DATE DUE

WATER RESOURCES DEVELOPMENT

U.S.A.:	BUTTERWORTH INC.
	WASHINGTON, D.C.: 20014: 7300 Pearl Street
ENGLAND:	BUTTERWORTH & CO. (PUBLISHERS) LTD.
	LONDON: 88 Kingsway, W.C.2
AUSTRALIA:	BUTTERWORTH & CO. (AUSTRALIA) LTD.
	SYDNEY: 20 Loftus Street
	MELBOURNE: 473 Bourke Street
	BRISBANE: 240 Queen Street
CANADA:	BUTTERWORTH & CO. (CANADA) LTD.
	TORONTO: 1367 Danforth Avenue, 6
NEW ZEALAND:	BUTTERWORTH & CO. (NEW ZEALAND) LTD.
	WELLINGTON: 49/51 Ballance Street
	AUCKLAND: 35 High Street
SOUTH AFRICA:	BUTTERWORTH & CO. (SOUTH AFRICA) LTD.
	DURBAN: 33/35 Beach Grove

WATER RESOURCES DEVELOPMENT

PLANNING, ENGINEERING
AND ECONOMICS

EDWARD KUIPER

C.I., S.M., P.Eng., F.ASCE, MEIC.

*Professor of Hydraulic Engineering, University of Manitoba.
Consultant to: Manitoba Hydro Electric Board; Department
of Agriculture, Canada; International Joint Commission.
Formerly: Chief Engineer of Manitoba Water Resources
Investigation; Senior Engineer, Saskatchewan River
Reclamation, and Winnipeg Flood Control Studies.
Project Engineer, Harlingen Navigation Works.*

WASHINGTON

BUTTERWORTHS

1965

Suggested U.D.C. number: 627·4

Suggested additional numbers: 532·53 : 551·482·2 : 627·8

Printed in Great Britain by
Spottiswoode, Ballantyne & Co. Ltd., Colchester and London

PREFACE

This book deals with the problem of developing a river system to the greatest benefit of the people that live in its drainage basin. Chapter 1 discusses the different aspects that have to be taken into consideration when planning the water resources development of a drainage basin. Chapters 2–5 deal with basic engineering subjects: hydrology to determine available river flows; hydraulics to analyse the flow in the river channels; river morphology to understand the behaviour of rivers; hydraulic structures to make possible the lay-out of river dams and other engineering works. Chapters 6–10 deal with the main components of river development: flood control, water power, irrigation, navigation and water supply. The discussions in these five chapters centre mainly around the question: how to make the best use of available water supplies; or conversely: how much water control and supply is required to obtain a certain objective. Chapter 11 discusses the economic analysis that must be applied to alternative water development plans, in order to find the one that will yield the greatest benefits.

It has been the objective of the author to provide sufficient information: (1) to understand the underlying principles and interrelationship of the main elements of river development; (2) to make possible the preliminary design and cost estimates of the required engineering works; (3) to provide guide lines for the selection of the most desirable water plan.   Once the development plan has been adopted, the task of detailed engineering design begins. This calls for a vast amount of engineering knowledge that goes beyond the scope of this book, and leads into the handbooks and professional papers that are listed in the bibliography.

Limited as the scope of this book had to be, it is believed that it can be of value to students at the undergraduate and post-graduate level, to introduce them to the basic concepts of river development; to practising hydraulic engineers who may have superior knowledge on particular subjects, but who would like to broaden their vision; and to economists, geographers and administrators who may have some degree of responsibility for the making of decisions regarding the development of water resources.

Winnipeg, Manitoba   E. KUIPER
May, 1965

v

CONTENTS

CONTENTS

CHAPTER 1

PLANNING

Before we begin to consider detailed aspects of water resources engineering, let us devote some discussion to water resources planning. It is obvious that planning must precede design and construction, not only in discussion but also in actual practice, or else there would be no assurance that the development of our human and physical resources takes place in the most beneficial way. If we pursue this thought a little further, we may observe that in creating an engineering project, we must perform the following, distinctly different, tasks.

(1) To define the purpose of the engineering project. To formulate its useful end, which makes the creation of the project a desirable objective.
(2) To plan the project in accordance with its established purpose. To investigate alternative proposals, and to select that project that will most effectively fulfil this purpose.
(3) To design the project in the most efficient manner and in accordance with appropriate criteria of safety.
(4) To construct the project according to its design, while applying suitable standards of workmanship.
(5) To operate the project, thus bringing the useful end of the engineering plan into concrete existence.

It may be noted that there is a certain amount of interdependence between these five tasks, in the reverse order in which they are listed. The successful operation of a project depends on its proper construction. The project can be properly constructed only when it is adequately designed. The correct design will only be effective when the project is properly planned. The most desirable plan can only be conceived when the purpose of the project is constantly kept in mind. Let us see how the engineering profession is carrying out these tasks.

(5) The operation of engineering projects is generally done with skill and knowledge, so that the desired results are obtained.
(4) The construction of engineering works is generally accomplished with a good deal of efficiency, while the workmanship is of a high calibre.
(3) The design of engineering structures has reached a high level of perfection. Occasionally, the highest degree of efficiency may not have been obtained due to a lack of imagination. Adequate criteria of safety are nearly always applied.
(2) The planning of many engineering projects is done with knowledge, insight and perception. However, it happens too often, particularly in the field of water resources development, that not all pertinent facts are taken into consideration. Instead, irrelevant aspects are introduced, unwise political pressure may be applied, the clear concept of resources development becomes obscured, and the eventual plan falls short of its ideal.

(1) The formulation of the purpose of engineering projects is a task that is sadly neglected by the engineering profession. We seldom venture beyond the point of stating that the purpose of the bridge is to relieve congested traffic, that the purpose of the hydro plant is to provide electricity and that the purpose of the floodway is to prevent future damage. However, we must admit that well-organized traffic, a dependable flow of electricity and harnessed rivers, are no ends in themselves. These are merely the means towards the ultimate end: 'the benefit of mankind'. But what constitutes the benefit of mankind?

ENDS AND MEANS

At one time in the history of engineering it was not so difficult to perceive 'the benefit of mankind'. Some hundred years ago, on the North American Continent, a small population, living under primitive circumstances, was faced with the task of conquering a vast continent. In this conquest, the engineers assumed a major role of responsibility. They provided efficient transportation, they built homes and factories, they generated power, they irrigated the land, they produced essential goods, and they provided a certain measure of leisure and comfort. Last but not least, they played an essential role in the winning of the wars that had to be fought to preserve freedom. There is no question that all this has been to 'the benefit of mankind'. From what we know of life 100 years ago, it appears that it was for many a painful struggle, filled with hardships and insecurity. If we look around us now, in North America, it appears that most of us, with very few exceptions, have all the material possessions that are prerequisite for pleasant living. (The inadequacies that still exist here and there are not due to insufficient technology or productive capacity but rather to the lack of appropriate social and political measures.)

When we cast a glance into the future, our vision of what the engineers can do for the benefit of mankind is not as clear as it was when we reviewed the past. Is our obligation towards mankind fulfilled when we keep advancing our technology and when we keep producing more and more goods? This is indeed what our friends the industrialists and stockholders tell us. The more production, the more profits, the more prosperity. This is also what our friends the political scientists tell us. For a strong western hemisphere it is necessary that we keep increasing our standard of living. This is also what our friends the economists tell us. They claim that for a healthy economy, it is essential that the production of goods, per capita, keeps increasing by a few per cent per year. In other words, in a few decades from now, we the engineers will be called upon to produce twice as much as we do now; and if we extrapolate this thought we would be producing, within the next 100 years, ten times as much as we do now. Would this be 'to the benefit of mankind'? There is ample evidence around us that this is not a trivial question.

Our friends the psychologists point out that in order to keep such abundant products flowing smoothly from the factory, to the consumer, to the garbage can, it is totally insufficient to merely offer these goods for sale. Instead, it is necessary to create a desire for obtaining these goods. To accomplish this, the consumer must be manipulated into believing that acquisition of more and more goods is essential for the good life of the individual and for the well being of the nation.

Our friends the social scientists point out that in this process of manipulation a great confusion of values takes place. Friendship is identified with beer, love with diamonds, happy family life with a station wagon, dignity with a Cadillac, and hospitality with Coca-Cola. At the same time there is the friendly persuasion from all directions to consume more and more of these goods. Please buy a new split level bungalow, the latest model car, the latest model T.V., a twin-powered, upholstered speedboat. As a result, our young people, being inescapably exposed to this friendly but constant persuasion, are developing a profound materialistic outlook on life. The acquisition of wealth is pursued at the expense of moral and spiritual values. There is ample evidence that such a trend is already taking place in the more affluent societies of North America and Western Europe. There is no doubt that this trend will become more pronounced with increased emphasis being placed on 'healthy economy' and its consequent need for increased consumption. This confronts us with the issue: do we deliberately wish our young people to grow up in an environment where the prevailing incentives lead towards materialism?

Our friends the philosophers point out that it is basically corrupt to manipulate people into believing something that they would probably reject if they were given all the facts and if they could rationally think about it. This is the way dictators have managed to degenerate whole societies to the point of extinction. Moreover, they point out that there is no correlation between the acquisition of material wealth and a happy life. In fact, luxurious living will hinder rather than enhance the balanced development of the human individual.

Finally, there is the warning, from our friends the geographers, that at our present rate of production, some basic resources will soon disappear from the face of the earth. We must add to this the speculation and hope that within the foreseeable future all people on earth will have acquired the sort of living standards to which we aspire. Allowing for predicted population increases, and assuming that all nations adopt the 'healthy economy' philosophy that requires an annual production increase of a few per cent, this would lead to the conclusion that within 100 years the world consumption of steel, copper, aluminium, timber, coal, oil, and uranium, would be 20 to 50 times as large as it is now. It is evident that such a situation cannot last very long. Some resources would be exhausted after a few decades; in fact, oil may already have disappeared before the world economy would develop to that point. Other resources may last somewhat longer, but they will probably have all disappeared before another 100 years have passed by.

It has been argued that the above image is not as frightening as it seems, because technology will find substitutes for raw materials before they are exhausted. This sounds encouraging, but it is not the prudent engineering approach that we are used to. We dislike to adopt unproven and speculative theories for determining our course of action. At the present, the engineering profession would have no idea how to continue its technology without steel, copper, aluminium, timber, coal, oil, and uranium. Therefore we must admit that there is some degree of urgency to consider this issue: do we, as engineers, act in a responsible manner, if we co-operate in persuading and coaxing the world into a fantastic consumption spree of natural resources, that may

3

culminate, a few centuries from now, in some three or four billion people, used to luxurious living, finding themselves at last without means of survival.

So here we find ourselves, as engineers, confronted with conflicting thoughts. On the one hand, we like to see the future of our profession as one that is distinguished by a steadily advancing technology. We want to improve our means of transportation, our means of communication, we want to conquer space, we want to meet every challenge that looms on to the horizon. On the other hand, we begin to suspect that some of our technological activities may have social implications that are far from desirable. What would be the sense of giving to mankind an abundance of wealth, leisure and convenience, and then finding out that we have corrupted the human individual, in the process of doing so. We would accomplish the opposite of what we set out to do!

How are we going to solve this dilemma? One way out would be to close our eyes and ears, and pretend that the problem does not exist, or to claim that we have no responsibility in these matters. This would be the surest way to lower our status of engineer to the level of the technical robot who does what he is told to do by his master who manipulates the control panel. It is simply inconceivable that we could take that attitude, since we are so intimately associated with the problem and therefore so deeply committed to help find a solution.

We must face up to the problem of choosing the sort of environment that we want our children and their children to live in. Do we want them to live in material abundance with the probable loss of the things of the mind? Or do we want it the other way round? Are we going to advocate the most rapid depletion of our non-renewable natural resources? Or do we have some obligation towards future generations that would caution us to use no more of these resources than is required for our basic needs? Are we going to continue building our cities in such a way that they become a nightmare of asphalt, concrete, smog, billboards and traffic lights? Or do we want to live and work in an environment where space, beauty, and quietude prevail? It is easy to ask these questions. It is not so easy to find the answers, but unless we face up to this challenge, we will never be able to clearly define the ends that will make the creation of our engineering projects a desirable objective.

The engineering profession is thus faced with a gigantic task: to participate in the search for guide lines for our future society, so that we can steer a predetermined course rather than drift into the unknown. This search will not be easy. It will require much thinking, writing and discussion. It will require much consultation with our friends in other professions, who are in their particular fields more knowledgeable than we are. After we come up with findings and guide lines, it will require much public discourse and debate to test the proposals on their soundness of content, and to get them adopted as policy of the nation.

Difficult and time consuming as this task may be, there is a great urgency about it. Scientific and technological development moves ahead very rapidly. Our undirected industries, eagerly making use of the latest innovations and supported by the most sophisticated methods of public opinion control that have ever been devised, are restlessly stirring about. Without moral restraints, without noble objectives, they may develop swiftly along lines that turn out to be detrimental to society. Once the tide is flowing in a certain direction, it will be extremely difficult to stem it, and change its course. Therefore, we must turn to our task without delay, and try to do the best we can.

The author has attempted in the above paragraphs to emphasize the need for developing a clear concept of ultimate ends, prior to planning our engineering activities, as means towards the fulfilment of those ends. Only then shall there be a reasonable chance that engineering will indeed benefit mankind. The author has discussed in another essay (listed in Chapter 12, Bibliography, along with other references on this subject) his subjective views on ends and means in society, and how they may be formulated. In the present treatise, which must be an objective discussion of engineering problems, this subject is pursued no further.

WATER RESOURCES PLANNING

In accordance with the above discussion we shall, first of all, try to formulate the objective of water resources planning. This may be done in several ways, each with its own merits and shortcomings. Let us consider the following definition and subsequent qualifications: The objective of water resources planning is to make the most effective use of the available water resources to meet all the foreseeable short- and long-term needs of the nation. The words 'most effective' imply that the well being of all the people should be maximized, while at the same time the total physical effort is minimized. The words 'short-term and long-term' imply that water resources must be managed and conserved not only for this generation but for generations to come. The interpretation of the word 'needs' should go beyond physical needs of industry and agriculture, into the realm of spiritual needs. To use the words of the U.S. President's Water Resources Council (1962): Proper water resources management requires that full consideration be given to ... 'the inspiration, enjoyment and education of the people.'

The most appropriate geographical unit for water resources planning is usually the river drainage basin. In this way we can realize fully the advantages of multiple use, reconcile conflicting interests of water use, and achieve the best co-ordination between all interests concerned. These interests may very well extend into other areas of resources development, such as forestry, agriculture, and mining. It is not unlikely that it is found that the planning studies should be extended to groups of closely related drainage basins.

To discuss all elements of water resources planning in such a manner that a logical sequence of thought is maintained, while at the end a full understanding of the subject is acquired, is not a simple assignment. Many elements are interrelated and one would have to move back and forth over the subject material several times to bring them all into proper focus. The manner and content of discussion that has been selected for the present purpose follows somewhat along the lines of the report 'Basic Considerations in Water Resources Planning', prepared by the Committee on Water Resources Planning, ASCE, of which the author was a member when the report was being prepared. As a first aid to the following discussions let us briefly review what is involved in water resources planning, before we proceed to details.

5

(1) Basic data (before any plans can be made, facts and figures must be collected).

   (a) Stream flow (a collection of all data pertaining to run-off. This section becomes in effect the water resources inventory of the basin).

   (b) Geophysical (maps of the basin showing streams, topography, soils, geological formations, forest cover, and minerals).

   (c) Economic (how many people live in the basin and what do they produce).

   (d) Jurisdictional (what levels of government are involved and how do they operate).

(2) Economic base projection (an estimate of the future growth of the population and the economy in the basin).

   (a) For the near future.

   (b) For the far future.

(3) Water requirements (how much water is needed for different purposes, in accordance with the economic base projection).

   (a) Domestic and industrial use.

   (b) Irrigation.

   (c) Navigation.

   (d) Power.

   (e) Flood control (this is a problem of controlling excess flows rather than consumptive use).

   (f) Recreation.

(4) Development of plan (the above isolated studies are now integrated. Reservoirs, canals, hydraulic structures are laid out so that maximum efficiency is obtained).

   (a) Priority of water use (if the demand for water exceeds the availability, a priority of use must first be established).

   (b) Alternative plans (a variety of multi-purpose plans, all aiming at maximizing the total benefits, is prepared).

   (c) Economic analysis (all plans are subjected to a benefit–cost analysis).

   (d) Selection of plan (based on the benefit–cost analysis, intangible aspects, and social consideration, a choice is made).

(1) *Basic Data*

It is obvious that the planning of water resources development cannot be successfully conducted without the availability of a wide variety of basic data, relating not only to the locality of the actual engineering works, but to the entire region that will benefit from or be affected by the project. The basic information that must be collected includes: stream flow, geophysical, economic and jurisdictional data.

(1a) *Stream flow data.* Any plan of water development must take into consideration the amount of water that is available, the amount of water that is required, and how these two can be reconciled. However, a mere cataloguing of the recorded run-off figures in the drainage basin is insufficient to serve as a water resources inventory. These figures must be supplemented, processed and studied. These last three activities will be discussed in the following paragraphs.

It is more than likely that some stream flow records in the drainage basin have missing years, or have unequal length, or should be extended backwards in time. This may be accomplished by correlating stream flow at one station with stream flow at another station, or with precipitation. Thus it becomes necessary to collect not only run-off data but also such meteorological data as precipitation, evaporation, temperature, wind movement, isolation, and relative humidity. It is also possible that recorded stream flow figures must be corrected for water supply or irrigation depletion, or reservoir operations, in order to obtain hypothetical 'natural stream flow figures'. The latter figures may have more value in future regulation studies.

After the stream flow data have been completed they must be processed so that they reveal the sort of information that the planning engineer wants to obtain. Hydrographs will show the distribution of run-off over the year and over longer periods of time. Duration curves of flow on a daily, monthly or yearly basis will show availability of water as a percentage of time. Mass curves will show the deficiency of flows over selected low flow periods. Frequency curves of maximum annual flood peaks will show the danger of flooding. Statistical research may reveal cyclical trends in water supplies that can be applied to the future.

From the information thus obtained, the planning engineer may decide to study the records from a viewpoint of possible redistribution in area or in time. This leads to studies of reservoir capacity and dependable flow, the possibility of river diversions, the possibility of supplementing surface water supplies with ground-water supplies, the possibility of using ground water as a storage reservoir, the possibility of reducing natural losses, the re-use of water, and the development of new sources. It is more than likely that some of these studies will be closely interwoven with the subsequent planning studies. In other words there may be some basic data, some preliminary studies, a promising plan, followed by more data collection, more analysis and more detailed planning.

Since the objective of this part of the study is to ascertain the availability of water in the drainage basin for different purposes, such as domestic use, and irrigation, the investigation must include sediment movement and water quality. It is necessary to collect data, to process and study these data, and to determine how proposed water development plans are affected by, and may in turn affect, sediment concentration and water quality.

The effects of sediment movement in rivers and the accumulation of sediments in reservoirs are among the most difficult problems that face engineers engaged in water resources developments in most parts of the world. The loss of storage capacity in reservoirs because of sedimentation gradually reduces the effectiveness of water resources developments and may, in time, seriously limit existing water supply facilities. More data are needed on the volumes of suspended sediment in rivers and the manner in which these sediments collect in reservoirs. More sediment surveys in reservoirs are needed to facilitate studies of how reservoirs may be operated more effectively so as to postpone the day when the capacity will be exhausted. Current methods must be improved, and new and better methods developed, for the removal and disposal of sediment deposits and for retardation of the production of sediment in contributing drainage areas.

Ground-water data are costly to obtain and it is often not economically possible to obtain them uniformly over an entire continental area. However, where ground-water reservoirs have been or are to be utilized, careful chronological records of water elevations must be obtained as well as yields from test wells. This is particularly true where there is danger of encroachment of sea-water or other water of poor quality. When ground-water supplies are being used to supplement surface water supplies, one must look carefully into the long-term results. If ground water is used as a storage reservoir that is deplenished during times of low stream flows and equally replenished during times of high stream flow, the long-term effect may be beneficial. However, when ground-water supplies are consistently deplenished, the result may be a harmful lowering of the water table. Moreover, the water pumped up from ground water in one place would eventually equal the decrease in stream flow somewhere else. It is obvious that such ground-water supplies should not be considered in long-term water development plans.

Analyses of the bacteriological and chemical characteristics of surface and ground waters are needed for almost all uses including municipal and domestic water supply, industrial consumption, irrigation use, and the disposal of wastes. The requirements for domestic water supply are well known, but the requirements for industries vary widely and are constantly changing with the developments of new products and new processes. Similarly, the requirements for maintenance of water quality vary widely in response to waste loads and the associated uses being made of the water. The requirements for agriculture are also exacting and this fact should always be recognized by the planners of new projects in underdeveloped countries.

It is obvious that there is plenty of water in the world if quality could be neglected. Poor quality may occur in five general forms as follows (although other forms, such as deoxygenation, may exist): bacteriological contamination, inorganic chemicals in solution; suspended sediments; thermal degradation and radiological contamination. Bacteriological contamination is largely an act of man and probably affects only approximately 5 per cent of the water withdrawn for all purposes in the United States and Canada. A high level of control of bacterial pollution can be accomplished through sewage treatment, and it is likely that in a few decades this source of pollution will render only a slight percentage of the total available water unfit for either industrial or agricultural consumption.

Contamination by inorganic chemicals may result from action by either nature or man. The most common example of natural chemical contamination is sea-water, but many ground-water basins and land-locked lakes suffer various degrees of chemical contamination. The possibility of reclaiming water containing naturally dissolved chemicals is receiving wide attention and Federal subsidy. However, it is unlikely that there will be an appreciable increase in total supplies available from these sources for many years.

Chemical contamination by man is a more serious matter and is created primarily by industrial use. Furthermore, the contaminated water is often discharged into rivers or ground waters that contain satisfactory water in their natural state. Of course, all industrial use, such as cooling water, does not have an unfavourable effect on water other than thermal, and such water is often re-used. It will become increasingly necessary to prevent industries from contaminating natural streams and ground-water basins through construction of treatment facilities. Thus, water will be available for other industries or agriculture as well as for increased domestic demands, although such re-use will be subject to economic and other limitations.

8

(1b) *Geophysical data* are those related to the configuration of the ground, the structure and composition of the earth's crust. These data fall into the following general classifications: topographic maps; geologic maps; sub-surface geologic explorations; agricultural soil surveys.

Topographic maps are needed for determining the drainage limits of river basins; for hydrologic studies; for location of dam sites; location and determination of the capacity of reservoirs; locations of canals, tunnels, aqueducts, levees and similar facilities; determination of head available for power development; location of areas for agricultural development; plotting of hydrologic data; and plotting of cultural, soil, geologic, agricultural, economic, and other data.

Geological maps are useful for water resources planning. Maps indicating regional geology will give indications of the probable composition of the basic soils, the location of water-bearing formations, the suitability of rock at dam sites, and the possibility of tight reservoirs. They warn of the danger of earthquakes and instability resulting from the presence of fault zones—important features in the planning of dams. However, in the final planning of structures, there is no substitute for field inspection and for the detailed exploration of sites for dams and for sources of construction materials by drill holes and test pits.

Soil surveys and regional soil maps are needed for land areas that are to receive irrigation water for the first time. Such surveys should include analyses of the soils in order to determine their suitability for crop production and studies of the physical structure in order to determine the rate at which water will pass through the soil. Such surveys cannot be economically undertaken on a national basis, but must be confined to areas selected for potential irrigation or water replenishment development.

(1c) *Economic data*, necessary to determine the need, cost, justification, and best over-all plan for water resource developments do not differ greatly from the economic data needed to evaluate most civil engineering projects. Typical items are as follows: population statistics; industrial development and future potential; land values and potential enhancement; agricultural statistics—crop productivity with and without irrigation; market value of crops; flood damages; transportation facilities; cost to ship various commodities; sources and value of energy; recreation needs and potential; fish and wildlife resources; pollution damages; hazards to navigation; erosion damages; damages caused by poor drainage; availability of construction materials; cost of construction materials; and wage rates.

(1d) *Jurisdictional data.* Water resources planning and development may take place under the auspices of private corporations, local, provincial or federal governments, regional authorities, or even international commissions. Each of these jurisdictions can be expected to have different social and physical objectives, each must recognize different legal restraints, and each has its individual economic limitations. It follows therefore that the relative political setting may influence the planning effort.

Generally speaking, as successively higher levels of government are engaged in the task, more emphasis is given to long-range resource development considerations, and to multi-purpose development. Private development, on the other hand, is usually characterized by single-purpose objectives. Emphasis is of necessity on profits rather than prudent resources management and social needs.

9

The need for successively higher levels of government or regional authorities arises when joint action is needed, for what cannot be accomplished by individual action. Since water resources planning and development requires extensive investigations and large capital expenditures, it is no wonder that, in many countries, the federal governments are major participants in the initiation of long-range resources conservation policies and become major investors in the construction of the nation's water projects.

Since political boundaries seldom agree with the boundaries of drainage basins, we may often expect conflict of interest between the several jurisdictions that are involved. Where such conflicts occur, legal considerations may be of paramount concern. Pertinent questions are: What earlier expressions of policy have been laid down? What jurisdiction will have management control? What new institutional arrangements will be needed and what will be their legal status? If engineers are to maintain a leading position in the water resources planning field, they must be able to recognize such problems, so that they can obtain timely and sound legal advice.

(2) *Economic Base Projection*

In the preceding section we have discussed the need for making an inventory of the available water supplies. In the present section we shall discuss the need for making an estimate of the future population and economic activity, since they are the principal factors in determining the future water requirements. It is necessary to make estimates of the near future as well as the far future. The estimates of the near future are required to develop the actual engineering plans that are contemplated for early construction. The estimates of the far future are required to develop an ultimate water plan, against which all intermediate plans must be tested for their effectiveness to provide the best long-term resources development.

(2a) *Economic base projection for the near future* must at least extend over a few decades. This is because of the magnitude and complexity of most river basin projects. Complete development may include several years for planning studies, several years for discussion and authorization, and several years for design and construction. Hence there may easily be a period of 10–20 years between the beginning of the planning studies and full operation of the project. Most estimates of future trends of population and their economic base are projections of past and present trends. It is generally assumed that there will be no wars or severe depressions, and that the economy of the country will continue to grow roughly at the same rate as it has in the past. It must be realized, of course, that all these assumptions create a substantial margin of possible error. However, since man is unable to predict what is going to happen in the future, we must accept the extrapolation of past trends (perhaps modified by logical reasoning) as the best alternative. To avoid the necessity of predicting the exact rate of growth of population and economy, which could create a false impression of precision, it would have merit to use a range of values, incorporating high, medium, and low levels of growth for all projections.

If future population growth figures are available for the entire country, but not for the drainage basin under consideration, one could initially take the present population ratio and apply that to the estimates for the country. There may be valid reasons to modify the resultant figures upwards or down-

10

wards because of circumstances that apply only to the drainage basin and not to the country as a whole. In fact, there will be a good deal of interdependence between the size of future population and the contemplated plans for resources development in the drainage basin. Here is another example of the need for moving back and forth in planning studies before any scheme can be finalized. For the purpose of estimating the future economic base level, it may be necessary to divide the drainage basin along political boundaries, rather than appraise it as one geographical unit. The above remark about interdependence applies even more so to future economic activity and future development plans. For instance, low-cost water-power development may stimulate mining and industrial development, which in turn may stimulate agricultural development. The latter activities may demand significant quantities of water which may affect the water-power development that triggered off the whole sequence. It is therefore evident that we should not make one inflexible economic base projection, but that such estimates must be re-appraised and may have to be adjusted as a result of the development plans that evolve.

(2b) *Economic base projection for the far future* should extend at least over the useful life of the projects that are considered; that is, a period of 50–100 years. It is obvious that any extrapolation of past trends, over such a long future period of time, is of a highly speculative character. It is therefore important that we try to apply logical reasoning, pertinent to this problem of projection. We may inquire into world-wide population trends and population distribution. We should consider probable future emigration and immigration policies of the country under consideration. In some heavily populated countries, the possibility of population control may be considered. If this becomes a reality, the extrapolation of past trends into the next 100 years would lead to completely erroneous results. In other areas of the world that are relatively sparsely populated, like northern Canada, the extrapolation of past trends could grossly underestimate the real growth. We must also keep in mind that population growth is related to economic activity, which in turn is related to resources development. Over such a long period as 100 years, the interplay between these factors becomes of major importance and may have to be given more consideration than the extrapolation of past trends.

We must also make due allowance for the likely increase in standard of living of the presently under-developed countries of the world. When this happens, there may come a greatly increased demand for food products with a consequent world-wide rise in food prices. This will probably result in the development of land reclamation projects that are presently considered uneconomical. This, in turn, will call for greater allocations of water for irrigation. Fortunately, it may be expected that in the foreseeable future the production of nuclear energy and the conversion of sea-water to fresh water will become so cheap, that water can be pumped from the sea to low-lying agricultural lands on the coastal plains. Thus, more water can be allocated from the headwaters of the river systems towards the high-lying agricultural lands.

It seems appropriate at this point of the discussion to ask why we should indulge in such highly speculative day-dreaming. It was noted earlier that the purpose of long-range water planning is to develop an ultimate water plan that may provide guidance for our present decisions. When we build a dam or excavate a canal, these works are likely to function for a period of some 100

11

years. During that period other dams, canals, and hydraulic works will be built. To obtain maximum efficiency, these works must function in a co-ordinated manner. This goal can only be attained if we are now farsighted enough to consider the future development of the country as far as we can look down the road.

It may be possible to develop an ultimate water plan by a different approach. Instead of first estimating the size of the future population and their economic base, and then preparing a resources development plan to suit their needs, we could turn the issue round: appraise first all available resources in the region and then estimate how many people and what economic activity could be supported by these resources. If such an approach is to be successful, we must take into consideration not only the available water resources, but all resources of the region, including space for living and recreation. After having developed a tentative plan to make the most effective and continued use of all available resources, we could then roughly determine how many people could live in the area without competing with one another for the use of the same resources. The last step would be to prepare a water development plan that would meet the demands of such a regional development.

Such an approach to water planning would serve a particularly useful purpose in regions where the resources are not yet fully developed and in countries where the government would be prepared to control the population of the region. This may require the adoption of migration or immigration policies and perhaps population-control measures. This leads us into politics and philosophy, which is beyond the scope of the present discussion.

(3) *Water Requirements*

After having estimated the probable future population and their economic activity, we must now determine the associated water requirements. It should be noted at the outset that initial estimates may have to be reviewed and modified at a later stage in the planning process. First, because the demand for water is to some extent a function of the cost of providing water. Since in the beginning we do not know yet the cost of water development plans, we cannot estimate precisely the water requirements. Second, because some water requirements may be in conflict with others. If these conflicts cannot be resolved, the original estimates must be revised. Third, because most water requirements are only partly consumptive. Therefore it may be possible, by appropriate re-use of water, to achieve a total water requirement that is substantially less than the sum of the individual requirements. Since most of these aspects are not fully revealed until a certain amount of planning has taken place, it is evident that a good deal of moving back and forth over the subject material is required. In the following paragraphs we shall pretend to conduct an inquiry for making the first tentative estimates. What questions would the planning engineer begin to ask with respect to possible water uses and control requirements, before he would even begin to think in terms of engineering plans?

(3a) *Domestic and industrial water requirements.* Assuming that we have an estimate of the population and the type and size of industries, it is a fairly simple matter to estimate the associated water requirements. Per capita requirements of water for household use range from a few gallons per day in primitive communities, to 60 gal./day in communities with a high standard of living where air-conditioning and watering of lawns is required. Per capita requirements of entire cities, including municipal, commercial, and industrial use, may range from 40–400 gal./day, depending on standard of living and type of industries. When the industrial requirements become relatively important, they should be estimated separately. To make such estimates one would have to consult pertinent literature such as the reports by the United States Senate Committee on National Water Resources (1960).*

It should be noted that the total domestic and industrial water requirements are usually small, compared with the available water supplies in the drainage basin. Moreover, only some 5–10 per cent of the total intake is consumptively used, the remainder returning to the river system. Although the total requirements are relatively small, it must be emphasized that they are of the highest priority and that a good quality of water is needed.

Associated with the domestic and industrial water requirements are the waste disposal requirements. The stream flow requirements to dilute municipal and industrial effluent so that adequate sanitary river conditions are maintained may be 10 or even 100 times larger than the pure-water intake. The exact amount depends largely upon the degree of treatment that industries and municipalities will apply to their waste.

With respect to waste disposal requirements the following may be noted. First of all, when sources of pollution are organic in nature and sufficiently far apart a good deal of natural sanitation in the river will take place, with the result that the same water can be used again for carrying off waste. Second, for the purpose of waste disposal, water does not have to be as pure as for irrigation or water supply. Therefore, a lake or stream with alkaline or saline water, unsuitable for other purposes, may very well be used for pollution abatement.

A quantitative discussion of water supply requirements for domestic and industrial purposes, and waste disposal requirements for different degrees of treatment, will be found in Chapter 10: Water Supply.

(3b) *Irrigation requirements.* The first question that may be asked on this subject would be: Is there any need for irrigation? This leads to an inquiry of the production of crops under natural moisture conditions. It is not unlikely that in many years good crops have been grown. However, during dry cycles severe crop losses may have been suffered. Such experience should be carefully recorded so that subsequent economic analysis can be applied. It is important to know how frequent and how extensive crop losses have occurred, and what effect they had on the economy of the area. Would considerable relief be attained if a small area was irrigated where fodder could be grown to keep the livestock in the area alive during drought periods?

Inquiries should be made with respect to future land use changes. In densely populated areas, a substantial percentage of agricultural land may be required for residential and industrial development, thus diminishing the irrigable areas. On the other hand, an increase in population will call for increased food production, which in turn will call for intensified irrigation. Since irrigation is a type of water use that may be given increasing priority in the future, it is important to appraise all economic, social and other factors that will determine the relative priority that irrigation should have with respect to other water

* For bibliographic references see Chapter 12.

uses. It is quite conceivable that eventually the use of water for irrigation will come second only to domestic water use.

After a clear picture has been obtained of present and future demands for agricultural products, the problem arises to determine to what extent irrigation could be applied to increase the present production. This calls for an inquiry into the soil and moisture conditions of the area. How much land is suitable to be irrigated from a viewpoint of location, topography, fertility and drainage? How much water should be brought to the land to ensure a good production of crops? This may range from 1–8 ft. of water over the irrigated land, depending on the climatic conditions and the type of crop.

It should be noted that irrigation requirements often constitute a major portion of the available water supplies. Moreover, irrigation water must be of even higher quality than domestic water since plants seem to be more sensitive to mineral content in water than are human beings. As a result, the problems in some regions may be one of determining how much land can be irrigated with the available water supplies, rather than how much water is required for the irrigable land.

In most reclamation districts, part of the water that is used for irrigating the land returns via ground-water flow to the stream channels. This return flow may range from 20–70 per cent of the original stream flow diversion. Whether or not the return flow can be used again for irrigating other land depends on the mineral content of the soils of the reclamation district. More discussion on these aspects will be found in Chapter 8: Irrigation.

(3c) *Navigation requirements.* To explore the navigation possibilities in a drainage basin, one must first know the present and future quantity of goods that would be transported by water if it were attractive from an economic viewpoint as compared with possible alternative means of transport. After having established that there will be a worth-while market for water transport, the problem becomes a technical one of determining whether and by what means navigation can be economically provided. There are several means available, each requiring different quantities of water.

The first method is by regulation of the existing river channels. By means of river training works, perhaps supplemented by dredging, a uniform and continuous channel may be created and maintained, accommodating tugboats, barges, and other vessels of limited dimensions. On some sections of the river system, the natural minimum river flow may be sufficient to provide the minimum allowable depth of flow. On the other, steeper sections, however, the natural minimum river flow may be insufficient to maintain navigation requirements. In such cases, water may be released from upstream reservoirs to provide the minimum allowable depth. There are few examples where this method has been applied successfully. First of all, reservoirs are usually located in the upper part of the drainage basin and navigation channels in the lower part. The great distance between the two makes operations ineffective. Secondly, to increase low-water stages on a sizeable river requires the release of large quantities of water. In most cases there is more profitable use of this water.

A second method of providing navigable waterways is by canalization of the river channels. A series of dams may be used to convert the natural stream into a string of navigable pools. The dams must be by-passed with locks. This

14

method may be more costly than channel regulation, but it also may require little or no release of water from upstream reservoirs.

A third method is by providing an entirely artificial canal. This canal may run parallel to a steep stretch of river, where canalization would become too costly, or it may connect different river systems. The water level in the canals must be practically horizontal and differences in elevation are overcome with shiplocks. A small water supply may compensate for lockage, leakage, and evaporation.

In summary of the appraisal of navigation facilities: first, an estimate must be made of the potential need for navigation; then, several methods must be explored to provide the navigational facilities, each associated with a certain quantity of water use and a certain cost. This basic information must then be subjected to economic analysis and to inquiry as to the relationship and desirability of the proposed use of water for navigation with other water requirements of the region. This subject will be discussed in more detail in Chapter 9: Navigation.

(3d) *Power requirements.* The generation of hydroelectric power is an important use of water. In Canada, nearly 50 per cent of the total amount of electrical energy is generated by water-power plants. One of the attractive features of water power is its non-consumptive use of water, so that it can be readily integrated with other water development aspects. With proper planning in a river basin, a high percentage of the available stream flow may be used through a series of successive plants from the head waters to the mouth of the rivers. On some rivers and in some years, a utilization of more than 95 per cent may be obtained. To estimate stream-flow regulation requirements for hydroelectric purposes requires rather complicated studies. This subject will be discussed in detail in Chapter 7: Water Power.

It was noted in the above paragraph that the generation of hydroelectric energy is an important water use. Not only because it makes energy available at lower cost than the next best alternative source of generation, but also because it conserves our non-renewable coal and oil supplies. This is an aspect that may be given more consideration in the future when these latter resources become less abundant. On the other hand, we may also expect in the future greatly increased water requirements for domestic, industrial and irrigation purposes. If conflicts arise, it may very well be that priority will have to be given to the latter, since there is no replacement for water for people and plants, whereas electrical energy can also be produced by other means.

(3e) *Flood control requirements* do not involve any consumptive or non-consumptive use of water. Nevertheless they must be appraised at an early stage in the planning process, since flood control measures can often be integrated with other water development measures. The appraisal of the situation will include a review of the history of flooding throughout the basin, an analysis of hydrologic data to determine the flood potential, and an economic study of past and potential flood damages. After this has been accomplished, tentative plans may be prepared to cope with the flood situation. Flood control and damage prevention measures may be divided into the following general categories: channel improvements; river diversions; dikes; reservoirs; and flood plain regulation.

The first three methods may often be evaluated on their own merits, since they will probably not affect other potential water development projects.

The possibility of reducing flood flows by storage reservoirs should be carefully appraised from the viewpoint of co-ordinating this purpose with the conservation of water for other purposes. Regulation of the use of a flood plain requires action by regulatory agencies.

The reclamation of land by diking and drainage is also a water development aspect that must be appraised. However, in most cases no measures are involved that would affect other water development plans. Such reclamation projects could be considered on their own merits, without integrating them into a basin-wide development plan.

(3f) *Recreation requirements.* It was noted earlier that the enjoyment of people is an important objective in water resources planning. This may be achieved by reserving beautiful areas of rivers, lakes and beaches for recreation; or by making new reservoirs available for the same purpose. It is not yet the practice of public agencies to construct reservoirs for recreational purposes only. Instead, recreation is looked upon as an incidental by-product of the reservoir that may be given some intangible credit. However, recent experience indicates that we should regard recreation a little more seriously. In 1960, in the U.S.A., 450 reservoirs had 100 million visitor-day attendance (for boating, fishing, swimming, water skiing, wildfowl shooting, and so forth). If recreation benefits would be evaluated at say one dollar per visitor-day, the total benefits would represent a substantial portion of the total annual cost of the reservoirs. Moreover, the attendance figures are growing at a rate of 10 to 15 per cent per year. It may therefore be expected that recreational objectives will be given more and more consideration in the future, when planning water resources development.

After recreation is recognized as an equal partner in a multi-purpose project, operating guides for the regulation of reservoirs and lakes, that are more compatible to recreational use, will result. Detrimental effects of drawdown such as is required for flood control and power generation will be minimized. The sale of shoreline property to industrial, business or other private interests should only be considered after all public recreational interests have been allowed for. Legislative measures may have to be adopted to prevent municipal and industrial stream flow pollution wherever waters are used for recreational purposes. Conversely, regulations may have to be laid down to prevent the public from polluting reservoirs that are used for domestic water supply purposes.

(4) *Development of Plan*

Having prepared an inventory of the available water resources in the drainage basin, and having estimated the different water requirements in the basin for the near future as well as the far future, the planning engineer is now ready to prepare an over-all plan of water resources development. If there are conflicting water requirements, the engineer will first establish a tentative priority of water use. Then he will develop a number of alternative plans that attempt to provide the most efficient use of the available water resources to meet the short- and long-term needs. These plans are worked out to the point that a reasonably accurate estimate of their cost and benefits can be made. Finally, a comparative analysis of the economic and other factors will be made, which will result in the selection of the optimum plan. These aspects of planning will be discussed in the following paragraphs.

(4a) *Priority of water use.* It is not unlikely that, in a given drainage basin, sooner or later the water requirements will exceed the available water supplies. Assuming that the importation of water from outside the drainage basin is impossible or too costly, and that the depletion of ground-water supplies is undesirable, a choice must then be made with respect to the priority of water use. It is generally assumed that domestic water supply has the highest priority of all types of water utilization, possibly followed by industrial and agricultural requirements.

It has been pointed out by Hirshleifer (1960) that water requirements are not absolutely determined by natural forces but rather depend on the economic balance of the community. Where water is cheap and abundant, it is liberally applied to domestic, industrial and irrigational uses and the 'water requirement' of that region appears to be high. Where water is expensive and scarce, elaborate facilities are constructed to minimize intake, and to re-use quantities withdrawn. As a result, the 'water requirement' of that region appears to be low. It would therefore be erroneous to establish, in advance of the planning study, arbitrary water requirements for different purposes, and then to saturate, in sequence, domestic use, industrial use, and irrigation use. Instead, an attempt must be made to equalize the marginal value of incremental water use for each different purpose.

When establishing priority of water use, we should also consider whether or not there is a substitute for the water. For instance, water power can be substituted by thermal or nuclear power; navigation can be substituted by land transport. However, there is no substitute for water for irrigation. It is likely that over the years there will be a trend to eliminate the less economical, non-essential uses of water. In the far future, in certain regions, only the essential and highly economical uses such as domestic and industrial water supply and waste disposal may be left over. In such cases, it is important to make the best possible estimate of the useful life of a project function that may be eliminated in the future. This affects the cost of the project and is pertinent to the economic analysis of the over-all plan.

(4b) *Alternative plans.* We have now come to the heart of the planning study: the most efficient lay-out of the actual engineering works. Such an objective is difficult to obtain and the proper solution may not be found without considerable search and study. The investigation of one plan is not enough. The process of conceiving new plans must be repeated until all viewpoints have been considered, and until all reasonably prospective combinations have been tried. It may be possible, however, to eliminate a large number of initial proposals with a minimum amount of formalities. A quick calculation may reveal that the cost of some structure is prohibitive, or that some proposal is not compatible with established priorities of water use. After this initial process of elimination, it may be expected that a number of promising alternative engineering plans will remain for serious consideration.

While these planning studies are being conducted it has merit to prepare interim reports that outline the different plans that are to be investigated and that record the results of the different engineering studies (hydrology, hydraulic computations, structural design, etc.) that are being made. These interim reports serve several purposes. First, any planning engineer who has to sit down and put his thoughts on paper, will realize more clearly what is involved in his plans and what will be required for the subsequent investigation. Second, his report becomes a guide to other engineers who did not participate in the conception of the plans, but who will be engaged on the subsequent investigation. Third, it enables the specification of supplementary field surveys that are required to complete the collection of basic data. Fourth, it preserves for posterity the engineering knowledge and experience that is gained during the investigations.

When the river basin investigation is of an unusually large scope, it may be advisable to separate the study into distinct phases. The first phase could be regarded as a reconnaissance study that takes into consideration a great number of alternative plans, that deals only with major factors, and that is conducted by a small group of experienced engineers. The result of this phase could be a report that recommends and specifies the more detailed investigation of a small number of the more attractive alternatives. This would constitute the second phase of the planning study. After the alternative river-basin plans have been established and the field surveys completed, the preliminary design of engineering structures may begin. It must be kept in mind that the purpose of the present investigation is the selection of the most feasible water development of the area. Therefore, the engineering works need not be designed with any greater precision than is needed to obtain a reasonable cost estimate, so that the alternative plans can be compared on the basis of their benefits and costs.

The preliminary design may be considered completed when an estimated cost is available for all the different plans under consideration. The capital cost includes the direct cost of structures together with such indirect costs as land or right-of-way costs, damage to other interests, engineering, contingencies, and interest during construction. The annual cost includes interest, depreciation, taxes and insurance, payments in lieu of taxes, and cost of operation, maintenance, and major replacements.

The end result of this part of the study could be a final engineering report that presents a number of alternative water development plans with their associated annual costs.

(4c) *Economic analysis.* A mere estimate of the average annual cost of different engineering plans is not always sufficient to be used as a basis for selecting the most feasible project, since different plans may yield different benefits. For this reason, the benefits of each alternative plan must be estimated as well.

Benefits may be classified as tangible and intangible. Tangible benefits can be expressed in dollars, whereas intangible benefits, such as the enhancement of scenery, or the increase in security, cannot be evaluated in dollars. In presenting the findings of an engineering study, intangible benefits, when considered of sufficient importance, should be described so that they can be given due weight in the final analysis.

The practices followed in evaluating tangible benefits vary according to legal requirements, locality, governmental precedence, and so forth. The general principles usually involved may be stated as follows. Flood control benefits are measured in terms of damages prevented and improvement in land use capability. Irrigation and reclamation benefits are measured in terms of increased productivity. Hydroelectric and navigation benefits are measured in terms of what it would cost to produce the same services (electricity, transportation, and so forth) by the least costly alternative means (thermal plants, railroad or highway, and so forth). This procedure may give a fair approximation, but does not necessarily produce a complete estimate of the absolute value of the services. Water supply and waste disposal benefits are measured in terms of what it would cost to provide the same water requirements by the least costly alternative means, or damages attendant on pollution in the absence of waste control, respectively. In some cases this principle may be successfully applied; in other cases it can lead to absurd results if the alternative is clearly too costly to be realistic. This subject of estimating the benefits of water development projects is discussed in more detail in Chapter 11: Economic Analysis.

(4d) *Selection of plan.* After having determined the annual costs and the annual benefits of a number of alternative water development projects, the problem arises to select the most feasible project. It is now generally accepted, at least in theory, that this should be the project with the greatest excess of overall annual benefits over annual costs.

It is quite possible that there are smaller projects with a larger benefit–cost ratio. However, going from such a smaller project to the one selected on the basis of greatest net benefits, would mean adding more benefits than costs, which is entirely justified. Although the smaller project promises a greater rate of return on the investment, it could lead to an underdevelopment of natural resources.

It is also possible that there are larger projects with a benefit–cost ratio that is still larger than unity. However, going from the project with the greatest net benefits to such a larger project would mean adding more costs than benefits and would, therefore, not be justified. Before the final project selection is made on the basis of the greatest net benefits, it would be wise to dwell for a moment on some aspects that may have been overlooked.

It may be asked with what degree of accuracy the computations have been performed. There are three areas in which large errors can be made. First, in deciding on the annual charges that must be applied to the capital cost, it is difficult to estimate several years in advance the proper interest rate for economic analysis, or to ascertain for financial analysis the rate at which money can be obtained to finance the project. An error of 1 per cent or 2 per cent in the estimated rate of interest would be possible. When the total annual charge is, for example, 8 per cent, a 2 per cent error would mean a 25 per cent error in the estimated annual costs of this item. Secondly, the estimates of the benefits are also open to some uncertainty. In the economic analysis of flood control projects, the third source of error could be the estimate of flood frequencies. A seemingly slight upturn or downturn of the frequency curve may make a large difference in the computed average annual damages and benefits, particularly when the damages continue increasing rapidly with increasing flood magnitudes.

When such inaccuracies, which cannot be avoided because of the limited amount of basic data and foresight, aggravate one another, it will be appreciated that the resultant benefit–cost ratios do not ordinarily warrant more than one figure behind the decimal point, and even then a benefit–cost ratio should not be considered as more than merely indicating an order of magnitude. It should be noted that such errors are not too harmful when the purpose of the economic analysis is to select the most feasible scheme. It may be assumed that any error in interest rate, or damage estimate, or flood frequency will affect every project benefit–cost ratio to nearly the same extent. However, when the benefit–cost ratio is used to justify the construction of a project, assumptions should be made that maintain a conservative benefit–cost ratio.

It must be recalled that the economic analysis has been based on direct and tangible benefits. It was noted earlier that it is conservative practice to base the analysis on these benefits only, but in the final analysis, when a selection between projects must be made, it would be quite correct to consider both tangible and intangible benefits. It is not impossible that several alternative projects are so close to being the most economic that slight differences in net benefits become insignificant in view of the probable errors in estimate. In such a case, an appraisal of the intangible benefits associated with each project would provide much better guidance for the final selection than placing complete confidence in precisely quoted benefit and cost figures. For this reason, it is advisable to include in the economic analysis of water development projects a clear qualitative description of expected indirect and intangible benefits.

The economic analysis has thus far been applied to one particular river-basin development or, maybe, only one phase of river-basin development. From different, alternative project proposals, the most feasible one may be selected, taking into account the relative magnitude of net benefits, and, if necessary, allowing for intangible benefits of each. In the case of projects to be financed from public funds, let the problem now be considered from the viewpoint of public investment, and let it be recognized that there is not only this particular river development project, but that there may be many more across the country, all of them competing for the limited amount of available public funds. The question then arises, what project should be begun first? In order to assure the greatest flow of future economic values, as soon as possible, the government may well develop first those projects across the country that have the highest benefit–cost ratios. Such a policy could lead to a gradual stage-development of one particular drainage basin plan, that would be spread over a period of several decades.

CHAPTER 2

HYDROLOGY

The applied science of hydrology, in its broadest sense of the word, is concerned with the circulation of water from the sea, through the atmosphere, to the land, and thence via overland and subterranean routes, back to the sea. For this reason many textbooks, devoted exclusively to hydrology, contain chapters not only on precipitation and run-off but also on climate, soil physics, river morphology, open-channel hydraulics, and regulation of reservoirs. In this chapter, hydrology will be treated in a more limited sense, as being the applied science that provides the hydraulic engineer with the stream flow data that are needed for the planning, design and operation of water development projects. Subjects such as river morphology, open channel flow, and reservoir regulation will be discussed in subsequent chapters.

Even when hydrology is limited in the above sense, it remains a vast subject that requires hundreds of pages of discussion to be covered completely. However, the aim of this book is to discuss water resources engineering in all its aspects. In this context, hydrology assumes a role of limited importance. Admittedly, the hydraulic engineer cannot do without stream flow data, but the availability of stream flow data in no way assures an efficient water development plan. Moreover, the applied science of hydrology is largely based on empirical methods and relationships. Several comprehensive textbooks and a vast amount of professional literature is available on this subject. To study this material and to become familiar with its content presents no unusual academic difficulties. Therefore the present chapter will be limited to acquainting the hydraulic engineer with the objectives and the more important methods in engineering hydrology. Anyone who is interested in special applications is advised to consult the literature that is listed in Chapter 12: Bibliography.

When the hydraulic engineer collects stream flow data for the planning, design and operation of water development projects, he makes an important assumption: he assumes that the characteristics of the recorded flows of the past will apply to stream flow behaviour in the future. In most cases this is a reasonable assumption, to a degree. If one has observed, for instance, at a certain gauging station, a minimum stream flow of 9,870 cusec, during the past 40 years, it is likely that the minimum stream flow during the forthcoming 40 years will also be about 10,000 cusec, plus or minus a few thousand cubic feet per second. However, to use the exact figure of 9,870 cusec in power or irrigation studies for determining the scope of development, without considering the consequences of substantial deviations from this figure, would be a bit unrealistic. In the applied science of hydrology it would be a good policy to keep the refinement of computations commensurate with the chance nature of the basic data and with the fact that we are primarily interested in an intelligent estimate of future stream flow conditions and not in a precise analysis of the past. Admittedly, the precise analysis will improve the intelligent estimate, but even the most precise analysis of the past is only a vague indication of what may happen in the future.

Before we start discussing such hydrology topics as flood frequency curves or unit-hydrographs, it may have some merit to review what function hydrology performs in water resources engineering. Most engineers will perform their tasks more efficiently when they know what purpose has to be achieved and how their work is related to that of their colleagues.

FUNCTION OF HYDROLOGY

Water resources inventory (the collecting and processing of all data pertaining to the availability of water in the drainage basin).

Data collecting (the systematic collecting of stream flow, precipitation, ground water and other pertinent records).

Missing records (analytical studies to complete or to supplement the above records).

Data processing (the transformation of the raw data into hydrographs, duration curves, mass curves, etc., so that a quick appraisal of the available water resources can be made).

Project planning (to assist in project planning the hydrologist will often make special studies of available records; for instance, dependable flow for an irrigation project or evaporation losses of a proposed reservoir).

Design of structures (for the design and construction of hydraulic works, the hydraulic engineer must determine design flow conditions such as the spillway design flood, or a cofferdam design flow, or a culvert design flow. To obtain such design flows, a statistical analysis of recorded flows must be made. In exceptional cases a synthetic study of extreme river flow conditions is required).

Economic analysis (to appraise the relative merits of a proposed project, a benefit–cost analysis is required. To determine the benefits of a water development project, statistical data are needed; for instance, duration curves for roughly determining the benefits of irrigation and power projects; flood frequency curves for determining the benefits of flood control projects).

Project operation (the efficiency of operating reservoirs for water supply, power or flood control purposes can be greatly increased when river flows can be forecast. To this end, a study must be made of the relationship between antecedent moisture conditions, precipitation, and run-off).

In the following paragraphs we shall discuss first of all data collecting and processing. After that follow four of the main topics in hydrology that are used for water resources engineering. These four topics are: flood frequency curves, unit-hydrographs, maximum possible flood, and coaxial relationships.

DATA COLLECTING

Stream flow records are collected by the U.S. Geological Survey and by the Canada, Water Resources Branch. Meteorological records such as precipitation, wind, and temperature are collected in both countries by the Weather Bureau. Measuring the discharge of a river can be done in several ways. The most common method is with the Price current meter. From a bridge or from a cable car, or from a boat attached to a cable, this instrument is lowered into the river and the velocity is measured. These velocity measurements are repeated two or more times in one vertical and in 10–20 verticals across the river. By multiplying the average velocity in one vertical with a representative depth and width and adding the results of all verticals, the total discharge of the river is obtained. This technique is extensively described in U.S. Geological Water Supply Paper No. 888 (1945). During times of ice break-up, or when the river is in flood and full of debris, or when surveying in unknown territory, the above method may not be feasible. To get a rough approximation of the discharge, one could take maximum surface velocity measurements from the shore, somehow determine the width and depth of the river and multiply velocity × width × depth × 0·5. These and other methods are described by Kuiper in the discussion of a paper by Matthes (1956).

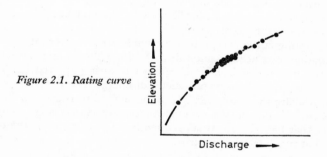

Figure 2.1. Rating curve

Another way of determining the discharge of natural streams is by applying an open channel formula or a weir formula. For an open channel formula one must know the slope of the water surface over a distance at least 10 times the width of the river, plus the average cross-section of the river, plus the rough-ness coefficient of the river. This method is mostly applied as a post-mortem on extreme flood peaks that passed by before the metering party arrived. The slope can be obtained from high-water marks; cross-sections and roughness coefficients are already known or can be determined afterwards. To apply a weir formula, one must look for rapids or falls where the river has a critical depth of flow. The difficulty is to estimate the elevation of the overflow crest. Such flow measurements are crude approximations at best, but they may be better than having nothing.

At regular metering stations, actual discharge measurements are only taken a few times per year. For the intervening days the discharge is derived from water level recordings and by using a rating curve, as shown in *Figure 2.1.* The rating curve shows the relationship between discharge and river stage for one particular station. When the river stage is observed daily by a local gauge reader or by an automatic instrument, these readings can then be transformed into discharge readings via the rating curve. Occasional discharge measure-ments with the Price current meter serve the purpose of ascertaining that the rating curve remains correct. There may be a variety of circumstances that make it difficult to apply this method. First, if the reach of river is alluvial with a shifting bottom, it is likely that there is no fixed relationship between discharge and stage. In such cases the rating curve must be continually corrected. Second, when the river discharge suddenly increases or decreases during times of flood, the river surface slope may increase or decrease which invalidates the stage–discharge relationship. For such cases one may have to use a rating curve with a loop, as will be discussed in Chapter 3: Hydraulics. Third, when there is a lake or a river junction downstream of the gauging station. The fluctuating lake levels or the tributary discharge may affect the river stage at the gauging station and thus invalidate the single relationship. It may be possible to devise a bundle of rating curves with the lake stage or tributary discharge as a parameter. Another solution for the second as well as the third case would be to observe not only the stage but also the slope of the river and use it as a third variable in the stage–discharge relationship. A fourth cause that may upset the rating curve at a gauging station is an ice cover over the river. As a result of the surface friction the river stage will become higher for a given discharge. Since the winter gauge readings correspond roughly with the upper surface of the ice, while the river discharge takes place underneath the lower surface of the ice, it follows that the thickness of the ice is a variable in the winter rating curve. To prepare a reliable set of winter rating curves for a gauging station, one needs a good deal of winter flow measurements.

Once a rating curve or bundle of rating curves is prepared for a gauging station it may be possible to reconstruct a discharge record that antedates the first actual discharge measurement. In undeveloped territory, for instance, it is not unusual to find someone who has kept a continuous stage record of the lake or river that he lives by. If subsequently a rating curve of the outlets of the lake or that river station is prepared, the original stage record can be translated into a discharge record. This may be a most valuable addition to the sparse stream flow data that are usually available in undeveloped territories.

If the stream flow data of a gauging station are discontinuous, it may be possible to complete them by correlating the known flows of this station with the same daily flows of another station close by, where the records are complete. If a satisfactory correlation is formed, this relationship is then used to estimate the stream flows on the missing days. The simplest method to do this is by a plotting of the flows of the two stations on rectangular or logarithmic graph paper. The scatter of the points will give a measure of the reliability of the correlation. It may be desirable to make separate graphs for the different seasons. This technique can be applied to mean annual, mean monthly, or daily discharges. In the last case it may be necessary to offset the discharges by one or more days, to allow for the lag of flow between the stations.

If there are no nearby stations to correlate the flows with, and if the gap in the record is only short, it may be possible to estimate the missing flows by plotting the hydrograph of all known flows, in conjunction with graphs of precipitation and temperature, and sketching in the missing gaps, giving due weight to the drainage basin characteristics as indicated by the available record.

If there are only a few records, or no records at all, and no nearby stations to compare with, the problem is much more difficult and it becomes necessary to make a comparative hydrologic study between the drainage basin under consideration and a similar drainage basin where both precipitation and run-off data are available. If conditions in the two regions are fairly uniform and comparable, records may be transferred on a unit-square-mile basis. If no comparable drainage basin with adequate records can be found, a rough estimate of run-off may be made by estimating, successively, the amount of precipitation, the amount of infiltration, and the shape of the hydrograph of the run-off. To make these estimates one must rely heavily upon empirical data published in hydrologic literature.

DATA PROCESSING

Before the raw stream flow data can become meaningful to the hydraulic engineer, they must be processed. The simplest way to attach some meaning to the stream flow record of a gauging station is to list the minimum, the average, and the maximum flow figures. This is usually done in all publications of stream flow figures. To detect other characteristics of the stream flow behaviour, it is advisable to plot hydrographs, mass curves, and duration curves.

Hydrographs

A hydrograph, as shown in *Figure 2.2*, is simply a graphical plot of the stream flow with time on the horizontal scale and discharge on the vertical scale. A day-by-day hydrograph reveals the distribution of the stream flow over the year. A prairie river for instance, as shown in *Figure 2.2(a)*, may have the bulk of its run-off during the spring and early summer and extremely low flows during the winter. A hydrograph of mean monthly flows, as shown in *Figure 2.2(b)* (which is much easier to plot and nearly as useful in most water

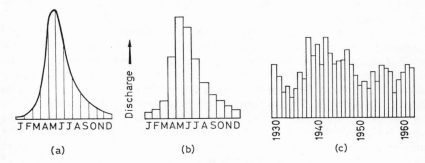

Figure 2.2. Hydrographs of rivers

development studies), shows the same characteristics. A hydrograph of mean annual flows, as shown in *Figure 2.2(c)*, may give an indication of long-term wet or dry periods in the drainage basin. Another way to show the long-term variation of stream flow is to plot the 10-year moving average against the terminal year of the 10-year period. This smooths out any large year-to-year fluctuations, and demonstrates more clearly the long-term changes. The same purpose can be accomplished by plotting a mass curve, shown in *Figure 2.4*, and discussed in the following section. Whether or not wet and dry periods occur with a predictable regularity has been a subject of debate for many years.

One of the more recent papers is by Williams (1961) who believes that proper statistical analysis may reveal predictable trends.

Before we leave the subject of hydrographs it should be pointed out that area-under-the-hydrograph represents volume of water, since the vertical ordinate (cusec) times the horizontal ordinate (time) represents cubic feet. Hence by planimetering, or by counting squares and by applying the appropriate conversion factor, one can quickly determine how much water passed the gauging station between two dates. For easy conversion of units it is convenient to remember that 1 cusec-day (1 cubic foot per second, flowing for the duration of one day) equals approximately 2 acre-ft. (an area of 1 acre, covered by 2 ft. of water). This makes 1 SFM (1 second foot month, representing 1 cubic foot per second flowing for the duration of 1 month) equal to 61 acre-ft. (assuming the duration of one month equal to 30·5 days).

Mass Curves

The mass curve is a graphical tool, to detect long-term trends in river flow. It is also a convenient device to determine storage requirements that are needed to produce a certain dependable flow from a reservoir. To illustrate the technique of doing this, a hydrograph of river flow is shown in *Figure 2.3(a)*. Let us assume that the two years 1930 and 1931 include the lowest flow period on record. The problem is to find the storage capacity required to increase the natural minimum flow of 2,000 cusec to a dependable flow of 20,000 cusec. In this simple example, this problem can easily be solved without resorting to a mass curve. All we have to do, is to draw a horizontal line at the ordinate of 20,000 cusec in *Figure 2.3(a)*, planimeter the shaded area, apply a conversion factor, and find that the required storage capacity is 90,000 SFM. Let us now solve the same problem by using the mass curve shown in *Figure 2.3(b)*.

The mass curve is the summation of the hydrograph. The abscissae are in the same units of time as the hydrograph. The ordinates represent the total volume of water that has passed from zero time up to that point. The slope of the mass curve at any point represents change of volume per change of time; in other words the rate of flow at that moment. Hence the mass curve is steep when the river flow is large, and flat when the flow is low. A small key diagram showing the value of slope in terms of cubic feet per second is convenient in appraising a mass curve. The slope of a line joining any two points of the curve represents the uniform rate of discharge that would have yielded the same total incremental volume in the same period. For instance, in *Figure 2.3(b)*, going from A to C along the mass curve, represents the same volume of water as going from A to C along the straight line. This feature of the mass curve enables us to determine readily the amount of required storage capacity. Let us assume that adequate, but yet unknown, storage capacity is available at point A (where the 20,000 cusec slope is tangent to the mass curve). From that moment on the release from the reservoir is 20,000 cusec, but the inflow into the reservoir is less and therefore the reservoir level goes down. At any time between A and C, the length of the ordinate intercepted between the straight line and the mass curve measures directly the total amount by which the reservoir capacity has been reduced. The maximum ordinate is reached at point B, and measures 90,000 SFM on the vertical scale. In other words, if the reservoir capacity had been 90,000 SFM to start with, a flow of 20,000 cusec could have been maintained from *A* to *B*. After *B*, the inflow into the reservoir is greater than 20,000 cusec. If we would still release no more than 20,000 cusec, the reservoir would gradually be filled and would be full again at *C*. It is interesting to note the corresponding features in *Figures 2.3(a)* and *2.3(b)*.

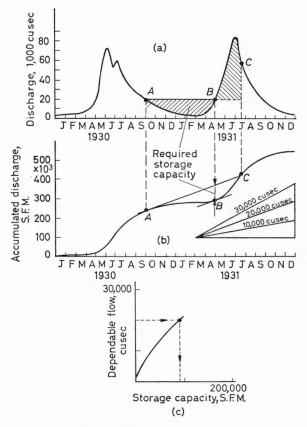

Figure 2.3. Mass curve of stream flow

Table 2.1 shows how the required storage capacity can be computed in the same manner as was done graphically in *Figure 2.3(b)*. Some hydraulic engineers prefer to do the initial exploratory work in graphical form, and after they have found the critical low flow periods in a 30- or 40-year record, to determine required storage capacities more precisely by computation. In Table 2.1, col. (1) shows the month; col. (2) shows the average monthly flow in 1,000 cusec; col. (3) shows the accumulated value of the mass curve in 1,000 SFM; col. (4) shows what draft from storage capacity is needed to provide an outflow of 20,000 cusec; col. (5) shows the accumulated draft from storage, in 1,000 SFM. The highest figure in col. (5) represents the required storage capacity.

27

The above procedure can be repeated for different dependable flow figures. *Figure 2.3(c)* shows a curve that relates the dependable flow from the reservoir as a function of the required storage capacity. Assuming that the low flow

Table 2.1. *Storage Requirements*

Date (1)		Flow (2)	Mass (3)	Draft (4)	Storage (5)
1930	Sep	25	240	0	0
	Oct	15	255	−5	5
	Nov	12	267	−8	13
	Dec	6	273	−14	27
1931	Jan	5	278	−15	42
	Feb	3	281	−17	59
	Mar	2	283	−18	77
	Apr	7	290	−13	90
	May	40	330	+20	70
	Jun	70	400	+50	20
	Jul	60	460	+40	0
	Aug	40	500	+20	0

period 1930–31 is governing for this reservoir site, the curve in *Figure 2.3(c)*, in the region shown, is nearly a straight line, since the pivot point of the straight lines in *Figure 2.3(b)* is close to *A*, while the maximum ordinates are close to point *B*.

In the above example a cycle of only two years was shown. This was done for the sake of simplicity of discussing the basic concepts of the mass curve. It was pointed out at the beginning of this section that in such simple cases, one can just as well analyse the hydrograph itself to find the storage capacity that belongs to a certain dependable flow. However, when we have to deal with

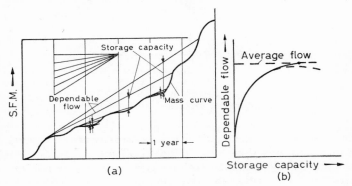

Figure 2.4. *Mass curve of stream flow*

cycles of many years, the latter method would become cumbersome, while the mass-curve method remains simple. In *Figure 2.4(a)* is shown the mass curve of natural stream flow at a reservoir site for a cycle of six years. By taking the whole range of slopes, from the minimum recorded flow to the average flow, and finding for each slope the section of mass curve that yields the highest required-reservoir-capacity-ordinate, one can prepare a dependable flow curve as is shown in *Figure 2.4(b)*. The beginning of this curve is the minimum recorded flow with no storage capacity. The end of the solid curve is the average flow with a large but still finite storage capacity. However, this upper part of the curve loses some of its meaning if one realizes that a full reservoir is required at the beginning of the period. It could take many years, after the completion of such a huge reservoir, to fill it up. This could be an economic liability of such a scheme.

It should also be pointed out that creating very large reservoirs may increase the total evaporation losses from the stream. As a result the average river flow will reduce. It is obvious that such aspects must be carefully investigated. This calls for specialized hydrological studies. If one were dealing with a potential reservoir site in a dry and hot region it is conceivable that the dependable flow curve would eventually decline, as shown with the dashed curve in *Figure 2.4(b)*, due to excessive evaporation and perhaps seepage losses, with increasing reservoir capacity.

It may very well happen that the water resources planning engineer is not only interested in the 100 per cent dependable flow, as discussed in the above

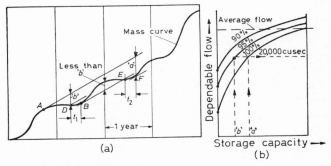

Figure 2.5. Partial dependable flow curves

(assuming that the record of the past is representative for the future), but that he would also like to know what flows are dependable during 90, 95, and 99 per cent of the time. To obtain such figures, the following procedure can be followed. In *Figure 2.5(a)* the mass curve of reservoir inflow (natural stream flow) is plotted. If storage capacity 'a' is provided, the dependable flow is 20,000 cusec, 100 per cent of the time, as discussed before. Let us now assume that we provide only storage capacity 'b' and that we release again 20,000 cusec. As a result, the reservoir will be empty at D. From D to B we can only release the reservoir inflow which is less than 20,000 cusec. After B the flow is again larger than 20,000 cusec. In this fashion we scan the whole mass curve and total all the time intervals, like t_1, that the flow was less than 20,000 cusec. We should watch out for years that the storage capacity 'b' is not entirely replenished, as is shown between B and E. Instead of following the above procedure we must now draw a line from B, parallel to the 20,000 cusec slope, till it intersects the mass curve at E, where the reservoir is again empty. The subsequent deficient period is t_2, from E to F. After having totalled up all the deficient periods, this figure is divided by the total time T of the number of years under consideration. Let us assume that the outcome is 5 per cent. This means that with a storage capacity '*b*', a flow of 20,000 cusec is dependable during 95 per cent of the time. This point is plotted in *Figure 2.5(b)*. The same process is repeated for different flows and different storage capacities. The resultant points are all plotted in *Figure 2.5(b)*. After enough points have been obtained, a bundle of curves can be drawn in as shown in the figure. It may be noted that the points where the curves intersect the vertical ordinate of no-storage capacity, can be obtained from the duration curve of the natural river flows.

A slightly different way to appraise the long-term trend of river flow is by the residual mass curve, as shown in *Figure 2.6*. These curves are prepared by

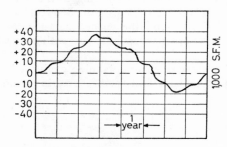

Figure 2.6. Residual mass curve

accumulating the positive and negative departures from average run-off. A period of above normal run-off is indicated by an upward slope, while below normal run-off corresponds to downward slope. A horizontal line joining any two points represents a period of average run-off. This feature makes the residual mass curve useful in selecting representative short-term periods of stream flow for detailed project analysis. It may be pointed out that the residual mass curve is basically the same as the ordinary mass curve. The horizontal line in the first one is the same as the straight line that connects the first and last point in the second one. The vertical deviations from these two lines are the same in both diagrams. The only advantage of the residual mass curve is that it is easier to appraise by eye.

Duration Curves

It was discussed earlier that a rough appraisal of a stream flow record can be obtained by listing the minimum, the average, and the maximum flow. However, for detailed studies we should know more precisely how often low flows or high flows occurred during the period of record. This can be accomplished by preparing a duration curve in which magnitude of discharge is plotted against the percentage of time that discharge is exceeded. A duration curve can be prepared for any period of time. One can prepare a duration curve of daily flows, mean monthly flows, or mean annual flows.

One way to prepare a duration curve is as follows. The total range of discharge, say from 0–100,000 cusec, is divided into 20 compartments of 5,000 cusec. One starts scanning through the selected period of record, day by day for a duration curve of daily flows (which is a tremendous amount of work), or month by month (for a duration curve of mean monthly flows). For every item in the record, a mark is made in the appropriate compartment. When all items are entered, one could plot the results as shown in *Figure 2.7(a)*, which shows the so-called frequency distribution of the sample. The compartment with the largest number of items is called the mode of the sample.

When the frequency curve of *Figure 2.7(a)* is accumulated, compartment by compartment, starting with the low value, the total frequency curve, or duration curve of *Figure 2.7(b)*, is obtained. The vertical ordinate still shows the discharge, the horizontal ordinate represents the total number of items, or, more

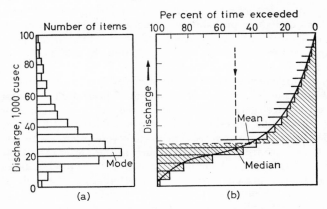

Figure 2.7. (a) Frequency curve, (b) duration curve

conveniently, 100 per cent of the time. In plotting the points of the duration curve, one must realize that the total number of items in, say, the first four compartments (0–20,000 cusec) must be plotted right on the 20,000 cusec value. After all 20 points are plotted, a smooth line can be drawn through the points. With this duration curve we can now select any discharge and find the per cent of time this discharge is exceeded. The stream flow value that is exceeded 50 per cent of the time is called the median flow. The average value of all items is called the mean flow. Since nearly all natural stream flows have an asymmetrical distribution (the high values deviate much more from the average than the low values) it follows that the median flow and the mean flow are usually not the same, as may be seen in *Figure 2.7(b)*. The median flow can be read directly from the curve. The mean flow can be found by drawing a horizontal line, such that the two shaded areas are equal in size, or by computing the arithmetic average of the sample.

It may be of interest to point out that the duration curve of mean annual flows may look entirely different from the duration curve of daily flows. Let us assume, for instance, that the river flows fluctuate from a low in the winter to a high in the summer, but that the total volume of run-off every year is nearly the same. This would result in a normal duration curve for the daily flows, but in a nearly horizontal line for the duration curve of the mean annual flows. The mean flow values of both curves are the same, but not the median flow values. Therefore, if one wants to estimate, say for reservoir filling purposes, the magnitude of flow that has an even chance of being exceeded for the next few years, one should consult the duration curve of mean annual flows, and use the median flow figure.

FREQUENCY CURVES

Until now we have discussed data processing (hydrographs, mass curves, duration curves) that has a wide application. From now on, in this chapter, we shall discuss a few special problems that have more limited application. First of all the frequency analysis of flood flows. A frequency curve of maximum annual flood flows is a graph showing the relationship between the magnitude of flood flows and their probability of occurrence. Hence they are basically the same as the duration curves, discussed above. The difference is that now we only analyse the flood flows and that we plot the results on special graph paper.

Frequency curves are used for selecting design flow conditions for hydraulic works and for economic analysis of flood control projects. In the first case, one must make a choice of the degree of required safety, as well as prepare the frequency curve. The selection of the degree of safety will be discussed in Chapter 5: Hydraulic Structures. There are several methods to analyse flood flows and to prepare frequency curves, such as the Hazen, the Goodrich, the California, and the Gumbel method. Each has its own advantages and disadvantages. In the present chapter we shall discuss primarily the Hazen method which is the oldest but still a widely used method. It is hoped that this discussion will suffice to make clear the essential features of frequency curves. To make a very important study, it would be advisable to look into the other methods as well.

In analysing flood flows we are mostly interested in flood peaks (the maximum discharge). However, in some cases, for instance when we have to route the flood flow through a reservoir, we may be interested in the flood volume (the total amount of water contained in the flood hydrograph) as well. Moreover, we may have to determine the correlation between peak and volume. In the following paragraphs we shall discuss first the frequency analysis which can apply to peaks as well as volumes, and afterwards the correlation between the two.

Frequency Analysis

In order to visualize the concept of a frequency curve, the 100-year period of record of the Cumberland River at Nashville, Tennessee, from 1850–1949 will be chosen as an example. In Table 2.2 the beginning and end values of the period of record are listed in col. (1). These 100 items are now rearranged and plotted on a graph so that they appear in ascending order of magnitude from the left to the right as shown in *Figure 2.8*. The relative position of the points and their magnitudes are also shown in Table 2.2, cols. (2) and (3). If a smooth line were drawn through the skyline of the items in *Figure 2.8*, this line would be commonly called a frequency curve, although the correct name is duration curve or total frequency curve. It is of interest to point out that the frequency curve of *Figure 2.8* is basically the same as the duration curve of *Figure 2.7(b)*. We could have obtained the latter duration curve also by plotting the items one by one, in ascending order of magnitude, but that would have been too much work for a duration curve of daily or monthly flows. However, if we only have a few dozen items, like the maximum annual flood peaks, this method is easier and more precise.

Table 2.2. *Frequency Curve of the Cumberland River at Nashville*

Year	Magn. (1)	Rel. Pos. (2)	Magn. (3)	Ratio (4)	Freq. (5)
1850	157,000	1	203,000	1·66	0·5
1851	127,000	2	194,000	1·59	1·5
1852	127,000	3	186,000	1·52	2·5
1853	123,000	4	170,000	1·39	3·5
1854	152,000	5	170,000	1·39	4·5
1947	101,000	98	73,400	0·60	97·5
1948	155,000	99	72,400	0·59	98·5
1949	91,500	100	69,000	0·57	99·5
		Mean	122,200		

Col. (1) lists the magnitude of the maximum annual average daily river flow for the given year, in cubic feet per second.
Col. (2) lists the relative position of the items in col. (3).
Col. (3) lists the items of col. (1), but rearranged in descending order of magnitude.
Col. (4) lists the items of col. (3), expressed in the ratio to their mean value.
Col. (5) lists the frequency of exceedence of every item, in per cent. For example: every year, there is a chance of 2·5 per cent that a peak flow of 186,000 cusec will be exceeded.

Assuming that the frequency curve of *Figure 2.8* is representative for the behaviour of the flood flows of the Cumberland River at Nashville, the following observations can be made with respect to the probability of occurrence

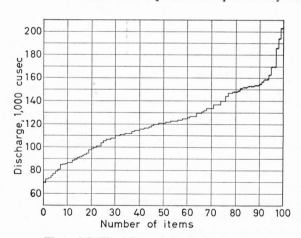

Figure 2.8. *Flood flows of Cumberland River*

of these flood flows. During 50 per cent of the time the flood flows are larger than 121,000 cusec. During 10 per cent of the time the flood flows are smaller than 87,000 cusec. During 10 per cent of the time, the flood flows are larger than 153,000 cusec. This last statement can also be phrased as follows:

During every year there is a probability of 10 per cent that a flood peak will occur which is larger than 153,000 cusec. Emphasis should be placed on the words 'larger than'. Unfortunately, it is common language, even among hydraulic engineers, to call 153,000 cusec the 10 per cent flood, conveying the impression that there is a probability of occurrence of 10 per cent that the flood peak will be 153,000 cusec. This, of course, is incorrect. A flood of exactly 153,000 cusec may never occur. If it is necessary to make reference to a certain range of floods, for instance in connection with flood damages, it may be observed, for example, that the flood flows from 110,000–130,000 cusec have a frequency of occurrence of $65 - 30 = 35$ per cent. In other words every year there is a 35 per cent chance that the peak flow will fall between 110,000 and 130,000 cusec.

It has been found convenient to distort the vertical and horizontal scales of *Figure 2.8* in order to make extrapolation and reading of low frequencies easier. From *Figure 2.8* to *Figure 2.9*, the vertical scale changed from linear to

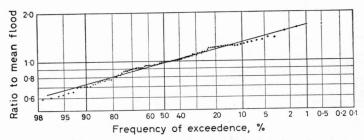

Figure 2.9. Frequency curve of Cumberland River, Nashville, Tennessee. (Records 1850–1949: Maximum 1927, 1882, 1937, 1862; D.A. 12,900 sq. miles; M.F. 122,200 cusec; F=63·0, R=1·7)

logarithmic, the peak flows are plotted in ratio to the mean (in order to enable comparison between different river stations) and the horizontal scale changed from lineal to a probability scale (in order to stretch the range of the low and high frequencies which form the most interesting part of the curve). The result of these scale distortions is that the plotted points fall on a straight or almost straight line. Fundamentally, however, the frequency curve of *Figure 2.9* has the same meaning as the curve of *Figure 2.8*.

Since there is never an infinite number of items available, it will never be possible to determine the curves accurately, but the next best approach is to draw smooth lines through the scattered points that are available. Strictly speaking, there is not even a series of points available, but only a series of columns, and the first question that arises is: which points on the top of the columns must be selected for drawing the frequency curve, or in other words: how are the points to be plotted?

In order to discuss this problem, reference is made again to *Figure 2.8*, which was obtained by placing all the available items from left to right in ascending order of magnitude. Is the frequency curve to be drawn through the upper middle of the bars or through the upper left corners? This would seem to be a trivial question because the difference in *Figure 2.8* would not be noticeable to the eye. However, once the points are plotted on logarithmic probability paper, it may be seen that it does make a noticeable difference. The highest point, for instance, is assigned a time value of 0·5 per cent in the one case, and 1 per cent in the other case.

Plotting the points in the middle of the bars, as shown in *Figures 2.8* and *2.9* was proposed by Hazen and is based on the assumption that every one of the 100 items represents 1 per cent of a true sample of the series. Plotting the points at the upper left corner of the bars, is known as the California method. It is based on the assumption that a subsequent flood has an equal chance of exceeding the highest point or falling into one of the intervals. Both methods are used at the present time. The California method is more conservative from the safety-of-design point of view and the Hazen method is more conservative from the standpoint of justification of flood-control projects.

The plotting positions of the points in the Hazen method are found by the following formula:

$$f = 100 \cdot \frac{m - 0\cdot5}{n}$$

in which f is the frequency of occurrence in per cent, m the relative position of the points when listed in decreasing order, and n the total number of items. An example of applying this formula to an actual sample, may be found in Table 2.2 and *Figure 2.9*.

Once the flood peak data are plotted, the question arises: how is a frequency curve drawn through the plotted points? The simplest and most widely used method is by eye. More complicated methods, but not necessarily more accurate ones, are based on theoretical frequency curves. These methods are developed by Foster, Hazen, Gumbel, and others. Foster's and Hazen's methods are based on the following idea. All distributions follow a basic law of probability, hence they can be expressed in mathematical terms. In order to determine the shape of a frequency curve, it is necessary to know two parameters: C_v, the coefficient of variation to indicate spreading of the curve and C_s, the coefficient of skew to indicate the skewness of the curve. Once these coefficients are known, the shape of the curve is determined. These two coefficients in their turn can be computed from the basic data, and thus it is possible to determine for a certain series of stream flow data the type of frequency curve that should theoretically correspond to the data.

Unfortunately, the computed values for C_v and C_s are extremely sensitive to the highest items in the series. The value of C_s, especially, will increase rapidly with a mild increase in value of only the one highest point on record. In fact, Hazen states in the last part of his book *Flood Flows*, that C_s can be used as an indication whether or not the highest items are truly representative for the series. However, that means that C_s will be of considerably less value for drawing the true frequency lines. Other students in the theory of probability curves claim that the computed value of C_s is not reliable when the series contain less than 150 items. Since most studies deal with record periods of about 50 years or less, the usefulness of the parameter C_s becomes doubtful.

In a study of 100 frequency curves of North American Rivers it was concluded by Kuiper (1957) that frequency curves can generally be drawn as straight lines, when plotted on logarithmic probability paper. Records of 22

35

river stations with a period of record of 60 years and longer were analysed. Twelve out of the 22 series of plotted points can be represented by a straight line. Nine out of the remaining 10 series approached a straight line fairly well. In most cases the whole group would fall on a straight line if only a few points were changed. In only one case would a straight line be a poor fit, and then only with respect to the lower third of the points.

Demonstration of the misleading appearance of a number of points on a seemingly curved line is provided by the following experiment. A straight line was drawn. From this line, 1,000 representative values were chosen; that is, the discharge values were read at the frequency: 0·05, 0·15, 0·25, 0·35, etc., up to 99·95. The 1,000 items that were thus obtained represented a true straight frequency distribution. From this group, 50 items were selected, one by one, and by pure chance. This was repeated until 20 groups of 50 items each were obtained. These 20 series were plotted on frequency paper. Five series fitted the original straight line well; nine series fitted the line fair; six series fitted the line poorly; two of them would not fit a straight line at all!

It is recognized that such an experiment is no conclusive proof that frequency lines have to be straight. The basic series in this example had a straight-line distribution to start with, which is an unknown quantity in an actual case. However, it is definitely demonstrated that a curved appearance of the plotted points can easily be a chance deviation from a true straight line.

In many cases a frequency curve of maximum annual flood peaks at a given river station is based on the flood peaks recorded at that particular station. Little or no attention is paid to frequency curves at river stations upstream or downstream, or in similar drainage basins. However, it can be demonstrated that definite relationships exist between such frequency curves and a study of these relationships is of help in drawing frequency curves of flood flows of river stations.

The peak flow of a flood depends upon many variables. First of all, there is the factor of chance. In one year, the flood peak may be large; in the next year, it may be small. This factor of chance is the principal cause of the variability of peak flows that can be found at any stream gauging station. However, chance is not the only factor that determines the variation between the extreme low and high peaks. The characteristics of the drainage basin also play a role.

In the following paragraphs, the various factors that influence the magnitude of the peak flow will be discussed. The purpose of the discussion is to find out to what extent these factors have a bearing on the inclination of the frequency curves of the flood peaks. This discussion will first be in general terms. Subsequent paragraphs will deal with specific cases, and it will be seen how the findings of this section apply to the frequency curves of stations on the North American Continent.

One of the main characteristics of a drainage basin is its size. What is the influence of size upon the inclination of the frequency curve? Assuming that all other drainage basin characteristics (climate, soil, topography) remain the same, the increase in size of a drainage basin will reduce the inclination of the frequency curve of its maximum annual flood flows. The reason is that in a larger drainage basin there are more independent causes for a flood. The mean flood will increase approximately in proportion with the drainage area, since it is made up of average conditions. Extreme floods, however, increase less than proportional since it is unlikely that extreme conditions coincide throughout the drainage basin. The situation is somewhat analogous to two roulette wheels. On one single roulette wheel with the numbers 0 to 36 inclusive (37 items in total) the average value of a great number of turns will be 18. The chance of getting a value of 34 or more is 3 in 37 or 1 in 12·3. The ratio of 34 to 18 is 1·89. On two roulette wheels, that are spinning at the same time, and where the values will be totalled up, the average value will be 36, or double what it was on one wheel. What are now the chances of getting a value of 68 or more, which has again a ratio of 1·89 to the mean. A simple calculation reveals that the chance is 15 in 1,369, or 1 in 91, instead of 1 in 12·3.

From the above discussion, it follows that frequency curves have a tendency to become flatter when the drainage area increases, provided that all other drainage basin characteristics remain the same. This tendency towards a decrease in ratio between the extreme floods and the mean flood is caused by the absence of complete correlation between the peak flows on the tributaries and main stem of a river system. One part of a drainage basin may produce a large flood but that does not necessarily imply that the other parts produce large floods also. The larger the drainage basin, the less correlation there is between the production of floods in the various sub-basins, and the smaller the ratio between the extreme floods and the mean.

The next characteristic that we shall discuss is topography. Consider two drainage basins, identical except for the fact that one has a topography that is twice as steep as the other. What effect will this have upon the frequency curve of the flood flows? Steep slopes mean a quick run-off of the precipitation and therefore a hydrograph with a relatively sharp peak. On account of the principle of the unit-hydrograph, it can be expected that a change in its shape will affect all floods to the same proportion. Hence, all flood peaks in the drainage basin with the steeper topography are a certain ratio larger than the flood peaks in some other basin, provided that the run-off coefficient remains the same. In other words, topography has no direct bearing on the steepness of the frequency curve. It was assumed in the above that the run-off coefficient remained the same in both cases and at all flows. However, it can be expected that for low flows, the run-off coefficient is smaller than for high flows, and that the difference is smaller with steeper topography. This secondary effect of changing run-off coefficient will be discussed later.

To study the effect of climate, let us consider two drainage basins, identical except for the fact that one is located in a region with a wet climate, like the Atlantic Coast of the United States; and the other in a region with a dry climate, like the Canadian Prairies. The wet climate will result in a relatively large figure of average annual rainfall and it can be expected that the magnitude of the mean flood is also relatively large. However, it has been learned from experience that the extreme rainstorms, and therefore the extreme floods, may very well be of the same order of magnitude in both places. There are even dry regions where the extreme rainfalls are larger than in wet regions. In terms of flood magnitudes, this means that the dry regions can be expected to produce a relatively large ratio between extreme floods and the mean flood. In terms of the frequency curve, it means that the drainage basin in the dry region will show a relatively large inclination.

The fourth and last important characteristic is the composition of the soil. Consider two drainage basins, identical except for the fact that one has an impervious soil, resulting in a high run-off coefficient, and the other has a loose porous soil, resulting in a low run-off coefficient. What effect will this have upon the inclination of the frequency curve? In the case of the impervious soil, the run-off coefficient may range from a low of, say, 60 per cent during the moderate rainfalls to a high of 90 per cent during heavy rainfalls of short duration. This means that the peak flows of the floods are approximately proportional to the size of the rainstorm.

In the drainage basin with the pervious soils, it can be expected that during moderate rainfalls, the infiltration capacity of the ground is so large that most or all of the precipitation will be absorbed in the soil and hence that very little water will reach the streams. The run-off coefficient may be as low as a few per cent. During heavy rainfalls, however, especially if they are of a short duration, the water that infiltrates in the ground constitutes only a fraction of the total amount of precipitation, and as a result a relatively large proportion will reach the streams. The run-off coefficient may become as high as, say, 50 per cent. This means that in the case of the pervious soil, the peak flows increase much more rapidly than the size of the rainstorm and, as a result, the ratio between the extreme floods and the mean flood is larger than in the case of the impervious soils. In other words, pervious soils tend to produce steeper frequency curves.

The effect of the drainage basin characteristics, as discussed in the previous paragraphs, can be found in the frequency curves of actual river flows. In order to facilitate the analysis, it would be desirable if the inclination of the frequency line could be expressed in some numerical parameter. One of the simplest ways of doing this would be to quote the point where the frequency line intersects the vertical line, representing the frequency value of 1 per cent. This parameter will be called R, which is then equivalent to the ratio of the 1 per cent flood to the mean flood. The problem is now to determine how R is influenced by the size of drainage basin, the climate, and the soils. It would be desirable if these independent variables could also be expressed in numerical parameters.

The first variable, the size of the drainage basin, can be expressed easily in numerical terms; namely, in square miles. The second and third variables are more difficult to express. Fortunately, they both produce similar results with regard to R and the mean flood. When the climate is dry, the mean flood is bound to be low and R can be expected to be high. When the soils are porous, the mean flood will also be low and R high. Therefore, it would seem logical to relate R to the mean flood instead of taking both the climate and the soils into consideration. The mean flood in itself is not a desirable indication because it also depends upon the size of the drainage basin. In order to overcome that difficulty the mean flood will be divided by the area of the drainage basin to the 0·8 power. Thus a parameter F, 'flood coefficient', is obtained that has from experience proved to be fairly constant when the characteristics of the drainage basin are constant, and when only the size of the basin varies. It can then be expected that, other variables being constant, R will be low when F is high, and vice versa.

Thus there are two independent variables, size of drainage area D.A. and flood coefficient F; and the one dependent variable, the ratio of the 1 per cent

Table 2.3 A–E. *Relationship of Inclination of Frequency Curves with Drainage Basin Characteristics*

TABLE 2.3A

Fig.	No.	River	Station	D.A.	F	R
	62	French Broad	Ashville	950	72·6	2·6
	63	Tennessee	Knoxville	8,900	65·7	2·4
	64	Tennessee	Chattanooga	21,400	69·0	1·8
16	65	Tennessee	Florence	30,800	64·4	1·7

TABLE 2.3B

Fig.	No.	River	Station	D.A.	F	R
4	6	Pembina	Neche, N.D.	3,200	3·0	4·4
	1	R.R. of the N.	Fargo, N.D.	6,800	2·9	4·4
2	2	,, ,, ,, ,,	Grand Forks, N.D.	15,600	4·6	3·7
	3	,, ,, ,, ,,	Emerson, Man.	36,200	4·7	3·6
	5	,, ,, ,, ,,	Winnipeg, Man.	104,500	3·3	3·0

TABLE 2.3C

Fig.	No.	River	Station	D.A.	F	R
5	11	Qu'Appelle	Tantallon, Sask.	17,500	0·4	5·8
	69	Arkansas	La Junta, Colo.	12,000	4·7	4·3
	29	Minnesota	Mankato, Minn.	14,900	6·2	3·6
	95	Snake	Neeley, Idaho	14,100	10·9	2·8
	39	Des Moines	Keosanqua, Iowa	13,900	20·0	2·5
15	61	Cumberland	Nashville, Tenn.	12,900	63·0	1·7

TABLE 2.3D

Fig.	No.	River	Station	D.A.	F	R
	66	Arkansas	Salida	1,200	10·7	2·0
17	67	,,	Canon City	3,100	7·7	2·8
	68	,,	Pueblo	4,700	7·1	3·2
	69	,,	La Junta	12,200	4·7	4·3
18	70	,,	Holly	25,100	2·9	4·5
	71	,,	Wichita	40,200	1·9	5·0

TABLE 2.3E

Fig.	No.	River	Station	D.A.	F	R
9	28	Minnesota	Montevideo	6,200	3·1	5·4
	29	,,	Mankato	14,900	6·2	3·6
7	25	Mississippi	St. Paul	36,800	8·3	2·5
8	26	,,	Le Claire	88,600	15·1	2·1
	27	,,	Keokuk	119,000	15·4	2·0
20	76	,,	Vicksburg	1,144,500	18·1	1·7

Data from Kuiper (1957): 100 frequency curves on the North American Continent flood to the mean flood R. Since there are two independent variables, it can be expected that the effect of one variable may obscure or even dominate the effect of the other one. For this reason the analysis will first of all be concerned with examples where either F or D.A. is kept more or less constant, and afterwards with examples where both vary at the same time.

The first example concerns the Tennessee Basin, where F is fairly constant and where the effect of D.A. upon R can be observed. It can be seen that R decreases when D.A. increases. The data are assembled in Table 2.3A. The second example concerns the Red River of the North Basin where F is also fairly constant. It can be seen from Table 2.3B that R decreases when D.A. increases. The third example concerns a variety of river stations, all with approximately the same D.A., but with different value of F. It can be seen from Table 2.3C that R decreases with an increase of F. The fourth example, shown in Table 2.3D, concerns the Arkansas Basin, where F decreases with increasing drainage area. Although R is supposed to decrease with increasing drainage area, the variable F apparently dominates to such an extent that R increases instead of decreases. The last example, shown in Table 2.3E, concerns the Mississippi Basin, where F increases with increasing drainage area. Here, a rapid decrease of R with increasing D.A. can be noted because the increasing F and the increasing D.A. have the same effect upon R.

If an engineer is faced with the problem of drawing a frequency curve of maximum annual flood flows for a river station, and if he has decided that the frequency curve can be drawn as a straight line, and if he has access to flood frequency data at other places in the drainage basin or in similar drainage basins, he would be well advised to study all these available flood frequency data. There is a good possibility that he will find definite trends in the inclination of the frequency lines under consideration, in which case he could use these trends as guidance for his estimates.

Peak–Volume Relationship

In some cases the hydraulic engineer is not only interested in the peak of a flood of given frequency, but also in its volume. For instance, in calculating a spillway design flow, one determines first the reservoir inflow hydrograph. This hydrograph is routed through the reservoir and the resultant maximum spillway flow becomes the spillway design flow. The study of frequency curves of maximum annual flood peaks, as discussed in the foregoing paragraphs, only provides the peak of the flood, and is therefore insufficient to cope with such problems as spillway design floods.

There are three different methods to remedy the situation. First, one could establish the basic shape of flood hydrographs at the river station under consideration, by analysing recorded floods, in a fashion similar to the one discussed under unit hydrographs. The ordinates of the resultant 'basic hydrograph' are then multiplied by such a coefficient that the peak becomes equal to the peak flow found from the frequency study. This method is simple, but may lead to an overestimate of the reservoir inflow hydrograph, since extreme floods tend to be relatively more peaked than the ordinary floods from which the 'basic hydrograph' was derived.

A second method is to route all recorded floods through the reservoir and to make a frequency study of the resultant outflows. This frequency curve is then extrapolated to the desired frequency, and the magnitude of the spillway design flood is thus found. This method has the danger that the recorded (ordinary) floods may undergo a much greater percentage of peak reduction in the reservoir routing than the extreme floods will. The frequency curve of the outflows would consequently become more suppressed than it should, and as a result the spillway design flow is underestimated.

The third method is more accurate, but also requires more work. A study is made of the correlation between the peak and the volume of maximum annual floods. Every flood is plotted in hydrograph form, the base flow is sketched in,

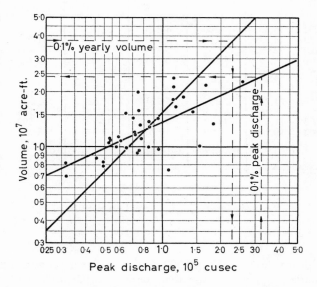

Figure 2.10. Flood volume–peak discharge relationship
(Saskatchewan River at Nipawin)

and the volume of the remaining flood flow determined by planimeter. The values of peak flow are plotted versus the corresponding values of volume as shown in *Figure 2.10*. Two regression lines are then computed as shown in Table 2.4. One line gives the most likely volume for a given peak, while the other line gives the most likely peak for a given volume. To understand the statistical theory behind the meaning of these regression lines, reference is made to textbooks on statistics and correlation, such as by Ezekiel (1945).

Application of the third method will be illustrated by the following example. The design flood for the dikes of the Saskatchewan Delta reclamation project was selected at a probability of exceedence of 0·1 per cent per year. A study was made of frequency curves of maximum annual flood flows in the Nelson Basin, resulting in the selection of a frequency curve at Nipawin, at the head of the delta. Since the delta provides a large amount of storage capacity, it was necessary to determine also the volume of the design flood. Therefore a correlation between peaks and volumes was established and a frequency curve of maximum annual flood volumes was made up. It was then possible to prepare a design flood with a peak of 320,000 cusec (0·1 per cent on the peak–frequency curve) and a corresponding volume of 24,000,000 acre-ft., with a shape resembling the basic hydrograph. In addition, another design flood was

Table 2.4. *Computation of Regression Lines: Peak–Volume*
(*Saskatchewan River at Nipawin*)

Year	Peak Flow in 1,000 cusec	Volume in 1,000 acre-ft.	x Log (vol.)	y Log (peak)	x.y	x^2	y^2
1913	75·1	15,875	7·200	4·875	35·109	51·850	23·773
1914	94·3	14,329	7·156	4·974	35·598	51·211	24·745
1915	260·0	22,793	7·357	5·414	39·842	54·137	29·321
1916	143·0	23,769	7·376	5·155	38·025	54·405	26·577
—	—	—	—	—	—	—	—
—	—	—	—	—	—	—	—
1948	167·6	21,767	7·337	5·224	38·334	53·843	27·292
1949	33·3	8,167	6·912	4·522	31·261	47·776	20·454
1950	77·3	11,020	7·042	4·888	34·423	49·592	23·894
1951	74·3	19,944	7·299	4·870	35·555	53·287	23·724
			276·618	191·404	1,358·300	1,962·755	940·847

$$x = 7·092 \qquad \bar{y} = 4·907$$

$$y^1 - \bar{y} = m_x(x - \bar{x})m_x = \frac{xy - n\bar{x}\bar{y}}{x^2 - n\bar{x}^2} \qquad y^1 - 4·908 = 0·935(x - 7·093)$$

$$x^1 - \bar{x} = m_y(y - \bar{y})m_y = \frac{xy - n\bar{x}\bar{y}}{y^2 - n\bar{y}^2} \qquad x^1 - 7·093 = 0·483(y - 4·908)$$

prepared with a volume of 38,000,000 acre-ft. (0·1 per cent of the volume–frequency curve) and a corresponding peak of 230,000 cusec. Both floods were routed through the delta area. In the upper part of the delta the first flood resulted in the highest stages, which were adopted for design purposes. In the lower part of the delta, the second flood with its larger volume resulted in the highest stages.

UNIT-HYDROGRAPH

It was noted earlier that, for design and other purposes, the hydraulic engineer must have knowledge of the nature of flood flows. If there are insufficient data available to apply a frequency analysis as discussed in the previous section, or if the engineer wants to verify the frequency analysis, some other approach must be followed. For instance, one could assume a hypothetical rainstorm over the drainage basin, estimate the amount of run-off, estimate in what shape the run-off will come down the river, and thus find the peak flow. One of the many difficulties in such a procedure is to estimate the shape of the stream flow hydrograph. To facilitate this, the unit-hydrograph method has been devised.

The principle of the unit-hydrograph method was developed by Sherman (1932), and is based on the following precipitation–run-off behaviour: a uniform rainfall of short duration over a small drainage basin produces a certain hydrograph; when a larger rainfall of the same duration falls over the same drainage basin, the resultant hydrograph has the same base width, and all its vertical ordinates are larger in the same proportion. Therefore, if one could detect the basic shape of such a hydrograph, one could draw any resultant stream flow hydrograph at that station for a given amount of run-off. This analysis of the basic shape is done as follows.

Going through the records of the gauging station and through the meteorological records, one selects all hydrographs that were caused by short, uniform rainfalls over the drainage basin. Let us assume that *Figure 2.11(a)* is one of those hydrographs. First of all, the base flow, that results from ground water or earlier rain storms, is sketched in the hydrograph. All vertical ordinates above the base flow line are now enlarged or reduced, in the same proportion, to such

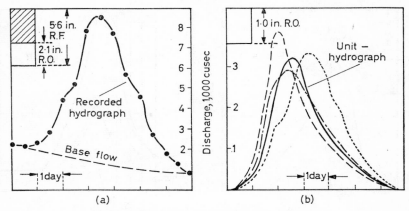

Figure 2.11. Development of unit-hydrograph

an extent that the resultant hydrograph has a volume of 1 in. of water over the drainage basin. This is repeated for all available hydrographs. The resultant graphs are all plotted on one sheet of paper as shown in *Figure 2.1(b)*. Through this bundle of graphs, one sketches now by eye an average graph, which is called the 'unit-hydrograph' and which can be defined as follows. It is the hydrograph, with a volume equal to 1 in. of water over the drainage basin, produced by a uniform storm, of unit length. An example of a unit-hydrograph with its notations is shown in *Figure 2.12(a)*.

There are some limitations to the use of the unit-hydrograph method. First, the size of the drainage area should not be larger than a few thousand square miles. For larger drainage basins, the precipitation is usually not evenly distributed; moreover, the effect of channel storage upon flood peak celerity begins to play a role, so that the base width of the hydrograph is not the same any more for storms of different intensity. Second, it has been found from experience that the unit time of rainfall duration should be no more than about a quarter of the time lag between the beginning of rainfall and the peak of the hydrograph. In small drainage basins where the time lag is one day or less, one would have to have the rainfall recorded in hours. Such records may not be available. In larger drainage basins where the time lag is several days, the critical run-off in which one may be interested may be caused by snow melt that is extended also over several days!

Assuming that we have an applicable situation, and assuming that a unit-hydrograph has been prepared, let us discuss an example of applying the unit-hydrograph. From a study of rainstorms we have found an exceptionally large three-day storm that occurred somewhere else, but that could occur just as well over the drainage basin under consideration. We wish to know what peak flow that storm would produce in our basin. This storm is now centred (on paper) over the drainage area. A study is made of the probable infiltration losses (this subject will be discussed in a following section under coaxial relations) for each of the three days, and thus the amount of run-off, in inches of water over the drainage basin, can be found. Let us assume we

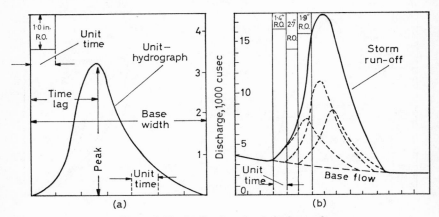

Figure 2.12. Application of unit-hydrograph

have 1·4 in. of run-off during the first day, 2·7 in. during the second day and 1·9 in. during the third day. We now estimate the probable base flow that could be expected prior to the rainstorm, as shown in *Figure 2.12(b)*. Upon the base flow we plot three individual hydrographs, offset by one day each, representing the run-off from the three-day rainstorm. These three hydrographs are obtained by multiplying the ordinates of the unit-hydrograph by the factors 1·4, 2·7 and 1·9, respectively. All ordinates are now totalled and thus the peak of the flood is found.

If all drainage basin characteristics are suitable for the application of the unit-hydrograph method, but no basic data on short uniform storms with their resultant run-off are available, one could develop a synthetic unit-hydrograph from data from similar drainage basins. If such data are not available either, empirical relationships as developed by Snyder (1938) could be used.

MAXIMUM POSSIBLE FLOOD

When the spillway of a large dam is designed, the hydraulic engineer wants to minimize the risk of failure, and therefore he wants to know what extreme flood conditions may occur at the reservoir site. Instead of relying on the frequency analysis of flood flows, he may want to investigate the situation with a storm–run-off analysis. If he wants to take no risk at all with his spillway design flood, he should determine the maximum possible flood at the reservoir site. If a small risk can be tolerated he may apply less severe criteria and determine a 'maximum probable flood'. A discussion of the virtues of frequency analysis, the maximum possible flood, and the maximum probable flood for spillway design floods is contained in Chapter 5: Hydraulic Structures. For the present discussion it suffices that the engineer may be interested in the maximum possible flood.

The maximum possible flood is defined as the absolute limit which nature may reach when all the maximum possible meteorological and drainage basin conditions are added together in critical sequence. It was noted that this maximum possible flood is not necessarily identical with the design flood, but that its magnitude should be known, in order to make an intelligent estimate of the appropriate design flood. In the following paragraphs the sequence of steps required to determine the maximum possible flood is discussed. This general discussion is followed by a résumé of the determination of the maximum possible flood of the Missouri River at Garrison.

The first step is to analyse the various types of floods in the drainage basin under consideration, and to determine which type of flood (e.g., spring snow melt or summer storm) has the largest flood-producing capacities. In some cases the choice of type of flood may be selected after a brief study; in still other cases one may have to carry the study of the different possible types of flood to the very end before it becomes known which produces the largest river flows.

The second step is to determine the maximum possible amount of precipitation (rain or snow or a combination of both) that can occur for various durations of time and over areas of various size in the drainage basin under consideration. Such a study is carried out by meteorologists and takes into account the geographical location of the area, the sources of atmospheric moisture and thermodynamic relationships. A discussion on this subject with selected data for 150 great storms may be found in a paper by Bernard (1944). Further information on area duration and depth values for selected storms may be found in U.S. Weather Bureau Bulletin No. 23 (1947).

The third step is to determine the minimum infiltration losses that may take place while the rainfall or snow melt water runs over the ground towards the stream channels. These infiltration losses may range from zero (where the ground is rock or frozen soil) to a substantial part of the gross precipitation. Such infiltration losses should be determined from a study of rainfall and run-off in the same or in a similar basin. In simple cases the result of such a study may be a selected rate of infiltration in inches per hour. In more complicated cases, such as the spring run-off from snow melt, the infiltration is a function of several variables. If sufficient data are available a coaxial relation may be established between the volume of run-off as a dependant variable, and soil moisture conditions, snowfall, rainfall, and melting conditions as independent variables. This subject will be further discussed in the next section. In other cases, where information about the drainage basin characteristics is very meagre, one may have to resort to the application of a run-off coefficient.

45

However, this should be avoided as much as possible since the run-off coefficient of extreme floods tends to be much larger than for observed moderate floods.

The fourth step is to determine at what rate the volume of water, determined above, will run off into the river system. For this purpose the unit-hydrograph principle is followed. The drainage basin under consideration is divided in sub-basins of homogeneous characteristics and of limited size (up to a few thousand square miles). For each sub-basin, unit-hydrographs are determined. With the volume of water from the third step and the unit-hydrographs from the fourth step, a series of extreme hydrographs is obtained, representing the outflow from the sub-basins.

The fifth and last step is to route the sub-basin hydrographs down the river system to the location under consideration, making due allowance for the time of occurrence of each contributing hydrograph. When the problem is one

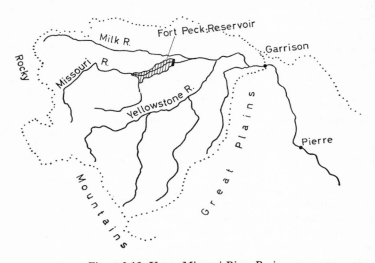

Figure 2.13. Upper Missouri River Basin

of determining a spillway design flood, it should be noted that the last reach in the flood routing sequence should be the reservoir, and that the peak of the outflow hydrograph becomes the spillway design flow.

To illustrate the foregoing, the determination of the maximum possible flood of the Missouri River at Garrison by the U.S. Corps of Engineers will be discussed. As a first step, early spring floods (resulting from snow melt from the Great Plains), late spring floods (resulting from snow melt in the mountains) and summer floods (resulting from rainstorms) were analysed. The Fort Peck Reservoir, shown on *Figure 2.13*, was assumed to control floods originating in the upper Missouri Basin. It was found that the most severe flood would be an early spring flood.

In the second phase of this study, meteorologists determined the maximum amount of snowfall that could possibly occur over the drainage basin. It was found that the maximum possible snow accumulation during the winter would be equivalent to 10 in. of water over the basin.

In the third step, an estimate was made of the various losses. It was assumed that a wet fall had saturated the ground, that a cold winter kept the ground in frozen condition, and consequently that no infiltration of melting snow into the ground would take place. The evaporation of snow during the winter was estimated at 2 in. The losses to run-off due to depressed areas which would not immediately contribute to the surface flow towards the river system was estimated at 1 in. The losses due to some snow drifting into places where the melting is delayed, beyond the critical period of run-off, was estimated at 0·5 in. As a result of these deliberations, the assumed snow melt run-off during the flood period became 6·5 in. of water over the drainage basin.

In the fourth step, the rate of run-off was determined. It was assumed that extremely high temperatures would prevail after 1 April, and that maximum south-westerly winds would prevail during the melting period, causing the melting of the plains snow cover to progress from west to east, so as to produce a high degree of synchronization of run-off from the tributaries. On the basis of these assumptions, it was found that all snow would melt in 19 days. The drainage basin between Garrison and Fort Peck reservoir was divided in 15 sub-basins, ranging in area from a few hundred to 10,000 square miles. For each sub-basin unit-hydrographs were developed and applied to the 19-day snow-water release.

It is of interest to note at this point, that in an alternative study of this situation an extreme rainstorm was assumed to have coincided with the snow melt. However, due to the drop in temperature that would necessarily accompany the storm, the melting of the snow was so much delayed that the addition of the rain did not result in a significant increase of the peak flow of the flood hydrograph.

In the fifth and last step the flood hydrographs of the sub-basins, plus the flood releases from the Fort Peck Reservoir (resulting from corresponding flood situations in its drainage basin), were routed down the river system towards Garrison. The peak of the maximum possible flood hydrograph was found to be 1,000,000 cusec and its volume 18,000,000 acre-ft.

COAXIAL RELATIONS

It was noted in the previous sections that in preparing hypothetical stream flow hydrographs from hypothetical rainstorms over the drainage basin, it is necessary to have a relationship between the amount of rainfall and the volume of resultant run-off. To obtain such a relationship is not always an easy matter, particularly in drainage basins where the major floods are caused by a combination of snow melt and rainfall. In such cases the volume of run-off depends at least on five major and more or less independent variables. First, the amount of rainfall during the snow melt period (which may range from a few days to a few weeks). Second, the concentration of the rainfall during this period. Third, the amount of snow that has accumulated during the winter. Fourth, the antecedent moisture conditions of the soil (which govern the rate of infiltration). Fifth, the rate of snow melt during the melting period (which determines over what period of time the water will be released). It would be ideal if a coaxial relationship could be established as shown in *Figure 2.14*.

To prepare such a diagram, one must solve two distinct problems: first, to find the proper index formula for the five variables, second to find the proper influence of each index. To find the proper index formula for the rainfall is simple, it can be the total rainfall in inches. The concentration index is not very difficult either, it could be the number of hours or days during which the rain falls. The snow index is a little more difficult. It could be the total snow fall, but that would not allow for evaporation losses of the early winter snow. One solution would be to make a separate study of the water content of the snow just before the melting period begins, and enter that figure as the index. The moisture index is the most difficult of all. What variable factors in the drainage basin determine the infiltration capacity during the melting period? One may have to go back one full year and see how wet the summer and autumn were. More weight must be given to rain just before the freeze-up than rain in the early summer. Perhaps the stream flow during the winter, indicating ground-

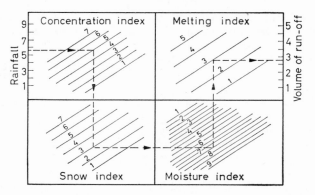

Figure 2.14. Coaxial relation between volume of run-off and independent variables

water levels, is a good index of soil moisture conditions. The melting index is also difficult to formulate. It depends on temperature, time of the season, wind, humidity, and sunshine. How to express all these variables in one figure that can be entered in the coaxial relationship? Here is ample opportunity to apply one's ingenuity. There is little point in suggesting hard and fast rules since the conditions and the available data change from basin to basin. After one has thought of formulae to compute the index values, the next step is to find out how valid the formula is. This is done as follows: Suppose there are three prospective formulae to compute the same index. Compute the index value for each formula, for all years on record, and plot index values versus volumes of spring run-off. See (or compute statistically) which formula yields the index with the best correlation. This is the one that should be adopted.

After having determined the index formulae, the next step is to determine the proper effect of each index upon the spring run-off. In other words, we now have to determine the proper slope, curvature and spacing of the sloping lines in *Figure 2.14*. This can be done by statistical analysis, which is rather complicated, or by trial and error. In the last case, one sketches in a bundle of curves in each of the four boxes. For the period of record, the run-off is determined with the coaxial relations and compared with the actual run-off. Any change in lines that improves the correlation between the computed and the actual values is an improvement of the coaxial relationship. After a while one develops a feeling where changes might have to be made in order to improve the diagram. Before long a satisfactory result may have been obtained.

It is of interest to point out that such coaxial relations are also very useful in stream flow forecasting for flood control or reservoir regulation. One might have to estimate a few factors that are not yet known, such as rainfall during the melting period, but having done so, a probable amount of run-off can be determined. This must then be applied to the unit-hydrograph, and the resultant flood routed down the river to the point of interest.

At one time the author attempted to use the coaxial relationship as a means of preparing a synthetic frequency curve of maximum annual flood flows. Frequency curves of all five variables were determined from recorded data. The frequency curves were combined, making due allowance for correlation between the variables, as is described in Chapter 6: Flood Control. Finally, the frequency curve of volumes of run-off was transformed into a frequency curve of flood peaks. When it was all over, it was felt that so many assumptions had to be made along the way, that the final result had no superior degree of reliability over the frequency curve of the recorded flood peaks.

CHAPTER 3

HYDRAULICS

This chapter has been written for the hydraulic engineer who is, or is going to be, engaged in water resources development studies. Therefore, more emphasis has been placed on practical aspects (such as determining the diameter of a conduit, the width of a spillway, or the dimensions of a canal) than on theoretical aspects (such as viscous resistance, boundary layer, or turbulence characteristics).

To acquire a feeling for what water will do under given circumstances is not an easy task. When given a series of connecting canals and spillways of different slopes and dimensions, there are few hydraulic engineers that can qualitatively predict under what surface profile the water will flow, without going first through a painstaking analysis. Yet to develop quick insight into how water will flow is a desirable objective. To achieve it, one needs first of all to understand the basic principles of flow, which will be discussed in this chapter. More important, however, is extensive and continued practice with a wide variety of flow situations.

HYDROSTATICS

Water may be defined as a substance that, in a state of rest, cannot maintain internal shear stresses. From this definition we can draw the following conclusions: (1) At any point, the pressure is the same in all directions; (2) At any point, the pressure (p, in lb./ft.^{2}) is equal to the specific weight of water ($\gamma = 62 \cdot 4$ lb./ft.^{3}) times the depth under the water surface (h, in feet), plus whatever pressure prevails above the water surface; (3) The force exerted by water pressure upon an element of plane is perpendicular to that element of plane.

With the above observations we can now determine the diagram of water pressure against a vertical wall, as shown in *Figure 3.1*. At A the pressure is zero. At B the pressure is:

$$p_B = \gamma . h = 62 \cdot 4 \times 90 = 5{,}620 \text{ lb./ft.}^2$$

The total pressure diagram consists of the triangle ABC. Hence, the total force against the wall is:

$$F_H = \tfrac{1}{2} . \gamma . h^2 = \tfrac{1}{2} \times 62 \cdot 4 \times 90^2 = 252{,}000 \text{ lb./lineal ft. of dam}$$

Since the resultant force of a pressure diagram acts through the centroid of that pressure diagram, and since the centroid of a triangle is situated at one third from the base, it follows that the force of 252,000 lb. acts at 30 ft. from the base of the dam, in a horizontal direction, and to the right.

To determine the force of water on inclined or curved planes, it is convenient to determine the horizontal and the vertical components of that force
separately, and to apply the following reasoning: to find the vertical component

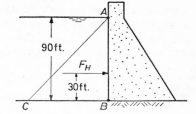

Figure 3.1. Force on vertical wall

of the force acting upon plane AB, consider all vertical forces acting upon the
element of water $ABCD$, shown in *Figure 3.2(a)*. The weight of that element of
water acts down, the reaction of the plane acts up. There are no vertical
forces along the boundary of the element since there can be no internal shear
stress. Hence, the vertical force F_v of the water upon the plane AB equals the
weight of the element of water, vertically above the plane. The line of action of
that vertical force must go through the centroid of that element of water. To
find the horizontal force acting upon plane AB, consider all horizontal forces

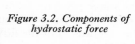

*Figure 3.2. Components of
hydrostatic force*

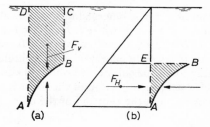

acting upon the element of water ABE, shown in *Figure 3.2(b)*. To the right
acts the force of the water pressure against AE, to the left acts the horizontal
reaction of the plane. Along EB there can be no horizontal force. Hence, the
horizontal force F_H of the water upon the plane AB equals the horizontal force
of water upon the projection of the plane AB upon a vertical plane.

To illustrate the above principles, let us consider the forces acting upon the
Taintor gate shown in *Figure 3.3*. The actual gate AC is part of a circle, with
B as centre and pivot point. The line AB is horizontal. The water level in the
reservoir is even with point A. The horizontal component of the force of
water against the gate: F_H, is to the right and equal to the pressure diagram
ADE:

$$F_H = \tfrac{1}{2}.\gamma.h^2 = \tfrac{1}{2}.62 \cdot 4.10^2 = 3{,}120 \text{ lb./lineal ft. of gate}$$

The vertical component of the force of water against the gate: F_v is upwards
and equal to the weight of the hypothetical volume of water ACF:

$$F_v = 62 \cdot 4 \times \tfrac{1}{2}\left[\pi \cdot 20^2 \cdot \frac{60}{360} - \frac{20 \times 20 \cos 30°}{2}\right] = 1{,}132 \text{ lb.}$$

51

The line of action of F_H is through the centroid of ADE. The line of action of F_v is through the centroid of ACF. However, there is no need to compute the location of these points since we can reason as follows. On every element of gate the total water force is perpendicular to that element and hence going through point B. Hence the sum of all elemental forces, which is the total force: R, must also have its line of action through B. The angle of this line with the horizontal is 1,132 vertical to 3,120 horizontal. Since the resultant force goes through B, there is no force required to keep the gate in the position shown.

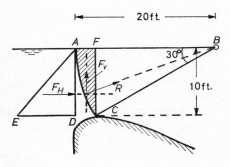

Figure 3.3. Forces on Taintor gate

When an object is submerged in water, as shown in *Figure 3.4*, the resultant vertical force of water pressure from above is equal to the weight of the shaded volume $ABDEC$. The resultant vertical force of water pressure from below is equal to the weight of the shaded volume $ABDGC$. Hence the buoyant force acting upon the submerged object is equal to the weight of the displaced volume of water; the direction of the force is vertically upward; and its line of action goes through the centroid of the object. It follows that when the object

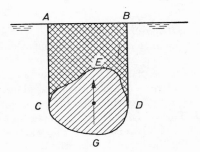

Figure 3.4. Buoyant force

weighs more than its buoyant force, it will sink; if it weighs less it will float up; if it weighs the same it will be in suspension. If the object weighs less, it will float at the surface and be submerged to the extent that the weight of the displaced volume of water is the same as the weight of the object.

To illustrate this principle let us consider the example of *Figure 3.5*. A reinforced concrete caisson is 40 ft. wide, 50 ft. long, 25 ft. high and weighs 450 tons. An air pump keeps the water out of the interior of the caisson. How deep will the caisson float under its own weight, and what air pressure is required? To answer the first question we can write:

$$450 \times 2{,}000 = 62 \cdot 4 \times 40 \times 50 \times d$$

$$d = 7 \cdot 2 \text{ ft.}$$

Hence the caisson will sink 7·2 d in the water under its own weight, and the required air pressure to keep the water out is

$$7 \cdot 2 \times \frac{62 \cdot 4}{144} = 3 \cdot 1 \text{ lb./in.}^2 \text{ (above atmospheric pressure).}$$

We could now ask what weight upon the deck of the caisson is needed to sink it to the bottom, and what is then the required air pressure? We can write:

$$450 \times 2{,}000 + W = 62 \cdot 4 \times 40 \times 50 \times 20$$
$$W = 1{,}600{,}000 \text{ lb.} = 800 \text{ tons}$$

Hence a weight of 800 tons is needed to sink the caisson to the bottom. The required air pressure in that position is

$$20 \times \frac{62 \cdot 4}{144} = 8 \cdot 7 \text{ lb./in.}^2$$

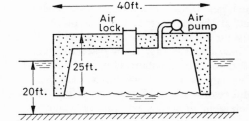

Figure 3.5. Floating caisson

BASIC EQUATIONS

After having discussed briefly the main principles of hydrostatics, the remainder of this chapter will be devoted to flowing water. Nearly all problems in hydraulics can be solved by one or a combination of three basic equations: the continuity equation, which states that whatever quantity of water passes through one section must also pass through another section; the energy or Bernoulli equation, which states that without friction losses the total energy of an element of water at one place must be equal to the total energy at another place; and the momentum equation, which states that force times a small element of time during which it is applied equals the resultant change in momentum of the element of water upon which it is applied. These three basic equations will be discussed in the following paragraphs.

Continuity Equation

In steady flow conditions (i.e., flow conditions where discharge is constant) we may observe that the discharge Q at one section must be equal to the discharge Q at another section, although the two sections may differ in wetted cross-sectional area A. Since water is, for practical purposes, an incompressible fluid, this means that we can write:

$$V_1 A_1 = V_2 A_2 = Q$$

This simple statement, that velocity times cross-sectional area must be the same in all sections, is known as the continuity equation. It is hardly necessary to amplify this principle with a numerical example.

Energy Equation

A small element of water in a flowing body of water may have five different forms of energy (capacity to do work). First, potential energy because of its elevation above a datum level; second, pressure energy because of its depth under the water level; third, kinetic energy because of its velocity; fourth, heat energy because of its temperature; and fifth, atomic energy which plays no role in the present problem because it does not change from place to place. The first three forms of energy are interchangeable. For instance, an element of water can change its potential energy for pressure energy and then back to its original potential energy. However, the change from the first three types of energy into heat energy is a one-way street. It is possible, but the reverse is not possible. Therefore, we call this an energy loss, from the viewpoint of the first three types. This energy loss takes place through boundary roughness, which generates turbulence, which generates internal friction, which generates heat, at the expense of the first three forms of energy. Therefore, when we consider a flowing body of water where the energy losses are nil or negligible, we may observe that from place to place the sum of the first three types of energy must be constant. How shall we express this in a formula? The dimension of energy is FL (Force times Length). The dimension of the weight of an element of water is $F = (\gamma . V = m . g)$. Hence, the dimension of the energy of an element of water per unit of weight is L, which is a convenient measure. We can now express the three forms of energy as follows:

$$\frac{\text{potential energy}}{\text{unit of weight}} = \frac{FZ}{F} = Z \quad \text{(elevation above datum)}$$

$$\frac{\text{pressure energy}}{\text{unit of weight}} = \frac{pV}{\gamma V} = \frac{p}{\gamma} \quad \text{(pressure head)}$$

$$\frac{\text{kinetic energy}}{\text{unit of weight}} = \frac{\frac{1}{2}mV^2}{mg} = \frac{V^2}{2g} \quad \text{(velocity head)}$$

Since, without energy losses, the sum of the three must be constant from place to place, we can write:

$$Z + \frac{p}{\gamma} + \frac{V^2}{2g} = H \quad \text{(total head)}$$

which is known as the Bernoulli equation. The sum of the first two terms, $Z + \frac{p}{\gamma}$, is known as the piezometric head. To illustrate the meaning of these terms reference is made to the water release conduit shown in *Figure 3.6*. Let us follow an element of water on its journey from position 1 to position 7.

In position 1, at the surface of the reservoir, the pressure head and velocity head are zero, and hence the elevation of 200 ft. above datum equals the total head of 200 ft. In position 2 the velocity is still negligible and so is the velocity head; hence, the pressure head is now 50 ft. and the elevation 150 ft. In position 3, the elevation is still 150 ft., but now there is a substantial velocity, say 25 ft./sec. (we shall see later how this velocity can be computed). Hence the velocity head is 10 ft. Hence the pressure head is 40 ft. In other words, from 2 to 3 the pressure energy is decreased, due to the fact that an exchange of pressure energy into kinetic energy had to take place. From 3 to 4 there is an exchange among all three forms of energy. The potential energy decreases to 45 ft. Let us assume that due to the narrowing of the conduit the velocity is twice as large: 50 ft./sec. This makes the velocity head 4 times as large: 40 ft. Hence the pressure head is now 115 ft. It may be noted in the figure that the

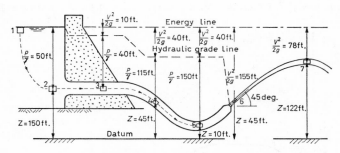

Figure 3.6. Bernoulli equation illustrated

line that shows the total head is called the energy line, while the line that shows the piezometric head is called the hydraulic grade line. From position 4 to 5 there is no change in kinetic energy (hence the hydraulic grade line remains parallel to the energy line). The only exchange is 35 ft. of potential energy into pressure energy. In position 6, immediately outside of the jet, the elevation is again 45 ft. Since the jet is surrounded by atmospheric pressure, the internal pressure must also be atmospheric (if it was higher the jet would expand, if it was lower the jet would contract, until the pressure became atmospheric). Therefore the velocity head is now 155 ft. and hence the velocity is 100 ft./sec. It should be noted at this point that we have thus a means of computing the discharge in the conduit. Assuming no energy losses, the velocity at point 6 is 100 ft./sec., regardless of the opening of the jet. Hence the discharge is 100 ft./sec. times the opening, say 50 ft.², which results in 5,000 cusec. Assuming a cross-sectional area of the conduit at point 3 of 200 ft.², the resultant velocity is found to be 25 ft./sec.; and assuming a cross-section area of the conduit at point 4 of the 100 ft.², the resultant velocity becomes 50 ft./sec. It is interesting to find out how high the jet would rise, again assuming no energy losses through friction with the air. From the laws of mechanics we know that in point 7 the horizontal component of the velocity is the same as in point 6, but that the vertical component is reduced to zero. Since the jet has an angle of 45 deg. with the horizontal, the horizontal component of the velocity is 71 ft./sec. Hence in point 7 the velocity head (or kinetic energy per unit of weight) is 78 ft., and therefore the elevation above datum is 122 ft.

In the above example it was assumed that there were no energy losses and therefore that the energy line was horizontal. In practice there are energy losses (in fact, in the above example, the energy losses would be substantial). Nevertheless, the principle of the Bernoulli equation can still be used. For instance, between the sections 1 and 5 we can write:

$$\left(Z + \frac{p}{\gamma} + \frac{V^2}{2g}\right)_1 = \left(Z + \frac{p}{\gamma} + \frac{V^2}{2g}\right)_5 + \Delta H$$

wherein ΔH represents all energy losses between the sections 1 and 5. These energy losses consist mostly of friction losses. In the section on conduit flow we shall discuss how these energy losses can be computed.

An interesting application of the Bernoulli equation is the venturimeter, shown in *Figure 3.7*, which is a locally narrowed section of conduit to measure

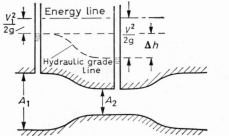

Figure 3.7. Venturimeter

discharge. Two stand pipes indicate the piezometric heads at a normal section and at the narrowed section. From consideration of the Bernoulli equation it follows that the difference in water level Δh equals the difference in velocity heads, assuming no energy losses between the two sections. After having measured the cross-sectional areas A_1 and A_2, and given a certain reading of Δh we can write:

$$\frac{V_2{}^2 - V_1{}^2}{2g} = \Delta h \quad \text{and} \quad V_1 A_1 = V_2 A_2$$

Hence we have two equations with two unknowns. We can solve for V_1, multiply by A_1 and find Q.

In *Figure 3.8* is shown a tank with an orifice, having an opening A_0. Since the flow lines of water cannot make a sharp turn at the edge of the orifice, the jet is somewhat contracted and has a cross-sectional area A_c smaller than A_0. Applying the Bernoulli equation to an element of water within the jet, and neglecting friction losses, we may observe that its velocity head is H, which is the distance of the centre of the orifice below the water surface in the tank. Hence the discharge from the orifice is:

$$Q = A_c \cdot \sqrt{2gH}$$

However, it is difficult to know A_c, unless one goes out and measures the jet! Hence the equation is written more conveniently in terms of A_0 with a correction coefficient. Moreover, we should allow for small friction losses. Therefore we write:

$$Q = C_d . A_0 . \sqrt{2gH}$$

wherein C_d is a coefficient that ranges from 0·6 for a sharp-edged orifice to 1·0 for a well-rounded opening.

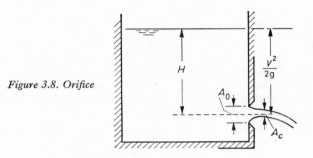

Figure 3.8. Orifice

Momentum Equation

In the mechanics of rigid bodies, the momentum equation is usually written in the following form:

$$F . \Delta t \quad \text{(impulse)} = m . \Delta V \quad \text{(change in momentum)}$$

However, in hydraulics, where we are dealing with a steady flow of water, it is convenient to rewrite the equation as follows:

$$F = \frac{m}{\Delta t} . \Delta V$$

Since mass divided by a small element of time is the same as discharge times density, we may write:

$$F = \rho . Q . (V_2 - V_1)$$

which is the momentum equation for hydraulics. Like all force–acceleration equations, this is a vector equation that should be applied to an x and y system of axis. The following examples will illustrate the use of this equation.

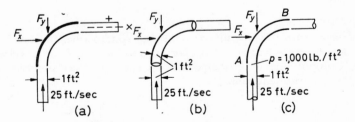

Figure 3.9. Application of momentum equation

In *Figure 3.9(a)* is shown a vane that deflects a jet of water with a velocity of 25 ft./sec. and a cross-sectional area of 1 ft.² Assuming that the experiment takes place in a horizontal plane, and that the friction losses are negligible, the exit velocity is also 25ft./sec (same kinetic energy). We can write the momentum equation as follows:

$$F_x = 1 \cdot 94 \times 25(25 - 0) = 1,210 \text{ lb.} \rightarrow \text{vane upon water}$$
$$= 1,210 \text{ lb.} \leftarrow \text{water upon vane}$$

$$F_y = 1 \cdot 94 \times 25(0 - 25) = 1,210 \text{ lb.} \downarrow \text{vane upon water}$$
$$= 1,210 \text{ lb.} \uparrow \text{water upon vane}$$

In *Figure 3.9(b)* the jet is carefully aimed into a 90 deg. tube with a cross-sectional area of 1 ft.^{2} Upon analysis, it turns out that nothing has changed from *Figure 3.9(a)* and hence the forces are again 1,210 lb. in the x and y directions. In *Figure 3.9(c)* the discharge is entirely within a pipe, with a cross-sectional area of 1 ft.^{2} Moreover, there is a pressure in the pipe of 1,000 lb./ft.^{2} Let us now consider what forces are acting upon the element of water in the elbow of the pipe, and what changes in velocity take place.

$$F_x - 1,000 = 1 \cdot 94 \times 25(25 - 0)$$
$$F_x = 2,210 \text{ lb.} \rightarrow \text{elbow upon water}$$
$$= 2,210 \text{ lb.} \leftarrow \text{water upon elbow}$$

$$F_y + 1,000 = 1 \cdot 94 \times 25(0 - 25)$$
$$F_y = 2,210 \text{ lb.} \downarrow \text{elbow upon water}$$
$$= 2,210 \text{ lb.} \uparrow \text{water upon elbow}$$

Assuming that the elbow of the pipe would be lying freely on a horizontal support, the tensile force in the metal at A as well as B would be 2,210 lb. Of this total force, 1,000 lb. is due to the pressure in the pipe, and 1,210 lb. is due to the change in momentum of the water.

CONDUIT FLOW

To understand the flow of water through conduits, and to solve such problems quantitatively, one must first of all be able to sketch qualitatively the energy line and the hydraulic grade line and then to compute the energy losses. The first aspect has been discussed before, the last aspect will be discussed in the following paragraphs.

Energy Losses

Experience with the flow of water through conduits has indicated that the energy loss for turbulent flow is proportional to the square of the velocity and inversely proportional to the diameter of the conduit. These observations have led to the empirical equation:

$$h_f = f \cdot \frac{L}{D} \cdot \frac{V^2}{2g}$$

This equation is known as the Darcy–Weisbach formula. The factor $2g$ has been entered for the sake of convenience so that the last term becomes the velocity head. The term h_f is called the friction loss and is expressed in feet. The term f is called the friction factor and should have been a constant if the above statement about proportionalities had been precisely correct. However, as it turns out f is a variable, dependent upon the roughness of the conduit, the diameter of the conduit, the velocity of flow and the viscosity of the fluid. This relationship is rather involved and can be found in most standard textbooks on hydraulics. For the present case, where we are dealing mostly with water, mostly with fairly large conduits in dams, and mostly with velocities that exceed 5 ft./sec., it is sufficient to make reference to *Figure 3.10* where f can be approximately determined as a function of the nature of the conduit material and the diameter of the conduit.

To illustrate the use of the equation and *Figure 3.10*, let us determine the friction loss in 200 ft. of conduit with a diameter of 10 ft. and a discharge of

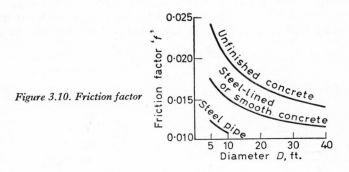

Figure 3.10. Friction factor

1,500 cusec. The conduit is made of smooth concrete. According to *Figure 3.10* the friction factor f is about 0·015. The velocity in the conduit can be computed at 19·1 ft./sec. Hence the friction loss becomes:

$$h_f = 0{\cdot}015 \times \frac{200}{10} \times \frac{19{\cdot}1^2}{64{\cdot}4} = 1{\cdot}7 \text{ ft.}$$

In addition to the friction losses in a uniform reach of conduit, there are energy losses due to sudden changes in diameter, bends, valves and so on. These sort of losses are known as 'minor losses', although in certain cases they may become the major part of the total losses. All minor losses are caused by local disturbance of the flow pattern. Such disturbances cause intense turbulence, which causes internal friction which causes energy loss. It has been found from experience that these minor losses, like the friction losses, are proportional to the velocity squared. Hence the minor losses are most conveniently expressed in terms of the velocity head of the flow. The following paragraphs list some of the more important empirical data on minor losses.

At the entrance from a reservoir to a conduit, or at the transition from a large diameter conduit to a small diameter conduit, the local energy losses may range from zero to 0·5 $\dfrac{V^2}{2g}$, if V is taken as the velocity in the conduit, downstream of the entrance or of the transition. The zero value applies when the entrance or transition is gradually and carefully rounded and smoothed out. The value of 0·5 applies when the entrance is sharp edged and when the transition is considerable and abrupt.

59

At the exit from a conduit into a reservoir, or at the transition from a small diameter conduit to a large diameter conduit, the local energy losses are given by the formula:

$$\Delta H = k \frac{(V_1 - V_2)^2}{2g}$$

In this formula V_1 is the velocity upstream of the widening (hence the high velocity), and V_2 is the velocity downstream of the widening (hence the low velocity; in case of the reservoir V_2 is zero). k is a coefficient that indicates the abruptness of the transition. For a sudden widening $k=1$; for a gradual widening under a slope of 10 longitudinal to 1 perpendicular, $k=0.2$.

The energy losses in the bend of a conduit depend on the angle and the sharpness of the bend. A reasonable value for the energy loss is 0·1 times the velocity head. Gates and valves have associated energy losses that vary over a

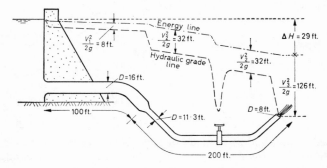

Figure 3.11. Energy losses in pipeline

wide range. An open gate that hardly disturbs the flow pattern may have losses that are as low as 0·1 times the velocity head. A needle valve or a partially closed butterfly valve may have energy losses that are several times the velocity head. For a more precise estimate of these kind of losses one should consult the handbooks on this subject, such as Abbett (1956) or Albertson (1960), or write to the manufacturer of the hydraulic equipment.

Let us now apply the above information about energy losses to the example of *Figure 3.6*, and compute what the actual energy losses would be. Let us assume that the conduit is steel lined and that the dimensions are shown in *Figure 3.11*. The losses at the entrance and first bend are estimated at $0.2 \dfrac{V_1^2}{2g}$. The friction losses in the wide conduit are:

$$h_f = 0.013 \times \frac{100}{16.0} \times \frac{V_1^2}{2g} = 0.08 \frac{V_1^2}{2g}$$

The losses in the transition from the wide to the narrow section of conduit are estimated at $0.1 \dfrac{V_2^2}{2g}$. The energy losses in the valve, the two bends and the nozzle shall be assumed at $0.5 \dfrac{V_2^2}{2g}$. The friction losses in the narrow conduit are:

$$h_f = 0.014 \times \frac{200}{11.3} \times \frac{V_2^2}{2g} = 0.25 \frac{V_2^2}{2g}$$

The sum of all these energy losses is represented by ΔH in *Figure 3.11*. It may also be seen that the sum of ΔH and the velocity head of the jet, outside of the nozzle, equals 155 ft. To solve for the unknown discharge, it would be convenient if all energy losses were expressed in terms of V_3. Since the ratios of V_1 to V_2 to V_3 are known, this can easily be done. We can now write the following equation:

$$\left(0.2 \times \frac{1}{16} + 0.08 \times \frac{1}{16} + 0.1 \times \frac{1}{4} + 0.5 \times \frac{1}{4} + 0.25 \times \frac{1}{4}\right) \frac{V_3^2}{2g} + \frac{V_3^2}{2g} = 155$$

or $V_3 = 90$ ft./sec., $\Delta H = 29$ ft., and $Q = 4,500$ cusec. It may be seen that the theoretical velocity of 100 ft./sec (neglecting all energy losses) has been reduced to 90 ft./sec due to friction in the conduit and local losses at bends and the valve.

Pumps and Turbines

If water has to be lifted to a higher level, we use a pump to accomplish this. In terms of the Bernoulli equation we could state that we add energy to the flow of water. If a potential fall of water is available, we can distract energy

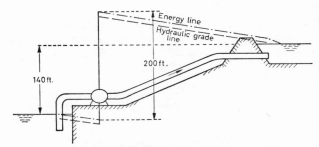

Figure 3.12. Pumping from reservoir into canal

by means of a turbine; or to put it in a different way: we convert energy of the water into electrical energy rather than let nature take its course and convert it into thermal energy. Let us try to express the flow of water over a certain height into terms of energy and power. If 1 lb. is lifted vertically over h ft., the amount of work required is h ft. lb. If 1 lb. is situated h ft. above datum it has a potential energy with respect to that datum of h ft. lb. If Q ft.^{3} of water is lifted up h ft., the amount of work done is $Q \times h \times 62.4$ ft. lb. If this is being done every second, the amount of power (rate of doing work) applied must be $Q \times h \times 62.4$ ft. lb./sec. Since 1 h.p. is defined as 550 ft. lb./sec, it follows that a flow of Q cusec over a height of h ft. is equivalent to

$$\frac{Q \times 62.4 \times h}{550} = \frac{Q \times h}{8.8} \text{ h.p.}$$

If we pump water up we must apply more motor power than the theoretically required power. Hence we must divide this expression by the efficiency e of the pump. If we lower water into a turbine to generate electrical energy, we expect to receive less than the theoretically available power. Hence we must multiply the above expression by the efficiency of the turbine. This leads to the following expressions:

For pumps, required motor capacity

$$= \frac{Q \times h}{8 \cdot 8 \times e} \text{ h.p.} \quad (e = \text{efficiency of pump})$$

For turbines, available generator capacity

$$= \frac{Q \times h \times e}{8 \cdot 8} \text{ h.p.} \quad (e = \text{efficiency of turbine})$$

The use of the first expression will be illustrated in the following example, shown in *Figure 3.12*. Water is being pumped from a reservoir into an irrigation canal that is 140 ft. higher, at a rate of 7 cusec, through 3,000 ft. of 1 ft. diameter pipeline. The problem is to determine the required h.p. of the pump, assuming an efficiency of 80 per cent. The friction factor f can be estimated at 0·015. The velocity in the pipe can be calculated at 9 ft./sec, which gives a velocity head of 1·25 ft. Hence the total friction losses in the pipeline are:

$$0 \cdot 015 \times \frac{3,000}{1} \times 1 \cdot 25 = 56 \text{ ft.}$$

The entrance, exit and elbow losses can be estimated at 3 times the velocity head, which makes the total energy losses in the pipeline 60 ft. Hence the pump must overcome a total head of $140 + 60 = 200$ ft. and the required capacity of the motor for the pump becomes:

$$P = \frac{7 \times 200}{8 \cdot 8 \times 0 \cdot 8} = 200 \text{ h.p.}$$

OPEN CHANNEL FLOW

We may divide open channel flow into steady and unsteady flow. Steady refers to time. If the discharge in a channel at any one point does not change with time, we are dealing with steady flow. If it does change with time, we have unsteady flow; for instance, when a wave is travelling in a channel or when the discharge is gradually increasing.

We may divide open channel flow also into uniform and non-uniform flow. Uniform refers to distance. If the wetted cross-sectional area of an open channel is the same from one location to another, we are dealing with uniform flow. If it does change from place to place, we have non-uniform flow; for instance, when the slope of the channel steepens, when the channel contains an obstruction, or when the flow is backed up by a dam.

In the following paragraphs we shall deal first with steady, uniform flow. The main problem here is to determine the discharge of a channel when its characteristics are given, or vice versa. In order to solve such problems we will derive the Manning formula. The next topic will be steady, non-uniform flow. This subject will be divided into two parts. The first group of problems will be concerned with situations where friction losses are relatively unimportant. In other words, the non-uniformity of flow takes place over short distances; for instance, channel transitions and channel obstructions. The main problem here is to determine the water surface profile. This will be done by using the Bernoulli equation and by introducing such concepts as the specific head. The second group of problems in steady, non-uniform flow will be concerned with situations whereby friction losses do play an important role. This group includes backwater problems. In order to solve these, we will discuss the general shape of backwater curves, and the various methods to compute surface profiles.

Uniform Flow

If a marble is released on an inclined plane, it will start rolling and it will roll faster and faster until it reaches a certain velocity at which it will keep

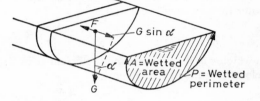

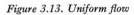
Figure 3.13. Uniform flow

rolling. The force that accelerates the marble in the beginning is the gravity component in the direction of the plane. The faster the marble rolls downhill, the more wind friction it generates. Finally comes the moment when this wind friction equals the gravity component. From then on the movement of the marble is steady.

Keeping this example in mind, we may look upon steady, uniform flow as a state of flow whereby the gravity component of the fluid in the direction of flow is equal to the friction force exerted by the boundary upon the fluid.

Let us consider an element of fluid, taken from the open channel. We may write: gravity component, $G \sin \alpha$ = friction force, F. Let the slope of the open channel be called S. We may then substitute S for $\sin \alpha$, provided α is small. The expression for the gravity component thus becomes: $\gamma . A . dx . S$. We know from laboratory experiments that the friction force is proportional to the wetted area of contact, proportional to the specific weight of the fluid and proportional to V^2, provided that the fluid is in turbulent motion. We may therefore write in general: friction force $=$ constant $. P . dx . \gamma . V^2$. Equating now the gravity component to the friction force, we obtain: $\gamma . A . dx . S =$ constant $. P . dx . \gamma . V^2$ or: $V = C\sqrt{RS}$. This equation is known as the velocity equation of Chezy, a French engineer who developed this formula in the last part of the 18th century. In this formula, C is the constant of Chezy and has to be determined by experiment; R is the hydraulic radius and equal to wetted area divided by wetted perimeter; S is the slope. Since we are dealing here

63

with uniform, steady flow, the slope of the bottom of the channel, the water surface, and the energy line are the same and S may represent any of the three. However, we apply the formula also when the flow is slightly non-uniform, and then it becomes important to realize that S in the Chezy formula, being associated with friction losses, represents the slope of the energy line S_e.

Many attempts have been made to determine the magnitude of the coefficient C in the formula $V = C\sqrt{RS}$. It was soon discovered that C is not a constant, but varies with the characteristics of the channel. One of the best-known early empirical formulae for the value of C was developed by Kutter in the middle of the 19th century:

$$C = \frac{41\cdot6 + \dfrac{1\cdot811}{n} + \dfrac{0\cdot00281}{S}}{1 + \left(41\cdot6 + \dfrac{0\cdot00281}{S}\right)\dfrac{n}{\sqrt{R}}}$$

in which n is another coefficient but now more constant than C itself. Because of the fact that this coefficient increases when the roughness of the channel bottom increases, this coefficient n is called coefficient of roughness. Gauckler, Hagen, and Manning (1891) proposed that C varies as the sixth root of R. Later investigators presented:

$$C = \frac{1\cdot49}{n} R^{1/6}$$

in which n is almost the same coefficient as in Kutter's equation. The complete velocity formula then becomes:

$$V = \frac{1\cdot49}{n} R^{2/3} S^{1/2}$$

and is widely known as the Manning formula. Later investigations in boundary resistance by Karman and Prandtl (1934) revealed that a fundamentally better expression for C is:

$$C = 33 \log 12 \frac{R}{k}$$

In this equation k is the diameter of imaginary balls, closely arranged on the channel bottom and giving the same resistance as the bottom under consideration.

At the present time there is not enough reason to abandon the Manning formula for application to open channel flow. The complicated nature of Kutter's equation is not justified by its slightly greater accuracy and in the Karman equation a great deal remains to be learned about the proper value of k. For artificial open channels with fixed boundaries, an appropriate n value can be found in handbooks such as Chow (1959). For river channels, particularly with movable beds, the selection of n becomes very difficult. This subject will be further discussed in the section 'River Channel Roughness'.

Non-uniform Flow

We will deal first with situations whereby the non-uniformity takes place over a relatively short distance. Relatively short in the sense that boundary friction losses can be neglected. If we exclude also the cases with local turbulence losses, then we have for practical purposes no energy losses at all, and the energy line is horizontal. In order to determine the profile of the water surface we may apply the Bernoulli equation. Let us consider the example of *Figure 3.14*. From section 1 to section 2 we may write:

$$z_1 + h_1 + \frac{V_1{}^2}{2g} = z_2 + h_2 + \frac{V_2{}^2}{2g} \quad \text{(Bernoulli equation)}$$

and
$$V_1 h_1 b_1 = V_2 h_2 b_2 \quad \text{(continuity equation)}$$

Strictly speaking we have to multiply the velocity head with a coefficient since the square of the average velocity V is not quite the same as the average of the squared velocities over the cross-section. However, this error, which is normally small, will be neglected.

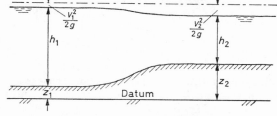

Figure 3.14. Non-uniform flow

It is evident that once the configuration of the channel and the flow condition in one of the sections are given, the flow condition in the other section can be computed.

For example:

    section 1:   9 ft. wide and 6 ft. deep
                  discharge is 207 cusec.

    section 2:   7 ft. wide and bottom 1 ft. higher.

What is the depth of flow at section 2?

We may write:

$$6\cdot0 + \frac{3\cdot8^2}{2g} = h_2 + 1\cdot0 + \frac{V_2{}^2}{2g}$$

$$207 = h_2 \times 7\cdot0 \times V_2 \quad \text{or} \quad h_2 = 4\cdot6 \text{ ft.}$$

or the problem could have been stated as follows:

    section 1:   9 ft. wide and 6 ft. deep.

    section 2:   7 ft. wide, bottom 1 ft. higher
                  and water surface 5 in. lower.

What is the discharge in the channel?

$$6\cdot0 + \frac{V_1{}^2}{2g} = 1\cdot0 + 4\cdot6 + \frac{V_2{}^2}{2g}$$

and
$$6\cdot9 V_1 = 4\cdot6 \times 7\cdot0 \times V_2$$

or

$$V_1 = 3\cdot8 \text{ ft./sec} \quad \text{and} \quad Q = 207 \text{ cusec}$$

3

The above problem is rather straightforward and may be solved in a simple way. However, there are many non-uniform flow conditions that are rather complicated and their analysis is greatly facilitated by the use of the so-called

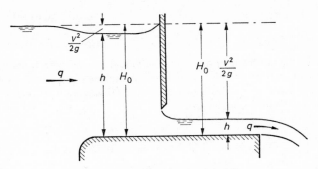

Figure 3.15. Alternate depths of flow

specific head diagram. In order to make this concept clear let us consider the flow in a short open channel, leading from a reservoir, and obstructed by a vertical gate as shown in *Figure 3.15*.

When the gate is in the position shown, we have the same discharge q on the left-hand side of the gate as on the right-hand side. We assume no friction

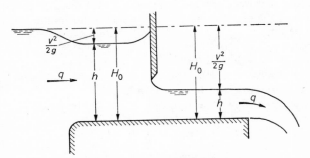

Figure 3.16. Alternate depths of flow

losses and therefore the total head on the left is the same as the total head on the right. Evidently there are two conditions of flow possible for the same total head, namely, a condition of tranquil flow (sub-critical flow) whereby the depth of flow is relatively large and the velocity head relatively small; and a condition of shooting flow (super-critical flow) whereby the depth of flow is relatively small and the velocity head relatively large.

Now we are going to raise the bottom of the channel, as shown in *Figure 3.16*, but we keep the discharge the same and therefore we have to raise the gate more than we raise the bottom. For a given discharge q we can evidently produce any depth of flow we like, and every time there are two alternate depths of flow that belong to the same specific head H_0. We should not confuse specific head with total head. Total head is velocity head plus depth of flow plus elevation of the bottom above datum. In the above we like to consider only the velocity head plus the depth of flow. This quantity is called the specific head. We may say then, that for every specific head and a given discharge, there are two alternate depths of flow.

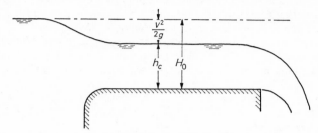

Figure 3.17. Critical depth of flow

Now let us go to the extreme. We remove the gate and raise the bottom so much that we produce again the same q as in the earlier experiments, as is shown in *Figure 3.17*. We have now the smallest possible specific head H_0 for the given discharge q. If we would further reduce H_0 we would reduce q! The depth of flow in this situation is called the critical depth h_c. We can establish the following relationship between h_c and H_0.

In general:
$$H_0 = h + \frac{V^2}{2g} = h + \frac{q^2}{2gh^2}$$

H_0 will have a minimum value for a given q when:
$$\frac{dH_0}{dh} = 1 - \frac{q^2}{gh^3} = 0 \quad \text{or} \quad h = h_c = \sqrt[3]{\frac{q^2}{g}}$$

combining the first and last equation:
$$H_0 = h + \frac{h_c^3}{2h_c^2} = \frac{3}{2} h_c \quad \text{or} \quad h_c = \frac{2}{3} H_0$$

Let us now represent in the form of a diagram shown in *Figure 3.18* what has been discussed in the above:

Figure 3.18. Specific head diagram

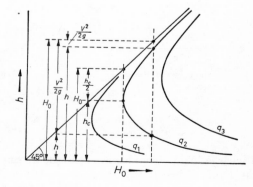

If we have a specific head diagram available for the range of flow conditions under consideration, we can take the proper q curve, look up the given H_0 value and find the two alternate depths of flow; or we may know one depth of flow and wish to know the alternate depth of flow, etc.

In the diagram as shown in *Figure 3.18*, each curve represents the relationship between h and H_0 for a constant value of q. At times it may be convenient to have a curve that represents the relationship between h and q for a constant value of H_0. Such a diagram is shown in *Figure 3.19*.

It is evident from the above diagrams that the maximum discharge for a given specific head is obtained when the depth of flow is two-thirds of the specific head. This is the flow condition we encounter at weirs, where the

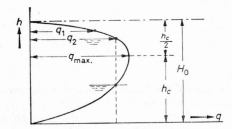

Figure 3.19. Discharge diagram

specific head is determined by the sill of the weir and the elevation of the energy line upstream (for instance, the surface of a reservoir). Since the natural free flow over a weir may be assumed the maximum possible for that particular specific head, we can then easily compute the depth of flow over the weir as two-thirds of the specific head.

It is of interest to note that in one of the earlier formulae, the critical depth h_c is a function of the discharge per unit width: q, only. In other words, when

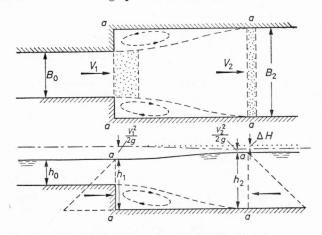

Figure 3.20. Energy loss at channel widening

the discharge in an open channel is given, we can determine immediately the critical depth. When the slope, cross-section, and roughness of the channel are also given, we can determine the normal depth of flow in the channel with the Manning formula. When this depth is greater than the critical depth, we have tranquil flow and the slope of the channel is called mild. When the normal depth of flow is smaller than the critical depth, we have shooting flow and the slope of the channel is called steep.

In the above we have discussed flow conditions whereby the non-uniformity took place over a short distance. In some situations the local turbulence losses may not be neglected, such as at an abrupt channel widening. In order to find the energy loss in such a case as is shown in *Figure 3.20*, we may apply the momentum equation: $F. \Delta t = \Delta(m.V)$.

Consider the free body of water *a-a-a-a*. During time dt it will change position as indicated by the shaded areas. Thus the free body has lost the momentum $V_1 \, dt h_0 B_0 \rho V_1$ and gained the momentum $V_2 \, dt h_2 B_2 \rho V_2$. We may now write the momentum equation as follows:

$$(\tfrac{1}{2}\rho g h_1{}^2 B_2 - \tfrac{1}{2}\rho g h_2{}^2 B_2) \, dt = V_2 \, dt h_2 B_2 \rho V_2 - V_1 \, dt h_0 B_0 \rho V_1$$

Since $V_1 h_0 B_0 = V_2 h_2 B_2$ we may write:

$$\tfrac{1}{2} g B_2 (h_1 + h_2)(h_1 - h_2) = V_2 B_2 h_2 (V_2 - V_1)$$

or

$$h_1 - h_2 = \frac{2h_2}{h_1 + h_2} \cdot \frac{V_2}{g} \cdot (V_2 - V_1)$$

Since $\dfrac{2h_2}{h_1 + h_2}$ is close to unity, we may approximate:

$$h_1 - h_2 = \frac{V_2(V_2 - V_1)}{g}$$

Since

$$h_1 + \frac{V_1{}^2}{2g} = h_2 + \frac{V_2{}^2}{2g} + \Delta H$$

We obtain

$$\Delta H = \frac{(V_1 - V_2)^2}{2g}$$

Backwater Curves

We will now discuss the large group of hydraulic problems where the non-uniformity in flow condition extends over a relatively long distance so that the

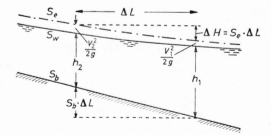

Figure 3.21. Elements of backwater curve

normal boundary friction losses have to be taken into consideration. We will first derive the general shape of backwater curves and then discuss the various methods of numerical computation. In *Figure 3.21* is shown the general case of a non-uniform flow condition. We apply the Bernoulli equation from section 1 to section 2, and include the loss in energy $\Delta H = S_e . \Delta L$.

$$h_1 + \frac{V_1^2}{2g} + S_e . \Delta L = S_b . \Delta L + h_2 + \frac{V_2^2}{2g}$$

$$\frac{\Delta h}{\Delta L} = S_b - S_e - \frac{1}{\Delta L} \Delta \frac{V^2}{2g}$$

or written in the form of a differential equation:

$$\frac{dh}{dx} = S_b - S_e - \frac{d}{dx}\left(\frac{V^2}{2g}\right) = S_b - S_e - \frac{d}{dx}\left(\frac{q^2}{2gh^2}\right)$$

Since $h_c^3 = \dfrac{q^2}{g}$ we may write:

$$\frac{dh}{dx} = S_b - S_e - \frac{d}{dx}\left(\frac{h_c^3}{2h^2}\right) = S_b - S_e + \frac{h_c^3}{h^3}\frac{dh}{dx}$$

or
$$\frac{dh}{dx}\left(1 - \frac{h_c^3}{h^3}\right) = S_b\left(1 - \frac{S_e}{S_b}\right)$$

Since $S_e = \dfrac{q^2}{C^2 h^3}$ and $S_b = \dfrac{q^2}{C^2 h_n^3}$ we get:

$$\frac{dh}{dx} = S_b \frac{1 - \dfrac{h_n^3}{h^3}}{1 - \dfrac{h_c^3}{h^3}} = S_b \frac{h^3 - h_n^3}{h^3 - h_c^3}$$

For a given channel, with a given discharge, h_c and h_n can be computed. We may then determine $\dfrac{dh}{dx}$ as a function of h and in terms of S_b. It should be kept in mind that in the above derivation x was taken in a positive sense in the upstream direction; and also, that a positive slope represents a lowering of elevation in the positive direction. Therefore, when we find that $\dfrac{dh}{dx}$ is positive and smaller than S_b, we are dealing with a slope that is in between horizontal and parallel to the channel bottom. A graphical key to the interpretation of the sense of $\dfrac{dh}{dx}$ is given in *Figure 3.22*. The general slope of backwater curves for mild and steep slopes is shown in *Figures 3.23* and *3.24*. Examples of backwater curves are shown in *Figure 3.25*. An explanation of these curves is given in the following paragraphs.

After having established the general shape of backwater curves on steep and mild slopes, as shown in *Figures 3.23* and *3.24*, it is of interest to make a few observations. When the flow in a canal with a mild slope is sub-critical (above critical depth) and the depth of the flow is not normal, the transition to normal

70

depth can only take place in upstream direction. There are no backwater curves that can lead from a non-normal depth to normal depth in downstream direction. Therefore, when we have an obstruction in a canal, like for instance a cofferdam, the depth of flow immediately downstream of the cofferdam must be normal. If it was more than normal or less than normal, there would be no possibility to get gradually back to normal in downstream direction. We may therefore conclude that if we have a disturbance or a discontinuity in a canal (like a change in slope, width, or roughness), and if the slope downstream of the

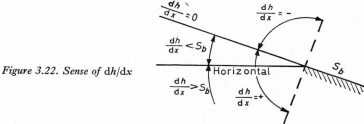

Figure 3.22. Sense of dh/dx

disturbance is mild, and if we know that the depth of flow at the disturbance is more than critical, then we must have normal depth of flow in downstream direction, starting immediately at the disturbance.

We can make a similar observation for steep canals with super-critical flow (below critical depth). It may be seen from *Figure 3.24* that the correction from a disturbed depth of flow to normal depth of flow can only take place in downstream direction. We may therefore conclude that if we have a disturbance in a canal, and if the slope upstream of the disturbance is steep, and if we know that the depth of flow at the disturbance is less than critical, then

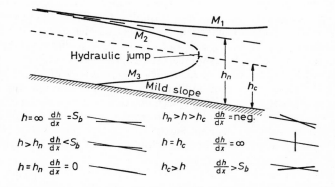

Figure 3.23. General slope of backwater curves on mild channel

we must have normal depth of flow in upstream direction, starting immediately at the disturbance.

These observations help us to sketch qualitatively the backwater curves in the different situations in *Figure 3.25*. For instance, in *Figure 3.25(b)*, the

71

slope downstream of the discontinuity is mild. The depth of flow must be more than critical since both slopes are mild. Hence the depth of flow at the discontinuity must be the normal depth that corresponds to the downstream slope. Having established this point we now consider it to be our starting point of the backwater curve on the upstream mild slope. Consulting *Figure 3.23*, we find that in between critical depth and normal depth we must have an M_2 curve. In *Figure 3.25(c)*, the slope upstream of the discontinuity is steep. Hence the depth of flow at the discontinuity must be the normal depth that corresponds to the upstream steep slope. This point is in between the normal depth and critical depth of the downstream slope. Hence, consulting *Figure 3.24*, we conclude that we have an S_2 curve in downstream direction. Similar

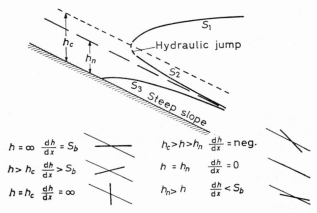

Figure 3.24. General slope of backwater curve on steep channel

reasoning can be applied to the *Figures 3.25(d)*, *3.25(e)*, *3.25(i)*, *3.25(j)*, *3.25(k)*, and *3.25(l)*. In the last four examples, the slope is kept constant but the width of the channel is changed, resulting in a change in critical depth as well as normal depth. In *Figure 3.25(a)*, we have a discontinuity from an upstream mild to a downstream steep slope. This is the only case where the above reasoning cannot be applied. Instead, we can look upon the steep slope as having the same effect upon the mild slope as an abrupt fall would have. It has been noted before that in such a case the depth of flow at the edge is equal to the critical depth. Hence upstream from that point we have an M_2 curve. Now considering the downstream situation, we have established the starting point of critical depth and the transition must be towards normal depth. Hence we have here an S_2 curve.

The situation in the *Figures 3.25(f)*, *3.25(g)* and *3.25(h)* calls for some further explanation. Upstream of the discontinuity we have a steep slope, and we would be tempted to conclude—hence a normal depth of flow. Downstream there is a mild slope, and again we conclude—hence a normal depth of flow. Right at the discontinuity these two unequal depths meet and the so-called hydraulic jump is formed, where the transition from super-critical to sub-critical flow takes place, as is shown in *Figure 3.25(f)*. However, in the section on the hydraulic jump (p. 88) we shall see that there is a relation between the depth and velocity of the super-critical flow on the one hand and the depth of the downstream sub-critical flow on the other hand. If the momentum of the incoming flow is too great, the hydraulic jump will be pushed to the right and an M_3 curve develops on the mild slope as shown in *Figure 3.25(g)*. If the momentum of the incoming flow is too small, the hydraulic jump will be pushed to the left and an S_1 curve develops on the steep slope as is shown in *Figure 3.25(h)*.

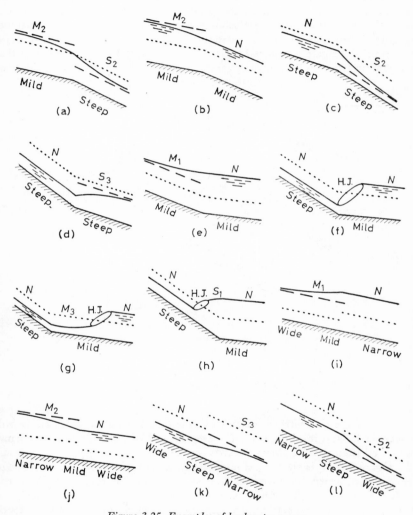

Figure 3.25. Examples of backwater curves

Now that we have gained some insight in where to expect normal depth and how to draw the connecting backwater curves in the vicinity of the discontinuity of an open channel, let us apply this knowledge to a somewhat more complicated situation. In *Figure 3.26(a)* is shown a channel with a mild slope that connects a reservoir with a falls. Halfway down the channel is a gate. What water surface profile do we expect? Let us first pretend that the gate does not exist. At the downstream end of the channel we must have critical depth. From that point upstream we must have an M_2 curve. If the channel is relatively long, normal depth will be reached. This normal depth must be such that the Manning equation is satisfied and that normal depth plus velocity head equals the height of the reservoir level above the entrance from the reservoir to the canal. For the data given in *Figure 3.26(a)* this works out to $h_n = 7.9$ ft. The corresponding discharge is $q = 67$ cusec/ft. Now we introduce the gate. This causes extra resistance and hence an M_1 backwater curve upstream. If the gate is lowered to 3 ft. from the bottom we get super-critical flow coming from underneath the gate and hence an M_3 curve in downstream

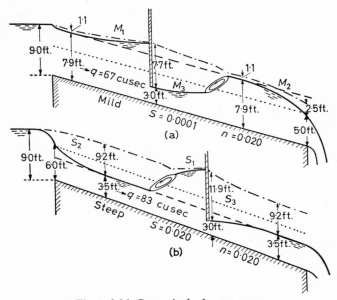

Figure 3.26. Composite backwater curves

direction. If the downstream distance to the falls is short, the M_3 curve will go over the falls; if it is long enough we get first a hydraulic jump, then normal depth, and then the M_2 curve towards the falls.

In *Figure 3.26(b)* is shown a similar situation, but now the channel is steep. Hence the discharge is governed by the outlet of the reservoir where the depth of flow is critical. With a velocity head of 3.0 ft., we find a discharge of 83 cusec/ft. From the reservoir downstream an S_2 curve towards normal depth of 3.5 ft. The gate causes backwater in the form of an S_1 curve and a hydraulic jump. Downstream of the gate, an S_3 backwater curve towards normal depth. The normal depth shoots over the edge of the falls.

After having determined the general shape of backwater curves, the next problem is to compute quantitatively the profile of the water surface. There are basically two different methods of doing this. In the first method we derive an equation for the entire water surface profile, and are thus able to compute the water level at any given point. It is obvious that this method is only applicable when the channel is uniform in slope, cross-section, and roughness coefficient for the entire stretch under consideration; a condition that seldom exists! The most widely used formula is the Bresse equation which is discussed later.

In the second method, we compute the water surface profile from section to section, using the earlier discussed principle that the slope of the energy line for any depth of flow in non-uniform flow conditions, is the same as it would be for uniform flow with the same discharge at the depth of flow under consideration. This second method is most widely used. When it is applied to uniform channels, an increase or decrease in depth of flow is assumed, and the length over which this occurs is found in a straightforward manner. When it is applied to non-uniform channels, sections of similar characteristics are selected, and the increase or decrease in depth of flow over any given section is computed by trial and error.

Bresse derived a general formula for the backwater profile by integrating the differential equation of the profile:

$$\frac{\mathrm{d}h}{\mathrm{d}x} = S_b - S_e - \frac{\mathrm{d}}{\mathrm{d}x}\frac{V^2}{2g}$$

resulting in:

$$x = \frac{h_n Z}{S_b} - h_n\left(\frac{1}{S_b} - \frac{C^2}{g}\right)\phi Z$$

In this formula:

x = horizontal distance in feet

h_n = normal depth

Z = ratio of actual depth to normal depth: $\dfrac{h}{h_n}$

S_b = bottom slope

C = constant of Chezy

ϕZ = complicated function of Z, summarized in Table 3.1. For more complete tables of this function, see appropriate handbooks.

Table 3.1. *Values of ϕZ as a Function of Z for M_1, M_2, and M_3 Curves*

Z	ϕZ	Z	ϕZ	Z	ϕZ	Z	ϕZ
0·00	0·00	0·95	1·47	1·03	1·06	1·40	0·30
0·10	0·10	0·96	1·55	1·04	0·97	1·50	0·26
0·50	0·52	0·97	1·64	1·05	0·90	1·60	0·22
0·60	0·64	0·98	1·78	1·07	0·79	1·80	0·17
0·70	0·78	0·99	2·02	1·10	0·68	2·00	0·13
0·80	0·95	1·00	~	1·15	0·56	2·50	0·08
0·85	1·07	1·01	1·42	1·20	0·48	3·00	0·06
0·90	1·23	1·02	1·19	1·30	0·37	5·00	0·02

To illustrate the use of the Bresse formula we will compute the water surface profile for the following situation. An open channel has a bottom slope of 0·0003, a width of 200 ft., a roughness coefficient of 0·025, and a discharge of 40,000 cusec. Due to a downstream disturbance (for instance a steepening of the bottom slope) the depth of flow at a certain section is 16·0 ft. Determine the water surface profile upstream of this section.

First of all, let us determine the normal depth and critical depth of this flow condition to find out what situation we are dealing with. From application of the Manning formula it is found that the normal depth of the channel is 25·5 ft. From application of the critical depth formula, it is found that the critical depth is 10·7 ft. It may therefore be concluded that the water surface profile is a backwater curve of the M_2 type, in other words a draw-down curve.

The C value in the Bresse formula is strictly speaking a function of the depth of flow. However, only a small error will be introduced by assuming C constant and giving it a value that corresponds to the average depth of flow between the first and last section, say 20 ft.

Hence
$$C = \frac{1\cdot5}{0\cdot025}\, 20^{1/6} = 100$$

The formula then becomes:

$$x = \frac{25\cdot5}{0\cdot0003}\, Z - 25\cdot5\left(\frac{1}{0\cdot0003} - \frac{100^2}{32\cdot2}\right)\phi Z$$

or
$$x = 85,000\, Z - 76,200\phi Z$$

and the computation of the water surface profile may be performed as follows:

Table 3.2. *Bresse Backwater Computation*

h	Z	ϕZ	x	Δx
16·0	0·63	0·67	+2,500	
				2,300
18·0	0·71	0·79	+200	
				3,200
20·0	0·78	0·91	−3,000	
				7,000
22·0	0·86	1·09	−10,000	
				18,000
24·0	0·94	1·40	−28,000	
				~
25·5	1·00	~	~	

The Bresse formula lends itself to an interesting manipulation. For small values of Z, that is, for slight disturbances of the normal depth, we can compute over what length Δx, the disturbance Δh is reduced to $\frac{1}{2}\Delta h$.

76

The formula becomes:

$$\Delta x = \frac{h_n}{S_b}(Z_1 - Z_2) - h_n\left(\frac{1}{S_b} - \frac{C^2}{g}\right)(\phi Z_1 - \phi Z_2)$$

which resolves approximately into:

$$\Delta x = 0 \cdot 25 \frac{h_n}{S_b}$$

To illustrate this equation with the foregoing example, let us assume that the normal depth of 25·5 ft. is distributed by a vertical lowering of 4 ft. This disturbance will be reduced to 2 ft. over a distance of:

$$\Delta x = 0 \cdot 25 \frac{25 \cdot 5}{0 \cdot 0003} = 20,000 \text{ ft.}$$

which checks fairly well with the distance of 18,000 ft. in Table 3·2, between the depths of flow of 22·0 and 24·0 ft. It is understood, of course, that this last formula should only be used for rough computations to establish the order of magnitude of a backwater effect, and not for the final analysis of a problem.

The trend of thought in applying the step method for backwater computations is fairly simple. Let us assume that we have the above example of a drawdown curve under consideration. At section 1 we know the magnitude of the disturbance from the normal depth, namely a depth of flow of 16·0 ft. instead of 25·5 ft. For this particular depth of 16·0 ft. we can compute the specific head and slope of the energy line. Somewhere upstream at section 2, at a yet unknown distance, we have a depth of flow of 18·0 ft. For this depth we can also compute the specific head and slope of the energy line. If we take now the difference in specific head between the sections 1 and 2 and divide that by the average slope of the energy line minus the slope of the channel bottom, we obtain the distance between sections 1 and 2.

When the channel is not uniform of slope, cross-section, or roughness coefficient, the above method cannot be followed. Instead, we divide the channel in sections of approximately equal characteristics and compute the backwater curve from section to section. Starting in section 1, we have complete information. In section 2, at a given distance upstream, we assume a certain depth of flow and compute the corresponding specific head and slope of energy line. The accuracy of the assumption is then checked by taking the average of the energy line minus the slope of the channel bottom, multiplying this by the distance between the sections 1 and 2, adding this value to the specific head at section 1 and comparing the result with the assumed specific head at section 2. After some trial and error computation, we find the correct water level at section 2.

The above discussion will be illustrated by the same example that was used for the application of the Bresse formula. We will assume first that the channel

77

is uniform and that we have the liberty of choosing certain depths of flow and computing the distances between these sections. The necessary computation may look as follows:

Table 3.3. *Backwater Computation With Fixed Difference in Water Level*

h	A	R	V	$\dfrac{V^2}{2g}$	H	ΔH	S_e	$\bar{S}_e$	$\bar{S}_e - S_b$	Δx
16·0	3,200	13·8	12·5	2·42	18·42		0·00136			
						1·49		0·00113	0·00083	1,800
18·0	3,600	15·2	11·1	1·91	19·91		0·00091			
						1·64		0·00078	0·00048	3,400
20·0	4,000	16·7	10·0	1·55	21·55		0·00066			
						1·73		0·00057	0·00027	6,400
22·0	4,400	18·0	9·1	1·28	23·28		0·00048			
						1·79		0·00043	0·00013	13,800
24·0	4,800	19·3	8·3	1·07	25·07		0·00037			
						1·38		0·00034	0·00004	36,000
25·5	5,100	20·3	7·8	0·95	26·45		0·00030			

For a better understanding of the backwater computation of Table 3.3 the different terms are graphically illustrated in *Figure 3.27*. The key to the computation is that the difference in specific height ΔH, equals the length of the reach Δx, times the difference between bottom slope and average energy slope $(\bar{S}_e - S_b)$.

Let us now assume that the channel in the above example is divided in sections of 2,000 ft. each and that we have to compute the water surface profile from section to section. The computational procedure of trial and error would then be as follows:

Table 3.4. *Backwater Computation With Fixed Length of Reach*
(Auxiliary Columns A, R, and v Are Omitted) *

h	$\dfrac{V^2}{2g}$	H	S_e	$\bar{S}_e$	$\bar{S}_e - S_b$	Δx	ΔH	H
16·0	2·42	18·42	0·00136					
				0·00113	*0·00083*	*2,000*	*1·66*	*20·08*
18·0	*1·91*	*19·91*	*0·00091*					
				0·00112	0·00082	2,000	1·64	20·06 √
18·2	1·88	20·08	0·00088					
				0·00080	0·00050	2,000	1·00	21·08 √
19·4	1·65	21·05	0·00072					
				0·00066	*0·00036*	*2,000*	*0·72*	*21·77*
20·4	*1·49*	*21·89*	*0·00061*					
				0·00067	0·00037	2,000	0·74	21·79 √
20·3	1·51	21·81	0·00063					

and so on...

* The figures in italic type denote the wrong trials that had to be repeated.

The above example in Table 3.4 is a bit unrealistic since one would not apply the fixed-length-of-reach method if throughout the entire channel the bottom slope, the width, and the roughness coefficient were constant. Moreover, if these channel characteristics are not constant, the computations are a little more complicated. Let us assume for instance that the first reach of 2,000 ft. is the same as before, that the second reach of 2,000 ft. has a roughness coefficient of 0·040, and that the third reach is again the same as before.

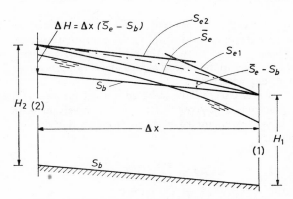

Figure 3.27. Elements of backwater computation

Having completed the first step, where we find a depth of flow of 18·2 ft., we must now recompute the value of S_e with the characteristics of the second reach and then we can proceed to estimate the depth at the end of the second reach. Such a procedure is demonstrated in Table 3.5.

Table 3.5. *Backwater Computation With Fixed Length of Reach*

h	$\dfrac{V^2}{2g}$	H	S_e	$\bar{S}_e$	$\bar{S}_e - S_e$	Δx	ΔH	H
16.0	2·42	18·42	0·00136					
				0·00112	0·00082	2,000	1·64	20·06
18·2	1·88	20·08	0·00088					
18·2	1·88	20·08	0·00222					
				0·00177	0·00147	2,000	2·95	23·03
21·7	1·32	23·02	0·00131					
21·7	1·32	23·02	0·00051					
				0·00049	0·00019	2,000	0·38	23·40
22·1	1·27	23·37	0·00047					

and so on...

From the above discussion it follows, that when we have to select suitable reaches of the channel for backwater computation, we should first of all watch out for changes in channel characteristics (width, depth, slope, roughness). We should locate our sections (the dividing lines between the reaches) at the discontinuity of the channel characteristics. At every section we should compute the downstream energy slope according to the downstream channel characteristics, and the upstream energy slope according to the upstream channel characteristics. Another criterion in selecting reaches is that the sections should be relatively close together when a rapid change in depth of flow is expected and farther apart when the depth of flow approaches normal depth.

If it so happens that the reaches with different characteristics have approximately the same length, we could simplify the computations, and still be fairly accurate, by choosing the sections in the middle of the uniform reaches. Then we do not have to recompute energy slopes at every section, and we still give due weight to all characteristics.

If several backwater curves have to be computed for a certain open channel (for instance, a stretch of a river upstream of a reservoir with a fluctuating surface level) it is convenient to prepare auxiliary curves or tables. In the Manning formula:

$$Q = \frac{1 \cdot 5}{n} AR^{2/3} S^{1/2} \qquad \bullet$$

we substitute k for $\dfrac{1 \cdot 5}{n} AR^{2/3}$ and obtain:

$$Q = kS^{1/2}$$

k is a function of the roughness coefficient, the shape of the channel cross-section and the depth of flow, but not the slope of the water surface. Therefore we can prepare graphs with the value of k given as a function of h, for each section of the channel under consideration. It becomes thus more convenient to compute S_e when Q and h are given.

Normally the k values in the above formula are computed on the basis of measured cross-sections and observed or estimated roughness coefficients. However, if such information is not available, k values may also be determined from observed water surface slopes and discharges and subsequently used for backwater computations, provided that the change in velocity head may be neglected, and that the backwater curves stay within or close to the observed conditions. The sequence of computations is then as follows: we plot recorded water surface profiles on a sheet of graph paper, measure the slope of the water surface at the selected sections, and since we know the discharges we can determine the k values for every section and every recorded stage. We plot then for every section, water level versus k values. When computing a backwater curve, we can determine the slope of the water surface for any given stage and discharge. Velocity heads have to be neglected since we have no means of computing the velocity. This is usually permissible on M_1 curves, where the velocity heads are small, but not on M_2 curves, where the velocity heads may be substantial. The computations in tabular form may look as shown in Table 3.6.

If we have to compute the backwater curve of a river with appreciable bank overflow, the following procedure may be followed.

River channel flow: $\qquad Q_1 = k_1 S^{1/2}$

Overbank flow: $\qquad Q_2 = k_2 S^{1/2}$

Total flow: $\qquad \overline{\Sigma Q = \Sigma k S^{1/2}}$

Table 3.6. *Backwater Computation With k Values**

Section	W.L.	k	S_e	$\bar{S}_e$	Δx	Δh	W.L.
1	844·0	$4·21 \times 10^6$	0·00021				
				0·00023	*10,000*	*2.3*	*846·3*
2	846·5	$2·80 \times 10^6$	0·00026				
	846·4	$3·60 \times 10^6$	0·00029	0·00025	10,000	2·5	846·5 ✓
				0·00041	*5,000*	*2·1*	*848·5*
3	848·1	$2·65 \times 10^6$	0·00053				

and so on...

* Figures in italic type denote the wrong trials that had to be repeated.

We compute k values for the river channel and the flood plain for every section, for different water levels and proceed in the trial and error computations as shown in Table 3.7 (assuming that we can neglect the velocity head):

Table 3.7. *Backwater Computation For River Channel and Overbank Flow*

Section	W.L.	k	Σk	S_e	$\bar{S}_e$	Δx	Δh	W.L.
1	273·6	$8·60 \times 10^5$ $1·35 \times 10^5$	$9·95 \times 10^5$	0·00022				
					0·00025	7,500	1·9	275·5
2	275·6	$7·10 \times 10^5$ $1·91 \times 10^5$	$9·01 \times 10^5$	0·00028				

and so on...

It would not be unusual in a situation with overbank flow that from section to section the channel length is different from the length of the flood plain, and consequently that the channel slope is different from the overbank slope while they were assumed the same in the above formula. This may be corrected by adjusting the k values accordingly. If we take the slope of the river channel as reference slope, and the slope of the flood plain is found to be steeper, then we should multiply the k value of the flood plain with a correction factor.

$$C = \sqrt{\frac{\text{channel length}}{\text{overbank length}}}$$

Thus we shall obtain the correct overbank discharge while using the channel slope. In other words, we have:

River channel: $\qquad Q_1 = k_1 S_1^{1/2}$

Overbank: $\qquad Q_2 = k_2 S_2^{1/2}$

81

but for computational convenience we pretend to have:

River channel: $\qquad Q_1 = k_1 S_1^{1/2}$

Overbank: $\qquad \dfrac{Q_2 = k_2 C S_1^{1/2}}{Q = (k_1 + k_2 C) S_1^{1/2}}$

River Channel Roughness

One of the great difficulties in predicting river stages or river surface profiles is the selection of an appropriate roughness coefficient. It is a well-established fact that the roughness coefficient of large rivers may range from values as low as 0·015 to values as high as 0·060. It has even been observed that the roughness coefficient on one reach ranged in value from 0·015 to 0·045. Since for a given discharge and slope the stage of the river is roughly proportional to the roughness coefficient to the three-fifth power, while for a given discharge and stage the slope of the river is proportional to the roughness coefficient to the second power, it will be appreciated that hydraulic computations on river channels may be subject to a large degree of error when they are not based on an adequate estimate of the roughness coefficient.

The most dependable way of determining this coefficient of roughness for a given reach of river and for a given rate of flow, is by making accurate field observations. Such observations, however, are too often not available for the reaches and the flows in which the engineer is interested. In these cases the determination will have to depend on the individual judgement of the engineer. The aim of this section is to improve this judgement by discussing the different variables that influence the overall roughness of a river channel.

It appears that the roughness of a river channel depends to a greater or lesser degree on the following characteristics: (1) The grain size of the bed material; (2) The small irregularities of the stream bed; (3) The major irregularities of the river channel; (4) The sediment transport; (5) The temperature of the water. These five items are discussed in more detail in the following paragraphs:

(1) The grain size of the bed material is obviously one of the factors that influence the total boundary friction of a moving body of water. If a river channel were lined with sand it would provide less resistance to flow than if it were paved with boulders! If the roughness coefficient of a channel would depend on the texture of the bed material only (which is hardly ever the case, as will be seen from the following points) the order of magnitude of the roughness coefficient would be as follows: sand, 0·020; gravel, 0·030; cobbles, 0·040; boulders, 0·050.

(2) The small irregularities of the stream bed such as ripples and dunes seem to play an important role in causing overall channel roughness. In an interesting laboratory test by the U.S. Waterways Experiment Station (1935), it was demonstrated that during a gradual increase of discharge over a bed of sand the roughness coefficient changed from 0·010 to 0·030 to 0·016 while the bed changed from smooth to ripples to relatively smooth. Laboratory experiments carried out by Brooks (1958) confirm these findings: while the bed condition was smooth, the roughness coefficient would be around 0·010; when the bed became covered with dunes, the roughness coefficient would increase to well over 0·030. Carey and Keller (1957) believe that this phenomenon of

appearing and disappearing dunes or sand waves is the main cause of the fluctuating roughness values on the Mississippi River. They ascribe the 'loop' in the rating curve to a lag in the rate of change of sand waves, which tend to become larger with increases in discharge. Thus, during the rising river the dunes (and hence the roughness coefficient) are somewhat smaller than they should be, while during a falling river they are somewhat larger than they should be.

(3) The major irregularities of the river channel, such as sand bars, islands, deepenings, widenings, bends, and constrictions, have an important bearing upon the overall roughness and may account for as much as 50 per cent of the total roughness coefficient. We may look upon these major irregularities as causing, not so much boundary friction, as local acceleration and deceleration of flow, with consequent turbulence losses.

The effect of the major irregularities upon the coefficient of roughness will not be as intimately related to the magnitude of discharge as was the case with the small irregularities, discussed in the previous section. Some variations, however, can be expected. For instance, when the flow increases from a very low to bankful stage, irregularities in cross-section may become relatively less important and the flow will become relatively less disturbed. When the flow increases above bankful stage, it may become very much disturbed on account of irregularities in local high banks or on account of bush and trees, producing eddies and cross currents that may disturb the flow far out in the channel. As a result of this the 'n' value may decrease up to bankful stage, and increase again above that stage, as has been observed on various types of rivers.

In connection with this subject, it is of interest to quote a discussion by Matthes, of a paper by Schnackenberg (1951) on the slope discharge formula. 'One troublesome factor in determining the roughness coefficient is the energy gradient. During a rapidly rising flood, slope steepening results from a marked straightening of the main axis of the current which effectively shortens the path of greatest flow. The main current then no longer follows the concave shores of bends but is found nearer the convex shores inflicting heavy scour there at the toes. Slope steepening, so caused, naturally affects the value of S, illustrating one more disturbing factor that affects 'n' which engineers cannot readily evaluate.'

(4) Sediment transport appears to have some bearing upon the resistance against flow in a river channel. Some of the earlier authors of papers on river engineering have presented sediment transport as an energy consuming task of the river. Therefore the roughness coefficient of the river would have to increase with an increase in sediment transport. This appears to be a wrong concept. We may look upon sediment transport as a result of the turbulent state of the river, and this turbulence as caused by the movement of a body of water within fixed boundaries. Hence the sediment transport is an incidental by-product of the movement of water, just like the whirling leaves are by-products of the passing train, and not a partial cause of the wind resistance of the train. In fact, it is believed by Vanoni (1946, 1960) that the reverse is the case, namely that the presence of sediment in the water reduces the total resistance against flow by damping the turbulence and thus reducing the

energy dissipation. This contention has been verified by laboratory experiments where the roughness factor was found to increase with a reduction in sediment concentration, while the bed configuration was artificially fixed.

In view of the discussion under 2 and 4, we may conclude that a sediment-carrying river acts in two ways to change its roughness coefficient: (a) by changing the configuration of the bed, (b) by damping out turbulence. Of these two factors, the first one is believed to be more significant than the second one. The total effect may be as much as a change in roughness coefficient from 0·045 to 0·015 as reported by Einstein and Barbarossa (1952) for the Missouri River at Ft. Randall. As a further illustration of the above discussion the following field measurements are quoted:

Mississippi River:

| Discharge in cusec | 84,600 | 197,000 | 365,000 | 1,010,000 | 1,964,000 |
| 'n' | 0·046 | 0·035 | 0·032 | 0·029 | 0·027 |

Tennessee River:

| Discharge in cusec | 12,800 | 30,500 | 65,400 | 195,000 | 273,000 |
| 'n' | 0·057 | 0·040 | 0·035 | 0·032 | 0·029 |

Beas River:

| Discharge in cusec | 3,300 | 6,940 | 15,130 | 47,710 | 100,800 |
| 'n' | 0·037 | 0·034 | 0·032 | 0·025 | 0·023 |

(5) The temperature of the water affects the viscosity of the fluid directly and the sediment transport indirectly. It can be demonstrated by computations that the change in viscosity within the normal range of river water temperatures has a very small effect upon the velocity. Since this effect is much smaller than the normal errors that can be expected in river flow computations, it can be disregarded for practical purposes. The indirect effect upon sediment concentration, however, warrants careful attention. Observations on the Colorado River, by Lane (1949), have indicated that the suspended bed material concentration during the winter was twice as high as during the summer for identical flow conditions. In view of the important influence of sediment transport upon channel roughness, it therefore follows that water temperatures should be taken into consideration when dealing with roughness coefficients in sediment-bearing river channels.

In the above paragraphs we have discussed the inter-relationship of the coefficient of roughness and various river channel characteristics in qualitative terms. It would be quite natural now to expect a set of firm rules that can be used in determining the coefficient of roughness for a given set of conditions, but unfortunately such rules do not exist yet—partly because the problem is complicated, to say the least, and partly because no systematic collection of complete river channel roughness data is available.

The best advice that can be given at present for estimating roughness coefficients is to make actual field measurements of the river reach under consideration. If no time or funds are available to do this (implying that the problem is not very serious or important, and therefore will not have to be solved with a great degree of accuracy) the engineer would be well advised to make a research of measured roughness coefficients on comparable streams. If such information is not available, the following notes may provide some guidance:

An all-round value for natural streams with fairly straight alignment, fairly constant section and during medium stages, is 0·030; for very low stages the value may come up to 0·040, or even higher; and for extreme high stages on alluvial rivers, the value may drop to 0·020, or even lower. Tortuous channels or channels with boulders and rock outcrops may have values up to and even exceeding 0·050. Extensive tables of roughness coefficients for various conditions have been prepared by Ven Te Chow (1959).

In order to overcome the lack of systematic information that makes it so difficult and hazardous to venture a roughness coefficient estimate, it is suggested that river channel roughness data be collected, as tabulated below, which aims at representing in numerical parameters all river channel characteristics that may have some bearing on overall roughness.

River: Saskatchewan River, Canada
Test reach: Sipanok Channel to bifurcation; 5·4 miles; 7 sections

Date:	17 Aug. '55	20 May '54	24 June '54	17 June '54	3 Sept. '54
Stage:	901·3	906·3	909·4	910·7	912·6
Discharge:	19,600 S	48,200 R	75,000 F	100,000 F	129,800 P
Slope:	0·00016	0·00015	0·00015	0·00015	0·00017
Channel width:	850	1,490	1,750	1,860	2,050
Hydr. radius:	8·6 A	9·7 A	12·6 A	13·3 A	13·9 A
Average vel.:	2·7	3·4	3·4	4·0	4·6
n (Manning):	0·029	0·025	0·029	0·026	0·025
Bed material:		0·30(0·78)			
Ripples:					none E
Dunes:					none E
Kolks:					10–200 E
Alignment:	1·05(4)	1·05	1·04	1·03	1·02
Uniformity:	2·1	1·7	1·7	1·7	1·8
Susp. sediment:	120	1,850(55)	790(31)	1,650(28)	1,700(36)
Temperature:	64	45	63	60	51

Remarks

The rest reach is located in the upper part of the Saskatchewan Delta. The river channel is fairly straight and contains 15 sand bars and 9 islands, the latter grown over with willows. The river is entirely alluvial and in its natural state. The sandy banks, covered with heavy trees, do occasionally cave in. The channel cross-sections are of irregular shape and contain deep and shallow parts. The channel elements are computed with Method C. The lowest and highest observed river flows are 2,000 cusec and 260,000 cusec, respectively. The stage data in the table refer to the gauge at Sipanok Channel, and are listed in geodetic elevation. The roughness data are obtained from P.F.R.A. hydrometric surveys.

Explanation of table headings

River

List the name of the river and the country in which it flows. If the name of the river is not generally known then the drainage basin in which it flows can be added in brackets behind the name of the river; e.g., Battle River (Saskatchewan Basin), Canada.

Test reach

List the location of the section of river that is under consideration. List also the length of the test reach and the number of cross-sections that were available for computing the average cross-section.

Stage

List the date on which the measurements were taken. List the stage of the river at some reference gauge. This will help to show any substantial shifts in the rating curve between measurements. These shifts may indicate trends of aggradation or degradation which, in turn, might influence the roughness.

Discharge

List the discharge of the river channel proper, for which the data in the table are observed, in cusec. List behind the discharge figure the letter S for steady flow, or R for rising flow, or P for peak flow, or F for falling flow.

Slope
List the average slope over the test reach. The test reach should preferably be a comparatively long reach of river, say from one to ten miles in length. The upper and lower gauges should be located on a straight stretch of river channel.

Channel width
List the channel width in feet. See notes on the computation of channel elements.

Hydr. radius
List the hydraulic radius in feet. See notes on the computation of channel elements. List behind the hydraulic radius figure the letter M when the channel elements are determined on the basis of cross-sections taken simultaneously with the discharge measurement. List the letter A when the cross-sections were taken before or afterward.

Average velocity
List the average velocity in feet per second. See notes on the computation of channel elements.

n (Manning)
List the roughness coefficient. See notes on the computation of channel elements.

Bed material
List the average grain diameter of the bed material in millimetres. List behind this figure, in brackets, the uniformity coefficient of the bed material as computed by the formula:

$$s = \left(\frac{D_{25}}{D_{75}}\right)^{1/2}$$

in which D_{25} is the grain size, of which 25 per cent is finer, and D_{75} is the grain size, of which 75 per cent is finer.

Ripples
List the height and the spacing of the bed ripples in feet, measured along lines parallel to the centre line of flow. For instance: 0·1–0·5. It can be estimated by lightly dragging a sounding weight over the river bottom from a drifting boat.

Dunes
List the height and the spacing of the dunes in feet. For instance: 3–20. It can be determined by a self-registering echo-sounding apparatus.

Kolks
List the depth and the spacing of the kolks (isolated deep scour holes, causing violent turbulence) in feet. For instance: 10–200. Determination as under previous item. It is recognized that the measurement of ripples, dunes and kolks is most difficult, especially in deep rivers at high discharges. However, since such measurements will add materially to the knowledge of river channel roughness, it is believed that they will amply justify the expense of taking them. When no ripples, dunes or kolks are present, the word 'none' should be listed.

Alignment
List the ratio of the total length of the centre line of flow to the length of the straight line between the beginning and the end of the centre line. List between brackets, in the first column only, the number of times that the flow reverses its direction of curvature.

Uniformity
List the ratio of maximum cross-sectional area to minimum cross-sectional area.

Susp. sediment
List the average suspended sediment concentration of the river flow in parts per million (1,000 p.p.m. = 1 g/l. = 0·1 per cent), and list between brackets the fraction of sand that the suspended sediment contains, expressed in per cent of the average suspended-sediment concentration. The sand fraction is assumed to range from 0·0625–2·0 mm.

Temperature
List the water temperature in degrees Fahrenheit.

Remarks

List any characteristics of the test reach that are not included in the table and that may have a bearing on the channel roughness (such as: the occurrence of islands and sand bars; the nature of the riverbanks; the type of inundated vegetation at high stages; whether the river bed is entirely alluvial, or contains rock outcrops or artificial structures). Describe the shape of channel cross-sections and the method of computing the channel elements (see following notes). List the range of river flows. List the station for which the stages are given. List the source from which the data are obtained. List any other information that seems pertinent. Include photographs of the test reach taken from the ground and from the air, from different angles, and at different discharges.

NOTE

If no actual measurement of one of the listed characteristics is available, it would be desirable to substitute this by an estimate. This will be indicated in the table by placing the letter E behind the estimated figure. If no observation has been made, and if no estimate can be made, the space in the table will be left blank. If possible, include some photographs of the test reach.

In order to compute the roughness coefficient for a test reach of appreciable ength, say from one to ten miles, it is desirable to survey a number of channel cross-sections that are spaced representatively throughout the reach. From these cross-sections and for a given river flow and surface gradient, it is possible to compute an average cross-sectional area, an average channel width, an average hydraulic radius, an average velocity, and finally the coefficient of roughness. There are several ways of making these computations, three of which will be discussed in the following paragraphs.

Method A: If all cross-sections are very similar, resemble an elliptical shape, and have approximately the same surface width, they can be superimposed on one another, bringing the cross-sections together to a common water surface. They are then reduced by eye to an average cross-section, which yields the average cross-sectional area and the average channel width. By dividing the average cross-sectional area by the average channel width, the average hydraulic radius is obtained. The error that is introduced by taking the surface width instead of the wetted perimeter is, for normal river channels, negligible compared to the other inaccuracies of measurement. In cases where the channel is deep and narrow, it would be advisable to determine the average wetted perimeter for computation of the hydraulic radius. The average velocity is computed by dividing the known discharge by the average cross-sectional area. The coefficient of roughness is computed by using the Manning formula.

Method B: If all cross-sections resemble an elliptical shape, but are not quite similar and have different surface widths, it becomes rather hazardous to sketch by eye an average cross-section. It then becomes more reliable to compute the average of all cross-sectional areas and the average of all surface widths, and from these figures the average hydraulic radius, and so on, as in Method A.

If similar computations have to be repeated for different river stages, it may be convenient to plot on graph paper a bundle of curves, each curve representing the wetted area of one particular cross-section in terms of the common stream profile elevation. This is also done for the surface width. The two bundles of curves are then averaged, and curves of average cross-sectional area and average channel width versus water surface elevation are obtained.

Method C: If some or all of the cross-sections do not resemble an elliptical shape, but are of irregular form (e.g., a deep channel adjacent to a shallow section with sand bars), the above methods may produce misleading results. In such a case, the following method can be used. For a given discharge and river gradient, every cross-section is divided into deep and shallow parts. For every part, the wetted area and the hydraulic radius are determined, and the value of $AR^{2/3}$ is computed. For every cross-section, the wetted areas are totalled and the $AR^{2/3}$ values are totalled. The outcomes are multiplied by the number of miles that the cross-section represents. The resultant figures of all cross-sections are then totalled and divided by the total mileage of the test reach. In this way, an average cross-sectional area and an average value of $AR^{2/3}$ are obtained. From these two figures, the value of R can be computed. By using the Manning formula:

$$Q = \frac{1 \cdot 5}{n} AR^{2/3}S^{1/2}$$

the value of n can be computed. By using the formula $Q=AV$, the average velocity can be computed. By using the formula $A=RW$, the average channel width can be computed.

If similar computations have to be repeated for different river stages, it may be more convenient to plot curves of A and $AR^{2/3}$ values for each cross-section and to average the bundles of curves, keeping in mind that certain curves may represent longer stretches of channel than others.

This method has the advantage that no undue weight is given in the computations to shallow parts of the cross-section. This can easily be seen by working out a simple example. Assume a river channel 1,000 ft. wide, consisting of two parts. One part is 500 ft. wide and has a hydraulic radius of 20 ft. The other part is also 500 ft. wide, but has a hydraulic radius of 5 ft. When computed with Methods A or B, the average hydraulic radius becomes 12·5 ft. When computed with Method C, the average hydraulic radius becomes 16·5 ft. These two answers would result in finding two corresponding coefficients of roughness that vary 20 per cent in value. It would seem that this is too much of a discrepancy to be disregarded. It is believed that in computing n values, due respect must be paid to the most desirable method of computation.

Hydraulic Jump

It was noted earlier, in the qualitative discussion of backwater profiles, that crossing the critical depth, going from super-critical to sub-critical flow, involves a hydraulic jump. This section will deal with the relation of water depths before and after the jump.

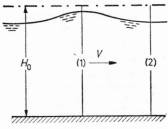

Figure 3.28. Wave in open channel

In order to gain a better understanding of the hydraulic jump let us discuss first the movement of a wave of small amplitude in an open channel as shown in *Figure 3.28*. The velocity of the wave in still water is V ft./sec to the left. However, for the sake of convenience, we assume that the water is flowing with a velocity V to the right and that the wave is standing still.

Going from section 1 to 2, and applying the Bernoulli equation, we can write:

$$\frac{V^2}{2g} + h = H_0 = \text{constant}$$

and when we differentiate with respect to x:

$$\frac{d\left(\frac{V^2}{2g}\right)}{dx} + \frac{dh}{dx} = \frac{dH_0}{dx} = 0$$

or

$$\frac{V}{g}\frac{dV}{dx} + \frac{dh}{dx} = 0 \quad \text{or} \quad \frac{V}{g}\,dV + dh = 0$$

Applying the continuity equation we may write $Vh =$ constant, and when we differentiate with respect to x:

$$V\frac{dh}{dx} + h\frac{dV}{dx} = 0 \quad \text{or} \quad V\,dh + h\,dV = 0$$

Combining the Bernoulli and the continuity equations, we obtain:

$$V = \sqrt{g \cdot h}$$

which is the general equation for the velocity of a wave in an open channel.

We have discussed earlier that in critical flow conditions the velocity head is equal to half the depth of the flow:

or

$$\frac{V^2}{2g} = \frac{h}{2} \quad \text{or} \quad V = \sqrt{g \cdot h}$$

In other words the velocity of water in critical flow equals the velocity of a wave in that same channel. We may therefore conclude that in super-critical flow a surface disturbance cannot be propagated in upstream direction. In sub-critical flow, however, surface disturbances will be propagated in upstream direction. It becomes evident now that when super-critical flow meets sub-critical flow this must take place in the form of a wave, continuously being built up from downstream, but unable to travel upstream and therefore standing still. This standing wave is called a hydraulic jump.

It is of interest to note that the ratio of V to $\sqrt{gh}$ is known as the Froude number. Thus:

$$F = \frac{V}{\sqrt{gh}}$$

It follows that for critical flow, the Froude number is 1, for sub-critical flow it is smaller than 1, and for super-critical flow the number is larger than 1. It has been found from experience that the Froude number of the upstream flow must be larger than 2, to produce an outspoken, surging, hydraulic jump. For values between 1 and 2 the hydraulic jump may become a series of undulating waves.

89

If we consider an element of water flowing through the hydraulic jump, we may observe that this element, on the upstream side of the jump, has a relatively high (super-critical) velocity, while on the downstream side it has a relatively low (sub-critical) velocity. We know from the laws of mechanics that, in order to slow down an object, we must apply a force in the opposite direction of the movement. In the hydraulic jump this retarding force is provided by the hydrostatic pressure of the higher downstream depth of flow. For a given discharge, a lower upstream depth of flow means higher velocity of flow, hence a greater momentum, hence a greater downstream depth of flow to slow down the greater momentum.

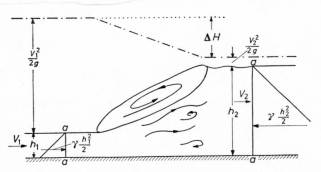

Figure 3.29. Hydraulic jump

The slowing down of each element of water in the region of the hydraulic jump implies that a corresponding force moves, against its sense of direction, over the length of the hydraulic jump. This implies that negative work is being done by these forces; in other words, that energy is being dissipated.

In order to establish the quantitative relationship between the upstream depth of flow and velocity on the one hand and the downstream depth of flow on the other hand, we shall apply the momentum equation and continuity equation. In order to define the amount of energy loss in the hydraulic jump we shall apply also the Bernoulli equation.

Let us consider the element of the hydraulic jump a-a-a-a in *Figure 3.29* and apply the equation:

$$F = \rho.Q.(V_2 - V_1)$$

or
$$\frac{\gamma}{2}(h_1{}^2 - h_2{}^2) = \frac{\gamma}{g} V_1 h_1 (V_2 - V_1) = \frac{\gamma}{g} V_1 h_1 \left(V_1 \frac{h_1}{h_2} - V_1 \right)$$

or
$$\frac{\gamma}{2}(h_1 + h_2)(h_1 - h_2) = \frac{\gamma \cdot V_1{}^2 \cdot h_1}{g h_2}(h_1 - h_2)$$

or
$$h_1 + h_2 = \frac{2 V_1{}^2 h_1}{g h_2}$$

or
$$h_2{}^2 + h_1 h_2 - \frac{2 V_1{}^2 h_1}{g} = 0$$

or
$$h_2 = -\frac{h_1}{2} + \sqrt{\left(\frac{h_1}{2}\right)^2 + \frac{2 V_1{}^2 h_1}{g}}$$

When we combine the last equation with the Bernoulli equation we obtain, after some algebraic manipulation:

$$\Delta H = \frac{(h_2 - h_1)^3}{4h_1 h_2}$$

Now that we have a relation between h_1 and h_2, we are able to determine the location of a hydraulic jump by the following procedure: let us assume we have a channel with uniform flow, where we introduce a gate that is situated as is shown in *Figure 3.30*. Starting at the gate and going in a downstream direction

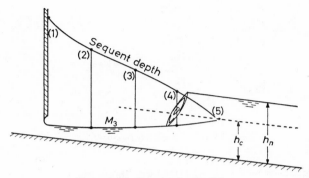

Figure 3.30. Sequent depth of flow

we compute the surface profile of the water from section to section. For every depth of flow h_1, we can compute the sequent depth of flow h_2, which is plotted in the same diagram. At the section where the sequent depth of flow equals the normal depth of flow, the hydraulic jump will occur. By trial and error we find that this condition is satisfied at section 4. It may be seen from this diagram that a greater depth of flow h_n would tend to move the hydraulic jump towards the gate, and vice versa. If the normal depth h_n would become larger than the sequent depth at section 1, the jump would become submerged. If the normal depth h_n would be close to the critical depth h_c, the hydraulic jump may become a series of undular waves near section 5, since by then the Froude number of the super-critical flow is nearly 1.

Hydraulic jumps are often found where the super-critical flow of a spillway chute meets the tranquil flow of the tail water below a dam. When dam and spillway are built on solid erosion-resistant rock, not much harm can result from simply dumping the spillway flow into the river channel. The only provision required is to give the lower end of the spillway chute the form of a ski jump so that the super-critical flow hits the river bottom some distance away from the toe of the dam, just in case some local erosion or dislodging of rocks would occur.

The situation requires much more care when the dam and spillway are built on erodible soil. A stilling basin is required at the end of the spillway chute to contain the hydraulic jump and to reduce the velocities to a few feet per second, before the spillway flow is discharged into the river channel. The main problem in the hydraulic design of the stilling basin is to ensure complete control of the hydraulic jump with the least costly structure. This design problem is discussed in Chapter 5: Hydraulic Structures.

Weirs

From a viewpoint of physical shape we may distinguish between broad-crested weirs (straight stream lines on top of the weir) and sharp-crested weirs (curved stream lines). From a viewpoint of flow conditions we may distinguish between weirs with a free discharge (overflow not influenced by the tailwater elevation) and submerged weirs (discharge a function of headwater and tailwater elevation).

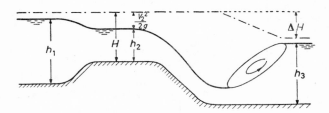

Figure 3.31. Broad-crested free-overflow weir

For the broad-crested weir with free discharge, as shown in *Figure 3.31*, we discussed earlier that the depth of flow is critical depth, which is two-thirds of the specific head. We may therefore write:

$$\frac{V_2^2}{2g} = \tfrac{1}{3}H \quad \text{and} \quad q = \tfrac{2}{3}H\sqrt{\tfrac{2}{3}gH} \quad \text{or} \quad q = 3 \cdot 1H^{3/2}$$

Due to friction and contraction the coefficient is usually somewhat less than 3·1. For sharp-crested weirs, the curvature of the streamlines begins to play a role, and the coefficient becomes 3·3. For a well-rounded spillway crest and a large depth of approach the coefficient may be as high as 4·0. For any given situation it is advisable to look up the value of the coefficient in a handbook.

It is of interest to discuss at what stage the free overflow weir becomes a submerged weir. The transition from the first type to the second type may be defined as the moment that the downstream water level has been raised so much that it begins to back up the upstream water level. Up to this moment, the above equation of the broad-crested weir is applicable; after this moment that equation does not apply any more. Let us see what different patterns of flow we may have when we raise slowly the downstream water level. Starting with the situation in *Figure 3.31*, we have first of all an outspoken hydraulic jump below the chute of the weir. When we raise the downstream water level the hydraulic jump is pushed upon the chute and becomes much less outspoken. When we raise the downstream water level above the sill of the weir the jump is drowned out, but the super-critical flow still clings to the chute, and the upstream water level is not yet affected. When the downstream water level is approximately as high as the water level over the weir, a change in flow pattern will take place. The critical and slightly super-critical flow plunging over the weir will no longer cling to the chute but will begin to ride the surface of the downstream channel. A series of undulating waves will form as shown in *Figure 3.32*. At or near this moment, we will notice that the upstream water surface will respond to a change in downstream water level. We have now reached the transition from free-overflow to submerged weir condition. The

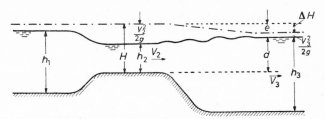

Figure 3.32. Broad-crested submerged weir

downstream water level at which this occurs can be defined approximately as follows. The energy losses downstream of the weir are for the submerged condition:

$$\Delta H = K \frac{(V_2 - V_3)^2}{2g}$$

(in which $K = 1 \cdot 0$ for an abrupt drop after the weir crest, and where $K = 0 \cdot 2$ for a smooth transition with a slope of 10 horizontal to 1 vertical between the weir crest and the downstream channel bottom). Therefore, when the downstream water level is more than $\Delta H + (V_3^2/2g)$ below the upstream energy line, the weir has a free overflow condition. When the downstream water level is higher, the weir is submerged.

After having established at what stage the submerged weir condition begins, the next problem is to develop a discharge formula. Such a formula should preferably express the discharge in terms of quantities that can readily be measured. Theoretically we can write:

$$q = h_2 \sqrt{2g(H - h_2)}$$

However, the water surface over the broad-crested weir is seldom a smooth, level, plane. Instead it is curved and warped, so that it is very difficult to measure h_2. Hence we substitute another approximate equation:

$$q = Cd \sqrt{2ge}$$

in which q is the discharge per ft. length of weir; C the correction factor that ranges from a value of $1 \cdot 0$ when the weir has an abrupt drop, to a value of $1 \cdot 4$ for a smooth transition—an average value of C equals $1 \cdot 1$; d the height of the downstream water surface above the sill of the weir, and e the difference in elevation between the upstream energy line and the downstream water level, as shown in *Figure 3.32*.

It should be pointed out that this formula has no great degree of accuracy, mostly because of the difficulty in estimating the correct C value. Here is one subject in hydraulics where further research is badly needed.

93

Flood Routing

Many problems arise in hydraulic engineering where the hydrograph of stream flow at one location in the river system is given, and where the resultant hydrograph at some downstream location has to be determined. In principle such problems fall in the category of unsteady flow. The technique of solving such problems is known as flood routing. We may divide the large variety of flood routing problems in three categories:

(1) Routing through reservoirs. In these cases, the modification of the hydrograph is caused by the temporary storage of water in the reservoir. The problem is solved on the basis of the relationship that inflow equals outflow plus storage. An example of hydrograph modification due to a reservoir is shown in *Figure 3.33(a)*. It will be noted that the peak of the outflow hydrograph falls on the recession limb of the inflow hydrograph, indicating the moment of maximum reservoir elevation.

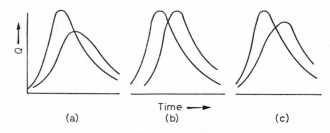

Figure 3.33. *Deformation of hydrographs*

(2) Routing through open channels without flood plain storage. In these cases we may consider the problem as one of the translation of a flood wave, and the hydrograph may retain its original shape, as shown in *Figure 3.33(b)*. It is possible that the velocity of the flood peak becomes greater than the velocity of the water.

(3) Routing through open channels with flood plain storage. These cases are a combination of 1 and 2 as shown in *Figure 3.33(c)*. In the following paragraphs these three cases will be discussed in more detail.

Let us discuss first the simple case of a reservoir without conduits and an ungated spillway of given dimensions. It will be assumed that the reservoir is filled initially to the spillway crest and that an inflow hydrograph of given proportions has to be routed through the reservoir shown in *Figure 3.34*.

When the flood wave begins to flow into the reservoir, the outflow is nil because there is no head of water at the spillway. Outflow will begin to take place as soon as a head has been created. This, obviously, requires that a certain amount of the inflow has to go temporarily into storage to create this head. In other words, inflow equals outflow plus storage. This relationship is controlled by the head versus discharge curve of *Figure 3.34(b)* and the head versus storage curve of *Figure 3.34(c)*. Given the inflow hydrograph, the storage curve, and the discharge curve, there is only one outflow hydrograph possible, shown in *Figure 3.34(d)*. At time t_1 the inflow equals the outflow. Hence the reservoir has reached its maximum level. Hence the outflow will

94

decrease after t_1. The shaded area between the inflow and outflow hydrograph between t_0 and t_1 represents the amount of water S_1 that has gone into storage, also shown on *Figures 3.34(a)* and *(c)*. The magnitude of the outflow Q_1 shown in *Figure 3.34(d)* should check with the discharge capacity of the spillway for the maximum reservoir elevation, shown in *Figure 3.34(b)*. From t_1 to t_2 the reservoir level is lowered again to spillway crest elevation. Hence the shaded area between the outflow and inflow hydrograph must also equal S_1. It may be

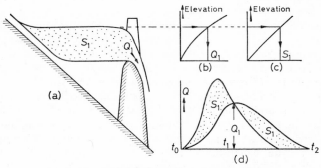

Figure 3.34. Elements of reservoir routing

assumed that the flow of water through the reservoir takes place with such low velocities that the reservoir has, for all practical purposes, a horizontal surface.

There are several ways to determine the outflow hydrograph quantitatively for given conditions. The quickest, simplest but least accurate method would be to sketch an outflow hydrograph by trial and error, such that the resultant Q and S from *Figure 3.34(d)* satisfy simultaneously the discharge and storage

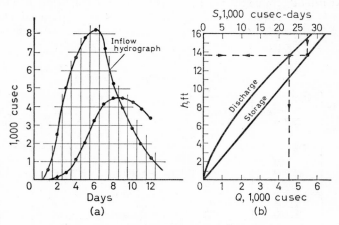

Figure 3.35. Numerical data of reservoir routing

curves of *Figures 3.34(b)* and *3.34(c)*. After one has gained some practice in estimating the shape of outflow hydrographs, this procedure may be followed in making a rough estimate of peak outflows or for the purpose of roughly

95

checking the more precise computations of someone else. To illustrate this method, let the following conditions be assumed: a reservoir has a surface area of 4,000 acres and, for practical purposes, vertical shores. The outlet of the reservoir is a spillway with a width of 30 ft. and a discharge coefficient of 3·0. The inflow hydrograph is shown in *Figure 3.35(a)*. The problem is to determine the outflow hydrograph. First of all we prepare *Figure 3.35(b)* showing discharge and storage as a function of the reservoir elevation above spillway crest. We then sketch an outflow hydrograph as a first trial and determine the area between inflow and outflow hydrograph from the beginning of the graph to the peak of the outflow hydrograph. This area may be determined by planimetering, or simply by counting the number of rectangles. Every rectangle represents 1,000 cusec-days or 2,000 acre-ft. With the resultant storage figure of 27,500 cusec-days (this is the last and 'correct' trial) we go to the storage-elevation curve and read an elevation of 13·6 ft., which corresponds with a discharge of 4,500 cusec. This in turn checks with the peak of the outflow hydrograph and confirms that the estimated outflow hydrograph was 'correct'. It should be noted that the outflow hydrograph thus found is only correct indeed when its shape is right. If we had assumed a wrong shape to start with we would have found a slightly incorrect maximum discharge.

The second method to be discussed is a computational method, based on the same principle as above, but providing a more precise answer. This method is more laborious than the methods found in most of the textbooks, but it has the advantage that one realizes what is being done. We prepare Table 3.8, where in cols. (1) and (2) are listed the dates and the average daily inflows into the reservoir. In col. (3) a trial value for the average daily outflow is entered. This outflow results in a change in storage, in cusec-days, listed in col. (4), which in turn results in a total storage value, at the end of the day, listed in col. (5). From this figure and the figure of the previous day, plus the storage-elevation curve, we can determine the average daily reservoir level, listed in col. (6). This elevation is now entered in the discharge-elevation diagram and the corresponding discharge is found and checked against the assumed outflow figure in col. (3). If the figures do not check, a second trial is made.

Table 3.8. *Flood Routing Computations*

Date (1)	Average Inflow (2)	Average Outflow (3)	Change in Storage (4)	Total Storage (5)	Average Water Level (6)
1	500	0	+500	500	0·1
2	2,500	100	+2,400	2,900	0·8
3	5,000	400	+4,600	7,500	2·6
4	6,700	1,100	+5,600	13,100	5·2
5	7,700	2,000	+5,700	18,800	8·0
6	8,200	3,200	+5,000	23,800	10·6
7	7,100	4,100	+3,000	26,800	12·6
8	5,200	4,500	+700	27,500	13·6
9	3,900	4,500	−600	26,900	13·6
10	2,800	4,300	−1,500	25,400	13·1
11	2,000	3,900	−1,900	23,500	12·2
12	1,200	3,400	−2,200	21,300	11.2

The third and last method of reservoir flood routing to be discussed is more commonly used. It is more sophisticated than the previous ones and requires no trial and error procedures. We start from the basic relationship:

$$\text{average inflow} = \text{average outflow} + \text{storage}$$

or

$$\frac{I_1 + I_2}{2} \Delta t = \frac{O_1 + O_2}{2} \Delta t + (S_2 - S_1)$$

Assuming a time unit of one day, and bringing all known terms to the left (including S_1 and O_1 which were computed for the previous day) we obtain the following equation:

$$2S_1 + I_1 + I_2 - O_1 = 2S_2 + O_2$$

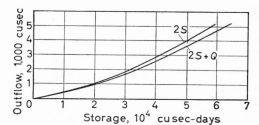

Figure 3.36. Flood routing curves

Since there is a relationship between S_2 and O_2 via the stage-storage and stage-discharge curves, we can determine O_2 when we know $2S_2 + O_2$. In order to facilitate this, we prepare an outflow versus outflow plus 2 times storage curve as shown in Figure 3.36. The necessary computations are shown in Table 3.9.

Table 3.9. Flood Routing Computations

Date (1)	I_1 (2)	$I_1 + I_2$ (3)	O_1 (4)	$2S_1$ (5)	$2S_2 + O_2$ (6)	O_2 (7)
1	0	1,100	0	0	1,100	50
2	1,100	4,700	50	1,050	5,700	250
3	3,600	9,600	250	5,450	14,800	700
4	6,000	13,300	700	14,100	26,700	1,400
5	7,300	15,400	1,400	25,300	39,300	2,400
6	8,100	16,100	2,400	36,900	50,600	3,500
7	8,000	14,200	3,500	47,100	57,800	4,300
8	6,200	10,700	4,300	53,500	59,900	4,500
9	4,500	7,900	4,500	55,400	58,800	4,400
10	3,400	5,800	4,400	51,000	52,400	4,000
11	2,400	3,900	4,000	48,400	48,300	3,500
12	1,500					

Col. (2) represents the inflow figures at the beginning of the day, taken from Figure 3.35(a).
Col. (3) represents the sum of the inflows at the beginning and at the end of the day, taken from col. (2).
Col. (4) represents the outflow at the beginning of the day and equals the figure in col. (7) for the previous day.
Col. (5) represents twice the storage at the beginning of the day and could be obtained from Figure 3.36 by using D_1 from col. (4). However, this figure can be more readily obtained by deducting col. (7) from col. (6) for the previous day.
Col. (6) represents $2S_1 + I_1 + I_2 - O_1$ and hence equals $2S_2 + O_2$.
Col. (7) represents O_2 obtained from Figure 3.36 by using $2S_2 + O_2$ from col. (6). After having entered the appropriate figure in col. (7) the same figure is entered in col. (4) as O_1 for the following day.

It should be pointed out that the selection of the time interval (in the above examples, one day) depends on the rate of change of the hydrograph and upon the desired degree of accuracy. In routing slowly changing river flows through very large lakes or reservoirs (upwards of one million acres of surface area) one may select a time interval of one month. For normal rivers and reservoirs, a time interval of one day may be appropriate. For rapidly changing flood flows and small reservoirs one may wish to use a one-hour period as time interval.

Let us discuss now the movement of a flood wave in an open channel with vertical boundaries. We cannot apply the equation $V = \sqrt{gh}$, since that equation was derived from Bernoulli's equation, assuming no friction. In the present case of a 'long' flood wave, the friction plays a dominant role. Applying the

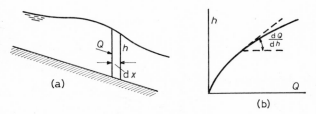

Figure 3.37. Velocity of flood wave

continuity equation to the element $Bh\,\mathrm{d}x$, shown in *Figure 3.37(a)* we may say that the water entering the element equals the water leaving the element plus the change in volume due to the change in water level, or:

$$Q = Q + \frac{\mathrm{d}Q}{\mathrm{d}x}\,\mathrm{d}x + \frac{\mathrm{d}h}{\mathrm{d}t}\,B\,\mathrm{d}x \quad \text{or} \quad \frac{\mathrm{d}Q}{\mathrm{d}x} + \frac{\mathrm{d}h}{\mathrm{d}t}\,B = 0$$

Now let us assume that the element moves with the velocity of the wave in a downstream direction and that the wave retains the same shape. Hence, at the location of the moving element there is no change in Q.

Hence
$$\mathrm{d}Q = \frac{\mathrm{d}Q}{\mathrm{d}t}\,\mathrm{d}t + \frac{\mathrm{d}Q}{\mathrm{d}x}\,\mathrm{d}x = 0$$

Combining the last two equations we obtain:

$$\frac{\mathrm{d}x}{\mathrm{d}t} = \text{velocity of wave} = \frac{1}{B}\frac{\mathrm{d}Q}{\mathrm{d}h}$$

in which $\mathrm{d}Q/\mathrm{d}h$ represents the tangent to the rating curve as shown in *Figure 3.37(b)*. We may further simplify by writing:

$$Q = CBh^{3/2}S^{1/2} \quad \text{or} \quad \frac{\mathrm{d}Q}{\mathrm{d}h} = CB\frac{3}{2}h^{\frac{1}{2}}S^{\frac{1}{2}} = \frac{3}{2}\,BV$$

or
$$\text{velocity of flood wave} = \frac{3}{2}\,\text{velocity of water}$$

The above coefficient of 1·5 applies to wide rectangular channels. For other shapes, such as triangular or parabolic, the coefficient may reduce to 1·2. Observations on several streams have confirmed that flood waves in confined channels move indeed with a speed substantially greater than the mean water velocity. The shape of the wave remains essentially the same, although there is a gradual subsidence of the crest and an elongation of the wave on both sides of the crest. If we are dealing with river channels with adjacent flood plains, rather than confined channels, the above coefficient may fall below 1·0 and hence the flood wave moves slower than the average velocity of the water. For a mathematical treatment of this subject, see the chapter on flood routing, in Rouse (1950).

It may be of interest at this point to dwell for a moment on the characteristics of the rating curve and the storage curve of a particular reach during the passage of a flood wave, as shown in *Figure 3.38*.

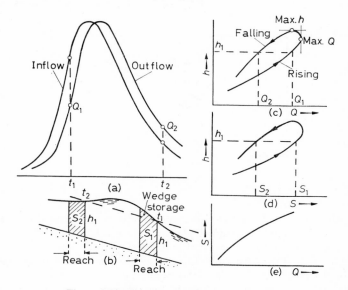

Figure 3.38. Elements of flood-wave movement

The hydrograph to the left in *Figure 3.38(a)* shows the inflow hydrograph for a certain reach of river, while the hydrograph to the right shows the outflow hydrograph. From *Figure 3.38(b)*, showing a profile over the flood wave, it may be seen that at time t_1 when the front end of the flood wave passes the reach under consideration, the surface slope is steeper than normal, while at t_2, when the rear end of the flood wave passes the reach, the surface slope is flatter than normal. As a result the total storage in the reach for a river level h_1 at t_1 is larger than the total storage for a river level h_1 at t_2. If we plot storage as a function of river level, we obtain *Figure 3.38(d)*. Another result of the change in surface slope is a larger velocity and hence discharge for a given stage h_1 during the rise of the flood than for that same stage during the recession of the flood, as shown in *Figure 3.38(c)*. It may also be noted that the maximum discharge in the channel is reached before the maximum stage is reached. If it so happens that the loop in the stage-discharge curve is nearly the same as the loop in the stage-storage curve, then the storage-discharge curve would be one

99

single line. In other words, the storage on the reach is then a single function of the outflow from the reach, as is shown in *Figure 3.38(e)*.

The problem of flood routing through river channels is basically the same as that for reservoirs. If the rise and fall of the river is so gradual that the wedge storage on the reach becomes negligible the total storage on the reach is a single function of the outflow and the flood routing computation is identical to the one shown in Table 3.9. The same applies when the loops in the stage-storage and stage-discharge curves tend to cancel one another.

There are two different ways to prepare storage-discharge curves for river channel reaches. One method is to use observed hydrographs at the beginning and at the end of the reach. By routing these hydrographs 'backwards' through the reach, the $2S+0$ and $2S$ versus 0 values can be obtained. The

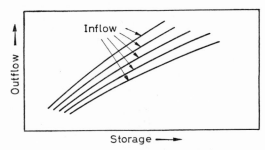

Figure 3.39. Storage-outflow relation

other method is based on hydraulic computations and topographic surveys. From the river channel and flood plain characteristics we compute a rating curve of the combined river channel and flood plain flow at the downstream end of the reach. From the field surveys we determine the total amount of storage in the river channel and flood plain, corresponding to different river stages, in terms of the gauge at the downstream end of the reach. We then combine the two curves and prepare the storage-discharge curve.

In cases where the storage-outflow relationship cannot be represented by a single line, the inflow must be introduced as the third variable, as shown in *Figure 3.39*. In order to prepare such a diagram, we perform first a provisional flood routing with a single line storage-outflow diagram. We know then approximately the shape of the hydrograph and its speed of moving through the reach. We then compute river surface profiles, allowing for varying discharge, and determine the relationship between outflow, total storage, and inflow. This is repeated for floods of different magnitudes. From this basic information we prepare a diagram as shown in *Figure 3.39*. With the aid of this diagram we then perform the flood-routing computation, similar to the one shown in Table 3.9.

Before we leave the subject of flood routing it may be of interest to discuss the possibility of determining the inflow into a reservoir or lake, in case the outflow is known. This situation may arise in hydrologic studies where the inflow into a lake has not been recorded but where a record of lake stages is available. From the lake stages and the outflow rating curve, we construct a continuous record of outflows. From the outflows and the stage-storage curve, we determine the inflow, as shown in Table 3.10.

Table 3.10. *Lake Winnipeg Net Inflow, Derived from Nelson River Rating Curve and Lake Winnipeg Stages*

Year	Month (1)	Outflow (2)	Stage (3)	Change in Storage		Net Inflow (6)
				Depth (4)	Flow (5)	
1927	Jan.	60,600	712·50			
	Feb.	55,000	712·50	0	0	55,000
	Mar.	51,900	712·50	0	0	51,900
	Apr.	55,100	713·35	+0·85	+85,000	140,100
	May	89,200	714·65	+1·30	+130,000	219,200
	June	120,000	715·55	+0·90	+90,000	210,000
	July	132,000	716·35	+0·80	+80,000	212,000
	Aug.	134,000	716·54	+0·19	+19,000	153,000
	Sept.	136,000	716·46	−0·08	−8,000	128,000
	Oct.	133,000	716·20	−0·26	−26,000	107,000
	Nov.	126,000	715·87	−0·33	−33,000	93,000
	Dec.	106,000	715·26	−0·61	−61,000	45,000

Col. (1) lists the months of the year 1927, selected from a study extending over 40 years. Since Lake Winnipeg has a surface area of 6,000,000 acres, the stages of the lake change slowly, and for the purpose of this study periods of one month were deemed to yield a sufficient degree of accuracy. If the lake or reservoir is smaller, periods of one week, or one day, or even a few hours, may have to be selected.

Col. (2) lists the computed outflows. This column was prepared from the lake level record and from an outflow rating curve for the different months of the year, allowing for ice conditions during the winter. This rating curve was based on field observations and supplementary computations. The figures listed in this column represent the average outflow during the month. Hence the lake stages that were used to compute this figure were the average lake levels during the months.

Col. (3) lists the lake stages at the end of the month. These figures were obtained by drawing an average line through the actually recorded lake levels in order to eliminate the temporary effects of wind set-up and other disturbances. The end-of-the-month stages were read from this line.

Col. (4) represents the change in lake level during the month. These figures are obtained by deducting the previous end-of-the-month lake stage from the present one.

Col. (5) converts the change in lake stage into cusec-months; 1 ft. on the lake being equivalent to 100,000 cusec-months.

Col. (6) represents the average net inflow into the lake. This figure is obtained by adding col. (5) to col. (2). The net inflow may be defined as the actual inflow plus the precipitation over the lake, minus the evaporation over the lake.

CHAPTER 4

RIVER MORPHOLOGY

When planning the development of a river basin, it is obvious that we should have thorough knowledge of the river system. Not only with respect to the water that it carries, but also with respect to sediment content and channel stability. An alluvial river, in the state of nature, maintains a delicate balance between its water discharge, sediment discharge, slope, meander pattern, and channel cross-section. Any disturbance at any point may have repercussions throughout the entire river system. If we build a dam somewhere, sediment will accumulate in the reservoir, the upstream river channel will aggrade, and the downstream channel will degrade. When we take water out of a river for irrigation the channel downstream as well as upstream may silt up. When we cut off river bends to aid navigation or flood control, the river upstream may degrade, while the river downstream may aggrade. In all such cases we should know in advance where and how fast these changes will take place.

To gain such knowledge is not an easy matter because no two rivers are alike and the behaviour of each river is controlled by many independent and interdependent factors. Because of the difficulty of the subject we shall devote a good deal of this chapter to a qualitative discussion of river behaviour. We shall begin by considering the evolution of a river from the viewpoint of the geologist.

NORMAL RIVER BEHAVIOUR

The ideal geographical cycle would be the upheaval of land masses and a standstill until they are worn down to a plain as near sea level as erosion can bring it. Actually this never occurs and many incomplete cycles are separated by a rising or sinking of the land surface. Nevertheless, such an ideal cycle can be conceived as a goal that the river constantly tends to approach, although it may never reach it.

During this process of wearing down the land masses, material is eroded from the upper part of the river drainage basins and deposited in the lower part and in the sea. The upper course of the river will be called that part of the river where erosion takes place consistently. The lower course of the river is that part of the river where deposition takes place consistently. The middle course of the river is the place where neither erosion nor deposition takes place. If the river had a steady flow, this place would be a point located at the mouth of the gorge of the upper course and at the apex of the alluvial cone of the lower course. However, due to the non-steady character of river flows, this place is normally a stretch of river that is alternately the site of deposition or of erosion.

During the beginning of the process of transformation from a recent geologic upheaval towards an ultimate plain near sea level, the changes in river profile take place rather rapidly and the streams are called young. During the middle of this process, the streams are not so active and are called mature. At the end, the changes are hardly noticeable and the streams are called old. It follows from this definition that the terms young, mature and old, have reference not so much to the length of the history of rivers in years, as to the amount of work which the streams have accomplished in comparison with that which they have before them. The meaning of these adjectives, young, mature and old, as applied to rivers, will become more familiar in the reading of Salisbury's (1908) discussion on stream erosion and alluviation.

The span of time that is involved in a complete transformation of the landscape may run into millions of years. Consequently, it can be expected that during a period of scientific observation, or even during the period of a civilization, no noticeable changes in some river profiles will take place. Strictly speaking, however, a river is never completely stable until it has reached a profile whereby the velocities are so low that no transportation of sediment can take place.

Transport of Sediment

It is evident that in this process of continuously reshaping the land surface by erosion and deposition, the transport of sediment by streams and rivers plays an important role. Since this will be the main topic of the following discussions, it may be of advantage to treat the subject first in general terms before proceeding to detailed descriptions.

Sediment is transported by a river as bed load and as suspended load. The bed load is composed of relatively coarse material, and rolls, bounces and saltates near the bottom of the river. The suspended load is composed of relatively fine material and is kept in suspension by the turbulence of the river. Actually, there is no sharp distinction between the two forms of sediment transport. The finer particles of what is designated as bed load may be found to saltate in the same manner as the coarser particles of the suspended load. The coarsest particles of a sample of the suspended load may even be found to be larger than the finest particles of a sample of the bed load. The distribution of suspended sediment tends to vary with the velocity distribution and with the turbulence of the flow. The finer particles may be fairly evenly distributed over a vertical plane, whereas the coarser particles usually have their largest concentration at the lower levels of flow. The suspended load may be sub-divided into 'suspended bed material load' and 'wash load'. The suspended bed material load includes all particles that are also found in the river bed material; for an alluvial river this is usually fine and coarse sand. The wash load includes all particles that are smaller than the river bed material, such as silt and clay. From a viewpoint of river channel characteristics (slope, depth, width, meander pattern) we are mostly interested in the suspended bed-material load. The wash load may be looked upon as an additive to the water that is picked up on the plains of the drainage basin, that remains in suspension at all times while being transported, and that is flushed through without participating in the formative process of the river system. It has only some importance when it finally comes to rest on flood plains, in river deltas or in reservoirs.

103

From various studies of sediment transport, it has become evident that the transport of bed load is some function of the tractive force which the flowing water exerts on the periphery of the channel and that the amount of sediment in suspension depends primarily upon the turbulence of the river. The tractive force is the component in the direction of flow of the weight of the water, and can be represented by $W.D.S.$ where W is the unit weight of the water, D the depth of flow, and S the slope of the river. The tractive force, or shear, is also, by definition, equal to the dynamic viscosity of the water times the velocity gradient at the point of consideration. The degree of turbulence can be expressed by Reynolds' number, represented by the depth of flow times the velocity of flow, divided by the kinematic viscosity.

From these relationships, it can be concluded that the sediment transport will in general increase with an increase in depth and an increase in river slope. However, the increase in transport of bed load and transport of suspended load does not necessarily have to be in the same proportion since they are a function of different variables. It is even conceivable that the transport of bed load will increase while the transport of suspended load will decrease. Consider, for instance, two river conditions with the same discharge, the same hydraulic roughness and the same average velocity, but with a different depth of flow and consequently a different slope and a different width of channel. The river condition with the greatest depth of flow will have the greatest turbulence and could, therefore, be expected to have the greatest suspended-sediment concentration. The river condition with the smallest depth of flow has the greatest velocity gradient, hence the greatest tractive force. It could, therefore, be expected to have the greatest bed-load transport. This speculation is confirmed by a conclusion reached by Leopold and Maddock (1953) after examination of field and laboratory data on the transport of sediment: 'At constant velocity and discharge, an increase in width is associated with a decrease of suspended load and an increase in bed-load transport.'

During every phase of geomorphologic activity, a river will try to obtain some form of equilibrium in which the slope is sufficient to carry the sediment that is brought into the river system. It will be of advantage to keep in mind that nearly all the energy that is spent by a river is consumed in external friction against the river bed and in internal friction through turbulence. According to Rubey (1933), only a very small part of the river energy, of the order of magnitude of a few per cent, is consumed in transporting the sediment. Consequently, this sediment transport should be regarded more or less as an incidental result of the agitated state of the river. Although the transport of sediment is incidental to the flow of water, it does play an important role in the behaviour of the river. When the supply of sediment from the drainage basin is larger than the river happens to be able to carry, some of it will have to be deposited in the river channel and the slope of the river will increase until the velocities become so great that a balance between supply and transporting capacity is reached. When the supply of sediment is smaller than the capacity, the river will, through its turbulent motion, pick up material from the river bed and add this to its incoming sediment load. If the river bed consists of alluvial material, this is easily done. When the river bed consists of bedrock, this process will take time, because the particles have to be abraded or dislodged by shock.

104

This natural process of obtaining equilibrium between sediment supply and sediment transport is obscured by a number of factors. One of them is variability of discharge. Its result upon river activity is described by Gilbert (1914): 'All streams vary in volume of discharge from season to season and from year to year. It is broadly true that streams give shape to their own channels, and among alluvial streams there are few exceptions. But through variation of discharge the same stream is alternately large and small, so that its needs are different at different stages. The formative forces residing in the current are so much stronger with large discharge than with small, that the greater features of channel are adjusted to large discharge, and this despite the fact that floods are of brief duration. Floods are of all magnitudes, and each flood presents not only a maximum discharge but a continuous series of changing discharges. At each instant the stream contains a system of currents of which the details depend not only on the discharge but on the shapes of channel, created by the work of previous discharges as well. So long as the discharge continues, its currents are eroding and depositing in such a way as to remodel the channel for its own needs, and so long as the work of remodelling continues, the loads and capacities at different cross-sections are different. With maximum discharge all the coarser grades of debris are in transit. With lessening discharge the coarsest material stops, but it stops chiefly in the deeps, because the change in bed velocity is there greatest. At the same time the coarsest of the suspended load escapes from the body of the stream and joins the bed load. With continued reduction of discharge the tractional load in the deeps becomes gradually finer and at last ceases to move. The tractional load on the shoal is then derived from the local bed. Soon the derivation becomes selective, the finer part being carried on while the coarser remains, with the result that the shallow channels on the bars become paved with particles which the enfeebled currents cannot move.'

Another factor which may confuse an essential state of equilibrium is the fluctuation of total run-off from year to year in different parts of the drainage basin. A description of the result on the Missouri River is provided by Straub (1935): 'The amount of detritus transported along the stream bed is essentially a function of the rate of discharge. There are usually two fairly well defined high-water periods each year. The first occurs in the early spring and has its greatest effect in the upper river; the other, the "June rise", usually has its greatest effect in the lower river. In case of considerably higher normal stage in one part of the river than in the other there is a dumping or picking up of bed load, depending upon the relative discharge past the two localities of the river. In the case of the tributaries a similar condition obtains. Frequently because of high rates of precipitation in one section of the drainage area more sediment is contributed to the Missouri River by the tributary than the Missouri is capable of transporting. This results in a rise in the river bed. The Missouri River therefore exhibits local variations of 5–10 ft. change in elevation of the stream bed in the course of less than a year, while changes in the general elevation throughout longer stretches of river amount to a foot or more.'

In addition to periodic changes in the volume of sediment transport, it can be pointed out that the actual mode of sediment transport may also differ from stage to stage. Gilbert (1914) made the following observations from laboratory flume tests: 'In another experiment a bed of sand was first prepared with the surface level and smooth. Over this a stream of water was run with a current so gentle that the bed was not disturbed. The strength of the current was gradually increased until a few grains of sand began to move and then was kept constant. After a time a regular pattern developed and the bed exhibited a system of waves and hollows, called dunes. The upstream face of the waves was eroded and the downstream face built out. In succeeding runs, the load was progressively increased. This caused increase of slope and velocity, with decrease of depth, and these changes were accompanied by changes in the mode of transportation. In the earlier runs dunes were formed, and these marched slowly down the flume. Then, somewhat abruptly, the dunes ceased to appear, and for a number of runs the channel bed was without waves and approximately plane. After further enlargement of the load, a third stage was reached in which the bed was characterized by anti-dunes, travelling against the current instead of with it. They travelled much faster than the dunes, and their profiles were more symmetric. The water surface, which showed only slight undulation in connection with dunes, followed the profiles of the anti-dunes closely. Some of these waves remained for two minutes or longer, but most of them not longer than a minute. A whitecap would form on the surface of the water when the larger waves disappeared.'

The last observation of anti-dunes and subsequent surface waves is seldom made on actual rivers. A description is provided by Pierce (1916): 'On the San Juan River sand waves appear at their best development on rapidly rising stages. The usual length of the sand waves, crest to crest, on the deeper sections of the river is 15–20 ft., and the height, trough to crest, is about 3 ft. The depth of the river may range from 4–10 ft. The sand waves are not continuous but follow a rhythmic movement. At one moment the stream is running smoothly for a distance of perhaps several hundred yards. Then suddenly a number of waves, usually from 6–10, appear. They reach their full size in a few seconds, flow for perhaps two or three minutes, then suddenly disappear. Often, for perhaps half a minute before disappearing, the crests of the waves go through a combing movement, accompanied by a roaring sound. On first appearance it seems that the wave forms occupy fixed positions, but by watching them closely it is seen that they move slowly upstream.'

These different modes of sediment transportation, causing different types of bed configuration, are closely related to the hydraulic characteristics of the river. Harrison (1954) reported, for instance, 'that in alluvial rivers the hydraulic roughness of the channel does not remain constant at all stages but varies with discharge. On the Platte River, the Missouri River and the Mississippi River, the roughness decreases with increasing discharges, within limits. This is consistent with observations of movable beds in flumes where it has been observed that at intermediate intensities of bed material load transport, ripples and bars tend to form on the bed; but as the bed load transport intensity increases, the ripples disappear and the bars become longer and flatter until they also eventually disappear altogether. It appears that the roughness of an alluvial river bed might be considered in two categories: first, that due to the granular roughness of the sand grains; and second, that roughness due to large scale irregularities of the bed surface such as bars and ripples. The variability of this "larger scale" roughness or "extra roughness" is evident on contour maps of the bed made at two different discharges in a short reach of the Missouri River at Omaha. It is reasonable to expect that as the discharge increases, increasing the intensity of bed load movement, the roughness due to the irregularities will decrease and the total roughness will approach the sand grain roughness as a minimum limiting value.'

In addition to causing different types of bed configuration, different sediment-transport intensities may also cause different types of channel alignment. A high sediment concentration, for instance, may lead to relatively steep slopes and may cause straight braided channels. A moderate sediment concentration may lead to moderate slopes and cause normal meandering channels. A low sediment concentration may lead to mild slopes and be associated with stable channels and an irregular meander pattern.

From the discussion in the above section, it would appear that rivers may behave in a rather complicated manner and that several variables are involved in the process. On closer inspection, the more important variables prove to be: Q, the discharge of the river; T, the flow of sediment in the river; d, the

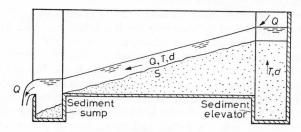

Figure 4.1. Sediment transport test

effective diameter of the sediment particles; S, the slope of the river; r, the ratio of mean depth to width of the river channel; and M, denoting the meander characteristics of the river, expressed in ratio of river mileage to river valley mileage.

In a laboratory flume as shown in *Figure 4.1* we could quickly establish some relations between these variables. Let us assume that we feed into the flume at the right-hand side a given discharge Q, with a given sediment load T, with a given diameter d. In the beginning of the test, the water and sediment will flow over the bare, horizontal bottom of the flume. The water surface profile will be a back water curve controlled at the left-hand side by the weir and with increasing depth of flow at the right. The velocities will be relatively low. We shall assume that they are too low to carry all of the sediment load T. As a result, some sediment will remain behind and start to fill up the bottom of the flume. This will continue until the situation, shown in *Figure 4.1*, has been established. The slope has now become so steep, that the velocity of the water is large enough to carry all of the sediment load T. By adjusting the exit weir in the correct position, we can produce a uniform depth of flow and hence a uniform slope. The discharge over the weir Q is the same as the discharge Q, entering the flume, and the sediment load T that falls in the sediment sump is the same as the sediment load T that enters the flume. We have now reached a steady, equilibrium condition whereby Q, T, d and S are in balance. With this slope, and only with this slope, the given Q is just able to transport the given sediment load. If we keep Q and d the same, but we would increase T, not all sediment could be transported any longer with the given slope. Some sediment would be deposited, first at the upper end of the flume, and then progressively downstream until the slope of the flume is so much steepened that the new T can be transported. If, starting with the original equilibrium conditions, we would keep Q and T the same, but we would increase d, the result would be similar. It is more difficult to transport coarse sediment than fine sediment and therefore some sediment would be deposited until the slope of the channel bed is sufficiently steepened to carry the new sediment load. If we keep T and d the same, but we would increase Q, the evident result is some scour at the upper end of the flume since there is now excess transport capacity. This scour will continue and extend throughout the whole flume until the slope is so much decreased that the new Q, T and d are in balance again. The criterion of balance is fulfilled when Q, T and d are quantitatively the same at the entrance and the exit, while the depth of flow is uniform throughout the entire flume.

In the above laboratory tests, we have arbitrarily selected Q, T and d as the independent variables so that S became the dependent variable. An interesting problem is to determine in nature which of these variables are independent and which are dependent. From a viewpoint of stream behaviour, it would seem that Q is definitely an independent variable. It would appear that d is a variable, dependent upon the nature of the mountains or plains, and also upon the nature of the erosional process. Hence, d would seem to be an independent variable from the viewpoint of stream behaviour. The variables r and M would seem to be dependent variables for alluvial rivers, provided these are defined as rivers that flow freely in alluvial sediments that are recently deposited under circumstances similar to the ones of the present time.

The two variables that are left—T, the sediment transport, and S, the slope of the river—have given cause to many disputes and misconceptions as to which is the independent and which is the dependent variable. In order to help clarify this point, let us conduct two more imaginary laboratory experiments. The first is with a short flume, which is going to be operated for a long period of time. At one end of the flume enters a certain flow of water, Q; a certain flow of sediment, T; with a certain gradation, d. After some period of time, the bed of the flume channel will have established a stable slope, S. There is no question that, in this case, T is the independent variable, and S, the dependent variable. The second experiment is with an extremely long flume, which is going to be operated for a short period of time. The flume has a thick bed of sand and a certain slope. At the upper end of the flume enters a certain flow of water, Q, and no sediment. After a short period of time, the flume channel bed at the lower end will still have the original slope S and carry a certain T, in accordance with this slope. There is no question that at that place and at that time, T is the dependent variable and S is the independent variable. It should be noted that the situation in the first experiment is stable. As long as the given Q, T and d are maintained, so long will the slope S remain. The situation in the second experiment is not stable at all. Initially, the sediment transport will adjust itself to the given slope, but it can only do so by deriving material from the channel bed, and by doing so, it will begin to change the slope of the channel and consequently its own magnitude.

When comparing these two examples with actual rivers, it is tempting to think first of the ideal geographic cycle, whereby the upper course of the river provides a certain Q, T and d, which will produce a certain S, r, and M in the lower course of the river. Through the wearing down of the mountains, there will, of course, be a gradual change in Q, T and d, and consequently in S, r and M, but essentially, during the whole process, the slope of the lower course of the river is the dependent variable and the sediment transport in the lower course is the independent variable. In the upper course of the river, the reverse situation exists. It is essentially the slope of the land, which is intimately associated with the slope of the river system, that produces a certain T with the given Q. Hence, the slope of the river is the independent variable and the sediment transport is the dependent variable.

It is in rare cases only that the upper part of a river drainage basin is eroded, and the lower part alluviated, in close resemblance with an ideal geographic cycle. In most cases, such a cycle is disturbed by either an upheaval of land masses or a change in climate. In both cases, the magnitude of the independent variables will change and consequently the regime of the river will change. The dividing zone between erosion and deposition, designated as the middle course of the river, may move upstream or downstream. Former alluvial deposits may begin to erode or river locations that were formerly degraded may now begin to aggrade. However, there will always be the tendency to establish a new definite river regime based on the new situation and in accordance with the laws of river morphology.

In summary of the above discussion, the interdependency of the variable factors involved in river regime can be presented as follows. In the upper course of a river system, the discharge Q is determined by hydrologic factors and can be regarded as an independent variable, from a viewpoint of river behaviour. The slope of the land, and therefore the slope of the rivers, is determined by geologic factors and can also be treated as an independent variable. The same holds true for the nature of the sediment particles. The sediment transport is determined by the discharge, the slope of the rivers, and the size of the sediment particles, and is, therefore, a dependent variable. The formation of the river channels, expressed in r and M, is determined by Q, S, d and T, and hence is a dependent variable. In the lower course of a river, it would seem logical to treat Q, T and d as given quantities, and hence as independent variables. It follows that S, r and M are then dependent variables.

The manner in which the independent variables and the dependent variables are related to each other is rather complicated and forms the subject of considerable research. At the present time, there are only approximate relationships established which allow the determination of S, r and M for a given Q, T and d. These relationships will be discussed in the last part of this chapter under 'Sediment Computations'. One complicating factor is that the hydraulic roughness of the river channel plays an important role in the relationship between the variables. It is found in laboratory experiments by Gilbert (1914), Brooks (1958) and others, and in the field by Harrison (1954) and others, that different magnitudes of sediment transport produce different types of topography of the river-channel bed; consequently, different magnitudes of hydraulic roughness, and therefore, different river surface slopes. Moreover, a change in r, or a change in M is also associated with changes in the river-channel roughness. Hence, the investigator should be aware that there may not be a simple relationship between say T and S, but that such a relationship may become highly obscured through simultaneous changes in hydraulic roughness, shape of the river cross-section, and meander pattern. Interesting discussions of some of these features are given by Leopold and Maddock (1953), Brooks (1958), and Rubey (1952).

Graded Rivers

When the upper course of a river, with a given Q, S and d, carries its optimum T and has shaped a corresponding r and M, or when the lower course of a river, with a given Q, T and d, has established its desired S, r and M, the river is more or less in balance. In geologic terminology, this condition is called graded. Mackin (1948) introduced the following definition: 'A graded stream is one in which over a period of years, slope and channel characteristics are adjusted to provide with available discharge just the velocity required for the transportation of the sediment load supplied from the drainage basin.'

An interesting discussion of the meaning of a graded river is given by Davis (1902): 'A graded river has reached a condition of essential balance between erosion and deposition. A mature river has this graded condition extended over all parts of the river system. A change at any one point thus involves a change of infinitesimal amount perhaps, all through the system. It is important to recognize that the maintenance of grade, during the very slow changes in volume and load that accompany the advance of the cycle, involves an appropriate change of slope as well. We have here to do with a slowly, delicately and elaborately changing equilibrium of river action, accompanied by a corresponding change in river slope. Once graded, a river will never depart perceptibly from the graded condition as long as the normal advance of the cycle is undisturbed. It is possible that the slope of a graded stream may have to be increased for a time after it has first been attained, for there is no necessity that the load should cease increasing just when its value has risen to equality with that which the streams can transport. If at any time in the cycle a change of climate should occur, new slopes would have to be developed by the streams in order to bring about a new balance between erosion and transportation. If, for example, the changes were from humid to arid conditions, all the valley floors would have to be steepened by aggradation. The incapacity of the Platte to deepen its valley leads Gannett to describe it as an overloaded river. This phrase is unsatisfactory, because it overlooks the fact that a river refuses to be consistently overloaded. It is ready to increase its capacity by aggradation. Like all streams with braided channels, the Platte is well graded although the quantity and texture of its load require it to maintain a relatively steep slope.'

An interesting demonstration that the upper course of a river, which is in the process of cutting downward into bedrock, can also be a graded river, is provided by Mackin (1948): 'The Shoshone River, East of Cody, has over a stretch of 50 miles, an average slope of about 30 ft. per mile. The profile is concave upward, similar to that of a river flowing on alluvial deposits. The stream is only mildly cutting down into the underlying soft sandstones and shales, much less than could be expected from its discharge and channel characteristics, as demonstrated by other streams with milder slopes which are vigorously cutting down in harder rocks. The down-cutting on the Shoshone is held in check by the required transportation of a large load of coarse rock waste continuously supplied by the headwaters. Its high gradient is perfectly adjusted to provide, with available discharge and with the prevailing channel characteristics, just the velocity required for transportation of the sediment. These velocities are so high that 12 in. boulders roll along its bed, and the river is such an effective grinding mill that boulders are reduced in diameter by one-half within a few tens of miles. Nevertheless the velocity is so definitely fixed that the river is not able to cut down appreciably at the present time.'

From the above discussion, it appears that the graded condition is more a matter of degree of balance than that of absolute balance. Due to the normal wearing down of the land masses, the continuous geologic warping and the periodic changes in climate, no river is in absolute balance. However, many rivers approach this balance so closely that for practical purposes they can be called graded. In particular, from an engineering viewpoint, which considers relatively short periods of time and mostly short stretches of river, natural streams are roughly in equilibrium.

It may be of interest at this stage of the discussion to point out that the appearance of a river valley with a meandering river, abandoned river loops, alluvial soils, and wide flood plains is in itself no indication whether the river is aggrading, in balance, or degrading. Sometimes these signs are taken as an indication for aggradation. However, it should be recognized that a river, in its process of meandering throughout the valley, periodically turns over the valley floor to a depth which is roughly equal to the depth of the river. The river cuts away the concave banks and builds up the convex banks. The latter deposits are composed of rather coarse material at the bottom and finer material on top. In due time, through the continued shifting of the river channel, these deposits are situated away from the active streams and are continuously being covered with deposits of silt and clay from occasional flood overflows. A boring taken at such a site may show 10, 20, or 30 ft. of stratified alluvial deposits. Let it be assumed that the cycle of completely reworking the valley floor has a duration of 100 years and that the depth of turnover is 30 ft. Then it becomes evident that a degradation of say 5 ft. in 100 years will have hardly any effect on the thickness of the alluvial deposits or on the general appearance of the flood plains.

Another misleading indication of whether or not a stream is in balance could be a short period of record with respect to river-bottom profiles or mean river surface profiles. Apart from the major changes in climate, stretching over periods of thousands of years, there are minor fluctuations in meteorological conditions over periods of five or ten years. A dry period will cause a relatively poor vegetal cover over the drainage basin. This will result in a relatively low Q and a relatively high T. Consequently, a lot of sediment will remain in the stream channels which are not adapted to these conditions. The return of the normal rainfall, or the occurrence of a subsequent wet period, will result in relatively high river stages, because the channels are, to a large degree, plugged up with sediment. Flood conditions will prevail for a number of years, creating the erroneous impression of aggradation. However, over a longer period of time, the river will restore its normal channel characteristics and may prove to be in balance.

As an illustration to the above discussions, it may be of interest to examine a few actual rivers. The first one will be the Platte River in the U.S. Middle West. This stream rises in the Rocky Mountains, flows over a distance of about 500 miles over plains country and joins the Missouri River near Omaha, Nebraska. The drainage area of the river system that lies in the mountains is small compared to the total drainage area. The transition from the mountains to the plains is abrupt. The plains are rather flat, are originally composed of non-alluvial materials and have a predetermined slope from a geomorphologic viewpoint. When flying over the drainage basin from west to east, the following observations can be made. Immediately east of the mountains, the plains are being alluviated by the streams coming from the mountains. Abandoned river courses and oxbows can be seen all over the landscape, and there are no river valleys. Further east, some land erosion begins to take place. The streams are cutting into the plains, and isolated gullies leading into shallow stream valleys can be seen. Still further east, the almost horizontal plain is more and more being dissected by long separated stream gullies and valleys with relatively steep banks. The streams in the tributary valleys are quite often vigorously meandering, whereas the main trunks have a straight braided character. In the lower part of the drainage basin, near the Missouri River, little is to be seen of the original plain in which the valleys are developed to such an extent that nearly all the surface has been reduced to slope. The depth of the valleys, with respect to the highest surrounding land, is a few hundred feet near the mouth of the river.

From a viewpoint of river morphology, it would seem that the following deductions could be made. The streams coming from the mountains have a rather steep slope and carry a sediment load that is in excess of what can be carried on the relatively flat plains. Consequently, alluvial fans develop on the most western part of the plain. These fans merge into one another and form a continuous aggrading alluvial plain. Further east, the discharge in the main channels is gradually increased through local drainage and finally the sediment-transporting capacity of these streams, with their predetermined slope of several feet per mile, becomes greater than the residual sediment load that is brought down from the alluvial cones near the mountains. Consequently, the river will begin to pick up local material. After degradation on the main streams has taken place for some time, valleys and gullies will be eroded backwards along the tributaries into the plains, thus gradually dissecting the country. If it were true that the original plain had a more or less even slope between the Rocky Mountains and the Missouri River, then it could be deduced that the above process will be more pronounced in easterly direction. The discharge of the main stem will gradually continue to increase through local drainage. Therefore, it could carry its sediment load with a continuously decreasing slope. However, the slope is predetermined, and hence the tendency to degrade will gradually increase in downstream direction. This could account for the observation that the lower part of the drainage basin is thoroughly dissected, whereas the middle section is only partly dissected.

Although it is not quite understood why, it is well known that relatively steep rivers develop braided channels and also that a small river needs a steeper slope to develop this braided aspect than a large river. This could explain why some of the tributaries of the Platte develop a meander pattern

112

with the predetermined plain slope and why the larger main trunks develop, with the same slope, a braided pattern. This braided aspect has sometimes been interpreted as proof that the Platte River is an aggrading stream. However, the appearance of the drainage basin as a whole would definitely seem to indicate degradation of the whole river system. In view of the fact that this degradation takes place over almost the whole drainage-basin surface, it becomes obvious that the quantitative degradation of the main trunk over a period of say the last 100 years is practically unmeasurable. Moreover, the balance of the river system may have been affected during this period by changes in climate, or other causes. It can be concluded, therefore, that neither the fact that degradation has not been measured, nor the fact that the river has a braided aspect, nor the fact that the river valley bottom consists of alluvial materials, necessarily indicates aggradation or equilibrium. On the other hand, the general appearance of the surface of the drainage basin and the relatively steep slopes of the streams would definitely seem to indicate consistent erosion throughout the drainage basin and in the river valleys. Consequently, the complete Platte River system should be classed as the upper course of a river. In the present case, the lower course of this river would be either the Missouri River or the Mississippi River.

Since any consistent changes in stream-bed profile are so small that they cannot be measured over a period of say 10 or even 100 years, it would seem perfectly logical to call such a stream in equilibrium, from an engineering viewpoint. Such an apparent equilibrium will be maintained as long as the natural circumstances remain the same. However, there is ample reason to believe that the activity of man will disturb this equilibrium, not only in this drainage basin, but at many more places. There are two possibilities. The first is that through abuse of the natural vegetative cover of the drainage basin, the natural sediment production of the area is increased. This will result, besides the loss of valuable land, in the aggradation of stream channels. The second possibility is that through soil conservation measures, the natural sediment production of the area is decreased. This will result in degradation of the stream channels. It may very well be that the benefits of soil conservation will outweigh the damages of stream-channel degradation, but it would seem that in planning such beneficial measures, due attention should be paid to their consequences.

The second example that will be discussed is the Saskatchewan River in Central Canada. This stream also rises in the Rocky Mountains, flows for a distance of about 700 miles over the Canadian Prairies and forms a huge delta area near the place where it enters Lake Winnipeg. Over the first 200 miles the slope of the river is several feet per mile. As a result, the tributaries of the river system, that emerge from the Rocky Mountains, do not form an alluvial plain at the foot of the mountains, as was the case with the Platte River, but begin to cut immediately into the foothills and the adjoining prairies. Evidently, the predetermined slope of the landscape is steeper than the equilibrium slope that belongs to the water and sediment discharge coming out of the mountains. In this stretch of river the bottom consists mainly of small boulders, pebbles and gravel. The valley banks become progressively higher till they are a few hundred feet above river level. There is evidence of active valley bank erosion along the entire stretch. Bank slides of one to two hundred feet high and several hundred feet long are not uncommon. The velocity of these slides does not generally exceed a few feet per year. Over the next 300 miles the river valley is a few hundred feet deep, nearly parallel to the prairie surface, and has a slope of one foot per mile. The river bed is composed of sand and fine gravel. In places the river has the appearance of a meandering river; in other places it appears more like a braided river. There is little or no evidence of valley bank erosion, and for practical purposes we can look upon this reach of river as being graded. Over the next 100 miles the surrounding landscape has again a slope of several feet per mile. The river valley becomes narrower and progressively deeper. The river bed consists of occasional boulder pavements, where rapids occur, with alluvial stretches in between. The valley banks are being actively eroded and huge bank slides of several hundred feet high and up to a mile long can be observed. At the end of this reach is a concentrated, boulder-paved rapids with a drop of about 50 ft. in 5 miles and below that is the Saskatchewan Delta, extending over two million acres, where all sediment is being deposited.

In this example we could look upon the entire reach of river from the headwaters in the mountains to the delta, as the upper course of the river. With the exception of a few hundred miles in the middle, where the river appears graded, there is consistent erosion from the beginning to the end. The middle course becomes the last rapid where the river valley terminates and where the delta begins. The lower course is the delta proper.

Channel Characteristics

Let us pursue the subject of channel formation in more detail. Why are river channels mostly relatively wide and shallow? A narrow and deep channel would convey the water much more effectively! Suppose a long stretch of natural wide river channel is replaced by an artificial narrow channel, to carry the same flow of water. Due to the more efficient shape of the channel, the velocities will be higher, and scour all around the perimeter will take place. The material of which the banks are composed will be extremely vulnerable to erosion, since the action of gravity will help to dislodge and move the particles. This material, eroded from the banks, will slide down towards the bottom of the channel and be transported together with the material that is eroded from the bed. Since the discharge is a given quantity and the velocity only varies within limits, it can be expected that the area of cross-section of the flowing water remains more or less constant. Due to the caving of the banks, the river will become wider and consequently shallower. The river will continue to widen until the velocities near the side slopes become too low for further erosion. This reduction of velocity is attained on the one hand through a lowering of the mean velocity in the river because of the less efficient cross-section, and on the other hand through a lowering of the ratio of side to mean velocity because of the change in shape of cross-section. When finally the banks become stable, the river channel is back to its original wide and shallow shape.

It is evident that the shape of the channel, all other factors being the same, will depend upon the nature of the bank material. Since that material is in alluvial rivers deposited by the same river at some earlier time, it can also be stated that the shape of the channel depends upon the nature of the transported sediment. If this is coarse sand, the banks will become uncohesive and the channel will become relatively wide. If it is fine sand and silt, the banks will become cohesive and the channel will be relatively narrow. It is also evident that the shape of the channel, all other factors being the same, will depend upon the slope of the channel. A steep slope produces relatively high velocities and will, therefore, result in relatively wide channels. Since the slope is a function of the sediment transport, it can also be stated that a large sediment transport will usually result in relatively wide channels. In general, the upper course of a river has a steep gradient and carries coarse sediment, whereas the lower course has a flat gradient and carries fine sediment. Consequently, it can be expected that, in general, a river is relatively shallow in its upper course and relatively deep in its lower course.

In addition to the slope and the shape of a river channel, its meander pattern also plays an important role in river studies. At the outset, it can be stated that it is not quite understood, and certainly not agreed upon, why a river meanders. It is not difficult to give an explanation that could be true for a certain case. For instance, it could be argued that a river flowing perfectly straight with a uniform movement of bed material is an unstable phenomenon. As soon as the symmetrical pattern is disturbed, due to turbulence or temporary one-sided deposition or other insignificant causes, erosion at one side of the channel, and deposition at the other, will start to take place and meandering is initiated.

However, if the above explanation is correct, why does not every river meander at every place? It is a fact that some rivers meander more actively than others and that some rivers do not meander at all. It is also a fact that some rivers meander at places, while there are long, straight stretches in between where there is no tendency to meander at all. Apparently, there is more to it than a simple uniform explanation.

Before an attempt is made to elaborate on the subject of meandering, a discussion of the nature of the movement of water and sediment in meander bends will follow. It can be demonstrated by drawing a flow net that the hypothetical flow of frictionless water around a bend will produce the highest velocities near the inside of the bend. In nature, there is no frictionless flow and it will be found that there is a certain velocity distribution over the cross-section of a straight channel, with the highest velocities in the middle near the surface and the lowest velocities near the perimeter. As a result of this difference in velocities, the fast surface water, because of its inertia, will move towards the outside of the bend and, consequently, the slow boundary water will have to move towards the inside of the bend. This second effect of the highest velocities accumulating near the concave bank offsets the first effect of the highest velocities being produced near the convex bank. Whether the one effect will dominate over the other depends upon the vertical velocity distribution. In a laboratory experiment of flow around a bend by Shukry (1950) the variation in velocity in a vertical plane was very small. Consequently, he found the highest velocities near the inside of the bend and a confused spiral motion. A river cross-section on the other hand may have a width of 10 to 20 times its depth. In the middle part of such a cross-section, one can expect a pronounced decrease in velocity from the surface to the bottom and therefore a pronounced movement of the high velocities towards the outside of the bend. It takes, of course, some distance for the fast water to reach the concave bank; hence, it

115

can be expected that at the beginning of the bend, the highest velocities are near the convex bank on account of the flow-net principle, and at the end of the bend near the concave bank on account of the inertia principle. Due to the inertia of the fast flowing water, it can also be expected that the highest velocities will remain for a while near the concave bank before they return to the middle of the river.

In the above paragraph, only the flow of water is taken into consideration. A most important role in the formation of river bends, however, is played by the movement of the sediment. It was noted earlier that the sediment that is transported by the river can be classified into two groups according to Einstein (1950). 'The first group is called "wash load" and includes the fine sediment particles. The concentration of wash load in a river depends upon availability in the drainage basin and not on the ability of the flow to transport it. Wash-load particles travel with the same velocity as the flow and are not deposited in the stream channels. The second group is called "bed-material load" and includes the coarse sediment particles. This material is at times deposited in the stream channels and at other times, picked up again and carried along as bed load or as suspended load. Hence, a river normally carries bed-material load at capacity level. Since the wash load represents usually the largest part of the total load, it is the most important factor in the studies of silting of reservoirs. For studies of stream-channel stability, however, the bed-material movement is important and wash load has practically no influence.'

From this classification it follows that the wash load will move through a bend without getting a chance to settle at any place. The bed-material load which is in suspension will partly settle out near the lower part of the convex bank where the velocities tend to be consistently lower than average. The bed-material load that rolls and saltates over the river bottom will tend to move towards the convex bank for the same reason that causes the spiral motion. Some further explanation of this last phenomenon may be of interest.

In flowing around a bend, the river has a transverse gradient, sloping upwards towards the concave bank. This gradient provides the centripetal hydrostatic force that makes every element of water describe a curved path. This gradient is consistent with the mean velocity of the water. Water particles that have a higher velocity tend to move towards the concave bank. If a hollow tube was laid on the bottom of the river across the middle of the bend and kept there, it would have a constant flow of water from the concave bank towards the convex bank. In the same way, the transverse gradient will act with a force towards the convex bank, upon any body of material that is laying still upon the bottom of the river. Hence, a grain of bed material that is not in motion is acted upon by the following three forces: one force, caused by the transverse gradient, acting in the direction of the centre of the bend; another force, caused by the flowing water, acting in the direction of flow; a third force, caused by the gravity, acting vertically downward. The last force has a component that has to be taken into account only when the surface of the bed is not horizontal.

When the particle moves through the bend over a horizontal bed with a low velocity, it will move towards the convex bank. The velocity of the water particles is already in that direction because of the spiral flow, while the particle is further deflected by the transverse gradient. After being deflected for a while, the particles will come into the region of still lower velocities near and downstream of the convex bank and some of them will be deposited. This will continue until such a side slope is built up that the gravity component balances the transverse gradient and spiral flow.

In addition to some deposition of bed load, there will also be deposition of suspended bed-material load near and downstream of the convex bank, due to the local decrease in stream velocities. Since the bed load is in many meandering rivers only a fraction of the suspended bed-material load, the latter form of deposition may be the more important of the two. There is, of course, a limit to the deposition on the convex bank. If the concave bank would not erode, deposition would decrease the cross-section of the river and therefore increase the velocities and side slope, until a state would be reached whereby no further deposition could take place. In a way, this is what happens in the formation of the proper channel shape in a straight stretch of river. The shape becomes such that no further erosion can take place with the existing velocity distribution. However, in a meandering river, the concave bank is usually composed of erodible material, hence deposition on the convex bank can continue to take place. It follows that the rate of erosion and the rate of deposition are inter-related. A river with cohesive banks or with little transportation of bed material will have a slower progress of its meanders than a river with erodible banks or a large transportation of bed material. Since in a normal river both the composition of the banks and the transport of bed material are a function of the sediment characteristics of the river, it would seem logical to state that the rate of meander progress is also a function of the same. This would seem to eliminate the old controversy as to whether erosion of the concave bank is the initial cause which permits deposition on the convex bank or whether the growth of the convex bank causes the concave bank to erode.

From laboratory experiments and field observations, it is found that eroded material from a concave bank has little opportunity to be deposited on the opposite convex bank. Instead, it moves downstream and is deposited mostly on the first convex bank downstream; i.e., at the same side of the river. This is probably because the most severe erosion takes place over the lower part and below the bend in the river, whereas the transverse gradient exists in the bend only. Moreover, it takes a certain length of travel before a particle is deflected appreciably. Hence, it seems logical that eroded material can be transported much easier towards the convex bank downstream, at the same side of the river, than towards the convex bank, opposite the site of erosion.

Coming back to the original cause of meandering, it is of interest to quote Matthes (1941): 'If a river flowing in an erodible channel could coordinate velocity with bed resistance simply by adjusting its width, then theoretically it could maintain a straight but wide and shallow channel down the axis of the valley. Its hydraulic gradient then would equal the valley slope and its bed load would travel over the full width of a very flat bed. ... (However), in ordinary streams having tough, resistant banks, the bed load collects in bars or shoals which tend to alternate in position along opposite shores and which force the low water channel to take a sinuous course, often causing it to swing alternately from bank to bank. ... In a meandering river, this action becomes greatly intensified through the yielding character of the banks.'

Mr. C. M. White, in a discussion of a paper by Inglis (1947) on river meanders, presented the following explanation of the cause of meandering: 'Many models tested at the Imperial College before the war proved that when water and solids were fed perfectly steadily and symmetrically to a straight channel, the bed load eventually congregated into long undulations or mounds of saw-toothed type, sometimes visible only with oblique illumination. Each mound was unstable when on the centre line of the channel, and went over to one side, which automatically caused the next to go to the other side, and so on. A very regular pattern of staggered shoals eventually developed, and the bank of the river tended to erode slightly opposite each mound. The resulting additional detritus was deposited on the next mound downstream, that was, on the same side, and once that deposition began, the whole system rapidly went into meander with a regular and predetermined wave-length. The wave-length changed only slightly as the process developed, and there was no fundamental change in action, except that the cross flows could now begin, though they were not easy to detect in a small model. In the primary stage there was no crossing of material or water whatsoever, only the formation of the shoals; in the secondary stage the banks were cut, and the detritus from the cuts was deposited on the same side; and it was not until the final or tertiary stage that true crossing of material and the diving flow could occur. At the Imperial College the whole sequence occurred under ideally steady conditions, and did not arise from fluctuations in feed, though the primary stage could be greatly accelerated by superimposing fluctuations of suitable periodicity.'

From experiments carried out by Friedkin (1945), it was concluded that 'meandering results primarily from local bank erosion and consequent local overloading and deposition by the river of the heavier sediments which move along the bed. Meandering is essentially a natural trading process of sediments from banks to bars. Sand entering at the head of an alluvial river travels only a short distance before it deposits on bars along the inside of bends and is replaced by sand eroding from concave banks. In turn, the sand from the caving banks of a bend travels only a short distance before it deposits on bars and is replaced by sand from caving banks. The rate of trading depends upon the rate of bank caving. In uniform materials and on a uniform slope, a series of uniform bends will develop. The radii of bends increase with increase in discharge or slope for the same alignment of flow into the bends. For the same flow and slope, the size and shape of bends depend upon the alignment of the flow into the bends. Cross-sections of a meandering river are deeper along the concave banks of bends because of the impingement of the flow against these banks. The depth of the channel of a meandering river depends upon the resistance of the banks to erosion. Resistant banks result in deep cross-sections and easily eroded banks result in shallow cross-sections. Every phase of meandering represents a changing relationship between three closely related variables: the flow and the hydraulic properties of the channel, the amount of sand moving along the bed, and the rate of bank erosion. These three variables constantly strive to reach a balance but never do even with a constant rate of flow. The bends of a meandering river have limiting widths and lengths. When a bend reaches this width, a chute forms and a new bend develops farther downstream. Distorted bends and natural cut-offs are caused by local changes in the character of bank materials.'

118

In general, there is a certain relationship between the meander pattern on the one hand and the discharge, sediment load, and slope of the river on the other hand. Large rivers have large meanders and small rivers have small meanders. The effect of slope is similar to that of discharge, because the greater the velocity of the flowing water, the less able it is to make sharp changes in direction. Since a steep slope is produced by a large sediment load, it would follow that an increase in sediment load tends to increase the meander bends. Inglis (1947) found that an increase in sediment load resulted in a small decrease in meander length and a large increase in meander width. In other words, the course becomes more tortuous.

However, there must be a limit to this tendency, because it is well known that very steep rivers tend to produce straight braided channels. A possible explanation of this phenomenon could be the following. A steep slope is normally an indication of a large or coarse sediment load. It was demonstrated above that these factors tend to produce a relatively wide cross-section. When the river becomes extremely wide, the effect of spiral flow diminishes rapidly because the transverse gradient becomes smaller. As a result, there is less tendency for the bed load to be deposited consistently on the convex banks. Consequently, the streamflow will wander casually from one bank to another without a pronounced meander pattern and the well-known braided river type is obtained.

There is also a lower limit to the tendency towards meandering. A possible explanation could be the following. An extremely mild slope of the river indicates a sediment load that consists mostly of clays and fine silts. As a result, very little deposition on convex banks will take place, whereas the opposite concave banks are highly resistant to erosion due to their cohesive nature. Hence, the river will become practically a stable channel.

Another point of interest on the subject of meandering rivers is the rate at which meandering takes place. It is evident that the erodibility of the banks plays an important role. The looser the material, the faster the meandering will take place. Another factor of importance is the total amount of bed material that is transported by the river. A large load will result in active sedimentation of convex banks and therefore in active erosion of the concave banks. In comparing the rate of meander progress of two rivers with a different sediment load, it must be kept in mind that these two rivers will also have a different slope and therefore a different meander pattern. Hence, the conclusion that the greater sediment load will result in faster meander progress cannot be made without further investigation of the type of meanders that are involved in the problem.

In most of the textbooks on geology or geomorphology, the activity of meandering rivers is presented as one of ever-widening meander loops, culminating in cut-offs between adjacent meanders, once they become too large. Lobeck (1939), for instance, states: 'It is to be noted that the development of meanders reduces the gradient of the stream by lengthening its course. During stages of high water, meandering streams cut across the neck of the meander spur and in this way shorten their courses. The arc of the meander, thus abandoned, is called an oxbow lake. Thus it is obvious that a mature stream maintains an approximately uniform length, as the reduction by cut-offs is in the long run equal to the added length produced by the enlargement of meanders. Each cut-off initiates a new meander which culminates in another cut-off. Therefore, all stages of meander growth may be represented at one time.'

Based on recent laboratory experiments and the analysis of aerial photographs of meandering rivers, it is at present believed by most river engineers that the above-described meander development is rather the spectacular exception than the common rule. In the normal process of meander development, the meander bends migrate downstream at practically the same rate, more or less maintaining the same shape, without developing cut-offs. Such an orderly migration, however, can only take place when the stream banks are composed of uniform material. When the bank composition is not uniform, the progress of the upstream arm of a bend may catch up with the slower progress of the downstream arm of the same bend and a cut-off may develop. There are many possible causes for non-uniformity in the composition of the river valley floor. Fluctuations in flood flows cause differences in gradation of sediment and deposition of these sediments at different places. Variations in vegetation on the flood plain cause variations in alluviation. The formation of an oxbow, which is likely to be filled with coarse sediment near the place of cut-off, and with clays and silts over the full length of its loop, is bound to disrupt the orderly migration of subsequent meanders. These, and other obstructions, may lead to the most fantastic meander patterns, but nevertheless it should be kept in mind that meanders will normally tend to migrate without widening and without developing cut-offs.

Sometimes if it stated that erodible banks in a meandering river add to the sediment load in the river and, consequently, that protection of these banks may reduce the sediment load in the river. This does not seem quite correct. Meandering is essentially an exchange of sediment from a concave bank to a downstream convex bank. In other words, it is a mode of sediment transport. In a graded river, this amount of sediment transport, plus the amount that is carried straight through, must be in balance with the total sediment load entering at the head of the section of the alluvial river that is under consideration. If it was not in balance, consistent aggradation or degradation would take place. It could be expected, therefore, that bank protection will only change the mode but not the volume of sediment transport. Instead of being partly exchanged from bank to bank, all the sediment will now continue to travel down the river channel. A slight change in the magnitude of the total sediment transport may be expected because of the change in shape of cross-section of the river channel due to the protective works. However, this change will be towards a relative deepening of the channel, and hence the sediment load could sooner be expected to increase than to decrease. In due time, this may lead to degradation of the river bed.

In connection with the above discussion, it may be of interest to point out that although the valley floor of a graded river cannot be a permanent source of sediment, it can certainly act as a 'detention reservoir', by the temporary storage and subsequent release of sediment. In order to visualize this, it should be realized that the production of sediment from a drainage basin is rather erratic and not always in proportion to the production of run-off. During a rainstorm, the sediment particles on the land are mostly washed away during the beginning of the storm. This may account for the many instances whereby the highest wash load concentration in a river is observed before the peak of the flood arrives. However, once water and wash load are in the river channel, this situation may reverse itself. The velocity of the flood peak, on its way down the river system, is quite often greater than the velocity of the water. Therefore, it could be expected that the flood peak will overtake the wash load concentration peak and will begin to run more and more ahead of it. Since the bed material transporting capacity is mostly a function of the water discharge, it follows that adjustments in scour and deposition will have to be made. Another cause for detention of sediment in the valley may be a period of successive dry years, as discussed in the above section on the Platte River. Then there is the possibility that similar floods from different parts of the drainage basin contain different sediment concentrations, or that a spring flood, running off from frozen soils, will contain a low sediment concentration while having a high sediment-transporting capacity because of its low temperature, as demonstrated by Lane (1949). In all such cases, the ratio of sediment to water discharge is not in accordance with the river-channel characteristics and, consequently, local temporary adjustments can be expected. It may be that the river valley will aggrade for years in succession. However, when the river is truly graded, these years will be followed by years of degradation. During these latter years, the river banks are indeed a source of sediment, but this should not be confused with the basic fact that the concave banks of a graded meandering river are no permanent source of sediment.

The flow of a river varies all the time. There are seasonal variations and there are also variations in annual peak flows. It would be rather confusing to keep all these variations in mind when discussing different aspects of river behaviour. It would seem more convenient to visualize one constant discharge which would produce the same effects as the actual fluctuating discharges. To this effect, the term 'dominant discharge' has been introduced by Inglis (1947). He states that 'at this discharge, equilibrium is most closely approached and the tendency to change is least. This condition may be regarded as the integrated effect of all varying conditions over a long period of time.' Blench (1957) remarks that 'the implication of the term is of a steady discharge that would produce the same result as the actual varying discharge. Obviously, the result must be specified; for example, the steady discharge that could produce the width between banks of an incised river would not be the same as could produce the slope or the meander breadth.'

There are more reasons why the usefulness of the concept of dominant discharge is limited. With respect to sediment transportation, for instance, the dominant discharge would be a few times the mean annual flow. However, such a flow can never transport the coarsest sediment that is transported during extreme flood flows.

A typical feature in river channel formation is described by Straub (1942) as follows: 'The route of strongest flow of a river ... alternates from one bend to another, from one shore back to the other. Between bends, where the depth is relatively great, the river traverses where the depth is relatively shallow. In the normal cycle of flow, during high-water periods, the bed within the bends is scoured deeper, while at the crossings the level of the bed rises. During the succeeding lower-water period, the bends are subjected to sedimentation and a rise in the stream bed, while the crossings are scoured to lower levels.' It is evident that this aspect of river behaviour would be lost if the fluctuating river flows are replaced by a steady flow.

Another interesting feature related to river-bed scour and deposition during floods is described by Lane (1953): 'During floods, the bed of the Rio Grande, in general, scours out at the narrow sections and most of the material thus removed is deposited in the next wide section downstream. This action causes the wide section to fill up somewhat, and at times promotes channel changes in the wide sections which may cause the stream to attack the bank.' The same remark can be made again, that this form of river behaviour could not take place during a steady dominant discharge.

With respect to the formation of natural river banks, the dominant discharge, instead of being close to the mean flow, would have to be close to an extreme flood stage in order to overflow the banks. But even then it is doubtful if similar results could be produced. Normally, bank sedimentation depends to a large extent upon the vegetation that grows upon the banks. This vegetation in its turn depends upon the frequency and severity of the overflow. Therefore, a steady discharge can never be expected to produce the same results as the actual fluctuating discharges.

Another feature that can hardly be simulated by a dominant discharge is an avulsion (a river breaking through its river bank) during a delta formation. Normally, this takes place during a sudden and unusual severe flood. If the river discharge was kept constant, the time of avulsion would be delayed considerably and the aspect of the delta would change accordingly.

It can be concluded that the concept of dominant discharge is often helpful in discussing river behaviour, but that it is advisable to question constantly the adaptability of the simulated presentation to the actual case.

The profile of a normal river from its origin to its mouth is found to be concave upward. There are several reasons for this shape, which will be discussed in the following paragraphs.

First of all, the bed material, which determines mostly the slope of the river, wears down in size while it is being transported. Due to this abrasio 1, an increasing portion of the bed material becomes suspended sediment while the rest becomes smaller in size. Because of this decrease in volume and in size of the bed-material load, the river will flow with a milder slope in downstream direction.

Secondly, the total flow of water becomes larger in downstream direction. Let it be assumed for the present discussion that one tributary adds a flow of water and sediment of the same magnitude as is flowing in the main stem of the river. Two rivers, flowing separately, are less efficient in transporting their sediment than when they are joined and flow in one channel. This seems logical and can further be demonstrated by the application of various bed-load transportation formulae. As a result of this increased efficiency, the one river below the junction will flow at a milder slope than the two rivers above the junction.

Thirdly, in some drainage basins, the concentration of sediment may decrease in the downstream tributaries. After all, the upper part of the drainage basin is, under normal circumstances, the principal source of the sediment, whereas the whole drainage basin provides run-off. As a result, the concentration of transported sediment in the main stem becomes increasingly smaller in downstream direction. Hence, the slope becomes increasingly smaller.

Fourthly, many rivers are actively aggrading their lower courses. In other words, they are continuously depositing sediment over a long section of river. This means that there is a gradual decrease of bed material load in downstream direction; hence there will be a gradual decrease in slope in downstream direction. In case the sediment is not uniform, but sorted, it can be expected, moreover, that the coarsest particles will be deposited first. Hence, the bed material load not only decreases in volume but also in gradation.

Finally, rivers usually have a relatively narrower channel (larger ratio of depth to width) in downstream direction. This is due to the more cohesive bank material that is produced through abrasion. As a result of this narrower section, the hydraulic efficiency of the river increases and consequently it will flow with a milder slope.

There are many instances whereby some of the above characteristics are hardly noticeable or are non-existent. It may be, for instance, that tributaries carry a relatively greater or coarser sediment load than the main trunk. As a result, the main river will flow with a steeper gradient below the junction and the profile may become concave downward. An example of such a situation is the Missouri River near the junction of the Platte River. Another cause for a concave downward shape of the river profile could be active degradation in a section of river channel where the discharge remains the same. The degradation will increase the sediment load in downstream direction, which may lead to an increase in slope in downstream direction.

It may be of interest, at this state of the discussion, to point out again that a small increase in sediment load will not necessarily result in an increase in slope. During hydrometric surveys on the Saskatchewan River, for instance, it was found by Kuiper (1954) that discharge measurements on the falling stage, taken during periods when the sediment concentration was relatively high, yielded points well below the normal rating curve. In other words, the hydraulic efficiency of flow increased with an increase in sediment concentration. This checks with the observations that Brooks (1958) made in a laboratory flume. The explanation appears to be that a higher sediment concentration is associated with a smoother bed configuration and hence with a lower friction factor. It seems likely that this tendency has a limited range of effectiveness. A considerable increase in sediment load would definitely have to result in an ultimate steeper slope of the river channel.

Degradation and Aggradation

During the ideal geographical cycle, presented in *Figure 4.2*, a river is constantly aggrading its lower course. If the climatological conditions remain the same while the mountains are worn down to a plain, the flow of water in the river system will also remain approximately the same. The flow of sediment, however, can be expected to decrease continuously. The profile *A–B* in *Figure 4.2* represents a young river. The sediment load is relatively great and the slope of the river is steep. The profile *C–D* represents a mature river. The slope of this river has become, all along its course, considerably milder. It is of interest to discuss how the sediment that is eroded along the upper course of the river is deposited along its lower course. Geologists have classified sedimentary deposits in a lake or in a sea into three groups. The bottomset bed is formed of the fine suspended sediment which can be carried over a long distance into the sea. The foreset bed is composed of the coarse suspended sediment and bed material and is deposited where the velocity of the stream is retarded when it reaches the sea. This bed is inclined steeply and may cover the formerly deposited bottomset bed. As the river mouth moves towards the sea, the current in the river upstream is retarded because its slope is reduced and as a result, the river deposits its topset bed, composed mostly of bed material. In the geologic terminology, the use of the name topset bed is mostly limited to the vicinity of the sea or lake. However, in a broader sense, it could also include the deposits further upstream.

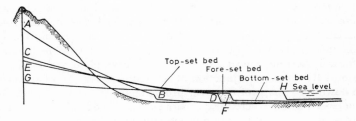

Figure 4.2. Erosion and deposition

In *Figure 4.2*, these three groups of deposition are shown as the river changes its profile from *C–D* to *E–F*. Of particular interest for the present discussion is the formation of the topset bed. Since the topset bed is formed from bed material, it follows that while the river is actively aggrading, there is a continuous decrease in bed material load in downstream direction. It was discussed in a previous section that this leads to a more pronounced curvature of the river profile. The ratio of the volume of the foreset bed to the topset bed plays an important role. When the sea is deep, this ratio is large. Hence, a small proportion of the coarse sediment will be used for aggradation and, as a result, there will be less tendency for an excessive curvature in the river profile. When the sea is shallow, the reverse will be the case.

It is evident that there must be a limit to the aggradation of the river bed along one particular course. The river may build up its bed and banks 10, 20, or even 30 ft. above the adjacent plains, but then an avulsion or break in the natural river bank is bound to take place and the river will choose a new course. In this fashion, it will form a cone-shaped plain, with its apex near the middle course of the river, where consistent degradation makes place for consistent aggradation. An example of such activity is the lower Assiniboine River. When the river is not flowing over a plain where it is free to wander about, but in a well-defined river valley, the same process of river-bed aggradation and consequent avulsion will take place, but now the wanderings of the river are restricted to the river valley. An example of such activity is the lower Mississippi River.

When the river becomes old, it may have a profile as is shown by the line *G–H* in *Figure 4.2*. The mountains are practically worn down; there is very

124

little sediment to transport and consequently the river will flow with a very mild slope. It can be seen from the figure, that the river, while it is still aggrading near its lower end, is now degrading into its former deposits. It depends upon the depth of the sea to what extent this will take place. When the sea is shallow, the mouth will progress a large distance and the mountains or hills at *G* will remain relatively high. When the sea is deep, the mouth of the river will stay close to its original place and, consequently, point *G* will sink relatively low.

It has been noted that, inasmuch as aggradation accentuates the curvature of a river profile, degradation may diminish this curvature. Due to gradual erosion of bed and bank material, the concentration of coarse sediment may continuously increase in downstream direction. As a result, an increasingly steeper slope may be required to carry the sediment. Degradation normally

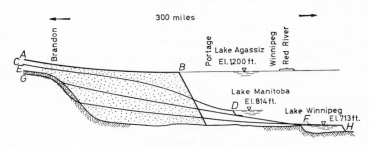

Figure 4.3. Morphology of Assiniboine River

takes place in the upper part of the drainage basin where mountains or hills are worn down. However, there are several reasons why degradation may also take place in the lower or middle part of the basin. A decrease in sediment load through the ageing of the river was discussed above. There is also the possibility of an increase in the flow of water through a change in climatological conditions. Finally, there is a whole range of possibilities through upheavals or depressions of the earth crust, or lowering of the controlling lake or sea levels.

An interesting example of degradation into former river deposits, because of lowering of the base level, is provided by the Assiniboine River between Brandon and Portage la Prairie, which is shown in *Figure 4.3*. About 10,000 years ago, the glacial Lake Agassiz had an elevation of about 1,200 ft. above sea level. The Assiniboine River deposited a delta of sand in the lake, extending from Brandon to Portage. The profile of the river at that time is represented by the line *A–B*. After the glaciers receded, Lake Agassiz drained to the north and its level dropped almost 500 ft. One of the remainders of the old lake is the present Lake Winnipeg. During the recession of the lake, the Assiniboine River began to degrade into its former delta deposits. During an intermediate stage, the river flowed into Lake Manitoba and the profile might have looked like the line *C–D*, which is concave downward between Brandon and Portage. The river profile at the present time is represented by *E–F*, which includes an almost straight line between Brandon and Portage. Along this stretch, the river is still actively eroding the valley banks and degrading its bed. The high valley banks are caving in and huge masses of sand, covered by trees and brush,

125

are sliding slowly but steadily into the river. The valley has a **V**-shaped cross-profile and recent alluvial deposits on its floor are thin. Retrogression of the erosive process in a westerly direction is checked by a sill of glacial till, five miles east of Brandon, where Currie's Rapids are found. The average slope of the river between Brandon and Portage is about two feet to the mile. Between Portage la Prairie and Winnipeg, the river flows over an alluvial plain on top of a self-formed ridge. At some places, the bottom of the river is higher than the elevation of the land adjacent to the natural river banks.

Ultimately, the river may establish a profile as sketched by the line *G–H*. Near Brandon, the river will degrade somewhat into the underlying glacial material. It may find enough coarse material during this process of erosion so that the river bed can be paved locally with gravel and boulders, thus forming an extension of the existing Currie Rapids. From below these future rapids to Lake Winnipeg, the profile will ultimately be in accordance with the small sediment load that is passing Brandon. As a result, the river gradient will be small, which may finally lead to degradation of the river bed near Portage la Prairie, where at the present time aggradation is a problem. It would be of practical importance if a prediction could be made of the time when the reversal from aggradation to degradation would take place. This will depend on the one hand on the rate at which the degradation between Brandon and Portage is decreasing, and on the other hand on the rate at which the mouth of the Red River is progressing into Lake Winnipeg.

One of the most evident places where river aggradation takes place is in a delta. Although the geography of a river delta may be very complicated, the principle of delta formation is rather simple. The river brings water and sediment into the lake or sea. Through deposition of the sediment, the river mouth is extended. Since the river needs a slope to carry its water and sediment, the lengthening of the river channel will result in a rise in river stages. There is a limit to how much a river can rise above the adjacent land, and after this limit is reached the river breaks through its own banks, chooses a new course, and begins to repeat the former procedure. The manner in which this process takes place will form the subject of the following paragraphs.

A clear description of what happens when a sediment-laden stream enters a lake or sea, how the streamflow decelerates, and how the sediment settles to the bottom, is given by Gilbert (1890): 'Suppose a river loaded with sediment enters a lake. Water flows to the shore, shoots out over the still lake water until its momentum has been communicated by friction to so large a body of water as to practically dissipate its velocity. The entire load consequently sinks and becomes a deposit. Coarse particles immediately, fine particles farther out. The river current is swifter in the middle than at the sides, which leads to deposition of suspended matter. A bank is thus produced at either hand. This constitutes an unstable condition. Whenever a current across this bank during flood no longer makes a deposit, there begins erosion. A side channel is produced, which eventually becomes deeper than the main channel and draws the greater portion or all of the water. Repetition of this process causes the river to build a sloping plain. This deposition causes a checking of current upstream. This in turn causes deposition until the profile of the stream has acquired a continuous grade from its new mouth sufficient to give it a velocity adequate to its load. In fact, the river carries forward the whole work at once.

Since the deposition begins at some distance from the mouth, the lessening load does not require a uniform grade and does not produce it. The profile is slightly concave upward.'

The essence of the above discussion is that deposition does not take place in a haphazard way all over the place but rather in the form of two banks along the main current of the river. When after recession of the flood the lake and river levels drop, these banks may become exposed. This will enable them to consolidate and possibly to grow some vegetation. When the next flood comes, the same cause for further deposition on the banks is still present and may even be increased due to the vegetation, while the causes for erosion are lessened through the consolidation and the presence of vegetative cover. When this goes on from flood to flood and no avulsion through the natural river bank takes place, the river mouth will extend like a finger in the lake.

Sometimes this form of river-mouth extension is very pronounced, and at other times it is hardly noticeable. There are several factors which affect this phenomenon. First of all, the type of sediment carried by the river is important. A large proportion of fine sediment will produce a relatively high sediment concentration in the overflow water and hence a rapid sedimentation of the banks. Moreover, fine sediment will result in cohesive banks that may delay the occurrence of an avulsion and thus promote the extension of a single mouth into the lake or sea. Secondly, a rapid progress of the river mouth will result in an excessive concave upward curvature of the stream profile. In other words, the slope near the mouth becomes relatively low. Hence, the mouth may protrude over a large distance into the lake or sea before the height of the river banks becomes critical for an avulsion. There are several possible causes for rapid progress of the river mouth, like a shallow lake or a large concentration of sediment. Since the slope of a graded river is determined by the bed-material load and not by the wash load, the circumstances would be ideal for finger-like extension when a river with a low concentration of bed-material load and high concentration of wash load would enter a shallow lake. The mouth would progress rapidly, the banks would be cohesive, and the slope would be small.

After an avulsion takes place, the river will be using both the old and the new channel for a certain length of time. However, there is a tendency for the old channel to silt up and to be abandoned for the following reasons. First of all, an avulsion often takes place in the outside of a river bend where the water is piled up during flood stages. This means that the new channel assumes a more or less straight course, whereas the old channel turns off to the left or to the right. Due to the tendency of the bed load to move towards the inside of a bend, the old channel will receive proportionally more sediment than water. In order to carry this sediment, the river needs a steeper slope, which it can obtain only by silting up its upper end. Even when the distribution of water and sediment over the old and the new channel is in proportion, then the new slope on the old channel still needs to be steeper because two separated channels are less efficient in transporting sediment than one combined channel.

In addition to the above causes for the actual silting up of the old channel, there is the tendency of the new channel to produce a local steep gradient at the site of the avulsion. This is first of all because of the local difference in elevation between the river channel and adjacent land, and secondly because the distance from the avulsion to the point where the sea or lake level is reached is usually less on the new channel, especially if deposits of the old river project far into the sea or lake. This steeper gradient in the new channel near the avulsion will cause a draw-down water profile on the river upstream. This will result in local scour and a further decrease in river stages near the entrance of the old channel, thus depriving this channel of more and more of its original flow.

The deposition of sediment in the old channel will take place in such a way that the remaining cross-section is in equilibrium, as discussed in the above. That means that deposition will not only take place on the bottom of the channel but also at the sides. The channel will become shallower and narrower. Finally, the entrance of the channel may become so narrow that it plugs up with debris or it may become so high that it becomes grown over with vegetation during low-water periods. After this has occurred, the channel will be abandoned and be filled up with marsh vegetation.

The rapidity of this silting-up process depends upon the proportion of excess bed-material load that enters the channel. If this is low, for instance, due to the entrance being located near the convex side of the main current, the river will silt up slowly. In the meantime, the new channel also silts up due to normal aggradation. Hence, the initial degradation upstream of the avulsion is checked and is reversed to aggradation. It may even be that the aggradation of the new channel catches up with the aggradation of the old channel and that both rivers become equal partners. The aspect that these two rivers are less efficient in their sediment transport than the former one river channel causes their slopes to be steeper. Consequently, the river channel upstream of the avulsion will aggrade above the level that it had at the time of the avulsion. After this has continued for a while, another avulsion may take place, eliminating the two old channels and replacing them by one new channel.

The formation and aggradation of the new channel, which initially flows through marshes and shallow lakes adjacent to the old channel, is somewhat similar to the formation of the old channel, as discussed in the above. However, one point of difference is of interest. In the old channel, it was assumed that the river bed consisted of normal river bed material. The new channel, flowing through low areas where formerly fine sediment was deposited, may find a bottom of cohesive material that resists scour. Hence, the formation of the cross-section may be different from what it would be if the river flowed over its own bed-material deposits. It could be expected that in the beginning the channel will be relatively wide. After the river digs into the clay deposits part way, it could be expected that normal meandering or casual shifting will be less pronounced. It could also be expected that the river slope can be steeper than would be in accordance with the transported bed material load, due to its resistance against scour.

From the above discussion, it follows that an avulsion may lead to the presence of two more or less stable and more or less active river branches. If this is repeated a number of times, the familiar picture of river arms in a delta, comparable to the veins in a leaf, is obtained. In downstream direction, the river channel will become increasingly smaller due to repeated bifurcating.

There is another reason why river channels in a delta decrease in size in downstream direction. During flood stages, when the river carries most of its sediment and when the formation of the channel takes place, water consistently overflows the river banks towards the adjacent marshes and lakes. As a result, the flow that remains in the channel will become increasingly smaller in downstream direction. Since a channel is formed in accordance with the volume of its flow, it follows that the channel capacity will also decrease in downstream direction.

The consequence of this decrease in channel capacity is that only one flood flow with a certain magnitude will cause uniform overflow over the delta. Floods of greater magnitude, which occur less frequently but which have a higher concentration of sediment and are of a longer duration, will flood more severely the upper part of the delta area. In this way, the sedimentation is kept in balance.

An interesting illustration of some of the above aspects is provided by the formation of the Saskatchewan River Delta. About 10,000 years ago, receding glaciers allowed the Saskatchewan River to take its present course. Near its lower end, where the river entered glacial Lake Agassiz, most of the sediment was deposited and the formation of the delta area, as it is known at the present time, was commenced. Deposits of coarse material were confined to channel locations, while the fine silts and clays settled down farther away from the river. After the channels were extended over a certain length into the lake, while possibly being split up into several branches, a situation developed whereby the river would somewhere break through its own formed banks and choose an entirely new course. In this fashion, the Saskatchewan River changed from one group of channels to another for thousands of years. Some of the oldest channels are completely covered by subsequent sedimentation. Some of the more recently abandoned channels are still plainly visible on aerial photographs and show up in the landscape as densely wooded ridges, enclosing low-lying marsh areas. The whole delta area, disregarding local ridges and depressions, is a gently sloping plain about 30 miles wide and 120 miles long, with a gradient of about one foot per mile near the upper end and one-half foot per mile near the lower end. The single river channel that enters the delta cone has a slope of about one foot per mile and may carry during normal flood stages a peak flow of 100,000 cusec, containing 0·3 per cent of sediment. The several river channels that leave the delta and flow into Cedar Lake, one of the remainders of the ancient glacial lake, have a slope of about one-quarter of a foot per mile and may carry during normal flood stages a combined peak flow of 50,000 cusec, containing 0·1 per cent of sediment. The difference in sediment load between the upper and lower end is used in raising the delta surface. The remainder of the sediment load is used in river-mouth extension into Cedar Lake.

One of the former main courses of the Saskatchewan River is the Old Channel, at present carrying only 20 per cent of the total flow. This branch must have carried, for at least a few hundred years, practically all of the flow in the river system. Due to continuous aggradation of its bed and banks, this river channel became unstable. In 1875, the river rose to a great height, due to an ice jam, and the water that overflowed the banks to the north followed a canoe portage. The velocities were so great that scour took place, and then rapidly a channel was eroded. After a few years, this New Channel carried more water than the Old Channel which began to silt up and narrow its bed. The sediment load carried by the New Channel has formed a local delta in Cumberland Lake, one of the largest remaining lakes in the delta area. This lake that had been by-passed until 1875 is now rapidly being silted up and it is expected that in a short time the river will have formed a well-defined channel connection with the outlets of the lake.

Another interesting example of delta formation is the Rhine Delta in Holland. About 20,000 years ago, the glaciers that covered part of Holland receded to the north and the combined river system of the Rhine, the Meuse, the Scheldt, and the Ems began to dump its sediment load into the sea, which had at that time an elevation of 200 ft. below the present sea level. During the following 10,000 years, the sea level rose 150 ft. and at the end of that period, through the opening of the English Channel, waves and currents of the North Sea formed a continuous ridge of beaches and dunes well in front of the active delta. The rivers now began to deposit their silts and clays in the quiet enclosed lake and broke through the belt of dunes at several places. The water in the tidal delta bay became almost fresh and a luxurious vegetation began to cover the mud flats. The continuous rising sea level, the occasional breaks in the dune belt, and periodic changes in river course caused a complicated pattern of sand deposits, clay strata, and peat formations, the latter sometimes being 10–30 ft. thick. At the present time, the major part of the original tidal bay is filled up with sediment and peat to an elevation above mean sea level. All of this land, forming the greater part of Holland, is reclaimed by dikes, and even some of the land below mean sea level is reclaimed. The present course of the rivers is fixed by training works, and changes in river regime are hardly noticeable. This is mostly because of dredging of coarse sediment for commercial purposes in the inland river channels and because of the fact that the main rivers have their outlets through the dune ridge where strong sea currents carry away the fine sediment and thus prevent river channel extension.

Further illustration of delta formation is provided by an interesting discussion of the Colorado River Delta by Vetter (1949): 'In the Colorado River Delta are thriving irrigated fields and also endless swamps and jungles and salt flats stretching as far as the eye can see. Through it all winds the Colorado River, not as a single confined stream, but in innumerable individual channels, now wide, now narrow, some deep, and some so shallow that even the lightest boat cannot penetrate them. For centuries, the flow of the river has changed from one group of channels to another, from east to west and back again, forming new channels and filling up old channels. The delta covers an area of approximately 1,800 sq. miles and may be described as a very flat cone, with its apex near Yuma, at an elevation about 100 ft. above sea level. Although several minor changes in the river delta occurred during the period 1890–1905, the main course of the river still was along the eastern portion of the delta. The situation was becoming unstable; however, the bed was building up and it was evident that the river was ripe for a major avulsion, or tearing asunder, and the only place the river could go was to the west. Then in 1905 came the break through to the north into the Imperial Valley, and for two years the entire flow of the river was diverted that way. When the break was finally closed in 1907, the old river bed had deteriorated and vegetation had grown up so it was not possible for the river to return to its old course. Instead, it overflowed the country towards the west, but no definite channel was established at once, the water being dissipated as sheet flow, gradually finding its way to Volcano Lake and from that into the Hardy. It was not until the great summer flood of 190? that the river established a definite channel between its old course and Volcano

130

Lake. The great silt mass carried by the river was deposited in Volcano Lake, gradually building up a silt cone which threatened to throw the river to the north and back into the Imperial Valley. Since there was a limit to how high the protective levees could be raised to prevent overflow into the Imperial Valley, a plan was developed for diverting the river into the Pescadero Channel, which hitherto had been simply a drainage channel located somewhat east of the Hardy. The plan was successful for a few years. Then the silt began to accumulate because the slope of the channel was too flat. In 1929, the river broke into the Vacanora Canal which had just been completed as an irrigation canal. The river followed this channel until the water was dissipated as sheet flow and started building up the Vacanora cone. In 1949, a definite channel had been formed leading from the apex of the Vacanora cone toward the west to the junction with the Hardy. For the first time, therefore, in more than 40 years, a continuous channel existed from Yuma to the Gulf.'

In connection with delta formation, it is also of interest to quote from a paper by Fisk and others (1954) concerning the development of the modern Mississippi River Delta: 'At Head of Passes, 100 miles downstream from New Orleans, the main channel of the Mississippi River divides into three principal distributaries, giving the birdfoot outline to the modern delta. As sediments have reached the sea, the most active deposition has occurred close to the mouths of the distributaries and has created bulges on the front of the delta platform. These submarine features resemble sub-aerial alluvial fans, with slopes radiating from the stream mouth as from the apex of a cone. The greater part of the sand and silt load of the river is deposited as distributary mouth bars on the upper part of the bulges, where the velocity of the stream is checked as it enters the sea. Most rapid bar building occurs during rising and falling high river stages; dredging is then necessary to maintain a navigable channel across the bar. Sediments accumulate on the bar crest and also at the ends of the submerged channels, where they form a continuous unit peripheral to the crest. The fine silts and clays are carried seaward from the mouths of passes for considerable distances as vast plumes of turbid water which have a maximum thickness of 15 ft. at the mouths and thin out seaward. A large proportion of the sediments which flocculate from the plumes are deposited around the toes of the distributary mouth bars, forming the basal part of the bulge. Much of the remaining load is deposited on the Gulf floor, but some is swept inshore by marine currents and reaches the interdistributary troughs. The passes are remarkably straight and have lengthened seaward across their stream mouth bars, scouring their channels in these sandy deposits. The channels are deepest upstream near Head of Passes and shoal seaward. They maintain a nearly constant width except at their mouths where they widen abruptly into a typical bell shape over the crest of a bar. During low stages of the river, salt water intrudes the channels and extends upstream beyond Head of Passes as a thick wedge, above which the fresh water moves seaward. It is during the low stages that the deeper parts of the channel thalweg may become filled with a coagulated mass of sludge consisting of clays, silts, and fine sands. Natural levees form during flood stages, when the river tops its banks and drops part of its load near the channel. At Head of Passes, the levees are 5–6 ft. high, approximately 1,000 ft. wide, and are covered with willows and oaks. Because the levees have developed concurrently with channel lengthening, they are smaller downstream along more recently developed segments of the channels; near the bar, they are barely distinguishable from the marshes and support only a growth of cane and marsh grasses. Their submarine counterparts are found as bars flanking submerged channels. Sediments swept over the natural levees during flood are carried into the shallow-water basins of the interdistributary troughs. Some of the floodborne materials may collect in the marshlands at the trough margins and some may be deposited in the central, deeper parts of the trough, or be carried by currents to the delta front. Wave and current action winnows fine particles from bay bottom sediments, leaving coarser particles to floor parts of the bays, or gathers some of the sands and silts to form spits and subaqueous bars.'

As a last illustration of delta formation, a description will be given of the Yellow River Delta in China. The presented material is obtained mostly from papers by Freeman (1922), Chatley (1938), and Todd and Eliassen (1940). The upper part of the Yellow River flows through hilly country covered with fine loess. The slope of the river in this part of the drainage basin is several feet per mile. The lower part of the river leaves the hills about 300 miles from the present seashore, and over this distance it has built a fertile delta-shaped alluvial plain, supporting a population of 100,000,000. The slope of the river on the delta cone is about one foot per mile. Because of this gentle slope, as compared to the steeper slope in the upper region, about 80 per cent of the transported sediment, which amounts to one billion cubic yards per year, is deposited on the flood plains, the remainder being carried into the sea. For about 3,000 years, incessant efforts have been made to exclude the flow of water and sediment from the cultivated lands. The river is controlled by dikes, supposedly built above flood level. However, the river bed is continuously aggrading and shifting, with the result that occasional dike failures, avulsions, and permanent changes in river course occur. In 1887, a flood covered 50,000 square miles of fertile and densely inhabited land, wiped out the existence of 2,000,000 people, and left a still greater number to perish of famine. It is of interest to quote H. van der Veen in his discussion of the paper by Todd and Eliassen (1940): 'To make a sweeping statement, it would have been infinitely better for the present generation if past generations had permitted the Yellow River to do its work unhindered, or if those in the past who were entrusted with its care had been far-sighted enough at least to permit the river to follow another course. This was not done, however, and therefore the present generation is faced with the fact that the river runs on a ridge where it is kept in bounds between dikes which are more or less far apart. Within the cramped space Nature goes on with its process of plain building in an endeavour to establish a slope sufficiently steep to create an equilibrium between the volume of silt that must be carried, and the carrying capacity of the current. It is evident, therefore, that, as long as such a slope has not been established, this narrow strip of plain, which by Nature's right belonged entirely to the River, will continue to rise higher until the final stage is reached.'

ARTIFICIAL INTERFERENCE

The aim of this section is to discuss in general terms the change in river regime that can be expected as a result of interference with a graded river. I has been noted that the interrelationship between the river channel character-istics S, r and M is complicated, and that no precise methods or formulae are known at the present time to determine the value of these variables for given values of Q, T and d. For this reason, it is also impossible to determine the change in value of S, r and M, for a given deviation of Q, T and d, for a known graded condition.

On first glance, this would seem to put an end to all speculation about changes in river regime, resulting from artificial changes in the flow of water and sediment. However, since the prediction of these changes, if only approxi-mate or even in qualitative terms, is frequently a desirable objective, an effort will be made to eliminate from such a prediction as many uncertainties as possible. It would be helpful if the interrelationship between the dependent and independent variables could be analysed in a simpler way than it would strictly require. This could be achieved by not dealing with S, r and M all at the same time, but with S first and with r and M afterwards. This would seem justified, in a way, because of these three variables the slope of a river channel appears to be the more important channel characteristic. In the following sections, the discussions will be concerned mainly with changes in S. For secondary changes in r and M, and consequent adjustments of S, the reader will have to consult the foregoing discussions on channel characteristics.

In the middle and lower courses of a river, where Q, T and d can be treated as independent variables, the relationship between these independent variables and the slope S is in general as follows. An increase in T, while Q and d remain constant, will cause an increase in slope, provided that the secondary effect of channel roughness adjustment does not play a dominant role. An increase in d, while Q and T remain constant, will also cause an increase in slope. An increase in Q, while T and d remain constant, will cause a decrease in slope. A simultaneous and proportional increase of Q and T, while d remains constant, will also decrease the slope.

In discussing the effect of artificial interference upon a river slope, two distinctions in place and two in time will be made: 1(a) Local effect—concerns the vicinity of the river structure; (b) Extended effect—concerns the graded part of the river system; 2(a) Temporary effect—concerns the stages of transition; (b) Final effect—concerns the ultimate graded state of the river.

From an engineering viewpoint, the local and temporary effects are of major interest. However, a better understanding of the problem is gained by also taking the extended and final effects into consideration.

Construction of Dams

When a relatively high dam is constructed in a river valley, a reservoir is created behind the dam and several problems arise. First, what is the mode of deposition of sediment in the reservoir. Second, what is the effect of the reservoir on the regime of the river upstream. Third, what is the effect on the downstream river regime. These three problems will be discussed in the following paragraphs.

When a river enters a reservoir, the velocity of the water will become less and less until it is a small fraction of what it was originally. When the river carries a load of sediment, which is kept in suspension by turbulence, it can be expected that due to the decrease in velocity, and the consequent decrease in turbulence,

133

the sediment load will gradually be dropped. Since coarse sediment particles require more intense turbulence to be kept in suspension than fine particles, it may be expected that the river, upon entering the reservoir, will drop the coarsest sediment first and progressively finer fractions farther downstream.

Through deposition of the sediment in the upper part of the reservoir, the river mouth is extended into the reservoir. Since the river needs a slope to carry its water and sediment, the river mouth extension will result in a gradual rise in river stages at any given point in the vicinity where the river flows into the reservoir. This is qualitatively illustrated in *Figure 4.4*. In situation 1, a small amount of sediment has been deposited and the transition from reservoir level to normal river level takes place via a short backwater curve. However, due to this backwater curve, reaching into the river, the first (coarse) sediment will be deposited in the vicinity of *A* and progressively upstream thereof. Flowing

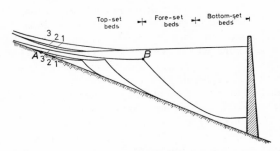

Figure 4.4. Delta formation in reservoir with fixed level

downstream, the gradient of the river, and therefore its velocity, becomes less and less, and consequently more and more sediment is deposited. This process will take place in such a fashion that river gradient and sediment transport are at all times in balance with one another. Since the rate of sediment deposition, due to slope reduction, must check with the rate of aggradation of the entire delta, some means is available to estimate the profile of the delta at progressive stages in the backwater area. It may be seen in *Figure 4.4*, situations 1, 2, and 3 that the backwater area, or the region of the top-set beds, becomes increasingly longer.

When the river reaches point *B* which indicates the end of the confined river channel, the velocities are suddenly reduced and the sediment, at least the coarser fractions, would theoretically settle out under the natural angle of repose. However, due to wave action and other disturbances, the actual slope of the so-called foreset beds will be somewhat milder.

Until now, the formation of the topset and foreset beds has been discussed from a two-dimensional viewpoint, as if the process took place in a laboratory flume between two glass walls. Actually, the sedimentation of the reservoir is three-dimensional process. The river flows in a wide valley which is inundated by the reservoir. As the flow in the natural channel proceeds downstream into the reservoir, the water depth increases until it eventually exceeds the height of the banks. Downstream of this point, part of the river flow spreads laterally over the flood plains and continues its progress. The bed-material load being relatively coarse and responding quickly to the decrease in transporting capacity, will deposit mainly within the river channel, while the finer wash load is carried over the banks and is deposited on the flood plains adjacent to the channel. Thus, the pattern of deposition that was sketched in the above paragraph will become partly obscured by the fact that the river flow in the upper reach of the reservoir is not evenly distributed over the cross-section.

The deposits of the wash load, at the downstream end of the foreset beds, are initially of a very low density. This may be illustrated by some observations made in the delta of the Little Missouri River, where it enters the Garrison Reservoir. At the lower end of the delta formation, a 6 ft. long, 3 in. diameter, clear plastic sampling tube would show clear water at the top, dirty water at three-quarter depth, a thin slurry half way, a thick sludge at one-quarter depth, and unconsolidated mud that felt like very soft butter at the bottom. The transition was so gradual that sounding of the reservoir bottom by hand was impossible; echo sounders produced an uncertain recording of the depth. Proceeding towards the upper end of the reservoir in a small boat with an outboard motor, the surface water was first observed as dark and clear. After a while, the outboard motor began to stir up a light-coloured mixture of mud in its wake. After another mile, the surface of the reservoir assumed the same colour. In the distance, some mud bars became visible that seemed quite solid. However, the motor boat could plough right through them, be it with an alarming decreasing speed. A 7 ft. oar could be stuck down in the mud without encountering substantial resistance. The gases from the outboard motor would bubble up as in a mud geyser.

Such deposits have initially a density of only 10–20 lb./ft.^{3} and practically no shear strength. While the process of consolidation takes place, the mass will deform under the influence of gravity and move farther into the reservoir as a slow, dense underflow. How long and how far such underflows will last is not known, but it is suspected that they result in a significant change of the initial deposition pattern.

Closely related to the underflow, discussed in the above paragraphs, is the so-called density current. When a river with a relatively large wash load enters a reservoir with clear water, there may be a measurable difference in the specific gravity of the river water and the reservoir water, the first one being the largest. As a result, the river water may continue to flow as a homogeneous mass over the bottom of the reservoir. Field measurements and laboratory experiments indicate that the interface between reservoir and river water may remain very distinct, till the density current arrives at the dam. Since the velocity of the density current is a function of the gravity component (the difference in weight between the turbid water and the clear water) in the direction of flow, it follows that the actual velocity of the density current is relatively low, say in the order of magnitude of 1 ft./sec. It follows that the cross-sectional area of the density current is several times as large as the cross-sectional area of the river flowing into the reservoir.

Considering the gradation of the suspended sediment, from sand to silt to clay; considering the different modes of sediment transport in a reservoir; and considering the different settlement patterns, it follows that the foreset slope of the delta formation is not one and the same from top to toe. Instead, it becomes gradually less and less until it merges imperceptibly with the bottom of the reservoir. In fact, the gravity flow of the finest sediment may extend all the way to the dam and finally come to rest in the very lowest part of the reservoir, as shown on *Figure 4.4*. Such deposits may be referred to as bottom-set beds.

Having distinguished between three essentially different types of deposition, namely, the topset beds, the foreset beds, and the bottomset beds, it may be noted that advance knowledge of such distribution of deposition is important, since each affects the useful life of the reservoir in a different manner. The top-set beds, inasmuch as they are situated above the full supply level of the reservoir, do not affect the storage capacity. However, they may be detrimental to upstream riparian interests. The foreset beds are the most deplorable ones since they reduce the live storage capacity of the reservoir. The bottomset beds are usually situated in that part of the reservoir that may be considered as dead storage. If gravity flows reach the dam, it may be possible to sluice part of these flows by releasing water from the reservoir. Such releases would have to be made at the right time and at the right place, and would only be practical when flow from the reservoir would have to be spilled anyway. Moreover, such provision would become superfluous if it can be ascertained that the dead storage capacity is large enough to accommodate all gravity flow during the foreseeable useful life of the project.

In the above paragraphs, the sedimentation pattern in a reservoir has been discussed in a qualitative sense and assuming a constant reservoir elevation. In most cases, this is an incorrect assumption. A reservoir performs the function of storing water during times of ample river flow and releasing water during times of low river flow. To perform this function, the reservoir level will necessarily fluctuate. The range of regulation may only be a few feet in a reservoir that serves primarily power interests; it may be from 10–50 ft. when the reservoir serves multiple purposes; it may be the entire depth of the reservoir when there is no power plant or irrigation outlet associated with the project. Whatever the reason, a substantial range of regulation will complicate the sedimentation pattern. During a high stage of the reservoir, at or near full supply level, which may last during several months of the high run-off season, sediment will be deposited in the upper part of the reservoir. During the subsequent draw-down period, part of the recently deposited sediments will be eroded and redistributed into a lower portion of the reservoir. During the following high water period, the previously eroded channels will be filled up again, and so on.

Although a precise computation of the resultant delta profile is out of the question, some qualitative remarks may be made. Profile a_1, on *Figure 4.5*, indicates the profile that would develop if the reservoir was constantly at the upper level of the range of regulation. Profile a_2 indicates the profile of deposition that would develop if the reservoir was at all times at the lower level. It is reasonable to assume that the actual delta profile that will develop while the reservoir fluctuates between the upper and lower limit, is somewhere in between the two profiles a_1 and a_2. This actual profile is indicated on *Figure 4.5* by the line b. The proximity of profile b to either a_1 or a_2 depends upon two factors. First the reservoir elevation duration curve. If the reservoir is at all times full or nearly so (e.g. for power), profile b will be close to a_1. If the reservoir is mostly near its lower limit (e.g. for flood control), profile b will be closer to a_2. The second factor that must be taken into account is the discharge of the river, which is related to the sediment-carrying capacity of the river, during the different stages of the reservoir. It is conceivable that the reservoir is only near its upper limit during a few per cent of the time, but during this short interval the river may carry a substantial portion of its total sediment load. Therefore, the reservoir elevation duration curve must be weighted with the sediment-transporting capacity of the river, prevailing during the different intervals of reservoir stage.

Another observation that may be made with respect to *Figure 4.5* is that the line that connects the points P_1 and P_2 is parallel, or nearly so, to the bottom slope. Therefore, it may be expected that profile b in the region between P_1 and P_2 has a slope of the same order of magnitude as the bottom slope. Whether or not the actual slope is somewhat more or somewhat less than the bottom slope, depends upon the factors that were discussed in the previous paragraph.

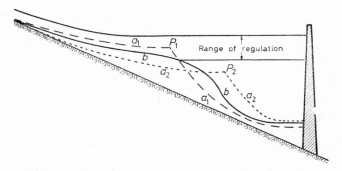

Figure 4.5. Delta formation in reservoir with fluctuating level

To the above image we must add the fact that the delta surface will be covered with vegetation. If the reservoir level would remain low for a few years in succession, this vegetation may extend relatively far downstream on to the delta. After the reservoir level comes up again, the vegetation will be inundated and will retard the flow over the delta surface. As a result, the deposition of sediment will take place farther upstream than would have occurred otherwise.

After having discussed the mode of deposition of sediment in a reservoir let us see what happens upstream of the reservoir. It was noted already that the temporary effect upstream is an aggradation of the river bed channel through deposition of sediment in the backwater reach of the reservoir, which will extend, with the advance in time, farther and farther upstream. This process is rather complicated since the relationship between sediment transporting capacity and water surface slope becomes highly obscure. In the upper reaches of the backwater curve, the coarsest particles will be deposited; in the lower reaches, the finer particles will be deposited. As a result the bed-material composition changes from place to place, thus upsetting the sediment transport capacity versus slope relationship that we might have established for the original river. Thus, any attempt at computing the rate of upstream aggradation becomes more difficult. Moreover, there is the probability that a change in bed-material composition and sediment transport may cause changes in the hydraulic roughness of the channel and changes in the cross-section of the channel. A decrease in sediment load in downstream direction may result in a larger friction factor, due to more pronounced formation of ripples and dunes on the river bed, which may lead to a greater depth of flow. The decrease in sediment load will cause a milder slope and lower velocities, which may lead to steeper river banks. These two factors combined may produce a narrower and deeper channel with lower velocities and hence a decrease in bed-material transporting capacity. As a result, the aggradation of the river bottom may proceed faster than expected.

So far, it has been assumed that aggradation of the river channel would only cause deposition of sediment in the river channel. This may be the case when the backwater does not overtop the natural river banks or when the river flows in a narrow valley without flood plains. However, when the backwater does overtop the banks and when the river has flood plains, the aggradation problem has a completely different aspect. The backwater curve will be different, the discharge in the river channel will continuously change, and sediment is going to be deposited upon the river banks.

In an attempt to find out by computation how such an aggradation situation will develop, it was assumed in an example that an alluvial river entered a lake, that the top of the natural river bank had the same elevation as the water level in the river before an artificial rise of 5 ft. in lake level. It was also assumed that the natural levees had a width of 1,000 ft. at each side of the river and that the downward slope was away from the river and so mild that all bed material, carried by the overflow due to the 5 ft. rise, would be deposited upon the levees. The first problem was to determine the backwater curve on the river channel. This was purely a hydraulic problem, but rather complicated because of the continuous change in river discharge. The solution of this problem had to be found by trial and error. After the backwater curve had been determined for the first step, the river was divided into reaches. For each reach, the mean discharge in the river channel was determined and also the corresponding sediment-transporting capacity. In order to find the sediment concentration in the overflow over the banks, sediment-distribution curves in a vertical plane had to be determined for every section. After this was done, the first step could be completed by computing for every section how much sediment was going to be deposited on the river bed and how much on the river levees. In this way, the initial conditions for the second step were found. Then a new back-water curve had to be determined, and new sediment-transporting capacities and sediment-distribution curves had to be calculated.

It was found that this computing procedure was extremely laborious and still open to question for the following reasons. If the deposition of sediment in the river was assumed to take place on the channel bottom only, the river would become shallow and the sediment-transport capacity would decrease rapidly, with more and more water going over the banks. However, when deposition of sediment was also assumed to take place at the river banks, the river cross-section could keep the same shape and the sediment-transport capacity would decrease less rapidly. Additional uncertainties were the possibility of a change in roughness of the channel and secondary changes in sediment-transport capacity, as discussed earlier in this section. For these reasons, this attempt at including overflow in aggradation computations was not pursued any further.

Fortunately, some conclusions could be made from the work that had been carried out. First of all, it was found that the backwater curve with overflow had only a fraction of the length of the backwater curve without overflow. As a result, the deposition of sediment took place in a short section of the river, located upstream of the place where the new lake level intersected the top of the river bank. Secondly, the overflow over the banks decreased the flow in the river rapidly. This decrease in river flow, combined with a decrease in slope due to the backwater profile, resulted in a very rapid decrease in sediment-transport capacity. In fact, the sediment, carried by the river, was dropped so rapidly that a bar in the channel was formed. Finally, it was found that the sediment distribution over a vertical plane played an important role in the formation of the levees. A slight change in this distribution would considerably increase the concentration near the water surface and hence the deposition on the levees.

Even if the above computations of aggradation could have been carried out with success for a while, a situation would ultimately develop whereby a natural river could not confine its flow between the banks any longer, and an avulsion would occur. Then the river will spread out over a great width of land with irregular topography and the mode of sediment transport will change completely. It can be expected that such a watercourse has in the beginning a very low capacity for transporting sediment and, consequently, that considerable deposition will begin to take place near the site of the avulsion. During the early stages of river-channel formation, the channel may be rather wide and braided because of the resistance to erosion of the underlying strata of clay deposits. In view of the discussions in the foregoing paragraphs, it seems impossible to make these modes of river behaviour subject to precise calculations.

The process of aggradation, as sketched in the above, may become further obscured by a number of circumstances. For instance, the controlling lake or reservoir level may not have a steady elevation, but may fluctuate. Moreover, the river flow is not steady, but varies from time to time. Hence, some flows are below bankfull stage whereas other flows are above bankfull stage. These flows will have a different effect upon the formation of the river gradient. There is also the aspect that the river may not have a straight course, but a winding or meandering one. During the process of aggradation, the meander pattern may change and thus cause a secondary adjustment of the stream gradient. Then there is the possibility that the vegetation of the river banks and valley lands changes due to repeated overflow. This in its turn may cause the mode of sediment deposition to be different from what is expected, on account of the present situation. Finally, the wash load carried by the river is not a function of the river flow, hence predictions of future concentrations are open to error. Moreover, it is rather difficult to estimate the location and the magnitude of wash-load depositions on the plains caused by overflow. Such an estimate is nevertheless important, because the volume taken up by these depositions has to be subtracted from the total volume that the bed-material load is going to fill up.

An interesting illustration of the above discussion is provided by Vetter (1953) in his description of aggradation on the Colorado River: 'In the Mohave Valley, down through the years, the Colorado River has acted as any other alluvial, meandering stream. It has moved from side to side in the valley, sometimes by gradual accretion, sometimes by avulsion. There is indisputable evidence that, at least during the past hundred years, there has been a general rise in the level of the valley because of deposition of sediment as the river moved from side to side. To this may have been added river-bed aggradation caused by the inflow of gravel from side washes. The mechanics of the river's behaviour seem to have been that it followed a certain course for a number of years, during which it would from time to time overflow its banks during floods. Since the river water was heavily laden with fine silt, suspended in an almost uniform concentration from top to bottom, that water overflowing the banks of the stream during floods would contain large volumes of fine silt. Since the banks were overgrown with tules and willows, as the silt-laden water penetrated this vegetation the silt was deposited along a fringe paralleling the river banks and natural levees were formed. As long as these natural levees kept pace with any general rise of the river bed, no avulsion would take place, but gradually the land adjacent to the river's banks became higher than the valley farther back. Eventually, an unusually large flood would descend the valley and the overflow at some point would be heavy enough to uproot the vegetation, thus permitting the river to course down a lower portion of the valley and an avulsion had taken place. The same levee-building processes would reoccur in its new course, until a new avulsion took place. The yearly floods would keep the river channel within the natural levees relatively free of obstruction and growth. With the closure of Hoover Dam (upstream of the Mohave Valley), this pattern of annually repeating floods changed. The flow of the river was now at a nearly constant rate. A substantial silt load was still carried by the water, but gradually its character changed. Its average size increased and its concentration near the surface decreased until the river, when looked upon obliquely, appeared blue and clear. The capacity of the river to transport the coarser material now available may have been less than that to transport the finer material of the past. In any case, deposits on the bed continued and occasionally overflows across the banks took place. The overflow came from the top layers of the river, which were now almost free of silt. As it penetrated the vegetation, no deposits were made, no natural levees were formed. A break was not again healed. Whenever a point of overflow had been created, the river continued to overflow at that point and kept on losing its water, but only a little of its silt. The result was that the remaining water was still less able to carry the sediment load and deposition on the river bed was accelerated. Soon all the water was passing through the vegetation and the river bed remained as a dry sandy ribbon where the river once flowed. Where a muddy river with a forever changing course had once meandered through an alluvial valley, but with a course which was at any one time well established, there now was a swamp with water covering an entire valley to a width of $1\frac{3}{4}$ to almost 5 miles, occupying innumerable sloughs and channels or simply filtering through the vegetation. For 15 miles a river had ceased to exist.'

After having discussed the temporary effect of dam construction in upstream direction let us discuss the final effect (the word 'final' to be interpreted in the sense as defined at the beginning of this section). This is a rise of the river-bed profile, parallel to the original profile as shown in *Figure 4.6*. Admittedly this will require thousands of years, but if we are prepared, for the present discussion, to reckon with such geological time elements, we must conclude that eventually the new river-bed profile must be parallel to the old one, and raised by an amount equal to the height of the dam, provided that the original value of the Q, T and d remain unchanged. It is an interesting although rather academic question, what happens to this parallel profile in the upper course of the river since it is evident that it cannot remain parallel and lifted up by the height of the dam, to the very origin of the river! If the dam is built on the lower course of the river this feature of the new and old profile being parallel will probably get lost in the middle course of the river where the slope is

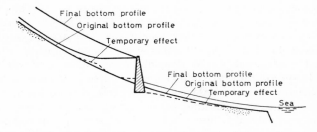

Final bottom profile
Original bottom profile
Temporary effect

Final bottom profile
Original bottom profile
Temporary effect
Sea

Figure 4.6. Upstream and downstream effect of reservoir construction

not a dependent function of T any more. If the dam is built in the middle or upper course of the river, upstream aggradation may slow down drainage basin erosion, thus reduce T and bring it in balance with a smaller S.

Now let us discuss what happens downstream of the reservoir. The temporary effect will be a degradation of the river bottom as shown in *Figure 4.6*. When the reservoir above the dam retains all the sediment, clear water will enter the formerly graded river bed below the dam and the river will start to pick up sediment. This form of erosion will take place over a stretch, long enough to enable the river to pick up its full former load of bed material. Since the result of this erosion will be a flattening of the slope and a paving of the river bed with coarse material, the length of river where degradation takes place will continuously increase. When after a certain length of time the reservoir will be silted up to the extent that some sediment begins to pass over the dam, degradation below the dam will slow down. Ultimately, the original river-bed profile will be restored, provided that the former Q, T and d will be restored. If the supply of sediment from upstream of the dam would be permanently stopped, and if the river bottom downstream would consist of erodible material, the ultimate profile would be an almost straight line between a point underneath the dam and the sea level, with a slope that would be insufficient to produce sediment-transporting velocities.

In connection with degradation below dams, it is of interest to quote from a paper by Lane (1934) on this subject: 'Since retrogression is the result of the stream recovering its normal load of solids after they have been reduced by the

141

abnormal conditions existing above the dam, the larger the load normally transported by the stream or the greater the proportion of it deposited in the reservoir, the more rapidly retrogression will take place. Since high transported loads are caused by steep slopes or fine material, these conditions tend toward high recession rates. A large pond has the same effect, since it removes a greater part of the solids. When the pond is filled, the degradation ceases and normal conditions tend to be re-established. For each type of bed-material and discharge condition there is a definite slope that is stable for a given supply of solid material carried. If all of the material is removed by the dam, this slope is flatter than if the removal is only partial. If, in the recession, sufficiently resistant layers are encountered to prevent the removal of more material under the existing slope and discharge conditions, the retrogression will cease.'

The process of degradation, as discussed in the above, becomes complicated to analyse, for several reasons. First of all the bed material is probably not uniform but may be composed of a range of grain sizes. In that case, computation of the transport rate of each grain-size fraction will have to be resorted to. The use of a constant dominant discharge in the river may be insufficient to remove the largest sediment particles that were deposited during extreme river flows. In that case, the use of river flows of different magnitude will have to be resorted to. In spite of these two refinements, it will be found that the degrading river bed becomes paved with the coarsest particles that are found in the bed material. In order to visualize this, the following discussion may be helpful.

Wash load is evenly distributed over a river cross-section and hence it is transported with the same velocity as the water in which it is suspended. Suspended bed-material load is transported with a velocity somewhat below the mean water velocity, since the highest sediment concentrations are found in the lowest velocity regions. Bed load is transported with a velocity well below the mean water velocity. It can be expected, therefore, that during the passage of a flood on a normal, graded river, the wash load is transported all the way from its place of origin towards the sea; that the suspended bed-material load is transported over a shorter distance, say one-half or one-quarter of the total length of the river; and that the bed load is transported over only a fraction of the total length of the river.

Actually, the movement of the bed material is still more complicated, because it does not move as one unit, but there is a continuous exchange between the particles in transport and the particles in rest on the bottom of the channel. This exchange of bed material is clearly demonstrated by two experiments, described by Einstein (1950): 'In the first experiment a flow is discharged continuously over a sediment bed and sediment is added at the upper end until deposition causes equilibrium to be established throughout the length of the bed. Then certain particles, marked so they can be identified, are added at the upper end. Visual observation shows that the bed-load particles move with a velocity that is comparable with the water velocity near the bed made visible by injecting dye. By assuming that the bed particles move at the same velocity as the flow in the bed layer, the time after which the marked particles should have reached the downstream end of the flume can be calculated. If 100 per cent is added to this time for a safety allowance, one might expect to find all the marked sediment particles safely in the deposit at the downstream end of the flume. On the contrary, however, if the flow is interrupted at that instant and the deposit inspected one may find only one or two of the marked particles there. Most of them have travelled only a small fraction of the distance and are found in the stream bed near the upper end of the flume. This result is not compatible with the assumption of an equilibrium condition unless an equal number of bed particles have been scoured from the bed during the same period. This possibility may be tested by a second experiment. Before the experiment is begun the water is drained from the flume and the bed allowed to dry. Dye is sprayed on a predetermined part of the bed area and thus all the sediment particles of that area are marked. Upon resumption of the experiment one observed that, gradually, all marked particles are eroded and replaced by others of the same type.'

In an ideally graded river, an observer would see, or he could measure, the following sequence of events during the passage of a flood. During the initial rise of the flood, the river stage is low and the water almost clear. There is practically no sediment transport. As the flood gains momentum, wash load and bed-material load are being supplied from the drainage basin to the river system. The wash load remains in supension. Almost all the very fine particles of the bed-material load remain in suspension. A few particles are exchanged in the river bed. These fine particles form an important fraction in the composition of the suspended bed-material load, whereas they form a very small fraction in the composition of the bed material at rest. The coarser particles of the suspended bed-material load are more frequently exchanged in the river bed. The coarsest particles are only moved during the peak of the flood. They mostly saltate, bounce, and roll near the bottom. These coarse particles form a small fraction in the composition of the suspended bed-material load, whereas they form an important fraction of the bed material at rest. When local adjustments of the river bed are not taken into consideration, it could be expected that the river bed keeps the same elevation during the passage of the flood. The erosion of bed material due to the increased turbulence of the river is balanced by the deposition of other bed material that is carried from upstream. After the flood has receded, the observer will see again a low river stage with almost clear water and almost no sediment transport. The river bottom has still the same elevation as it had before and during the flood, only it is now made of bed material formerly situated some distance upstream.

It was postulated in the above discussion that the river was ideally graded and that local adjustments in the river bed would not take place. In nature, these conditions are seldom found. In most cases, the river flow is not graded at all times. In other words, the supply of bed material from the drainage basin is not always equal to the bed-material load that the river can carry with its given discharge and channel characteristics. If, for instance, the supply of sediment falls short of what it should be, appreciable scour of the river bed may occur. However, this erosion of bed material will be restricted mainly to the location where the deficit of sediment load originates. After the river has regained its capacity load, it will keep the river bed downstream in balance. The effect of bends and narrow sections in the river channel upon local adjustments in river bed during floods has been discussed earlier. The fact that many observation sections are conveniently located at places where the river channel is narrow may account for the persistent belief that a river normally erodes its bed during times of flood.

In coming back to the subject of the present discussion 'paving of the river bed through degradation', two remarks in the above paragraphs will be given special attention. First, that the fraction of fine particles in the suspended bed-material load is much greater than in the composition of the river bed material. Second, that the fine particles, being higher up in the river flow, move faster than the coarse particles. When degradation takes place, the first phenomenon will cause the fine particles of the river bed to be picked up much more rapidly than the coarse particles. The second phenomenon will cause the fine particles to be carried away much faster than the coarse particles. As a result, the composition of the degrading bed will become increasingly coarser and the rate of degradation will decrease continuously.

An attempt has been made to evaluate this effect of paving by computation. For this purpose, sediment-transport curves were plotted for four individual bed-material fractions. The river was divided into reaches. For each step, the transporting capacity for every fraction was determined. On the basis of the incoming load into a reach, the outgoing load could be determined. After each step, the change in bed-material composition was determined. It was assumed that the upper layer of half a foot would be periodically worked over by moving ripples and would thus be available for the selection of eroded material. After the change in bed-material composition was determined at the end of each step for every reach, new sediment-transporting capacity curves for every new fraction of bed material had to be computed before the subsequent step could be commenced. This procedure was so laborious that it took one engineer two full weeks to compute a degradation of half a foot. Besides, there was no assurance that the answer was correct, for the following reasons. First of all, the assumption of half a foot layer of bed material being available for selection was open to question. Secondly, the change in bed-material composition and the change in bed-material load might have caused a change in hydraulic roughness, which was not included in the computations. Finally, the only available method of computing the sediment-transport capacity did not seem quite applicable to the present case. For instance, in one step it would be found that due to the high transporting capacity of the finest fraction, all this material would be removed at once from the available layer, half a foot thick. As a result, the bed composition for the following step in the same reach contained practically none of this fine material and, consequently, the computed transporting capacity of that fraction was found to be practically zero. Since there was still fine material coming from upstream, this had to be processed in the computations as a deposit in the reach under consideration. This was evidently a misrepresentation of the actual sediment movement.

The above form of paving will be found in rivers that degrade into their own former deposits. Another form of paving will be found when a river degrades into foreign material like glacial deposits. In such a case, the river may find material that cannot be removed at all, like pebbles and boulders. If these are found in large enough quantities, the pavement may become so effective that erosion stops completely.

A further decrease in the rate of erosion, as compared to the one that could be expected on account of the formulae only, may be caused by compaction of and colloidal activity in the former deposits. When such circumstances are present, the flow conditions at the beginning of erosion are no longer the same as the flow conditions at the end of the former deposition. In other words, as Matthes (1934) states: 'The ordinary tractive force values such as brought the bed-load material to their resting place are not able to get those same materials in motion again. The finer materials, such as silts and clays, cohere more firmly than do the coarser sands and gravels.' The erosion resistant qualities of consolidated clay deposits are well known. Harrison (1953) reports, for instance, of the delta in the Conchas Reservoir on the South Canadian River: 'The channel which has cut 10 ft. into the predominantly sandy delta exposed an old clay layer. It appeared that the layer had resisted further cutting even though the cobbles that were found on the bed must have been carried in and deposited when velocities on the layer were quite high. It is significant that a small layer of clay can retard degradation on a delta which is predominantly easily eroded sand.' The erosion-resistant qualities of sand deposits are less spectacular. However, on examining steep-cut river banks it will often be found that the lowest and oldest sand deposits have a much firmer and cohesive consistency than the loose upper layers. When the former layers become exposed to erosion, they will undoubtedly provide more resistance than the upper layers.

While the river is in its process of degradation, it not only has to remove sediment from the river bed, but also, after a while, the sediment that comes sliding down the banks. In case the river is eroding a valley of considerable depth, it can easily be understood that the sediment originally supplied by the river bottom is only a fraction of the sediment supplied by the valley banks. It would seem that the process of picking up this sediment will follow closely the mode of degradation that was visualized in the example. In an actual case, it would be necessary to investigate the grain sizes of the valley bank material in order to compute the corresponding transport capacities.

It could be expected that steepening of the river gradient through degradation will lead to more intense meandering of the river. This will cause a secondary adjustment of the river gradient and may modify the shape of the river valley. To what extent such phenomena will occur may have to be decided after investigating similar cases in the field.

In the above paragraphs we have discussed the effect of a relatively high dam, with its associated reservoir, upon the regime of a river. We shall now discuss the effect of a low dam resembling more a sill or overflow weir in the river channel. Before we do so, it may be advantageous to discuss first the different ways in which a sediment bearing river channel can flow over a fixed crest weir, as shown in *Figure 4.7*. If the depth of water over the weir is equal to the normal depth of flow in the river channel, we may expect a water surface and river bottom profile as shown in *Figure 4.7(a)*. If we would increase the depth of flow over the weir, as shown in *Figure 4.7(b)*, a backwater curve on the river channel is created. This results in lower velocities, hence less sediment transporting capacity, and hence a deposition of sediment. The final equilibrium condition will look like the situation in *Figure 4.7(b)* where the lower velocities in the backwater reach are compensated by a steeper bottom slope, to result in the same sediment transporting capacity. It may be noted that the river bottom in the uniform reach, upstream of the backwater reach, is now at a higher level than in *Figure 4.7(a)*.

145

If the depth of water over the weir is decreased, as shown in *Figure 4.7(c)*, the reverse phenomenon takes place. In the draw-down reach, the velocities are increased and hence scour takes place. The final equilibrium condition will resemble the situation in *Figure 4.7(c)*, where the river bottom in the uniform reach upstream of the draw-down reach is now lower than in *Figure 4.7(a)*.

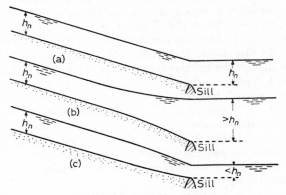

Figure 4.7. Bottom profile near sill

Now let us come back to the problem of the low dam in the river channel. If we are dealing with a low sill that causes practically no backwater, as is shown in *Figure 4.8(a)*, we may expect that the effect on the river regime is

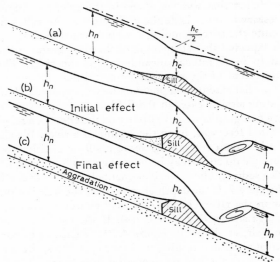

Figure 4.8. Aggradation near sill

negligible. There is no difference in sediment transport capacity until the low sill is reached. This obstruction will easily be overcome by the sediment particles due to the local increase in velocities in the vicinity of the sill.

Assuming a streamlined sill so that the deceleration losses downstream of the sill are negligible, we can compute the maximum height of the sill, whereby no change in river regime takes place, as follows. The height of the sill plus the critical depth of flow for the given discharge, plus the corresponding velocity head (equal to half of the critical depth) should equal the normal depth of flow plus its corresponding velocity head.

If the sill would be higher than computed above, there would be a noticeable backwater effect as shown in *Figure 4.8(b)*. Hence there will be aggradation upstream of the sill and consequently degradation downstream of the sill since there is the same Q with a deficiency in T. The final river profile will resemble the situation in *Figure 4.8(c)*.

It may be seen that the backwater curve upstream of the sill has become a draw-down curve, in accordance with the concept of *Figure 4.7(c)*. The river bed upstream of the sill is parallel to the old river bed and higher by an amount that is somewhat less than the height of the sill. The river bed downstream of the sill is back to normal since the original Q, T and d have been restored.

Consumptive Use of Water

Let us assume a graded river-bottom A–B–C, as shown in *Figure 4.9*. At point B, clear water is taken out of the river for irrigation. This water is lost through evaporation and does not return to the river. Consequently, the river downstream of point B needs a steeper slope to transport its unchanged T with a smaller Q. As a result, some deposition will take place below B. This will raise the water level and cause a backwater effect upstream of B. This will cause deposition of sediment in the backwater reach. The final bottom profile is

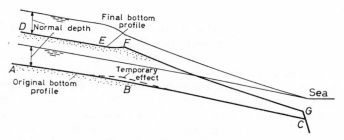

Figure 4.9. Effect of irrigation

represented by the line D–E–F–G. A local draw-down effect between E and F will remain. Note that the normal depth between F and G will be smaller than between B and C, because of the steeper slope. Hence, point G will theoretically be slightly above point C, because the sea level remains the same.

An illustration of the temporary effect of irrigation is provided by Hathaway (1948) in his description of the Arkansas River: 'From its source in the Rocky Mountains, the Arkansas River increases in size, with a large tributary system, to a point about 50 miles below the John Martin Dam. Throughout the next 200 miles, the drainage is limited to an extremely narrow strip along the river. Due to the use of water for irrigation, the normal flow below this

200-mile stretch is almost wholly dissipated and the river virtually dry for the greater proportion of the time. The result is a completely unstable channel. After a prolonged period of low flows, this channel is filled with sediments, and with extensive growth of vegetation. The channel capacity may be so reduced that flows of 2,000 or 3,000 cusec will cause flooding of adjacent lands. After a few days' flood flow, however, a redistribution of the bed materials occurs, a scoured channel has been produced, and the scoured material is transported downstream or deposited in the adjacent bed. Following several days of flow at 5,000 cusec in 1947, the channel had been scoured until it could safely carry about 7,500 cusec, and after the high flow receded it was found that the low water profile had been lowered by almost 2 ft.'

Dredging of Bed Material

It will be of advantage to recall first the permanent local effect of backwater upon a river-bottom profile, as shown in *Figure 4.7(b)*. The outlet of the channel bottom is fixed by a sill. The vertical distance between lake level and the top

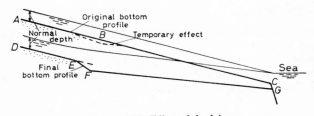

Figure 4.10. Effect of dredging

of the sill is greater than the normal depth of the channel. Consequently, there will be a backwater effect. This means that the velocities decrease in the backwater reach. Hence, the bottom slope must increase in order to keep the sediment transport constant.

Consider, next, a river with a bottom profile *A–B–C*, as shown in *Figure 4.10*. At point *B*, continuous dredging of bed material starts to take place. The river downstream of *B* can then transport the decreased *T* with the same *Q* at a milder slope. The initial result will be scour below *B*. This will lower the water profile, cause draw-down of the water profile upstream of *B* and therefore scour of the bottom upstream of *B*. The final profile of the bottom is represented by the line *D–E–F–G*. A local backwater effect between *E* and *F* will be permanent. Note that point *G* is slightly lower than point *C*, due to the increased normal depth.

Shortening of a River

Let it be assumed that a river with a bottom profile *A–B–C–D*, as shown in *Figure 4.11(a)*, is shortened by an artificial cut-off over the length *B–C*. The bottom and water profiles immediately after the cut-off is made, are represented by the lines *A–B–C–D* and *E–F–G–H* in *Figure 4.11(b)*. Due to the draw-down curve in the water profile from *F* to *G*, local erosion of the river bottom upstream of the cut-off will start to take place. This will cause a temporary overloading of the river downstream of the cut-off, with consequent deposition of sediment. The final bottom profile is represented by the line K–C–D. It has the same slope as the former bottom profile A–B–C–D.

It can readily be appreciated that the temporary effect may last for a very long period of time, since lowering of the profile A–B towards K–C over the full length of the graded part of the river upstream involves the removal of tremendous quantities of bed material. During this whole period, the river downstream of the cut-off is overloaded. Due to the consequent temporary aggradation, river stages and ground-water tables will rise. Due to the decrease

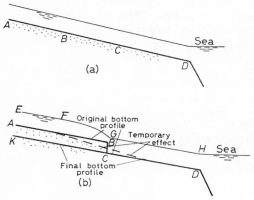

(a)

(b)

Figure 4.11. Effect of cut-off

in river length and loss in storage capacity, flood peaks will increase in magnitude. It is possible that these temporary, detrimental downstream effects will offset the final upstream beneficial effects of the cut-off. A quantitative investigation into this problem, before the actual river shortening is undertaken, would certainly be justified.

In the previous section, it was assumed that the change in river slope over the distance by which the river was shortened could be neglected. Let it now be assumed that a river is considerably shortened by a long series of artificial cut-offs or by a major avulsion of the lower part of the river towards the sea. The line A–B–C–D–E in *Figure 4.12* represents the bottom profile of the graded river in its original state.

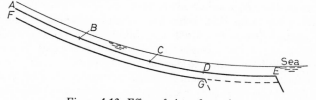

Figure 4.12. Effect of river shortening

Suppose the lower part of the river D–E is eliminated. The river stage at D will have to come down to sea level; a draw-down curve upstream of D will be formed; erosion of the river bottom will start to take place and slowly progress

149

upstream. The final profile of the river bottom is represented by F–G, which is the profile A–D, lowered over a distance equal to the average slope between D and E, times the distance D–E.

Suppose now that the river stretch B–C, which has the same length as D–E, is eliminated. In other words, the river is shortened to the same over-all length as in the previous example. The final profile is then again represented by the line F–G. This follows from the two following considerations: first of all, the curvature in the profile is a result of factors that remain the same; secondly, the lower end of the profile must coincide with the sea level. Thus, it can be seen that the river upstream of B will finally be lowered over a vertical distance, which is not equal to the difference in elevation of B and C, but equal to the difference in elevation between D and E. The temporary effect on the river upstream of B and downstream of C will be similar to the situation described under the previous section on minor shortening of a river.

Extension of a River

The reverse effect of shortening a river will be extending the total length of a river by a certain distance. Let the profile A–B in *Figure 4.13* represent the original bottom of a graded river. The river is suddenly extended by a distance equal to B–E. The final bottom profile is shown by the line C–D–E. The slope from D to E will continue to decrease in magnitude as compared to the slope upstream of D. Hence, the vertical rise in bottom profile will be equal to average slope between D and E, times the distance D–E.

Figure 4.13. Effect of river extension

In this example, it was assumed that the river, previous to the extension from B to E, was graded, and that during the temporary aggradation of the river bottom from A–B to C–D, the river mouth remained at E. In most practical cases, such simplifying circumstances are not present and the problem becomes more complicated. One reason may be that the river, instead of being truly graded, is in a state of aggradation, due to natural river-mouth extension. This would have the effect that the slope at point D in *Figure 4.13*, instead of being the same as at B, would become steeper.

In order to illustrate this, reference is made to *Figure 4.14*, where the process of aggradation is presented in a simplified form. The river carries Q, T and d. Abrasion of the sediment is assumed not to exist. The curvature of the stream profile is due only to aggradation of the river bed which causes a continuous decrease in T and hence a continuous decrease in S in downstream direction. After the river has reached the profile A–B, the increment in aggradation per unit of time is represented by the shaded area between the lines k and l. After the river has reached the profile C–D–E, the increment in aggradation during the same unit of time is represented by the shaded area between the lines m and n. Since Q, T and d remain the same, the two shaded

150

areas must be of equal magnitude. The distance from C to E is greater than the distance from A to B. Therefore, the thickness of the aggradation between A and B must be greater than between C and D. Hence, the transport at D must be greater than the transport at B. Consequently, the slope at D must be greater than the slope at B. Note also that the slope at C is the same as at A, because Q, T and d are the same. The slope at E is smaller than at B, because T has become smaller. The progress of the mouth per unit of time is smaller at E than it is at B.

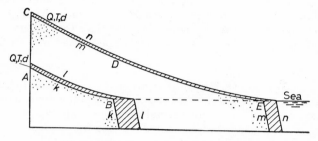

Figure 4.14. Aggradation near river mouth

River Regulation

The graded river, shown in *Figure 4.15(a)*, is going to be regulated between B and D. The river regulation will consist of bank protection and the construction of jetties which will narrow and deepen the channel in order to make it more suitable for navigation. The initial result of narrowing the channel will be a rise in river levels between B and C, backwater between A and B, and draw-down between C and D. The temporary effect on the river bottom will be deposition of sediment between A and B, mild scour between B and C,

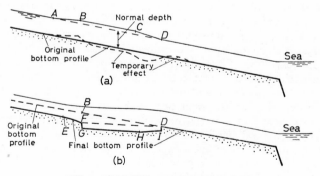

Figure 4.15. Effect of river regulation

severe scour between C and D, and deposition downstream of D, due to the overloading of the river. The final effect is shown in *Figure 4.15(b)*. The original bottom profile is represented by the line B–D. Over the regulated stretch of river B–D, the slope has become milder due to the increased efficiency in sediment transportation. The change in slope is from B–D to F–D. The change in bottom profile is from B–D to G–H–I. This is because the narrowing of the river was also compensated by a deepening of the river. Between H and I, the bottom slope decreases due to the local draw-down effect. Upstream of the regulated stretch, the slope remains the same but the bottom profile is lower. Between E and F, there is a local increase in bottom slope, due to the backwater effect.

In the above discussion, it was assumed that narrowing of the river channel would increase the sediment-carrying capacity of the river. However, it is conceivable that the contraction works are laid out in such a pattern that they cause considerable energy dissipation through acceleration and deceleration of the river flow. This may retard the velocities so much that the sediment-transporting capacity of the regulated river becomes lower than that of the uncontrolled river and thus cause a steeper slope instead of a smaller slope.

Dike Construction

Under certain conditions, the effect of dike construction can be the same or similar to the effect of river regulation as discussed in the previous section. In other words, dikes will narrow the natural watercourse, therefore, facilitate the transport of sediment and, consequently, decrease the slope and degrade the river bed.

However, there are other circumstances for which dike construction may have a reverse effect on the bottom profile. In order to demonstrate this, reference is made to *Figure 4.16*. The line $A–B–C–D$ represents the bottom profile of a graded river. Between B and C, the river has natural flood plains of considerable width, which are going to be diked. Before the construction of the dikes, the river overflowed its banks at high stages and deposited sediment upon the banks and plains. As a result of this natural process, the channel capacity decreased in downstream direction. Due to the return flow of clear water from the plains, the sediment concentration also decreased in downstream direction. After construction of the dikes, the river channel between B and C, and also between C and D, will have to carry greater river flows with a higher sediment concentration. The greater flows will tend to decrease the slope; the greater sediment concentration will tend to increase the slope. Hence, it depends upon the ratio of increase of flow of water versus sediment concentration if the river-bottom slope will decrease or increase. When considering the sediment concentration, particular attention must be paid to that part of the sediment that formerly overflowed the river banks and that will be deposited in the river bed after dike construction. Suppose the river carried coarse bed material only, none of which left the proper river channel during natural bank overflow, then the bottom profile, after dike construction, would change according to *Figure 4.15(b)*. Suppose the river carried very fine silt and clay only, none of which will be deposited in the river channel after dike construction, then the bottom profile would also change according to *Figure 4.15(b)*. However, if the river carried a large amount of suspended bed material, most of which was deposited on the river banks and flood plains during natural overflow, then the river is bound to become increasingly overloaded in downstream direction after the construction of the dikes. The original bottom profile $A–B–C–D$, in *Figure 4.16*, which had a progressively smaller slope in downstream direction due to the decrease in sediment concentration, will undergo the following transformation: $C–D$ will become *G–D*. The increase in slope is due to the increase in sediment concentration. *B–C* will become *F–G*. The slope is also increased, but not as much as from *C–D* to *G–D* because the increase in sediment concentration takes place between *B* and *C* and attains its full value only at *C*. *A–B* will become *E–F* and will retain the same slope as it had before.

The above paragraph can be summarized by saying that the construction of dikes may accelerate the natural river-channel aggradation of the lower course of a river, because the flood plains are then excluded from this process. An example of such a situation can be found in a publication by the United Nations (1951): 'Regarding the utility of dikes for flood control, opinion in

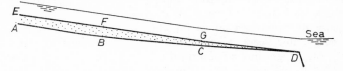

Figure 4.16. Effect of river dikes

India differs. For some time past, the Bengal engineers have expressed themselves against diking. They state that it appears to be common occurrence for rivers to change their courses at points close to the heads of their deltas, like the Ganges in 1600–1700, the Damodar in 1700, the Teesta in 1787 and the Kosi in recent times. They think it is a natural outcome that the new course of the river should be along the lower ground, skirting the edge of the land already raised, as has occurred in all the cases mentioned. It is concluded, therefore, that by restricting deltaic action at the head of a delta by constructing dikes, the problem is made more acute and the critical time is advanced because, as a result of dikes, the flood levels rise much faster than without them, and in places they are now as high as 15 ft. above the country along the Damodar.'

SEDIMENT COMPUTATIONS

The previous sections have dealt with river morphology and its connected engineering problems in purely qualitative terms. It was stated that a river under certain circumstances would 'tend to do this' or would 'temporarily do that'. Although qualitative insight into a river problem is a prerequisite for an adequate quantitative answer, it must be recognized that qualitative insight alone can hardly form the basis of sound engineering judgement. It may be true that a dam will cause upstream river-bed aggradation but such a statement will immediately be followed by practical questions about the extent and rate of sedimentation at particular points of interest.

It should be admitted without delay that the present status of sediment engineering does not permit the solution of sediment problems with a degree of accuracy comparable, for instance, to the accuracy with which the calculation of a backwater curve can be performed. Sediment problems are so much more complicated than hydraulic problems and the definite knowledge about sediment transport is so limited that predictions of river behaviour are quite often an intelligent guess at best. However, these predictions play perforce an important role in river-basin planning, and therefore it seems imperative that all possibilities of making the required guess more reliable be explored. For this reason some of the following sections contain discussions and calculations which are included with the object of gaining insight into fundamental problems of sedimentation.

The sediment problems that are of interest from an engineering viewpoint fall into three groups. The first group includes all problems that result from deliberate artificial interference with natural river regime. The nature of these problems has been briefly discussed in the previous section. The second group of problems may arise when man wants to live on, or make a living on, lands that are naturally the domain of the lower course of a river. Flood plains and delta areas produce wonderful crops, and natural river banks seem to provide the logical site for houses and barns. At the crossroads of traffic arteries, cities arise as close to the river bank as possible, in order to save on the cost of handling cargo. For years, maybe for centuries, no signs indicate that the river is a normal aggrading stream until this is suddenly realized through the occurrence of an extreme flood or through a consistent rise in river levels. By that time, it is practically impossible to evacuate the flood plains and the engineer is presented with flood- and sediment-control problems. The third group of problems is created when the natural sediment supply from the basin to the river system is increased or decreased because of climatic cycles or through the activity of man. The most common example, all over the world, is an increase in sediment load and consequent aggradation of river channels, caused by deforestation, overcultivation, and destruction of the natural vegetal cover.

Generally speaking, there are three principal methods of coping with rivers that carry too much sediment. The usefulness of each method depends, of course, upon the nature of the problem, local circumstances, and economic aspects. The first method is reduction of the sediment load through soil conservation. There is such a vast amount of literature on this subject, part of which can be found in Brown (1944), that a further discussion at this place is not needed. The second method aims at increasing the sediment-transporting capacity of the river channel by river training works. It has been noted that a decrease in channel width and consequent increase in channel depth will increase the velocity of flow. This will increase the transporting capacity and hence decrease the regime slope of the river. This may lead to a satisfactory solution of the problem. The third method includes all plans for controlled storage of sediment by reservoirs, on flood plains, or in delta areas. In connection with this subject, the following quotation by E. W. Lane in his discussion of a paper by Todd (1940) is of interest: 'Unless the inflow of sediment at the western side of the Yellow River Delta can be made to equal that flowing into the sea at the eastern side, a levee system alone will not be a permanent solution for the Yellow River floods, since the sediment storage which takes place will raise the bed of the stream and cause progressively higher flood levels until so great a height is reached that levee construction becomes very expensive and insecure. In this condition, if a break occurs, the elevation of the river bed will be so much higher than the land that the water cannot be forced back into the channel. Since it seems unlikely that in the plains section of the river the combination of reduction in sediment inflow by soil conservation and the increase in sediment outflow by river training works can reduce the excess of inflow over outflow to zero, some form of controlled sediment storage must form a part of any permanent plan for the control of the Yellow River floods. If this is the case, such storage will have to be carried on continuously and sooner or later the basins for the deposit of the sediment will be filled. It appears, therefore, that no set of works can in themselves be a complete solution of the flood-control problem, but sooner or later other works will have to be constructed and the control of the Yellow River will therefore be a continuing process of construction work. Another reason why continuous work will be necessary is that as the sediment continues to be carried out to sea, the mouth of the river will move outward. The river levees, therefore, will have to be extended seaward and those now near the sea will have to be raised. Although this process will no doubt be slow, it seems to be inevitable. In considering the best plan for controlling Yellow River floods, therefore, it should be realized that no solution will entirely eliminate the necessity of future construction, and therefore works which are obviously not a permanent solution should be considered along with those which may appear to be, but in reality are not, permanent solutions. When one considers the necessity of controlled sediment storage in any plan for permanent flood control, with the consequent expense and difficulties due to opposition of land owners, the advantages of soil conservation from the flood-control standpoint become apparent. Of course, such measures also would be a continual expense but it seems certain that a large programme of work to keep much of the soil in its place will be cheaper and more satisfactory than artificially storing the soil after it has reached the river.'

Measurement of Sediment Transport

It will be recalled that the sediment transported by a river is divided into three groups. The first group is called the wash load. This sediment fraction consists of the smallest suspended-sediment particles and can be measured by various devices. Normally, a river never carries wash load to its full capacity, nor is there a relationship between the discharge of the river and the wash-load concentration. When the sedimentation problem is concerned with the mean annual wash-load transport, and when many years of observations are available, this lack of relationship will not present serious difficulties. In other cases, it may cause a great deal of uncertainty.

The second group of transported sediment is called the suspended bed material load. This sediment is, by definition, abundantly available in the river bed. The concentration of this sediment in the river flow can be measured by various devices and it is a function of the discharge. When no observations are available, this function can be established by theory. Einstein (1950) has developed formulae for the transport of bed load that enable an estimate of the concentration of bed material at a short distance above the river bed. O'Brien (1933), Lane and Kalinske (1939), and other investigators have developed formulae for computing the concentration of suspended bed material at all points in the vertical when the concentration at one point is given. Combining these two methods, it is possible to compute the concentration of suspended bed material when the gradation of the bed material is given. However, Brown (1950) points out that small errors in the various assumptions that have to be made may lead to grave errors in the results.

155

The third group of transported sediment is called the bed load. This sediment is, by definition, also abundantly available in the river bed. A great number of investigators have presented formulae which enable computation of the bed load when the flow characteristics and the gradation of the bed material are given. Most of the formulae are based on laboratory experiments. Unfortunately, it is extremely difficult to check the accuracy of the formulae because no reliable instruments have been developed to measure the bed load in a river. Elzerman and Frylink (1951) describe field measurements that were taken on the Rhine. It is concluded by the authors that the modified Kalinske and Meyer–Peter formulae are in good agreement with the observations. However, some of the 40 measurements that are presented vary more than 100 per cent from the theoretical curve. If it is realized that every measurement represents the average value of 10 observations, repeated at several points in one cross-section, repeated at three cross-sections, then it would seem that either the formulae are still imperfect or that the sampling method is inadequate.

The three modes of sediment transport that were discussed in the above paragraphs have been the subject of extensive research and an impressive amount of literature has been published, presenting the results of laboratory experiments and mathematical analysis. Some of the more important publications can be found in the bibliography. Further references can be found in Brown (1950) and in a recent paper by Chien (1955).

Since a considerable amount of time and effort is required to assimilate this material, it is of interest to discuss to what extent the student of river morphology should concentrate upon the mechanics of sediment transport. This will, of course, depend to a large extent upon the ultimate purpose of the study. If this has the character of research, an extensive review of the literature is undoubtedly warranted. If the purpose is to solve practical engineering problems, it would seem to be more important to gain understanding of sediment transport in the broad sense of river regime than in the narrow sense of particle movement.

The engineer, faced with practical sediment problems, should also keep in mind that the result of much theoretical research work is not always applicable to practical problems. This finds its cause in the complexity of river behaviour as compared to the simplicity of the laboratory experiments, on which the results are based. It would be interesting to debate if the present science of sediment engineering would not benefit more from a systematic collection and analysis of field data than from further theoretical research work.

Observations of the sediment concentration of river flows form the basis of sediment computations. In measuring the sediment transport, there are two problems. The first one is to select the proper sites along the river system, the proper points in the cross-section, and the proper frequency of measurement. The second problem is to take the sediment sample with the proper device in the proper way. On the second problem, a number of excellent reports have been published by the U.S. Government (1952). A summary of these reports can be found in the last chapter of a publication by the United Nations (1953) on sediment problems. It may be noted at this point that the so-called depth-integrated sediment sampler is a device that is lowered from the surface of the river to the bottom and back to the surface with uniform speed and takes a continuous sample of river water during this journey. Thus, we obtain an average sediment concentration in that particular vertical of the river. The point-integrated sediment sampler is lowered to a particular point in the vertical whereupon the nozzle is opened, so that a sample of water enters the device. After a few seconds the nozzle is closed and the sampler is hauled up. Thus we have obtained the sediment concentration in one particular point of the cross-section. The bed-material sampler is a device that scoops up a sample of river bed material so that it can be analysed in the lab for grain size distribution. The bed-load sampler is a device that is lowered on to the bottom of the river and left there for a given time, to trap the passing bed load. When it is hauled up, its contents are weighed and form the basis for calculating the total bed-load movement. It was noted earlier that no satisfactory bed-load samplers have yet been developed. This is mainly due to the intermittent movement of the bed-load material as well as the local disturbance in flow pattern caused by the sampler itself.

With respect to the problem of determining the location of the sampling points, a brief discussion may be of interest. The most important factor in selecting the place and frequency of sediment sampling is the nature of the sediment problem. For instance, if the problem is to predict the rate of sedimentation in a reservoir that is being planned, the engineer will be primarily interested in the average annual volume of sediment transport at the site of the reservoir. For this purpose, it will be sufficient to take systematic daily sediment samples at that site for a number of years. If the bed material of the stream is relatively fine, so that the bed load becomes a negligible fraction of the total load, only depth-integrated samples will have to be taken. If the bed material is relatively coarse, so that the suspended-sediment load becomes a fraction of the total load, the measurement of the bed load only may suffice.

However, when the sediment problem is one of predicting changes in channel characteristics as a result of say the planned construction of river training works, the engineer will be interested not only in the volume of sediment transport but also in the composition of the sediment, and in the mode of transport along the whole section of river that is under consideration. For this purpose, it may be necessary to take systematic point-integrated sediment samples in a vertical, at several verticals in a cross-section, at several sites along the river, in conjunction with samples of the local bed material, possibly supplemented by bed-load measurements.

It could be stated briefly that the engineer has to determine first if his sediment problem is one that can be solved by simple volumetric computations or one that has to be solved by complicated analysis of channel equilibrium. In the first case, simple observations will suffice. In the second case, the engineer will have to make use of theoretical and empirical formulae and relationships, which can only be applied if a great deal of detailed sediment data is available.

As an example, it may be of interest to discuss briefly one of the sediment problems in the Saskatchewan Delta. Most of the sediment that is carried by the river is at present deposited in the Cumberland Lake area. As a result, the river channel below the lake is degrading. After a number of years, the lake will be filled up and it can be expected that the outlet will then begin to aggrade. The problem is first to determine after how many years the lake will be silted up, and second, at what rate aggradation below the lake will take place. The first problem is one of volumetric computations; the second is one of channel equilibrium. In order to obtain the necessary data, one group of stations has been set up to measure the volumes of transported sediment throughout the delta area. At these stations, depth-integrated sediment samples are taken daily during floods and twice weekly for the balance of the open water season. Another group of stations has been set up to study the distribution of sediment over the cross-section, the nature of the bed material, and the channel characteristics. Near one of these stations, the river is at present in equilibrium. In this section, the bed material load is computed on the basis of the channel characteristics and the bed material composition, and then compared to the measured load. Since the figures agree fairly well, the same method of computation is then used to predict the sediment transport for circumstances where only the channel characteristics are known and where the bed material composition is estimated. After having established the bed material transport for different channel characteristics, the computations of aggradation can be commenced.

Bed Material Load

In the preceding paragraphs, it was discussed that the bed material load is normally a function of the river discharge and that the total bed material load consists of the bed load which moves on or near the bed and the suspended bed material load which moves above the bed. Since it is difficult to measure the bed load, it is of interest to know the approximate proportion of bed load to suspended bed material load that can be expected under given circumstances. For that purpose, some computations have been made, based on the theory developed by Einstein (1950). In the following Table 4.1, the ratio of the total load to the bed load is given as a function of the diameter of the grain size and the depth of flow for an infinitely wide river with a slope of 0·00020

Table 4.1. *Ratio of Total Load to Bed Load*

Depth of Flow (ft.)	Diameter of Grains (mm)				
	0·84	0·42	0·21	0·11	0·07
7	1·1	1·2	2	20	300
11	1·2	1·4	3	200	2,000
14	1·3	1·6	4	400	4,000
16	1·4	1·7	6	800	6,000
18	1·5	2·0	12	2,000	12,000
22	1·7	3·0	40	5,000	22,000

NOTE: A factor 2·0 represents equal amounts of bed load and suspended bed material load.

It can be seen from this table that a decrease in diameter of bed material from say 1·0 mm to 0·1 mm gives a tremendous increase in ratio of total load over bed load. For a diameter of 1·0 mm, the suspended bed material load is only a fraction of the bed load, but for a diameter of 0·1 mm, the bed load is only a fraction of the suspended bed material load. It must be realized that the absolute amount of bed load generally increases with a decreasing grain diameter but the absolute amount of suspended bed material load increases so much more rapidly that the ratio between the two undergoes a considerable change. Some measurements on the Rhine, reported by Schaank and Slotboom (1937), indicated a bed load of 900 tons per day as compared with a suspended bed material load of 50 tons per day, or a ratio of total load to bed load of 1·05. The average grain diameter of the bed material was about 1·0 mm, the depth of flow about 14 ft., the slope about 0·00015, and the discharge about 10,000 cusec. Measurements on the Colorado River, reported by Grunsky (1930), and others, indicated a suspended bed material load of about 2,500,000 tons per day during a discharge of 50,000 cusec. The average grain diameter of the bed material was about 0·1 mm. Although no bed load measurements were reported, it can readily be appreciated that the bed load was likely a fraction of the total load. Measurements on the Yellow River, reported by Todd and Eliassen (1940), indicated a suspended bed material load of about 150,000,000 tons per day during a discharge of 400,000 cusec. The average grain diameter of the bed material is about 0·01 mm. No bed load measurements are reported. However, the bed load must have been a negligible fraction of the total load. This follows from considering the fact that the suspended-load particles occupied about 5 per cent of the total cross-section of the river and moved with the velocity of the water. The bed load, however, which is, by definition, the material that moves on or near the bed, could have occupied only a fraction of a per cent of the total cross-section of the river and must have moved with a much lower velocity

Average Annual Sediment Transport

Since there is no definite relationship between river discharge and wash load, the only way to determine a reliable figure for the average annual transport of wash load is to treat this problem the same way as the problem of determining the average annual run-off. Daily observations must be made for a duration of say 10 to 20 years. After such a period, the average annual transport can be determined and can be considered as fairly accurate. If such a period of observation is not available, the next best solution is to establish an average relationship between discharge and wash load from the available data and compute the average annual wash load in the same way as will be discussed in the following paragraphs for the bed material load. It must be realized, however, that the outcome of such a computation may be subject to considerable error.

In the preceding sections, it was discussed that the relationship between bed material load and discharge can be determined from observations, and also analytically, when the composition of the bed material and the hydraulic characteristics of the river channel are given. In case observations are available, the most precise procedure of determining the average annual bed material transport is as follows. From the point-integrated suspended-sediment samples, the concentration of bed material over the cross-section of the river will be known. In *Figure 4.17(a)*, the concentration of suspended bed material in one vertical for one particular discharge is shown. In the same vertical, velocity measurements have been taken so that the velocity distribution, as shown in *Figure 4.17(b)*, is also known. By combining these two graphs, the rate of bed material transport, still in that same vertical, can be computed

159

and is shown in *Figure 4.17(c)*. When these observations have been repeated at different verticals in the same cross-section, at the same discharge, the total transport of suspended bed material load over the cross-section can be determined. A less refined way of obtaining the same result is to take depth-integrated suspended-sediment samples in each of the verticals. In that case, the rate of transport in each vertical per foot width will be found by multiplying the sample concentration with the average velocity and the depth. A still simpler but less accurate procedure is to take one depth-integrated suspended-sediment sample near the centre of the flow and multiply the concentration of that sample with the water discharge over the cross-section.

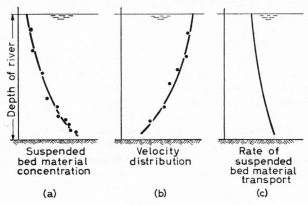

Figure 4.17. Bed material transport in one vertical

For practical purposes, the last method is the one that is most commonly used for taking daily sediment samples. In order to determine how reliable the one-vertical samples are, they are several times per year duplicated by the second method, and a few times per year by the first method. From the latter, more accurate, control measurements, it is possible, if needed, to determine a correction coefficient that may have to be applied to the daily, one-vertical measurements, in order to arrive at the correct daily sediment load in the river.

The discussion in the above paragraph was restricted to suspended bed material load. The bed load remains to be included. When the bed material is relatively coarse, the suspended bed material load will be a fraction of the bed load and it becomes important to determine this bed load independently, either by measurements or by applying one of the various bed-load formulae, or by both methods. When the bed material consists of medium-sized grains, the bed load may be of the same order of magnitude as the suspended bed material load. In that case, it is still important to determine the magnitude of the bed load. This may be done by computations only, since the necessity of checking the computations with difficult field measurements becomes less urgent. When the bed material is relatively fine, the bed load will be a fraction of the total load and may be included in the total load simply by extrapolating the curve in *Figure 4.17(a)* down to the bottom.

The result of the observations and computations, as discussed in the above, will be one figure for the total transport of bed material for one given discharge.

When this is repeated for different discharges, a relationship will be obtained as shown in *Figure 4.18(a)*. Let it be assumed that flow records have been kept for an ample number of years so that a daily flow duration curve can be constructed as shown in *Figure 4.18(b)*. The following step will be to combine then the *Figures 4.18(a)* and *4.18(b)*, resulting in *Figure 4.18(c)* which represents a duration curve for the transport of bed material. From these curves, the average annual bed material transport can easily be computed as shown by Table 4.2.

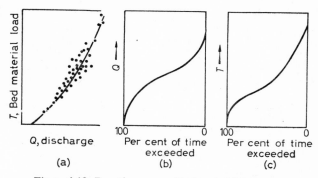

Q, discharge	Per cent of time exceeded	Per cent of time exceeded
(a)	(b)	(c)

Figure 4.18. Duration curve of bed material transport

Table 4.2. *Average Annual Suspended Sediment Load, South Saskatchewan River at Outlook*
(From Kuiper, E., Report Sedimentation South Saskatchewan Reservoir)

Q	% Time Exceeded	% Time Interval	Suspended Sediment Load	Average Suspended Sediment Load	Contribution to Total
500	100		50		
		50		525	300
5,000	50		1,000		
		30		6,000	1,800
16,000	20		11,000		
		10		21,500	2,100
25,000	10		32,000		
		5		59,000	2,900
35,000	5		86,000		
		3		158,000	4,700
48,000	2		230,000		
		1		365,000	3,700
60,000	1		500,000		
		0·6		750,000	4,500
76,000	0·4		1,000,000		
		0·3		2,000,000	6,000
104,000	0·1		3,000,000		
		0·1		6,000,000	6,000
148,000	0		9,000,000		
					32,000

$$\frac{32,000 \times 365 \times 2,000}{43,600 \times 90} = 5,700 \text{ acre-feet per year.}$$

NOTE: Sediment transport figures in the table are in tons per day. The weight of the sediment is assumed at 90 lb./ft^3.

Sediment Transport Graphs

It was noted earlier that the relationship between discharge, sediment transport, sediment size and river slope can be determined analytically by using the procedure developed by Einstein (1950). Due to the complicated nature of the problem, the application of Einstein's theory is an extremely involved and laborious process, whereby one may easily lose sight of what is being computed. Moreover, the final results may contain an error of a few hundred per cent, if not a factor 10. This is due to the many assumptions that necessarily have to be made while applying the theory. For this reason it would be much better to develop the relationships from field measurements than from the theory alone. However, adequate field measurements of sediment transport are costly and have to be taken over a long period of time before the results are reliable. Therefore, if no time or funds are available, the engineer must resort to the theory, and an answer with a possible error of one decimal place may be better than no answer at all.

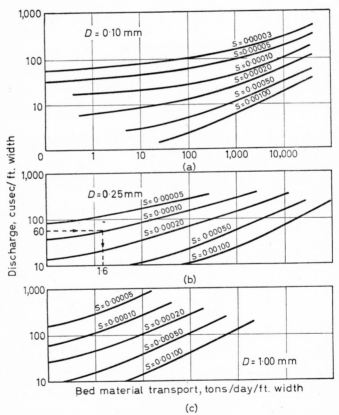

Figure 4.19. Sediment transport graphs for uniform bed material

Since the Einstein computations are rather involved, it may be convenient to have some graphs available that show the relationship between the different

162

variables for the most common values of river slope, discharge, and sediment size. Such graphical relationships are presented in the *Figures 4.19* to *4.23*. All graphs are developed with the Einstein procedure. It cannot be emphasized too strongly that these charts only serve the purpose of preliminary orientation.

Figure 4.19 shows on logarithmic scales the total bed material transport (bed load plus suspended bed material load) in tons per day per foot width, for uniform bed material, as a function of discharge and river slope. The grain size of 0·1 mm is about the lower limit for the application of the Einstein procedure. The majority of graded, alluvial rivers have bed material with the range of 0·1 to 1·0 mm. *Figure 4.20* is supplementary to the previous figure, and shows the ratio of total bed material load to bed load. The following

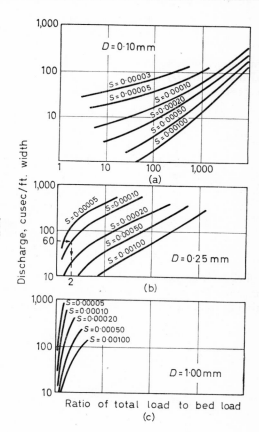

Figure 4.20. Ratio of total load to bed load for uniform bed material

example will illustrate the use of these charts. Assume a river has a width of 1,000 ft., a slope of 0·0001, a discharge of 60,000 cusec and uniform bed material with a diameter of 0·25 mm. From *Figure 4.19(b)* it can be read that the sediment transport, corresponding to a discharge of 60 cusec/ft. is 1·6 tons per day per foot width. This yields for the whole river 1,600 tons per day. From *Figure 4.20(b)* it may be seen that the bed-load fraction of this total sediment transport is 50 per cent.

The above-mentioned charts deal with uniform bed material in order to limit the work involved in their preparation. However, river-bed material usually consists of a mixture of various grain sizes. *Figure 4.21* has been prepared to

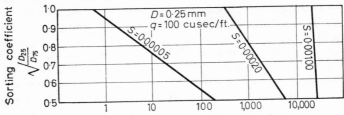

Figure 4.21. Effect of grain-size distribution

demonstrate the effect of the sorting coefficient (defined as the square root of $D25$ divided by $D75$, wherein $D25$ and $D75$ are the grain sizes for which 25 per cent and 75 per cent of the bed material is finer; hence the sorting

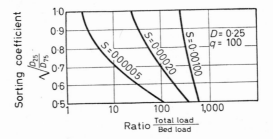

Figure 4.22. Effect of grain-size distribution

coefficient is always a number smaller than one) upon the total bed material transport. *Figure 4.22* is supplementary to the previous figure and shows the ratio of total bed material load to the bed load.

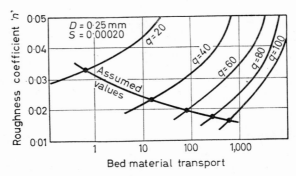

Figure 4.23. Effect of roughness coefficient

When applying the Einstein procedure, certain assumptions are made in the theory about the roughness coefficient of the river channels. *Figure 4.23*

164

demonstrates to what extent the actual bed material transport would deviate from the computed transport, if the actual roughness would deviate from the assumed roughness. The chart is only valid for a slope of 0·0002, and for uniform bed material with a diameter of 0·25 mm. The roughness values that are automatically assumed in the Einstein procedure are indicated with a dot.

In the above charts, sediment transport is given as a function of one single discharge. Actual rivers have fluctuating discharges. To compute the average annual sediment transport of a river it would be advisable to prepare a daily flow duration curve and to divide the flows in say 10 groups, from low to high. One could then take the average flow of each group and compute the corresponding sediment transport. It would be advisable to make the division such that the total sediment transport in each group is approximately the same. The total of the ten groups represents the average annual sediment transport.

CHAPTER 5

HYDRAULIC STRUCTURES

This chapter deals with the major hydraulic structures that are associated with reservoirs, such as dams, spillways and conduits. Other hydraulic structures such as shiplocks are discussed in Chapter 9: Navigation; hydroelectric plants are discussed in Chapter 7: Water Power; and dikes are discussed in Chapter 6: Flood Control. The operation of the reservoir depends largely on the purpose that the reservoir serves, and is discussed in the chapters on Flood Control, Water Power and Irrigation.

To create a reservoir, we must first of all build a dam across the river valley, to impound the water. This dam may be an earth dam, or a concrete gravity dam, or an arch dam, or a rock fill dam, depending on the foundation conditions and the availability of materials.

To use the water of the reservoir for the designated purpose, we must provide facilities to take the water out of the reservoir. For irrigation we may wish to lead the water into an open canal that has its level not much lower than the reservoir. For this purpose an outlet structure with gates and stilling basin has to be provided. For hydro development we want to bring the water to the power house that may be situated on the river bank, downstream of the dam. For this purpose we may provide conduits with gates, well under the level of the reservoir. For flood control we desire to release the reservoir content gradually into the downstream channel, after the flood danger is over. For this purpose we require conduits near the bottom of the reservoir with control gates and a stilling basin.

In addition to the above structures that are associated with the purpose of the reservoir, nearly all major dams are provided with a spillway. The capacity of the power, irrigation, flood control, or water supply conduits is seldom larger than a few times the average river flow, and mostly much less than that. However, extreme inflows into the reservoir may be 10 or even 100 times the average river flow. Since it is not unlikely that extreme river flows will be preceded by a relatively wet period, causing the reservoir to be full at the beginning of the flood, it follows that we must make provision to by-pass the extreme reservoir inflows or else the dam would be overtopped and washed away. This function of 'safety valve' is provided by the spillway. The selection of the spillway design flood and hydraulic design aspects of the spillway will be discussed in this chapter.

During the construction of the dam it may be necessary to divert temporarily the river flow around the dam under construction. This may be accomplished in several ways, as will be discussed in this chapter. A few solutions include the construction of diversion tunnels. After the dam is built, these tunnels or conduits may be used for power conduits, or spillway tunnels, or they may be plugged with concrete to prevent subsequent leakage or collapse.

To illustrate the general lay-out of the hydraulic structures associated with a dam and reservoir *Figure 5.1* shows, diagrammatically, the McNary Dam on the Columbia River, completed in 1956, and *Figure 5.2* shows the South Saskatchewan River Project, to be completed in 1968.

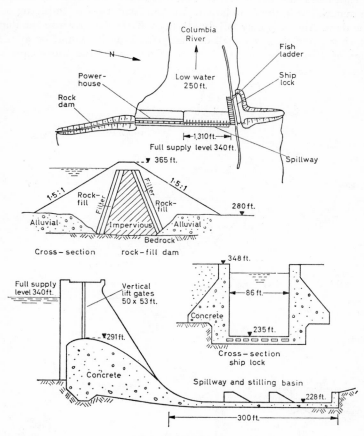

Figure 5.1. McNary Dam

DAMS

The design of dams and appurtenant structures should be performed with a relatively large degree of safety, because of the terrifying disaster that could result from failure of the dam. Unfortunately there are several dam failures on record, including earth dams, gravity dams and arch dams. In many instances the failure occurs rather suddenly. As a result, a huge flood wave, ten, twenty, thirty feet high, rolls down the river valley with the speed of an express train, erasing trees, buildings, bridges, and leaving behind in its wake a mass of tangled debris. Too often the death roll has been in the hundreds, and a few times it has exceeded a thousand. In one tragic case, the engineer responsible for the design of the dam concluded his statement in court by saying 'I envy the dead'.

There are several possible causes for the failure of dams. A gravity dam may overturn about its toe or slide over its foundation. An earth dam may have internal slides and may subsequently fail, due to insufficient shear strength in the dam material or in the foundation. All dams may fail due to piping underneath the dam. All dams may fail due to flood flows that overtop the dam. In the following paragraphs we shall discuss what criteria should be applied to the design of dams to prevent their failure from any of the above causes.

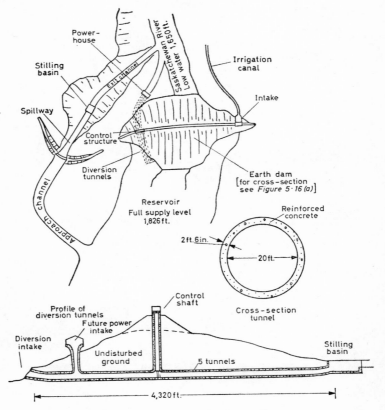

Figure 5.2. South Saskatchewan Dam

The choice of the type of dam depends largely on the foundation conditions and the availability of materials. Where solid bedrock is at or near the surface, a concrete gravity dam may be the logical choice. Where bedrock is a considerable distance below the surface, an earth fill dam is usually more economical. Where the river valley is narrow and has sound rock formations, an arch dam may be the best solution. Where large quantities of rock are found, or become available from channel and power house excavation, a rock fill dam may be considered. All other factors being the same (particularly the

168

factors of safety), the least costly dam is the most desirable. It is not unusual that the choice of the type of dam can only be made after a number of different, preliminary, dam designs and cost estimates have been worked out.

In designing the dam, we should first of all establish its height. It will be assumed that the full supply level of the reservoir has been determined from studies, described in the chapters on Flood Control, Irrigation, and Water Power. This full supply level may coincide with the crest of the spillway as shown in *Figure 5.3*. The first addition we have to make to this level is because of the passing of extreme flood flows over the spillway. In the section on spillways we shall discuss how the maximum flood level is determined. To this level we

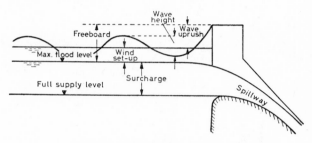

Figure 5.3. Definition of freeboard

must add an amount to allow for the set-up of the reservoir level, during the passing of the flood, due to winds blowing in the direction of the dam. These winds will also generate waves, and hence we must make a last addition that allows for the uprush of the waves against the slope or the face of the dam. The computation of the wind set-up and wave uprush will be discussed in the section on Freeboard.

Gravity Dams

The following forces must be considered in the design of gravity dams: (1) Weight of the dam; (2) Hydrostatic force; (3) Uplift force; (4) Ice force; (5) Earthquake force; (6) Reaction.

Let it be assumed that the height of the dam in *Figure 5.4* has been established and that a trial cross-section has been assumed. The first problem, then, is to determine the magnitude of the five active forces listed. The next step is to determine the sixth force, namely the reaction from the foundation upon the dam. The last and most important problem is to judge the stability of the dam, given the magnitude and the line of action of the six forces, acting upon the dam.

Determining the weight of the dam, or rather the weight of a unit of length of dam, is a simple problem, assuming that the cross-section and specific weight of the concrete or masonry are known. Determining the hydrostatic force upon the face of the dam is also relatively simple, and is discussed in Chapter 3.

The reservoir level to be assumed for the stability analysis of the dam depends on whether or not the ice force and the earthquake force are also considered. One design criterion is to assume the highest possible water level during the passing of the spillway design flood including the wind set-up, as shown in *Figure 5.3*. It is obvious that this condition cannot co-exist with ice pressure against the dam. Since such a reservoir level has a small probability of occurrence one may moreover ignore co-existence with the earthquake force which also has a small probability of occurrence. Another design criterion is to assume the reservoir at its normal full supply level, and this time co-existing with ice pressure as well as earthquake force.

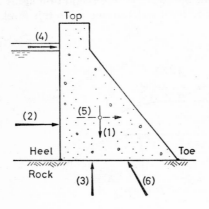

Figure 5.4. Forces on gravity dam

Determining the magnitude of the uplift force is more difficult. Let us discuss first what causes uplift. Suppose we had a solid concrete dam poured on to a solid rock foundation. We might be inclined to think that concrete poured on rock would be so well bonded that no water can seep through.

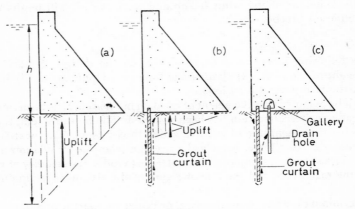

Figure 5.5. Uplift pressure under dams

Unfortunately, this is not so. Given enough time and pressure, water will seep through the tiniest cracks; it will even seep through solid concrete! The amount of seepage is extremely small, but this is immaterial; as soon as there is seepage there is a pressure gradient, and eventually the concrete dam on solid

170

rock would be subjected to an uplift pressure, as shown in *Figure 5.5(a)*. Such an uplift force is detrimental to the stability of the dam, from a viewpoint of reducing the resistance to sliding as well as increasing the danger of over-turning. It is therefore important to reduce this uplift force. This may be accomplished by the construction of a grouted cut-off wall, as shown in *Figure 5.5(b)*. If this cut-off wall is completely watertight (or at least much more watertight than the plane between the dam and the foundation) the water is forced to flow around and underneath the cut-off wall towards the toe of the dam. The hydrostatic gradient is now spread over three times the original length (assuming that the cut-off wall is as deep as the base of the dam is wide). As a result the uplift under the dam has now been reduced to one-third of what it was originally. It is evident that the uplift will remain reduced only as long as the cut-off remains effective. Some engineers have expressed some doubt as to the permanency of a cut-off. For this reason a gallery is often designed above the cut-off wall, so that supplementary grouting can be per-formed when required.

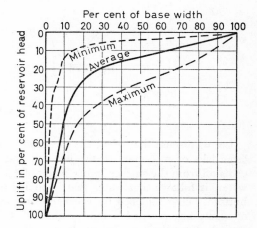

Figure 5.6. *Average uplift pressure for eight large dams*

A further reduction in uplift pressure might be obtained by a system of drainage holes, drilled from a gallery in the dam near its upstream face, to penetrate the foundation downstream of the cut-off so that water passing the cut-off area can escape to the gallery without building up the uplift intensity, as shown in *Figure 5.5(c)*. Such drain holes, however, are not completely reliable since they may become clogged by sedimentary or mineral deposits. However, the gallery enables inspection and the drilling of additional holes when required.

Over the years a good deal of measurements have been recorded about uplift pressure under gravity dams. Much of this has been presented in a report on uplift on masonry dams (1951) prepared by the ASCE. Some of this informa-tion is presented in *Figure 5.6*, showing the average uplift pressure at the base of eight dams, all provided with grouting and drainage, and ranging in height from 100–600 ft. The depth of the grout curtains ranged from 20–500 ft., depending on the height of the dam and the type of underlying rock formations. The depth of the drainage holes ranged from 20–100 ft.

171

For preliminary design of gravity dams it is recommended that grouting and drainage be included, and that the intensity of uplift pressure at the heel of the dam be assumed at one-half of the sum of maximum headwater plus the maximum tailwater. The uplift gradient shall then extend in a straight line to the maximum tailwater pressure at the toe of the dam, as shown in the example of *Figure 5.7*. For final design it is recommended that a thorough study of the pertinent literature be made.

The problem of ice pressure against dams is of special concern in Canada and the northern part of the United States. Unfortunately, very few reliable quantitative data are available. This finds its reason in the difficulty of field measurements and the non-similarity of laboratory research. At the present time little more can be done than discuss the problem in qualitative terms and quote some figures of ice pressure that have been more or less arbitrarily assumed by various agencies.

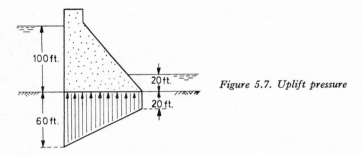

Figure 5.7. Uplift pressure

Let us discuss first of all the formation and behaviour of an ice sheet on a large lake or reservoir. In the late fall and early winter the water gradually cools off from the surface. Since water of 4°C is heavier than colder or warmer water, it follows that water that is cooled off at the surface will sink to the bottom, being replaced by warmer water from below. After the entire body of water has reached a temperature of 4°C, surface water can be further cooled off to the freezing point, without sinking to the bottom of the lake. Assuming air temperatures below the freezing point, we may now expect the formation of ice crystals at the surface of the lake. Since water expands by about 10 per cent when it freezes, these ice crystals will float at the surface. Under sustained freezing conditions, the ice crystals will bond together and form a solid sheet of ice. When the air temperature remains well below freezing, the ice sheet will grow in thickness by the continued addition of ice crystals on the underside of the ice sheet. If we would measure the temperature of the ice sheet, we would find a temperature of 0°C at the interface between water and ice, and increasingly lower temperatures higher up in the ice sheet, till we find approximately the prevailing air temperature at the surface.

Since the air temperature fluctuates not only from day to night, but also due to changes in weather, it follows that the ice sheet is also subject to changes in temperature. Since ice, like any other solid substance, contracts when it gets colder and expands when it gets warmer, it follows that an ice sheet is subject to internal stresses. During a crisp cold winter night we may hear the ice crack on a lake as if all over the lake people were occasionally firing a rifle. Such

172

cracks may be surface cracks only, that will not fill up with water from below. During the following day these cracks may close again, without causing much internal pressure in the ice sheet.

On large lakes and reservoirs, however, and during a sudden and severe drop in temperature, the surface of the ice sheet may contract so much that the entire ice sheet rips open with a deafening roar. The crevasse, which may be several feet wide, will fill with water, and will freeze solid in a few hours time. When some time later the air temperature rises again, the ice sheet, not being able to expand, will start building up internal pressure. The first result of this internal pressure is a buckling of the ice sheet, which may be observed by running a line of levels on the ice. Other evidence may be found in the formation of a layer of slush in places where the ice has buckled down and where the water, through a crack in the ice, is forced by hydrostatic pressure to flow on to the ice.

At this point it must be noted that ice does not quite behave as a solid substance. When subjected to pressure, it will slowly deform, as we know from the behaviour of glaciers. Therefore, it makes a good deal of difference how quick the air temperature rises. A rise of say 20° will be more effective in building up pressure, when it takes place in a few hours' time, than when it is spread over a few days. When sudden temperature rises do take place on large expanses of ice, the buckling of the ice sheet may become so severe that the sheet will break and form a so-called pressure ridge. Here, the two edges of the broken ice sheet crush into each other and form a continuous ridge of ice that may at places be as high as 10 ft.

With respect to the ice pressure on dams we may observe in a qualitative sense that the ice pressure will be greater when there is no opportunity for the ice sheet to expand. For instance, when the reservoir is small so that buckling becomes insignificant, or when the shores of the reservoir are steep and solid so that the ice sheet cannot slide up the sloping banks. It may be expected that the ice force against the dam becomes greater when the ice sheet becomes thicker. However, there appears to be a limit. No further increase may be expected beyond a thickness of two to three feet. One reason is that the rate of temperature rise in ice is rapidly reduced as the thickness of the ice increases. This effect compensates for the action of the increased thickness. Another reason is that in most cases the heating of the ice begins in the morning and ends in the evening. With too great a thickness of ice this length of time is insufficient to attain maximum pressure.

It may be of interest to quote the following design assumptions for ice force upon dams. In Canada, 10,000 lb./lin ft.; in Sweden, 20,000 lb./lin ft.; in Norway, at one time 30,000 lb./lin ft., but now 3,350 lb./lin ft. Many dams in the northern United States have been designed for ice forces as large as 60,000 lb./lin ft.

One of the few cases where ice pressure against a dam has been measured, is reported by Monfore (1954). The highest thrust measured at Eleven Mile Canyon Reservoir during three winters was 20,000 lb./lin ft. The ice sheet was 18 in. thick and the shores of the reservoir offered as severe a restraint to an ice sheet as any to be found.

Before closing this discussion on ice pressure it should be mentioned that the force of the ice sheet can be virtually eliminated by the application of a bubbler system. A horizontal perforated steel pipe is attached to the face of the dam near the bottom of the reservoir. During the winter, air is blown in this pipe which escapes in the form of small air bubbles into the reservoir. The air bubbles rising to the surface, along the face of the dam, set in motion a circulation of water from the bottom of the reservoir to the surface. The difference in temperature, combined with the associated turbulence, are sufficient to prevent the ice sheet from freezing to the dam even under the severest winter conditions. It is, of course, a matter of economics, whether or not such a bubbler system is less expensive than the added cost of the dam when it is designed to withstand the force of ice.

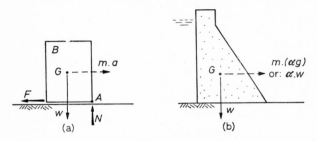

Figure 5.8. Earthquake force

The last of the five active forces that have to be taken into consideration is the earthquake force. This one may be visualized as follows: assume that block B in *Figure 5.8(a)* is pulled to the left with a force F. The resultant acceleration of the block is a. One can easily imagine that when F becomes large enough, there is danger that the block will overturn around point A. In

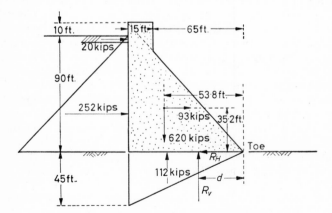

Figure 5.9. Numerical example, gravity dam

order to investigate the stability of the block, one convenient procedure is to assume a hypothetical force $m.a$ acting through the centre of gravity of the block and opposite to the direction of motion. After having introduced this

174

hypothetical force the stability of the block may be determined as if we were dealing with a problem in statics.

This principle is applied to a dam by assuming a hypothetical force, acting through its centre of gravity, as shown in *Figure 5.8(b)*. The magnitude of this force is *m.a.*, in which *m* is the mass of the dam, or the unit length of dam, and *a* is the horizontal acceleration of the earth crust that may be expected from earthquakes. Usually, the severity of earthquake acceleration is expressed as a fraction of the acceleration of gravity *g*. For instance, the most severe earthquakes ever recorded have been in the order of magnitude of $0.5\,g$. Earthquakes with a severity of $0.1\,g$ are considered disastrous; stone buildings will fall apart, bridges may tumble from their supports, landslides will occur, and cracks may open up in the earth. Most dams in seismically active regions are designed for an acceleration of $0.1\,g$.

The inertia of the water in the reservoir also produces an extra force on the dam during earthquakes. The determination of this force is very complicated. For preliminary design it is suggested that the selected earthquake acceleration be increased by 50 per cent ($0.10\,g$ would then become $0.15\,g$), that it be applied to the dam only, and that the extra water force be ignored.

After having determined the five active forces (weight, hydrostatic force, uplift force, ice force, and earthquake force) it is now a relatively simple problem of statics to determine the magnitude and the line of action of the reaction of the foundation upon the dam. This is illustrated by the numerical problem, shown in *Figure 5.9*.

For a section of dam 1 ft. long:
Sum of the forces in the horizontal direction equals zero:

$$252,000 + 20,000 + 93,000 - R_H = 0$$
$$R_H = 365,000 \text{ lb.}$$

Sum of the forces in the vertical direction equals zero:

$$112,000 - 620,000 + R_v = 0$$
$$R_v = 508,000 \text{ lb.}$$

Sum of the moments with respect to the toe equals zero:

$$20,000 \times 88.0 + 252,000 \times 30.0 + 93,000 \times 35.2$$
$$- 620,000 \times 53.8 + 112,000 \times 53.3 + R_v \times d = 0$$
$$d = 29.3$$

We must now appraise the situation, determine what criteria of safety have to be applied, and find out whether or not these criteria are satisfied. Some failures of gravity dams have been caused by inadequate resistance in the foundation to horizontal movement. Other failures have been due to overturning of the dams. Let us discuss first in general terms these two causes of failure: overturning and sliding.

In *Figure 5.10(a)* it is assumed that the vertical reaction acts in the middle of the base of the dam. It follows that the foundation pressure is evenly distributed over the base of the dam. The magnitude of the pressure can simply be computed by dividing the vertical reactive force by the area of the base of the dam.

Let us now assume that R_v remains the same in magnitude but shifts to the right. The result will be a redistribution of the foundation pressure. The total area of the pressure diagram remains the same since it represents the magnitude of R_v, but its centroid shifts to the right, coinciding all the time with the line of action of R_v. When R_v has arrived at a distance of $\frac{2}{3}b$ from the heel of the dam, the pressure diagram has become a triangle, as shown in *Figure 5.10(b)*, with the pressure at the heel being zero, and the pressure at the toe being twice as large as in *Figure 5.10(a)*. We must ascertain that the compressive stresses near the toe do not exceed the allowable values. Let us now assume that R_v shifts still further to the right, beyond the middle third of the base of the dam, as shown in *Figure 5.10(c)*. The pressure diagram must now include a negative area on the left side. In other words, there will be tension between the dam

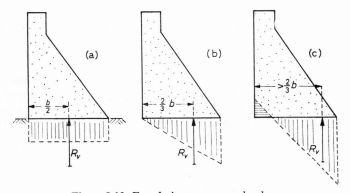

Figure 5.10. Foundation pressure under dam

and the foundation near the heel of the dam. This presents a situation that is potentially dangerous. Due to the tension small cracks may appear between the dam and foundation. Water will enter the cracks and thus increase the uplift pressure. The result will be a decrease in magnitude of R_v, but also a further shifting to the right of R_v. The decrease in magnitude of R_v will lessen the resistance to sliding. The shifting to the right of R_v will increase the compressive stress at the toe and the tension at the heel. When this tension becomes large enough, wider cracks may appear between the concrete and the rock and the downward force, exerted by the rock foundation upon the heel of the dam, will be lost. This will result in a shift of the positive pressure diagram to the right, further increasing the compressive stress at the toe. If by now the allowable foundation pressure is exceeded, the toe of the dam will start digging in and overturning can be expected. If the allowable foundation pressure is not exceeded, we must at least expect, due to the cracks between the heel and the foundation, full uplift pressure under the dam and therefore a reduced resistance to sliding. Thus, while the dam had an initial tendency to overturn, it could finally fail by sliding.

The conclusion to be drawn from the above discussion, is that the resultant of all active forces should intersect the base of the dam within the middle third so that no tension shall exist in any joint under any condition of loading.

In the above discussion reference was made to the possibility of sliding. To reduce the problem to its simplest proportions, we may compare the dam with an object that is pulled over the ground. As soon as the horizontal force exceeds the vertical force times the friction coefficient, the object will begin to move. Applying this principle to the dam, we may state that the dam will not slide as long as R_H is less than R_v times the friction coefficient. Laboratory experiments on specimens of concrete and rock indicate a friction coefficient of about 0·75. Since the actual concrete of the dam and the actual rock foundation can be made much rougher than laboratory specimens, there is a substantial factor of safety in using a value of 0·75.

Moreover, the rock foundation of the dam can easily be excavated in such a fashion that shearing resistance in the concrete or in the rock is added to the resistance to sliding. This provides a further factor of safety. One must be very

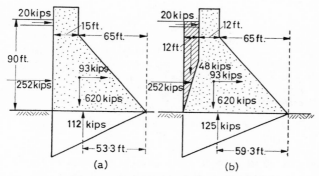

Figure 5.11. Stability of gravity dam

careful in exploring the sub-strata of the foundation to determine if there are any horizontal seams that could have less resistance to sliding than the dam on its plane of contact.

It may be of interest to point out that, from a viewpoint of stability against overturning and sliding, the design of the type of gravity dam with a vertical face may be improved by making the lower part of the face slanted, as shown in *Figure 5.11(b)*. In this example, the total volume of concrete and hence the cost of the two alternative dam designs of *Figure 5.11* is the same. The only difference is that some concrete from the upper part of the face of the dam in *Figure 5.11(a)* has been taken away, and added to the lower part of the face of the dam in *Figure 5.11(b)*. As a result, the vertical downward force upon the dam is increased by the weight of the water above the slanted face which equals 48,000 lb., whereas the increase in uplift under the dam is only 13,000 lb. Since all horizontal forces remain the same, there is now more resistance against sliding. From a viewpoint of overturning we may observe that the counter-clockwise moment of the added weight of water, with respect to the toe, is several times greater than the increase in clockwise moment of the uplift. Moreover, the centre of gravity of the dam is slightly shifted, increasing the counter-clockwise moment of the weight, and decreasing the clockwise moment of the earthquake force. Hence, the design in *Figure 5.11(b)* has more

177

resistance against overturning. It should be pointed out that complete analysis should include computations of the stresses at the heel as well as at the knick-point in the face of the dam, to ascertain that these do not become negative.

It has been found from experience and theoretical analysis that the compressive stress at the toe, as shown in *Figure 5.10(b)*, for dams that are higher than one to two hundred feet, becomes so large that only solid rock constitutes an adequate foundation. This is what has been assumed in the foregoing discussions. For gravity dams lower than about 100 ft. the foundation may consist of gravel or sand. To analyse the stability of such dams we must first of all apply the criteria of overturning and sliding as discussed before, keeping in mind that the bearing strength may not be higher than 10,000 lb./ft.^{2} and the friction coefficient not higher than 0·4.

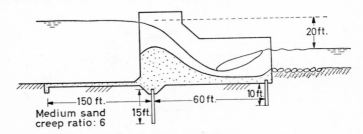

Figure 5.12. Dimensions of line of creep

Moreover, we must inspect the preliminary design of the dam upon the possibility of piping (this is the removal of soil from the foundation, by the water seeping underneath the dam). A small amount of seepage through earth foundations of gravity dams can be expected and will do no harm. Excessive seepage must be avoided. The amount of seepage and the stability of the soil can be determined by sketching a flow net and appropriate soil mechanics consideration. However, it has been pointed out by Lane (1935) that the resistance to percolation along the planes of contact between the dam and the foundation may be less than through the foundation. Thus the length of the so-called 'line of creep' becomes the criterion of safety against piping. Lane, after having studied a number of dams, came up with the following minimum creep ratios: fine sand 7; medium sand 6; coarse sand 5; fine gravel 4; coarse gravel 3; clay 2. The creep ratio is defined as the ratio of the creep distance to the head on the dam. The creep distance is the sum of all the vertical contacts plus one-third of all the horizontal contacts (these being less liable to have intimate contact). If the line of creep is less than what it should be, the design may be improved by the application of a vertical cut-off wall or an upstream blanket as shown in *Figure 5.12*, where the creep distance should be $6 \times 20 = 120$ ft., and is: $\frac{1}{3}.150 + 15 + 15 + \frac{1}{3}.60 + 10 + 10 = 120$ ft.

Earth Dams

A few decades ago, earth dams were only used for relatively small reservoirs, and were considered unsuitable for high head developments. However, the recent advances in soil mechanics and construction methods have resulted in earth dams being built to a height of several hundred feet, and these dams are now considered as safe as any other type of dam.

The main problem in the design of earth dams is to select such a composition of materials and such side slopes that, with the given foundation conditions, no failure of the dam or any part of the dam will occur. Since the function of the dam is to hold back water, and since all earth is pervious to some extent, we may expect that this problem of slope stability is intimately associated with problems of water movement through and under the dam. Let us discuss first the hydraulic aspects of an earth dam.

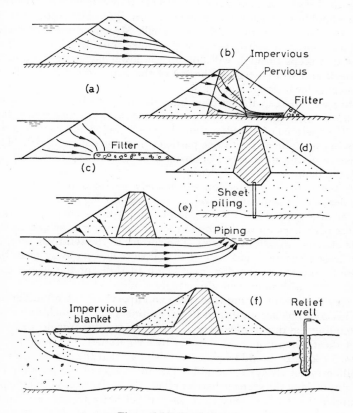

Figure 5.13. Earth dams

When a dam of homogeneous material, situated on an impervious and stable foundation, is permanently exposed to a head of water, as shown in *Figure 5.13(a)*, a certain seepage pattern through the dam will be established. This seepage pattern, and the place where the seepage line emerges on the downstream slope, is the same whatever material (sand, loam, clay) the dam is made of, provided the material is homogeneous and all dimensions similar in each case. The rate of seepage, of course, depends entirely on the type of soil that is used. For instance, through a dam composed of fine sand would seep one thousand times as much water as through a dam composed of clay, all other circumstances being the same. The emergence of the seepage lines on the lower part of the downstream slope of the dam has a tendency to make this slope less stable than it would be under dry conditions. Hence we must either design a relatively mild slope that will be stable while the seepage emerges, or we must somehow divert the seepage away from the downstream slope. The first solution would be rather costly. The second solution happens to be quite feasible and is widely applied in dam design.

One way to prevent the seepage lines from emerging on the downstream slope, is to design the dam with a relatively impervious core and relatively pervious shoulders, as shown in *Figure 5.13(b)*. The total amount of seepage through the dam will become relatively small. Moreover, most of the vertical drop of the seepage lines will occur in the impervious core. The small amount of water that seeps through will flow in the lower layers of the downstream pervious fill and will emerge at the toe of the dam, where a gravel and rock filter prevents local sloughing.

Another way to keep the seepage lines down is to provide the base of the dam at the downstream side with an adequate gravel filter, as shown in *Figure 5.13(c)*. The material of this filter should be well graded and well placed so that it cannot become plugged with fine particles from the dam material. Moreover, the capacity of the filter should substantially exceed the maximum seepage through the dam.

When the dam is situated on a relatively pervious foundation, such as alluvial deposits of sand and gravel, as is shown in *Figure 5.13(e)*, there may be danger of piping. This phenomenon occurs if the seepage pressure of the water that percolates upward through the soil beneath the toe becomes greater than the effective weight of the soil. The surface of the soil then rises, at some place of least resistance, and water and soil start flowing away from the dike. As a result, the toe of the dike may begin to slide, or the whole dam may begin to settle, with failure an ultimate possibility.

One way to prevent piping is to make an impervious cut-off underneath the dam. This could be accomplished by excavating a deep trench under the dam and extending the impervious core of the dam into this trench; or by driving a row of sheet piling under the dike; or by a combination of both, as is shown in *Figure 5.13(d)*. It should be noted at this point that the experience with steel sheet piling is not always favourable. The interlocks of the piles are far from watertight, making the seepage barrier relatively ineffective. When the pervious layers under the dam are very thick so that neither a trench nor sheet piling can reach the impervious sub-strata, the only effect of the trench or sheet piling is a lengthening of the seepage path. This may also be accomplished and probably at lower cost, by providing an impervious blanket in front of the dike, as shown in *Figure 5.13(f)*. The extent of this blanket should be such that the hydraulic gradient of the seepage water underneath the dike is reduced to a harmless magnitude. If, in spite of such measures, it is found that the seepage towards the toe of the dam is still dangerously large, it may be advisable to apply relief wells near the toe of the dam, as shown in *Figure 5.13(f)*. An interesting account of the successful application of relief wells, is provided by Middlebrooks and Jervis (1947): 'The foundation of the Fort Peck Dam consisted of an extremely tight, practically impervious stratum of clay overlying very pervious sands and gravels. Although a sheet pile cut-off was driven to shale, sufficient leakage occurred through the pile cut-off to develop high hydrostatic pressures at the downstream toe that produced a head of 45 ft. above the natural ground surface, which was considered dangerous. A board of consultants recommended the immediate installation of relief wells 100 ft. outside the toe on 50 ft. centres. The result was that the hydrostatic pressure at the toe was reduced from 45 to 5 ft. while the total flow from all wells averaged 10 cusec. It is concluded that the installation is entirely satisfactory.'

After having considered the composition of an earth dam from the viewpoint of seepage control, we shall now discuss the stability of the embankment and foundation. If the foundation is relatively solid, a possible failure of the dam would be restricted to a slide on one of the slopes of the dam. To prevent such

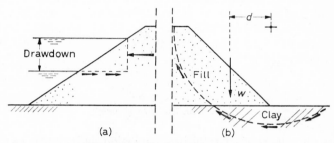

Figure 5.14. Analysis of earth dams

slides, a slope of about 3 horizontal to 1 vertical is sufficient, with the upstream slope being somewhat flatter than the downstream slope in order to take care of possible draw-down pressures. It often happens that the foundation of the dam consists of recent alluvial deposits such as silts and clays. A possible failure of the dam would then not necessarily be restricted to the slope but could include a slide in the foundation as well. To prevent such slides, the slopes of the dam must be lessened, and they may become as mild as 10 horizontal to 1 vertical.

The slope figures in the above paragraph have only been quoted to establish an order of magnitude. It is evident that the selection of the slopes for a particular dam must take into consideration all the circumstances pertaining to that particular site and that it must be based on adequate soil mechanics analysis. One of the methods used to check the stability of dam and foundation is to assume a potential sliding surface, and to analyse the equilibrium of the segment by taking moments about the centre of the arc, as shown in *Figure 5.14(b)*.

The circle method of stability analysis consists in drawing a circle through the slope and foundation, and taking moments of all the forces acting upon the segment about the centre of the circle. These forces are the weight of the entire segment of earth and the shear forces along the perimeter of the circle. This analysis must be repeated until the most dangerous circle is found. The shear force used in this analysis is the maximum shearing resistance which will be effective along a possible sliding surface. This shear resistance is partly due to internal friction and partly to cohesion. For clays it is not very well known how much shear resistance is due to the one or the other. The factor of safety against sliding is equal to the ratio of the maximum resisting moment to the overturning moment.

The actual computations may consist of dividing the segment into vertical elements and computing the overturning and maximum resisting moment for each element. The accuracy of such an analysis depends mostly on what values have been selected for the shear strength. Unless the designer knows clearly how the selected shear strength was determined, and how reliable it is, the analysis and the resultant factor of safety may create an unwarranted feeling of security. To perform such analysis one would have to make first a thorough study of the pertinent soil-mechanics literature.

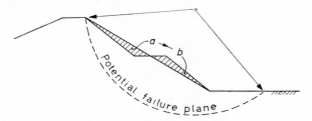

Figure 5.15. Stability of slope

While on the subject of safety against foundation slides, it may be of interest to point out that provision of a berm, without changing the volume of the dam, may increase the stability, as shown in *Figure 5.15*. Removing the amount of embankment '*a*', and placing it in the position '*b*', will reduce the moment of the segment of earth with respect to the centre of the potential failure circle. After having made such a change in design, one must, of course, check alternative potential failure planes. Care must be taken that the berm is well drained so that it cannot become a place of entrance into the dam for surface run-off.

Another method to check the stability of dam slopes is the sliding block analysis shown in *Figure 5.14(a)*. This method is particularly suitable for the upstream slope of the dam in the zone of drawdown. The forces involved are caused by the active earth pressure and hydrostatic pressure to the left, and the shear resistance to the right. The most critical condition must again be found by trial and error.

Fortunately for the safety of the completed dam, most failures occur during construction of the dam when the shear strength of the soil is at its lowest. Once the construction is completed, the soils will consolidate, and the shear strength of embankment and foundation will gradually increase with age.

To illustrate the above discussions, *Figure 5.16* shows three earth dams that are at present under construction. *Figure 5.16(a)* shows the South Saskatchewan River Dam in Canada. The river valley bottom contains a layer of fine to medium sand, up to 50 ft. in thickness. To reduce seepage, an impervious blanket extends 1,500 ft. upstream of the heel of the dam. To cope with the seepage that will still come through, a filter and relief wells have been designed at the toe of the dam. The layer of sand is underlain by clay and shale with a low resistance to shear. Therefore the shoulders of the dam had to be designed at a relatively mild slope. *Figure 5.16(b)* is of interest because this dam is situated in a narrow rock canyon where a gravity or arch dam might be expected. However, a comparative cost analysis of a multiple arch, thin arch, concrete gravity, rock fill, and earth dam indicated the last one to be the least costly.

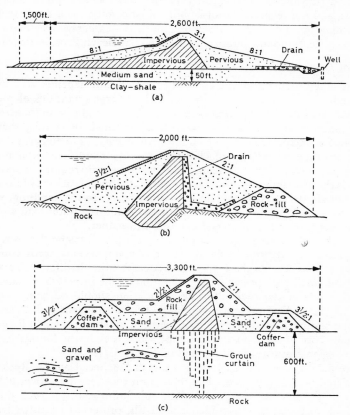

Figure 5.16. (a) *South Saskatchewan Dam (Canada), 210 ft. high;*
(b) *Mammoth Pool Dam (California), 330 ft. high;* (c) *Aswan Dam*
(Egypt), 370 ft. high

The impervious core is tied in with the underlying rock with concrete to obtain an even surface for the earth compaction equipment. The drain layer consists of coarse gravel and is dumped in horizontal layers along with the impervious and pervious material on each side. Since the drain isolates the downstream half of the dam from water, the downstream face has the relatively steep slope of 2:1. *Figure 5.16(c)* shows the Aswan Dam in Egypt as designed by Russian engineers. The most notable feature of the project is the foundation upon which the dam must rest, namely a layer of silt, sand and gravel from 600 to 700 ft. deep. This alluvial material will be made watertight by injecting a clay and cement mixture right down to bedrock. The designers are confident about the effectiveness of this grout curtain, so that an upstream impervious blanket has been omitted. The two cofferdams, required for the construction of the dam, will eventually be incorporated in the cross-section of the dam.

The upstream slope of an earth dam should be protected against wave action. This may be accomplished by covering the slope, over the range of the reservoir fluctuation, with a layer of rip rap, hand-placed stones, concrete, or asphalt mixture, depending on the severity of the wave action, the availability of materials, and the relative cost of labour. Most common is a 2 to 3 ft. layer of rock dumped on a 1 to 2 ft. layer of gravel. An extensive treatment of this subject may be found in a publication by the U.S. Beach Erosion Board (1961).

Rock Fill Dams

It was noted earlier that the choice of type of dam depends very much on the materials that are available. In cases where large quantities of rock have to be excavated for other purposes, a rock fill dam might be more economical than a concrete or earth dam. Rock fill dams must, of course, include an impervious element. This may be an impervious core, as in an earth dam, or it may be an impervious layer of clay near the upstream face, or it may be a concrete slab on top of the upstream face. One of the main requirements of a rock fill dam is the availability of sound rock for the fill. Sound rock may be described as rock which will not disintegrate in the quarry nor in the handling, which is strong enough to sustain the weight of the dam and the water load and which will not disintegrate rapidly when exposed to weather.

The design of a rock fill dam is relatively simple. A triangular rock section is dumped at the natural angle of repose. An impervious earth core at the upstream slope prevents the seepage of water. A suitable filter of coarse sand, gravel, and small rock prevents the migration of the impervious material into the large voids of the rock fill. The foundation should be solid enough to withstand the vertical pressure, and should have sufficient resistance against horizontal sliding. If necessary, a grout curtain must be provided between the impervious core of the dam and the impervious rock in the foundation. Some of these design features are illustrated in *Figure 5.17.*

Figure *5.17(a)* shows the Bersimis Dam in Quebec, Canada. Rock for the fill consisted of sound granite. The maximum rock size was governed by the capacity of the moving equipment. The minimum size was rock dust. The impervious core was a well-graded glacial till, protected by three transition zones consisting of coarse sand, gravel and rock ranging from 3 to 10 in. Stability analyses, assuming slide planes through the impervious core, were made to ensure that no failure would occur during construction or after sudden drawdown of the reservoir.

Figure 5.17(b) shows the Lemolo Dam in Oregon, U.S.A. A rock fill dam was selected in this case because the foundation was not considered suitable for a concrete dam and material for an earth dam could not be located within economic hauling distance. The concrete deck type was selected because material for an impervious compacted fill could not be found. The rock used in the fill is all basalt, obtained from spillway excavation and from a quarry. The placed rock was constructed of selected stones carefully placed and chinked. The surface voids were filled with gravel before the deck was placed. The concrete deck was reinforced with 0·50 per cent steel in both directions, and divided into 30 ft. squares by contraction joints, sealed with rubber water stops.

Figure 5.17(c) shows the Derbendi Khan Dam in Iraq. Comparative cost estimates between a rock fill and a concrete gravity dam were very close. Therefore designs were made and tenders were invited for both types of dam. After the bids were received, it was found that the rock fill dam was 15 per cent, or eight million dollars, cheaper than the gravity dam. The rock fill dam was designed with a central core because the low friction strength of the available

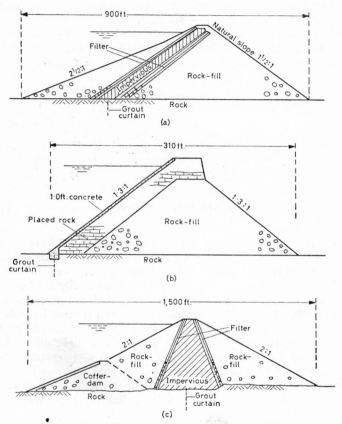

Figure 5.17. (a) Bersimis Dam (Canada), 230 ft. high; (b) Lemolo Dam (Oregon), 120 ft. high; (c) Derbendi Khan Dam (Iraq), 450 ft. high

core material would necessitate flatter slopes for the upstream shell of a sloping core type dam than for a central core type dam. The base of the impervious core was carried to sound and grouted rock throughout the entire foundation.

FREEBOARD REQUIREMENTS

The freeboard of a dam may be defined as the difference in elevation between the top of the dam and the maximum reservoir level that would be attained during the passing of the spillway design flood through the reservoir, as shown in *Figure 5.18.* This freeboard may be as little as a few feet for a gravity dam holding back a small reservoir, or it may be as much as 25 ft. for an earth fill dam holding back a large reservoir. The formulation of adequate freeboard requirements is important since over-topping of the dam by waves could be disastrous for the stability of the dam. An earth dam, for instance, whose downstream slope becomes saturated, may fail through progressive sloughing and sliding, starting at the toe of the dam. A gravity or rock fill dam, resting on an earth foundation or keyed into earth embankments, could fail through erosion at the toe or at the flanks of the dam.

The first problem in determining the freeboard of a dam, is to establish the freeboard criteria. This is mostly a matter of selecting the wind velocity that may prevail during the passing of the spillway design flood. Before the engineer

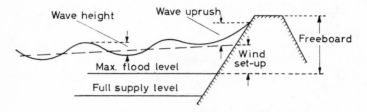

Figure 5.18. Definition of freeboard

makes this selection, he must explore the possibility of dam failure as a result of over-topping, and the scope of disaster as a result of dam failure. Some of the underlying philosophy in making this choice is discussed in the section Spillway Design Flood (page 196). After the wind velocity has been decided upon, the consequent wind set-up and wave uprush are computed. Unfortunately, this subject is still full of uncertainties and it must be realized that the final answer may easily contain an error of plus or minus a few feet. In terms of the dam, this may mean plus or minus several million dollars of construction cost! It is therefore obvious that from a viewpoint of safety as well as economy, the freeboard deserves to be treated with due consideration. The different problems involved in deciding upon the appropriate amount of freeboard is discussed under the following headings: (1) wind velocity; (2) wind set-up; (3) wave height; (4) wave uprush.

The meaning of these terms is illustrated in *Figure 5.18.* They are further discussed in the following paragraphs.

Wind Velocity

The selection of the wind velocity as noted earlier, depends upon the degree of safety that is desired. Let it be assumed, for example, that the safety of the dam should correspond to a probability of overtopping of less than 0·1 per cent per year. If the maximum flood level and the full supply level practically coincide (like for instance in a large reservoir with little inflow, regulated for recreational or navigational purposes) it would require the selection of a 0·1 per cent per year wind velocity in the direction of the dam under consideration. However, if the occurrence of the maximum water level is a rare event, the frequency of the required wind should be less. In simple cases, the problem may be solved by plain reasoning. For instance, a reservoir has its maximum water level during 40 per cent of the summer–autumn period, when the severe storms occur. During the remainder of the time, the reservoir level is considerably lower. Assuming no correlation between wind and high reservoir level, it follows that a 0·25 per cent wind in the direction of the dam would have to be selected (40 per cent of 0·25 = 0·1 per cent).

In more complicated cases where the extreme water levels cover a range of elevations, close to the maximum, it is necessary to combine the frequency of still water levels with the frequency of wind-set-up-plus-wave-uprush, in order to determine the proper elevation of the top of the dam. This technique will be discussed in Chapter 6: Flood Control.

After having determined the frequency of occurrence of the required wind, the magnitude of this wind has to be found. If the problem is very important, a competent meteorologist should be engaged. Otherwise, charts in hydrology handbooks and publications may be consulted. It will be found that maximum wind velocity not only depends upon the frequency of occurrence but also the direction and whether the wind blows over land or over water.

It appears from automatic lake level recordings that a full wind set-up can take place within a few hours. In fact it is believed that the wind set-up, for a given wind velocity, is largest before rotational movement in the body of water has been fully developed. It follows that the wind velocity to be selected should have a duration of a few hours only.

It has been observed that wind velocities over lake and reservoir surfaces are substantially higher than velocities recorded over adjacent land surfaces, the overwater winds being approximately 30 per cent higher than those recorded at comparable elevations above adjacent land surfaces. Accordingly, the wind velocity, obtained from Weather Bureau land stations, should be increased by approximately 30 per cent to obtain overwater winds. It should be pointed out that this is only justified when the figures, thus obtained, are applied to experimental relationships that are also based on overwater winds.

It may be found, for a certain location, after having studied all available information, that the following six-hour winds occur in the direction of the dam.

Overland	Overwater	Frequency
40 m.p.h.	52 m.p.h.	10 per cent per yr.
50 m.p.h.	65 m.p.h.	1·0 per cent per yr.
60 m.p.h.	78 m.p.h.	0·1 per cent per yr.

In one of the earlier examples, where the design wind had a frequency exceedence of 0·25 per cent per year, its magnitude would have been, according to the above table, about 74 m.p.h.

Wind Set-up

When the wind blows over a body of water, a friction force is being applied to the surface of the water, in the direction of the wind, as shown in *Figure 5.19*. As a result of this force, the surface water will begin to move in the direction of the wind and will begin to pile up against the windward shore. This will cause a return flow along the bottom of the lake from the windward shore to the lee shore. Considering a vertical element of the body of water in *Figure*

5.19, we may observe the following forces in balance: wind friction to the right, plus bottom friction to the right, against excess hydrostatic pressure to the left. This leads to the following equations:

$$\tau_w.\mathrm{d}x + \tau_b.\mathrm{d}x = \tfrac{1}{2}\rho.g.(h+\mathrm{d}s)^2 - \tfrac{1}{2}\rho.g.h^2$$

or

$$(\tau_w + \tau_b)\,\mathrm{d}x = \rho.g.h.\mathrm{d}s$$

or

$$\frac{\mathrm{d}s}{\mathrm{d}x} = \frac{\tau_w + \tau_b}{\rho.g.h}$$

In other words, the slope of the water surface is proportional to the wind stress and the bottom stress, and is inversely proportional to the depth of water. It has been found from field observation and experimentation that in normal

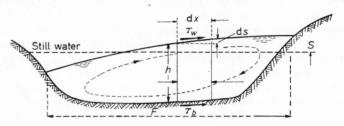

Figure 5.19. Wind set-up on lake

cases the bottom stress is only a small fraction of the wind stress. It has also been found that the wind stress is proportional to the square of the velocity of the wind. Since we are mostly interested in the wind set-up at the end of a lake or reservoir, we may write:

$$S = \frac{V^2.F}{C.D}$$

in which S is the wind set-up, in feet, above still water level; C is a coefficient; V the wind velocity in m.p.h.; F the fetch in miles, and D the average depth in feet.

The usefulness of this formula depends on the proper selection of the coefficient C. Observations carried out over a number of years on the Zuider Zee, indicated a C value of 1,600. Observations on Lake Okeechobee, indicated the same value. Although both lakes are relatively shallow, application of this coefficient to Lake Erie and Lake Ontario gave results that checked closely with observed values. In some literature, the coefficient C is listed at a value of 1,400, probably to stay on the safe side. In other publications, the coefficient is listed at 800. However, the wind set-up S is then defined as the difference in water level between the windward and leeward shores of the lake.

When the body of water is composed of sections with much difference in depth, it would be advisable to segment the body of water and apply the formula to each section individually, as shown in *Figure 5.20*. If there was a wall between the two sections of lake, the water levels would be as shown by the dashed lines. Since there is no wall, the water levels at the section boundary should match, and the total volume of water raised above still-water level must equal the total volume that is depressed below this level. This will involve some trial and error computations, but the problem is simple in essence.

When the body of water is very irregular in plan as well as in depth, it becomes difficult to determine the wind set-up. One may have to establish a correlation of observed wind values with observed wind set-up values. If the problem involves a reservoir that is still under design, it may be advisable to estimate the wind set-up on the conservative side, to carry out adequate field observations after construction of the reservoir, and after these have been observed and worked out, to adjust the regulation of the reservoir to the corrected reeboard val ues.

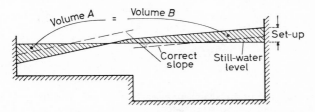

Figure 5.20. Wind set-up on lake with variable depth

In connection with irregular-shaped reservoirs, it may be of interest to point out that the wind set-up above still water level in a triangular-shaped reservoir would be somewhat larger than in a rectangular-shaped reservoir with the same length, if the wind is blowing towards one of the corners of the triangle, as shown in *Figure 5.21*. The slope of the water surface is only a function of wind velocity and depth and therefore the same in both reservoirs. It is evident that the volume of water depressed below the still-water surface level must equal the volume of water raised above this level. It therefore

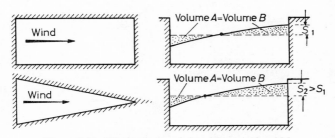

Figure 5.21. Wind set-up on lake with variable width

follows that the set-up S_2 in the triangular reservoir must be larger than the set-up S_1 in the rectangular reservoir. In *Figure 5.21*, with the wind set-up occurring at an apex of the triangle, the value of the set-up would be 1·3 times the value of the set-up in the rectangular basin.

Wave Height

For the present purpose we are interested in the height and period of waves generated in lakes and reservoirs. The height of a wave has obviously a direct

189

bearing upon the magnitude of the wave uprush. The period of the wave (the time that elapses between the passing of two wave crests) is also important,

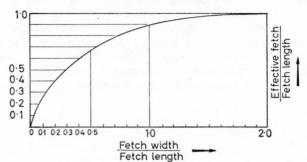

Figure 5.22. Effective fetch
(by courtesy of the U.S. Beach Erosion Board

since the magnitude of the wave uprush depends on the form of the breaker which, in turn, depends on the behaviour and timing of the backwash from the preceding wave.

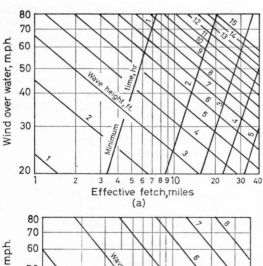

Figure 5.23. Wave height and wave period for deep water
(by courtesy of the U.S. Beach Erosion Board)

190

To determine the height and period of waves, reliance has to be placed on the results of field observations and laboratory tests. It has been found that the height and period of waves depends on the velocity of the wind, the length of the fetch, and the depth of the body of water.

To determine the length of the fetch is not as simple as it seems. It has been found that the width of the fetch is also important. As soon as the width of the fetch becomes less than twice the length, the waves will reduce in height, due to the 'narrowness' of the body of water. A procedure has therefore been established by the U.S. Beach Erosion Board (1954) to determine the so-called 'effective fetch' which is subsequently to be used in all experimental relationships. The ratio between effective fetch and maximum fetch length may be determined from *Figure 5.22.*

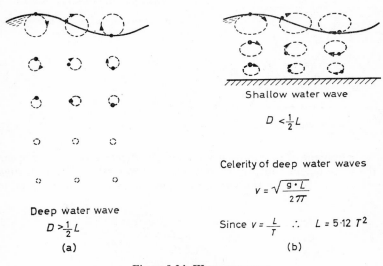

Shallow water wave

$$D < \tfrac{1}{2}L$$

Celerity of deep water waves

$$V = \sqrt{\frac{g \cdot L}{2\pi}}$$

Deep water wave

$$D > \tfrac{1}{2}L$$

Since $V = \dfrac{L}{T}$ $\therefore$ $L = 5 \cdot 12\ T^2$

(a)   (b)

Figure 5.24. Wave movement

Having selected the velocity of the wind over water, having computed the effective fetch length, and knowing the depth of the body of water, it is now possible to apply this data to empirical relationships in order to find the resultant wave height and period.

When the body of water is relatively deep (exceeding approximately one-half of the wave-length) *Figure 5.23* may be applied to find the wave height and period. *Figure 5.23(a)* also shows the minimum time required for the waves to fully develop. It may be seen that for lakes and reservoirs with an effective fetch less than 40 miles this minimum time is only a few hours.

When the body of water is shallow (less than one-half of the wave length) the waves generated by the wind, striking over the same stretch, will be smaller in height. This might be explained as follows: when a wave travels in deep water, the water particles describe a path as shown in *Figure 5.24(a)*, but when a wave travels in shallow water the water particles describe a path as shown in *Figure 5.24(b)*. The effect of frictional dissipation of energy at the bottom, for

191

shallow water waves, limits the rate of wave generation and places an upper limit in the wave heights which can be generated by a given wind speed and fetch length, as pointed out by Bretschneider (1959). If the fetch is long enough

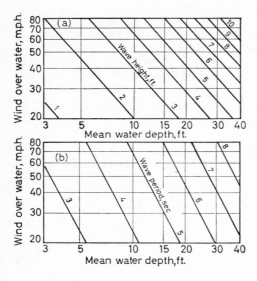

Figure 5.25. Wave height and wave period for shallow water and unlimited fetch
(by courtesy of the U.S. Beach Erosion Board)

for full development of the limiting wave height, *Figure 5.25* may be applied to find the wave height and period. When the fetch is relatively short, *Figure 5.26* may be used. However, this last diagram does not provide the wave period.

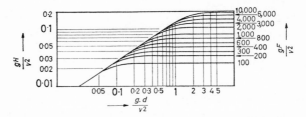

Figure 5.26. Wave height for shallow water and limited fetch:
g = 32·2 ft./sec²; d = depth of water, ft.; H = height of wave, ft.; v = wind/water, ft./sec; F = effective fetch, ft.
(after Delft Hydraulics Laboratory)

Wave Uprush

The ratio of wave uprush to wave height (both defined in *Figure 5.18*) is one of the most important items in determining the freeboard of a dike or dam. It is not uncommon to find an arbitrarily selected ratio of 1·5 applied to the wave height to find the wave uprush. However, it is now recognized that the correct value depends on several variables, such as the period of the wave, the slope of the dam, the roughness of the dam surface, the depth of the water at the toe of the dam, and the angle under which the waves approach the dam. These variables will be discussed in the following paragraphs.

192

It was pointed out earlier that the period of the wave is important since the magnitude of the wave uprush depends on the form of the breaker, which in turn depends on the behaviour and timing of the backwash from the preceding wave. *Figure 5.27* illustrates the effect of wave period upon the ratio of uprush to wave height, all other variables remaining constant. This figure is derived from laboratory tests carried out by Savage (1959). In the example shown, the slope of the dam was 1 in 4, and the face of the dam was artificially roughened with 10 mm bluestone. It may be seen that the ratio of the wave uprush R to the wave height H is more than 2·0 for very long (and high) waves (large values of T) while this ratio declines to a value of about 0·5 for very short (and low) waves. For example, an ocean wave with a height of 10 ft. and a period

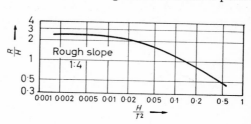

Figure 5.27. Effect of wave period
[from Savage (1959), by courtesy of the ASCE]

of 20 sec (and a speed of about 70 m.p.h.) would have a value $H/T^2 = 0·025$ and a corresponding value of $R/H = 2·0$. In other words, the wave uprush would be 20 ft. On the other hand, a shallow-water wave with a height of 3 ft. and a period of 4 sec (and a speed of about 14 m.p.h.) would have a value of $H/T^2 = 0·19$ and a corresponding value of $R/H = 0·8$. In other words, the wave uprush would be 2·4 ft.

The slope of the dam has also an important bearing on the uprush. On the one extreme, when the face of the dam is vertical, standing waves are generated, and the maximum height to which the water rises against the dam is theoretically H above the mean water level. In other words the ratio $R:H = 1$. On the other extreme, when the face of the dam has a very gentle slope, the approaching waves will break early, energy is dissipated, and the final uprush is diminished. It may therefore be expected that the ratio $R:H$ becomes very

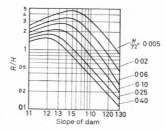

Figure 5.28. Wave uprush
[from Savage (1959), by courtesy of the ASCE]

small with a very gentle slope. In between these two extremes, the ratio $R:H$ may be substantially larger than 1, when the energy contained in the travelling wave is nearly all converted in the uprush of the wave. The relationship

7   193

between the ratio $R:H$ on the one hand and wave height, wave period and dam slope on the other hand is shown in *Figure 5.28*, which is derived from publications by Savage (1959). It may be seen that the relative uprush is much larger for long waves (with a large T value, and therefore a high speed) than for short waves (with a low speed). It may also be seen that the maximum value for each curve corresponds to different slope values. This is caused by the relationship between slope, wave period and backwash. Since *Figure 5.28* is based on laboratory experiments it would be advisable to increase the relative uprush by 10 to 20 per cent to allow for scale effect (the larger percentage for the larger waves and steeper slopes).

The roughness of the face of the dam has obviously a bearing upon the wave uprush, although not as important as one may think. A dam surface of hand placed masonry will reduce the uprush by only a few per cent, as compared to a

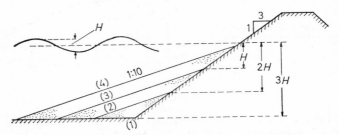

Figure 5.29. Effect of toe of dam
[after Saville (1958)],

perfectly smooth slope. A dam surface with a single layer of rip rap may reduce the uprush by 20 per cent. When the dam is entirely composed of rubble, like for instance a harbour breakwater, the uprush may be reduced to 50 per cent.

The depth of water at the toe of the dam has some bearing upon the uprush. When the depth at the toe is more than three times the wave height, a further deepening has no effect. When the depth becomes less, the waves may begin to break offshore from the dam, thus dissipating energy. This will result in a lower wave uprush on the dam. An example of such a reduction is derived from observations by Saville (1958) and illustrated in *Figure 5.29*. In situation 1, the depth of water in front of the dam is $3H$, and the ratio $R:H = 2 \cdot 0$. In situation 2, some soil has been placed on the toe of the dam under a slope of $1:10$, and the ratio $R:H$ is reduced to $1 \cdot 9$. In situation 3, the ratio is further reduced to $1 \cdot 8$ which is still not very significant. Finally, in situation 4, the ratio is reduced to $1 \cdot 0$. If still more soil was dumped in front of the dam, reducing the above-water slope also to $1:10$, the ratio $R:H$ would become $0 \cdot 5$. It should be emphasized that the above example illustrates the effect of the depth of approach in a qualitative sense only, since the basic data from which this example is derived, show a considerable degree of scatter.

The last important item to influence the amount of wave uprush is the angle under which the waves approach the dam. The foregoing discussions were all based on a perpendicular approach of the waves towards the dam. However, if the waves approach the dam under an angle, the uprush becomes less; the effect of the angle being similar to a lessening of the slope of the dam. One way of determining the effect of the angle would be to determine the corresponding slope reduction and treat it as such. Another way is to simply multiply the wave uprush with a factor equal to the sine of the angle between the dam and the direction of wave travel.

SPILLWAYS

In this section we shall discuss some of the more important hydraulic aspects of spillway design. The most important aspect of all, is to establish the discharge which the spillway should be able to pass. In other words, to determine the spillway design flood. Before we discuss this, let us devote a few words to the different methods by which a reservoir can be operated.

Reservoir Control

A simple way to design a reservoir would be to establish the spillway crest at the full supply level, as shown in *Figure 5.30(a)*. If the reservoir is to serve power or water supply interests, we may expect that the full supply is the normal level of the reservoir. Only during times of water shortage will the reservoir be drawn down to supply downstream requirements. If the reservoir is to serve flood control interests, it will normally be empty, but we may look upon the full supply level as the level that will be reached when the reservoir design flood (not to be confused with the spillway design flood, as will be discussed in Chapter 6: Flood Control) passes through the reservoir. For the sake of the present discussion let us conveniently assume that we are dealing with a power reservoir and that the full supply level is maintained during most of the time.

Having designed the reservoir as shown in *Figure 5.30(a)*, and assuming an extreme flood flow entering the reservoir, we now expect a rise in the reservoir level, and a flow over the uncontrolled spillway. If the extreme flood flow happens to be the spillway design flood, the reservoir will rise to the maximum flood level, shown in *Figure 5.30(a)*. The reservoir storage capacity between

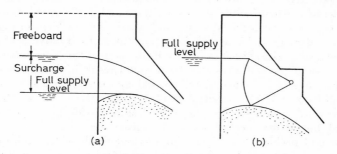

Figure 5.30. Reservoir control

the full supply level and the maximum flood level is sometimes called 'surcharge storage'. If the volume of the surcharge storage is a significant part of the volume of the spillway design flood, it is evident that a substantial reduction of the peak flow of the spillway design flood can take place between the moment of inflow into the reservoir and the moment of outflow over the spillway. As noted earlier, we must add the freeboard to the maximum flood level to obtain the level of the top of the dam.

An alternative way to operate the reservoir, assuming the same height of dam and the same maximum level of the reservoir, would be to establish this level as the full supply level, as shown in *Figure 5.30(b)*. When the spillway design flood, or any other flood, passes through the reservoir, the gate is opened to release as much flow over the spillway as enters the reservoir. As a result, the reservoir level remains constant at full supply level all the time. Since there is no flood storage capacity, the outflow peak of the spillway design flood is not reduced, and equals the inflow peak. Therefore, the crest of the spillway in *Figure 5.30(b)* must be lower than the crest of the spillway in *Figure 5.30(a)*, assuming that the length of the spillway is the same in both cases.

This second way of operating the reservoir has the significant advantages over the first one, of a larger storage capacity for water supply and a higher head for power. The disadvantages are: the cost of constructing and operating the control gates; the risk that the control gates will fail to open during extreme flood conditions; the loss of the flood-reducing feature of the first method which may be of downstream interest; the more costly stilling basin because of the larger design flow. Whether or not the advantages are more important than the disadvantages becomes largely a matter of economic analysis. In many reservoirs the features of the two methods of control are combined; that is to say, the spillway is controlled with gates, the full supply level is at the top of the gates, but the maximum flood level is higher than the full supply level, thus providing some surcharge storage capacity.

Spillway Design Flood

We have now come to one of the most important and also one of the most difficult problems in dam design, namely to determine the required spillway capacity. The first problem is to decide upon the amount of risk that can be tolerated. Should we accept no risk at all, like for instance in the case of the six Missouri River dams with a total storage capacity of nearly one hundred million acre-ft., where failure of the spillways and dams would cause a huge floodwave to race down the Missouri and Mississippi Valley, causing billions of dollars of damage and leaving behind thousands, if not millions of dead. Or can we take some risk, like for instance in the case of the Nelson River dams, where failure of the spillways and dams would result in limited damage and probably no loss of life, since no one lives along the river and since the power plants will be operated by remote control. If we can take some risk, how much can this be? A number of methods to deal with this question have been suggested, but no one method is entirely satisfactory.

The second problem is to determine the magnitude of the spillway design flood that will correspond to the desired degree of safety that was discussed above. Strictly speaking, the first and second problems are quite distinct from each other. We could look upon the first one as involving a policy decision, while the second one is largely an engineering problem. We shall first discuss the problem of defining adequate safety. It is suggested that this problem can be approached from three different viewpoints: an engineering, an economic, and a social viewpoint.

From an engineering viewpoint, it may be requested that an important structure should stand up under circumstances that can reasonably be foreseen. Unfortunately, there is no widely accepted definition of 'reasonably foreseeable' in terms of probability of occurrence. Since such a definition would be desirable, it is hereby proposed that 'reasonably foreseeable' be defined as having a chance of occurrence of 0·1 per cent per year. This may seem on the conservative side, but let it be noted that if the design flow conditions for a hydraulic structure were selected accordingly, this structure would have a 10 per cent chance of failure during the first 100 years of its existence; and if an engineering agency would build one hundred of such structures, the chances are that every 10 years one of these structures will fail! It is not very likely that any board of engineers would be willing to accept a much greater risk of failure, if it were only to keep up the reputation of the profession.

From an economic viewpoint, the most desirable degree of safety would be the one that results in the smallest sum of construction cost plus capitalized risk. To assume a more conservative design flood than the one thus determined, would result in an addition of construction costs that is larger than the corresponding capitalized reduction of average annual damages. The technique of finding the point of maximum economy will be discussed in Chapter 11: Economic Analysis. In many cases it will be found that the degree of safety that would yield maximum economy, corresponds to design flow conditions with a frequency of occurrence larger than 0·1 per cent per year. In such cases the previously discussed engineering considerations take precedence over the economic considerations. In other cases, where failure of the structure would result in relatively large economic damage, and where more rigid design flow conditions would only result in a relatively small increase in construction cost, it may be found that it is sound economy to assume design flow conditions with a smaller probability of occurrence than 0·1 per cent per year. In such cases, economic considerations take precedence over engineering considerations.

From a social viewpoint, one has to consider the possibility of the loss of human life, as a result of failure of the structure. It is almost superfluous to state that under given circumstances the probability of loss of life is decreased when more rigid design flow conditions are assumed. As an initial reaction to being faced with such a situation, one may suggest that when human lives are involved, the design flow conditions are to be assumed so rigid that the probability of the loss of human life is excluded. However, nowhere in the life of a community are unlimited funds available to provide complete safety for all individual members, for the simple reason that the productive capacity of the community is limited. It is therefore again necessary to compare the merits of increase in safety with corresponding increase in cost. This, of course, is a most delicate matter because the assignment of any monetary value to human life is wide open to criticism. One approach is to compare this problem with other social problems of human safety, such as fire protection, traffic control, or cancer research. It is therefore suggested that the incremental cost of the engineering structure and its corresponding increase in safety be determined, and compared to what alternative increase in safety these same expenditures could accomplish in other fields of social activity. A quantitative treatment of this problem will be discussed in the section on dikes, Chapter 6, page 226.

After having considered the required safety of a hydraulic structure from an engineering, an economic, and a social viewpoint, a conclusion may have been reached with respect to the probability of occurrence of the design flow conditions. For example, in designing the spillways for the hydroelectric projects on the Nelson River (where failure of the spillway would only result in limited economic damage and negligible danger of loss of life) it was decided that the spillway design flood should have a probability of occurrence of 0·1 per cent per year. On the other hand, in designing the spillways for the series of major dams on the Missouri River (where failure of the structure could result in a terrible disaster), it was decided to accept no probability of failure at all. In between these extremes there is an array of situations where the selected probability of occurrence of the design flow conditions may range from 0·1 per cent per year to zero.

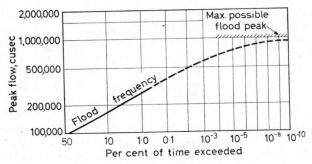

Figure 5.31. Extrapolation of flood frequency curves

The next problem is one of determining the magnitude of the design flood (in terms of peak flow and volume) that corresponds to the selected probability of occurrence. Before this last problem is dealt with, it may be of interest to discuss, first, the nature of extreme flood flows. If one has available the recorded data on maximum annual river flows at a certain station, over the past 10 years, one of these 10 figures is bound to be the highest. If one had available the peak flows at that same station over the past 100 years, it is likely that the highest of the 100 items would exceed the highest of the 10 items. If one could reconstruct the peak flows at that same station over the past 1,000 years, it is likely that the highest of the 1,000 items would exceed the highest of the 100 items. This thought could be pursued for a while, and it may be concluded that an increase in period under consideration corresponds to an increase in maximum flood flow. Or, in other words, a decrease in probability of occurrence per year, corresponds to an increase in flood magnitude. Could this statement be extrapolated to saying that a probability of occurrence of once in infinity corresponds to a flood with an infinitely large magnitude? No, certainly not since there is a physical limit to the flood producing capacity of a drainage basin. There is a limit to the amount of precipitation that causes floods, high as it may be. The run-off coefficient of the precipitation cannot exceed 100 per cent. The run-off cannot pass the river stations in less than a certain time limit

Therefore, there is a 'ceiling' to the flood potential, which may be called the 'maximum possible flood'. If enough time, knowledge, and research is devoted, the magnitude of the maximum possible flood can be determined quantitatively for any station on any river, as was discussed in Chapter 2, and is shown in *Figure 5.31*.

The following situation has now developed. On the basis of recorded river flows, a fairly reliable estimate can be made of the magnitude of flood peaks with a probability of occurrence of, say, larger than one per cent per year. When due consideration is given to the frequency curves of other river stations in the drainage basin (as was discussed in Chapter 2) the frequency curve may be extended to a probability of occurrence of say 0·1 per cent per year. On the basis of a comprehensive hydrologic study of the drainage basin, a fairly reliable estimate can be made of the magnitude of the maximum possible flood. In between, is an area of uncertainty where the relationship between flood magnitude and probability of occurrence is not known. Unfortunately, it is this very area that often has to provide the clue to determining the proper design flow conditions, which, as was noted earlier, may have a probability of occurrence ranging from 0·1 per cent per year to zero.

In spite of the uncertainties involved, one reasonable solution to the dilemma would be to extrapolate the frequency curve towards the ceiling of the maximum possible flood, as shown in *Figure 5.31*, and to use this curve for finding the peak flow that corresponds to the predetermined probability of occurrence. For instance, if the safety against failure should be 0·01 per cent per year, the corresponding peak flow would be 500,000 cusec.

It should be recalled at this point, that reservoirs can be operated in different ways, as shown in *Figure 5.30*. If the maximum flood level coincides with, or is close to, the full supply level, the spillway design discharge is practically the flood peak of the spillway design flood entering the reservoir. For such conditions the above method would give immediately the desired flow figure. However, when the surcharge capacity of the reservoir is substantial, it becomes necessary to determine not only the flood peak, but also the flood volume. The technique of doing this was discussed in Chapter 2. It should be pointed out here that the probability of occurrence of the final results becomes more and more uncertain.

A different method to arrive at the spillway design flood is to apply the procedure of computing the maximum possible flood, but to water down the components that make it up. Maximum possible precipitation is replaced by maximum recorded precipitation. Storms are rearranged over the drainage basin to represent 'reasonably severe' conditions. Instead of no infiltration, some infiltration losses are assumed. The unit hydrograph is given a somewhat lower peak and longer time of concentration. The resulting flood is then given a name, like 'maximum probable flood', and is defined as the largest flood that can reasonably be expected to occur. Since this method is usually applied without also computing the maximum possible flood, and without making a supplementary frequency analysis, the obvious disadvantage is that one loses one's orientation about the probability of occurrence of the 'design' flood. As a result it is no longer possible to separate the policy decision on the required degree of safety of the dam and the engineering computation of the corresponding spillway discharge. Instead, all decisions are left to the good judgement of one or a few engineers to produce a flood that is 'reasonably rare'. This seems to be an unsatisfactory state of affairs.

It should be noted that, over the years, numerous empirical formulae have been developed that give a relation between 'extreme flood flows' and certain drainage basin characteristics. The use of such formulae to determine the spillway design flood would be still less desirable than either of the above two methods.

Having determined the hydrograph of the spillway design flood at the entrance to the reservoir, we must now route this flood through the reservoir to obtain the maximum outflow from the reservoir. The technique of flood routing was discussed in Chapter 3. It was noted earlier in this chapter that the reduction of the peak flow depends upon the surcharge storage capacity in the reservoir. It should be noted that the maximum ordinate of the outflow hydrograph is not necessarily identical with the spillway design discharge. First of all, it is possible that part of the outflow takes place through conduits. If such auxiliary discharge capacity can be relied upon, it may be deducted from the maximum outflow. Secondly we must realize that due to wind set-up the reservoir level may rise and cause a larger discharge to flow over the spillway. Since we design the dam for this wind set-up it is logical that we design the spillway also for this water level and its corresponding discharge. It is suggested that the conventional hydraulic design of the spillway and stilling basin be based on the maximum reservoir level without wind, and that this design be checked for a discharge resulting from maximum reservoir level including the full wind set-up, to ascertain that no chute walls are overtopped, that the hydraulic jump is still contained in the stilling basin, and that no disastrous downstream erosion can take place.

It is evident from the above discussion that the subject of determining the spillway design flood is full of controversial problems. Therefore, a great deal of responsibility rests with the engineer who is in charge of the design of the project. He must familiarize himself with all aspects of the problem and he must select what he deems best. If he feels that he should not bear alone the responsibility for the final choice, then he must take the initiative of presenting the problem clearly, and inviting the opinion of others who are knowledgeable and in a responsible position to pass judgement.

Chute Spillways

It was noted earlier that the function of a spillway is to enable flood flows to pass the dam site without endangering the dam. There are several types of spillways. The most important are:

(1) Overflow spillways which are simply a lowered section of gravity dam over which the flood is allowed to pass, as shown in *Figure 5.1*. The crest is formed to fit the shape of the overflowing water under conditions of maximum discharge. If the foundation of the dam is solid rock, the lower part of the spillway may be designed as a flip bucket or ski jump. If the foundation is erodable, a stilling basin is required. This subject will be discussed in the next section.

(2) Free over-fall spillways are associated with arch dams, where the thickness of concrete is insufficient to fill the nappe under the overflowing water. Unless the foundation is solid rock, some protection is required to resist the impact of the falling water.

(3) Shaft spillways have a funnel-shaped entrance at reservoir level and a vertical shaft leading to a tunnel. This form of spillway is adapted to narrow canyons where room for some other type of spillway is too restricted.

(4) Chute spillways are usually associated with earth and rock fill dams and will be discussed in more detail in the following paragraphs.

The chute spillway presents several interesting design problems. It is considered good practice not to locate the spillway on top of the earth dam, since every earth structure is subject to some degree of settlement after construction. Such settlement could result in cracking or uneven settlement of the concrete spillway chute. When the design flood passes over the spillway, the water velocities on the chute may reach magnitudes of 50 ft./sec or more, depending on the height, the slope and the roughness. With such high speeds,

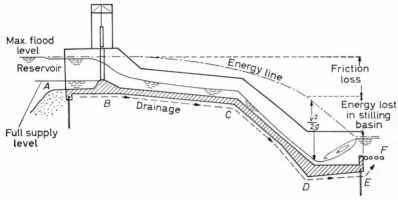

Figure 5.32. Chute spillway

the smallest unevenness on each side of a crack in the concrete could build up a high hydrostatic pressure underneath the slab and lift it up. Needless to say, once part of the chute is ripped up, the whole spillway might quickly follow and the entire dam may breach. Therefore, the spillway of an earth dam is invariably located in an abutment in undisturbed soil. In fact, when searching for a dam site one should look for the possibility of locating the spillway away from the main dam, into a small tributary creek or depression, as was done with the South Saskatchewan River Dam, shown in *Figure 5.2*. The profile of a chute spillway of an earth dam may look somewhat like *Figure 5.32*. The entrance channel at A is relatively wide so that no significant energy losses or eroding velocities will take place. The control of the spillway at B may be a low sill to induce critical depth to assure that the maximum discharge will be obtained for a certain rise of the reservoir level above the spillway crest. From B to C the spillway chute may have a mild slope because of the topography of the abutment. From C to D the spillway chute slopes down to the side of the river valley and reaches the bottom of the valley at D. The dissipation of the excess energy takes place in the stilling basin D–E. After the water flows again in a tranquil state, it returns to the river channel at F. In the following paragraphs some principles of designing a chute spillway will be discussed.

In many cases the width of the spillway, the chute, and the stilling basin are the same. However, this does not necessarily have to be so. The width of the spillway is related to the method of operating the reservoir, the cost of the dam, and whether or not the spillway is provided with gates. A discussion of this topic may be found in the sections on Spillways and Sluice Gates, Chapter 7, page 320. The width of the stilling basin is related to the tail water rating curve. A discussion of this topic will be found in the next section, Stilling Basins. If for the above reasons, or for reasons of economy, the width of the spillway, the chute, and the stilling basin are not the same, care must be taken that the transitions take place very gradually or else undesirable standing waves may develop, or the flow may separate from the side walls. Final design of spillways should always be checked in a hydraulic model. For preliminary design one could assume uniform width of spillway, chute and stilling basin, in feet, equal to the square root of the discharge, in cusec.

The spillway chute is usually made of reinforced concrete, from 12–18 in. thick. The side walls of the chute and stilling basin are designed as gravity or cantilever retaining walls. To prevent hydrostatic uplift under the spillway chute, a cut-off wall is provided under the spillway intake and a drainage system of filters and pipes is provided under the chute. This drainage system may release its flow via weep holes on to the chute surface. In cold climates it is possible that the weep holes freeze up, and provision must be made to have an outlet of the drainage pipes through the lower cut-off wall, at the end of the stilling basin.

When the stilling basin is in operation there is considerable hydrostatic uplift under the lower part of the chute and the upstream part of the stilling basin floor. This is because the hydrostatic pressure under the floor corresponds to the tail water level, while the pressure of water upon the floor is only caused by the relatively thin sheet of supercritical flow. To counteract this hydrostatic uplift the floor must be made sufficiently heavy, or must be anchored to the foundation.

Stilling Basins

The function of a stilling basin is to dissipate the kinetic energy of the supercritical flow on the spillway chute, before the spillway discharge is returned to

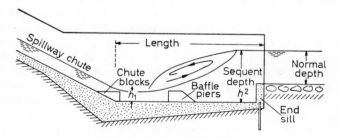

Figure 5.33. Stilling basin

the river channel. The means of energy dissipation is the hydraulic jump, discussed in Chapter 3. It may be recalled that there is a relation between the velocity and depth of flow, upstream of the hydraulic jump, and its so-called 'sequent depth' downstream of the jump, as shown in *Figure 5.33*. Let us discuss first under what ideal conditions the hydraulic jump can perform its function.

The length, the width and the depth of the stilling basin are related to one another. Let us approach the problem by discussing the depth first. From the design flow Q we can determine the normal depth h_n of the river channel and therefore the tailwater elevation. From the flow Q and an assumed width of the stilling basin we can also determine the depth of flow h_1. From the value h_1 and V_1 we can determine the sequent depth h_2. We then subtract h_2 from the tailwater elevation and find the elevation of the bottom of the stilling pool.

Unfortunately, we are not dealing with one discharge but with a range of discharges, from zero flow to the spillway design flow. For all these discharges the stilling basin should be able to contain the hydraulic jump. In order to

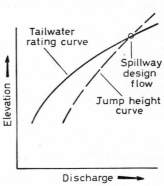

Figure 5.34. Stilling basin design curves

appraise the situation we prepare two rating curves: one for the tailwater elevation, resulting from the total discharge; and one for the jump height h_2, above the bottom of the stilling basin, resulting from the spillway discharge, as shown in *Figure 5.34*. In the ideal case, the two curves will coincide. However, this seldom happens. Where the jump height is higher than the tailwater, there is a danger that the hydraulic jump will shoot out of the stilling basin, which should be avoided. Where the jump height is lower than the tailwater, the jump will move upon the apron and be partly or completely drowned out. This will result in incomplete energy dissipation, and higher exit velocities, which should be avoided also. To correct the situation we may change the width of the stilling basin which will change the jump height curve, or the elevation of the bottom of the stilling basin, or the tailwater elevation by means of a downstream weir. A common corrective measure is the installation of baffles on the bottom of the stilling basin. These provide a force upon the hydraulic jump in upstream direction, in addition to the downstream hydrostatic force. As a result of having more resistant force, the jump will move to the left, where more momentum is to be dissipated; or the jump can remain at the same place, with a lower tailwater elevation.

The length of the stilling basin should equal approximately the length of the hydraulic jump. Experimentally, it has been found that the length of a hydraulic jump on a smooth horizontal floor is approximately seven times the sequent depth minus the supercritical depth. This length may be reduced by building chute blocks, baffle piers, and an end sill. Chute blocks are installed at the entrance of the stilling basin and have the function of breaking up the flow. Baffle piers are installed on the stilling basin floor and have the function of stabilizing the position of the jump and providing a force in upstream direction. End sills also stabilize the position of the hydraulic jump. Moreover, they lift the flow off the river bed, thus creating a bottom eddy which causes bed material to be deposited rather than eroded from behind the sill. As a result of such appurtenances the length of the hydraulic jump may be reduced to five times the difference in height between sequent depth and supercritical depth. For final design it is advisable to test the performance of the stilling basin in a hydraulic model.

As an added precaution to prevent downstream erosion of the river bed and possible undermining of the stilling basin, it is good practice to have a cut-off wall under the end sill and to pave the adjacent river bed with rip rap.

CONDUITS

The function of a conduit through a dam is to discharge water from the reservoir when its level is below the spillway crest. This may be desirable for several reasons, such as draining the reservoir to make room for flood storage, or leading the water to a power station. A conduit in a concrete gravity dam is

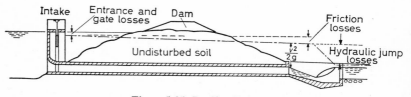

Figure 5.35. Profile of conduit

merely a hollow tube, saved during the construction of the dam and afterwards equipped with gates. Its design is relatively simple. A conduit for an earth dam presents more problems, as will be discussed in the following paragraphs.

The main elements of a conduit for an earth dam are shown in *Figure 5.35.* First there is the intake, that leads the water from the reservoir into the conduit. The elevation of the intake may be slightly above the bottom of the reservoir so that silt and debris are not carried into the conduit at low reservoir stages. As a further protection against debris entering the conduit, the intake is equipped with trash racks. To keep the hydraulic losses to a minimum, the entrance is given a smooth, rounded shape.

The second element of the conduit is the gate that regulates the flow through the conduit. The gate may be situated near the intake, as is shown in *Figure 5.35*, or it may be situated half way down the conduit, as is shown in *Figure 5.2.* The gate should never be located near the exit of the conduit, since that would place the lower end of the conduit under high pressure when the gate is closed. This could have disastrous results when a crack would develop in the wall of the conduit.

During closure of the gate, high velocities and therefore low pressures will develop underneath and behind the gate. To avoid cavitation and damage to the gate, provision must be made that open air has free access to the gate well, so that the pressure cannot drop substantially below atmospheric.

The conduit is usually constructed of reinforced concrete. To avoid the possibility of settlement and cracking, the conduit is preferably located in undisturbed soil in one of the abutments of the dam. The conduits may be constructed by the cut and cover method or by tunnelling, whichever is more economical. The thickness of the reinforced concrete conduit wall may range from 15–20 per cent of the diameter of the conduit. If deformation of the conduit is expected (due to settlement or due to rebound), with the associated danger of cracking, it may be advisable to place a steel penstock within the concrete conduit, with crawling space in between. If a crack develops in the concrete, no harm is yet done to the steel penstock. Even if the steel penstock would begin to leak, the leakage would drain away through the concrete conduit and failure of the earth embankment would be avoided. To prevent piping along the outside of the conduit, seepage collars must be provided and compaction of the surrounding fill must be carefully controlled. Conduits should never be supported by piles or piers since this may cause uneven settlement of the conduit and the surrounding soil and thus create an opening underneath the conduit through which piping may occur.

Between the exit of the conduit and the downstream river channel, a stilling basin must be provided to dissipate all excess energy. This stilling basin can be designed on the same principles as discussed earlier. Between the stilling basin proper and the conduit some transition section may be required to overcome difference in elevation or difference in width.

DIVERSION PROCEDURES

During the construction of the dam, measures must be taken to bypass the river flow. These measures are sometimes very costly and may run as high as 10 or 20 per cent of the cost of the project. In essence, nearly all diversion procedures have the following in common. During the very beginning of the project, the river is left in its existing course, while a temporary passageway of the river flow is built adjacent to the river. Upon completion of this bypass, the river is diverted into it, and will keep flowing through it until the project is finished. Then, the temporary bypass will be closed and the filling of the reservoir can begin.

There are numerous variations on this theme. When the dam is a relatively low concrete gravity dam, it may be possible to construct first, within a cofferdam that only closes off part of the river bed, the concrete overflow spillway of the project. When this spillway is finished the cofferdam is removed, the river is diverted over the spillway, the remainder of the river is coffer-dammed, and the project can be completed. When the overflow spillway is too high, it is possible to spare large openings in the concrete dam, at a low level, that are temporarily used for diversion. When the dam is completed, these openings are closed with gates and are filled up with concrete, as was done with the Kariba Dam. When the overflow spillway has an intermediate height, it is possible to follow a procedure as was applied to the McNary Dam, shown in *Figure 5.1*. First, the base slab and the piers of the spillway and the complete shiplocks were constructed in a cofferdam that closed off the north half of the river channel. Then, this cofferdam was removed and the river diverted through the unfinished spillway. The south half of the river was now cofferdammed and the power house constructed. The spillway was then completed by closing off each bay individually by steel bulkheads, so that the concrete for the overflow section could be poured in the dry. During this process the reservoir level gradually came up to the level of the spillway crest. After completion of all spillway bays, the gates were closed and the reservoir attained its full supply level.

If we are dealing with an earth dam and a chute spillway it may be possible to construct first the flood control or power conduits with the adjoining section of dam, while the river remains more or less in its existing course. After this is completed, the river is diverted through the conduits, and the other part of the dam can be built. It is not unlikely that the capacity of the conduits is governed by diversion requirements rather than their subsequent flood control or power function.

If, for some reason, any of the above solutions are not feasible, it may be necessary to resort to the costly procedure of diverting the river completely around the construction site of the dam by means of tunnels in the adjacent valley banks. In such a case, the tunnels have to be dug in the canyon walls from a point well upstream of the dam site, to a point well downstream of the dam site. After the tunnels are completed, a cofferdam is built just downstream of the tunnel intake, thus forcing the river flow into the tunnel. After this is accomplished, another cofferdam is built upstream of the outlet of the tunnels. The river section between the cofferdams is then dewatered and construction of the dam can begin. It is obvious that the size of the tunnels and the height of the upstream cofferdam are related. The smaller the tunnels, the higher the cofferdam must be, to create the head for the design discharge. It is a matter of economics to find the least costly combination. For the design discharge one may select an annual flood peak with a chance of exceedence of say one per cent per year. Assuming that no loss of life would be involved as a result of failure of the cofferdams, the problem of selecting the tunnel design flood can be subjected to economic analysis, as illustrated by the following numerical example.

Figure 5.36(a) shows a river valley where a gravity dam will be built. The river will be temporarily diverted through the tunnel *B–B*. The rating curve of the river at point *B* is given in *Figure 5.36(b)*. The cost of the cofferdams has been computed in terms of feet above the valley floor (which has an elevation of 100) and is given in *Figure 5.36(c)*. The cost of the tunnel has been computed and is shown in *Figure 5.36(d)*, in terms of the diameter of the tunnel. The frequency of exceedence of annual flood peaks is shown in *Figure 5.36(e)*. It is assumed that the cofferdams are needed for a duration of two years, and that failure of the cofferdams at any time during these two years would result in a damage of $10,000,000. The problem is to determine the most economic height of the cofferdam *A–A* and the diameter of the diversion tunnel *B–B*.

First of all, we assume a certain design discharge, say 20,000 cusec. With this given discharge, we must find the least costly combination of cofferdam and tunnel. We now assume a certain tunnel diameter, say 34 ft., and find that the cost of tunnel plus cofferdam is \$3,870,000. This is repeated a number of times until we find that a tunnel diameter of 36 ft. is most economical for the given discharge. This procedure is repeated for different discharges, and thus we find for every discharge, a corresponding, most economic tunnel diameter. We should now allow for the risk, or insurance premium, that is associated with each discharge. For instance, there is a chance of 10 per cent per year, or

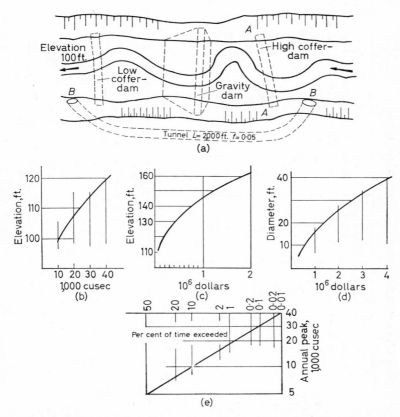

Figure 5.36. Pertinent data diversion scheme: (a) general plan; (b) rating curve; (c) cost of cofferdam; (d) cost of tunnel; (e) flood frequency curve

20 per cent in two years, that a flow of 10,000 cusec will be exceeded. We may equate this to a risk of loss, or to an insurance premium of 20 per cent of \$10,000,000 which equals \$2,000,000. We must now look for the design flow, where the total cost of tunnels plus cofferdam plus insurance premium is lowest. It may be seen from Table 5.1 that this is a flow of 15,000 cusec, corresponding to a tunnel diameter of 32 ft. and a cofferdam height of 27 ft.

After the project is finished, the diversion tunnel, or tunnels, may be used to supplement spillway capacity, or they may be used for power conduits, as is shown in *Figure 5.2*, or they may be plugged up at the entrance with concrete.

207

Table 5.1. *Computation of Most Economic Cofferdam Height*

Q	Tunnel Diameter	$\dfrac{\pi D^2}{4}$	v	$\dfrac{v^2}{2g}$	ΔH	Cofferdam Height	Cost Tunnel	Cost Cofferdam	Total Cost	Insurance Premium	Total Cost
10,000	22	376	26·6	11·0	61	61	1,700	1,900	3,400	2,000	4,350
	25	490	20·4	6·4	32	32	1,900	450	2,350		
	28	610	16·4	4·1	19	19	2,300	150	2,450		
15,000	30	707	21·2	7·0	30	34	2,550	550	3,100	400	3,450
	32	790	19·0	5·6	23	27	2,750	300	3,050		
	34	905	16·6	4·3	17	21	3,100	200	3,300		
20,000	34	905	22·2	7·6	32	39	3,150	720	3,870	140	3,840
	36	1,020	19·6	5·9	22	29	3,300	400	3,700		
	38	1,130	17·7	4·8	17	24	3,750	250	4,000		
25,000	38	1,130	22·1	7·6	27	38	3,750	700	4,450	40	4,490
	40	1,260	19·8	6·1	21	32	4,000	450	4,450		
	42	1,380	18·1	5·1	17	28	4,200	400	4,600		

Sample computation:
Assume $Q = 15{,}000$ cusec
Assume $D = 32$ ft.

$$\Delta H = hf + \frac{v^2}{2g} = f \cdot \frac{L}{D} \cdot \frac{v^2}{2g} + \frac{v^2}{2g}$$

$$= 0{\cdot}05 \times \frac{2{,}000}{32} \cdot 5{\cdot}6 + 5{\cdot}6 = 23 \text{ ft.}$$

Height of cofferdam $=$ T.W.L. $+ \Delta H$
$= 104 + 23 =$ EL. 127

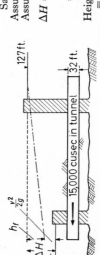

One of the most difficult engineering feats is the actual diversion of the river from its existing course into the tunnels, or dam openings. To accomplish the diversion, the river may have to be backed up 10–20 ft. or more, and while this is being done, very high velocities will develop. One method is to build a temporary bridge across the closure opening and to protect the river bed underneath and downstream with lumber or willow mattresses in case the river bed is erodible. From this bridge, rock is dumped until the closure is accomplished. Another method is to build a 'friction weir' across the river. This is a broad-crested weir of such dimensions that the river flows over it without being able to scour out a local channel. The dimensions of the weir depend on the size of the available material. During the construction great care must be taken that all parts of the weir are raised at an equal pace.

One of the most difficult river closures was accomplished on the Columbia River at the McNary Dam in 1951. The closure opening was 240 ft. wide and 60 ft. deep, carrying a flow of 110,000 cusec. A cableway across the opening placed 2,000 tetrahedrons of 12 tons each, until these emerged above water. Upstream of this pervious 'dam', successive layers of rock, gravel, sand and clay were placed until the closure was completed. During the closure, velocities up to 30 ft./sec were experienced. After the closure the river had been backed up by about 20 ft.

CHAPTER 6

FLOOD CONTROL

River-valley land has always had a great attraction upon early settlers; and for good reasons. In the early days, the rivers were the main arteries of traffic, and therefore it was convenient to live close to the river bank. Moreover the flood plains of rivers are usually fertile and have a relatively high moisture content, thus making them suitable for agricultural development. Finally it is much more pleasant to live under the shade of tall trees, with the view over a river, than to live high up on bald dry plains!

Whatever the historic reasons have been for the occupation of river-valley land, it must be accepted as a fact that many villages, towns and cities are situated in areas that are subject to flooding. The task of the hydraulic engineer is to devise ways and means to protect those communities from flooding; and not to bewail the fact that people have elected to live where they are.

The attempts of man to protect himself from flooding are as old as the history of civilization. The method of constructing large mounds, in order to have a place of escape, is thousands of years old, and is still applied! The method of protecting an entire area by building a dike, dates back to the early Middle Ages. In contrast, the more sophisticated method of reservoir construction was only developed in this century.

When discussing different means of flood control, the word 'control' should not be taken in its literal meaning. When a man brings the foundation of his house above flood level, he does not control the flood, but nevertheless he does protect himself from flooding. It is generally understood that flood control includes all measures which aim at reducing the harmful effect of floods. It is important to place the emphasis on the word 'reduce' since flood control measures very seldom, if ever, eliminate the hazard of flooding. Flood control measures may be divided into the following categories:

A. Engineering Measures
    (1) The construction of reservoirs
    (2) The construction of dikes
    (3) The diversion of flood flows
    (4) The improvement of river channels

B. Administrative Measures
    (1) Flood forecasting
    (2) Flood plain zoning
    (3) Flood insurance

RESERVOIRS

Storing flood water in the upstream part of a drainage basin is the most direct way (although not necessarily the most economic way) to reduce the flood hazard in the downstream part of the basin. There are three places where water can be stored: in the ground, in small reservoirs on creeks and minor streams, and in large reservoirs on the major stream channels of the river system.

To 'hold the water where it falls', in other words to store it in the ground, appears very attractive indeed. However, the beneficial effect of the type of land management that is required, is rather limited from a viewpoint of flood control. From careful observations on treated and untreated watersheds in the U.S., it has been concluded that the effect of terracing, contouring, and cropping will reduce small- and medium-sized floods to an appreciable extent, but has an insignificant effect upon the major floods. The effective decrease in run-off is a matter of tenths of inches. Since major floods are measured in several inches of run-off, it will be appreciated that for the purpose of the present discussion, land management as a means of flood control, will not be further pursued.

The choice between small reservoirs in the head waters and large reservoirs on the major stream channels has been subject to much controversy and mis- understanding. This subject will be discussed at the end of this section of reservoirs. It will be concluded that small reservoirs are a supplement rather than a substitute for large reservoirs. However, let us discuss first of all the functioning of reservoirs in general.

Detention Basin

The purpose of a flood control reservoir is to store water during times of extreme river flow conditions and to release this water when the critical flood conditions are past. The simplest form of a flood control reservoir is a so-called detention basin. The dam of a detention basin is equipped with an uncontrolled conduit and with a spillway. During normal summer flows the river discharge passes through the conduit and there is little or no water stored in the reservoir. When the river discharge increases to flood proportions, the capacity of the conduit is too small to handle all the flow and consequently some water will be stored in the reservoir. As a result the reservoir level will gradually begin to rise. Consequently the head on the conduit increases and therefore the dis- charge through the conduit increases. As long as the inflow into the reservoir is larger than the outflow, the reservoir level will keep rising, until the crest of the spillway is reached. A further increase in reservoir level now means a large increase in reservoir outflow capacity, namely through the conduit and over the spillway. As long as the inflow keeps increasing the outflow will have to increase, which means an increase in reservoir level, which means that some more water is going in storage, which means that the outflow is always somewhat less than the increasing inflow. After the inflow begins to decrease, the point is soon reached where inflow equals outflow, which represents the time of maximum reservoir level. After that, the outflow is larger than the inflow until the reservoir is drained and the river flow back to normal. The above sequence of events is illustrated in *Figure 6.1* for a detention basin and for the passage of a small flood and a large flood.

It may be seen from *Figure 6.1* that the reduction of peak flow in cubic feet per second is largest for the smaller flood, in spite of the fact that during the passage of the larger flood more water went into storage. The reason is that during the larger flood a lot of storage capacity was already consumed before

211

the peak flow arrived, and consequently there was not much storage left to reduce the peak of the flood. If a still larger flood would be routed through the reservoir, it would be found that the peak reduction is still less.

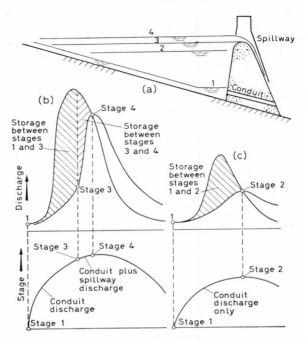

Figure 6.1. Effect of detention basin

Reservoir Control

Uncontrolled detention basins, as described in the above, have been built in the past. However, since the last few decades it has been found that the available funds and water resources can be used more efficiently when the reservoirs are controlled. This control may consist of a gate in the conduit and perhaps also a gate on the spillway crest. As a result, we may be able to use the reservoir more effectively from a viewpoint of flood control and, moreover, we may be able to use the reservoir also for water supply purposes without harming the flood control interests. In order to develop these thoughts in an orderly fashion, let us first discuss in general the areas of freedom that we have in designing and controlling a reservoir. We may distinguish the following variables: (1) Capacity of reservoir; (2) Diameter and control of conduit; (3) Length and control of spillway.

In the initial stages of our flood control study, the most economic capacity of the reservoir is an unknown quantity. In fact, we do not even know if a reservoir of any size will be economic! Let us therefore assume that we have prepared a storage capacity versus elevation curve for the dam site and that we have arbitrarily chosen one or a few reservoir capacities for preliminary investigation. We should qualify this last statement a little more precisely. We

212

can, namely, select arbitrarily either a full supply level, or a spillway crest level, or a maximum flood level, or the elevation of the top of the dam. Let us assume for the present discussion that we have selected the maximum flood level. This level corresponds to a certain capacity on the storage–elevation curve. This figure is often quoted as the storage capacity of a flood control reservoir.

Having established arbitrarily, and for the time being, the capacity of the reservoir, let us turn to the next variable: the control and diameter of the conduits. In *Figure 6.2(a)* is repeated the outflow hydrograph of the detention basin of *Figure 6.1(b)*. It may be seen that, although the storage capacity is substantial, the peak flow reduction is not very significant. The situation could be improved if more water was discharged from the reservoir during the

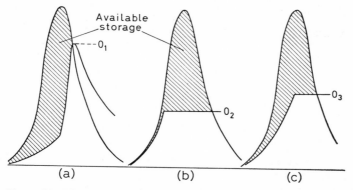

Figure 6.2. Effect of conduit size: (a) small conduit, uncontrolled; (b) very large conduit, controlled; (c) large conduit, controlled

early part of the flood, so that more storage capacity would be available during the peak of the flood when it is needed. Let us assume for a moment that we have an extremely large conduit. We could then simply let the river run through the conduit during the early part of the flood without filling up the reservoir, as shown in *Figure 6.2(b)*. When the river has reached a discharge O_2, we start closing the gate of the conduit. The reservoir now begins to fill. We keep adjusting the gate on the conduit, so that its outflow remains constant at O_2. The reservoir will have reached its maximum storage level when the recession curve of the flood hydrograph reaches a magnitude of O_2. It will be appreciated, by comparing *Figures 6.2(a)* and *6.2(b)* that with the same available storage capacity the second method of operation results in much lower reservoir outflows and therefore greater flood control benefits.

It is very likely uneconomical to build conduits as large as was assumed in *Figure 6.2(b)*. This was only done for the purpose of discussion. Instead, one should try to determine such a conduit size that an increment in size would result in a greater increment in cost than in flood control benefits, while a reduction in size would result in a greater increment in flood damages than a saving of costs. The technique of conducting such computations will be discussed in Chapter 11: Economic Analysis. A probable operation of the reservoir for such conditions is shown in *Figure 6.2(c)*. It may be seen that there is still a substantial reduction in flood peak as compared to the situation in *Figure 6.2(a)*.

Now that we have fixed the capacity of the reservoir and the diameter of the conduit, the only remaining variable is the spillway. We shall assume that from a hydrologic study of the drainage basin, the spillway design flood has been determined. In the hydraulic design of the spillway there are actually two areas of freedom: first of all the length of the spillway; secondly, the spillway can be gated or ungated. The design of an ungated spillway of given length would take place somewhat as follows. The spillway design flood is routed through the reservoir and the maximum reservoir level and spillway discharge are found. The maximum reservoir level is shown as stage 1 in *Figure 6.3(a)*. The maximum spillway flow forms the basis of the design of the stilling basin.

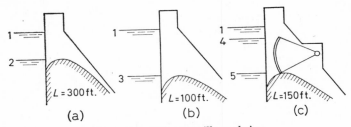

Figure 6.3. Different spillway designs

Maintaining the same maximum reservoir level at stage 1, we could also design a spillway of smaller length. In order to pass the same spillway design flood without exceeding stage 1, it is obviously necessary to lower the crest of the spillway, as is shown in *Figure 6.3(b)*.

The advantage of the design in *Figure 6.3(b)* over the design in *Figure 6.3(a)* is mainly its lower cost. In case we are dealing with an earth dam, the saving of a narrower spillway may be substantial. When dealing with a concrete gravity dam the saving would be relatively small since we need a gravity section in any case. The disadvantage is in its lesser effectiveness of peak reduction of medium and small floods. Such floods reach the spillway crest in *Figure 6.3(b)* earlier than they do in *Figure 6.3(a)*, and therefore they will not attain as high a reservoir elevation. A lower reservoir elevation means less flood water gone into storage, which in turn means a larger outflow.

This disadvantage of the lower spillway crest can be eliminated when the spillway is equipped with a gate, as shown in *Figure 6.3(c)*. The small and medium floods can now be stored more effectively in the reservoir, and the outflow would be at least as low as in *Figure 6.3(a)*. In fact, with a good flood forecasting system, the conduit gates and spillway gates can be manipulated in such a way that the small and medium floods are stored up to stage 4 in *Figure 6.3(c)*. Assuming that this elevation is higher than the maximum elevation that those same floods would have attained in *Figure 6.3(a)*, this would in effect mean a further reduction of flood peaks. Since flood forecasting is not always reliable, the width of the spillway in *Figure 6.3(c)* should be greater than in *Figure 6.3(b)*, so that when the flood turns out to be greater than expected the gates can be opened, and the reservoir levels maintained below stage 1.

A gated spillway would yield further advantages, when the reservoir is part of a flood control system, including other reservoirs. Many situations are possible whereby it would be advantageous to completely stop the outflow from certain reservoirs, if only for a few days, in order to reduce the peak flows in critical areas of the drainage basin. Assuming that the reservoir has already risen above the spillway crest, such a manipulation would only be possible when the conduits and the spillway are both equipped with gates.

Reservoir Design

It was noted in the foregoing discussion that there are several degrees of freedom in the design of flood control reservoirs. First, the capacity of the reservoir, which is associated with the height of the dam. Second, the diameter of the conduits. Third, the length of the spillway, with or without gates. To arrive at the most economical final design, the following procedure may be followed. A number of different reservoir capacities is selected. For every capacity, the most desirable size of conduit and spillway is tentatively estimated. The benefits of the different capacities are compared to their costs, and the largest one, showing a positive benefit–cost increment over the foregoing one, is selected. For this particular reservoir capacity, the size of conduits and spillway is now determined in more detail, until a satisfactory design has been attained.

Instead of arbitrarily selecting different reservoir capacities, to perform the foregoing analysis, a somewhat more sophisticated procedure, in regard to the subsequent benefit–cost analysis, would be to choose a series of river floods with even intervals of frequency of exceedence (for instance, 0.5, 1.0, 1.5, 2.0 per cent, etc.) and to design reservoirs that can cope with each of these floods. Each of these floods becomes then the reservoir design flood for its corresponding reservoir.

The reservoir design flood, which may be defined as the largest flood for which full control is provided, should be sharply distinguished from the spillway design flood. The latter may be defined as the maximum flood that can be passed through the reservoir without endangering the dam. Should the reservoir design flood be exceeded, some damage of downstream interests may be expected. Should the spillway design flood be exceeded, failure of the dam, which is infinitely worse, may be expected. It is not unusual that the frequency of exceedence of the reservoir design flood is in the order of magnitude of 1 per cent per year, whereas the frequency of exceedence of the spillway design flood may have the order of magnitude of 0.01 per cent per year.

Let us assume now that we have selected a reservoir design flood and that a storage–elevation curve of the reservoir site is available, as shown in *Figure 6.4*. How do we go about designing the size of reservoir, conduit and spillway?

First we plot the reservoir design flood in the form of a hydrograph and show the maximum discharge that can be released from the reservoir without harming downstream interests, as is shown in *Figure 6.4(a)*. This horizontal line is arbitrarily connected with the zero point on the hydrograph as shown. The shaded area in *Figure 6.4(a)* now represents the total amount of flood water that has to be stored in the reservoir. Going to *Figure 6.4(b)*, it is found that in order to store S_1 we need a reservoir level of E_1. The dam is therefore designed with its spillway crest elevation at E_1, as is shown in *Figure 6.4(c)*. A

215

conduit is designed at the base of the dam with such dimensions that the downstream bankfull discharge can be released while the reservoir is about 25 per cent full of capacity. A spillway width is tentatively assumed, and now an elevation–discharge curve can be prepared, as shown in *Figure 6.4(d)*. With this discharge curve and the storage curve available, the spillway design flood is routed through the reservoir and the maximum reservoir level is found. To this level is added the required freeboard, and thus the top of the dam is found.

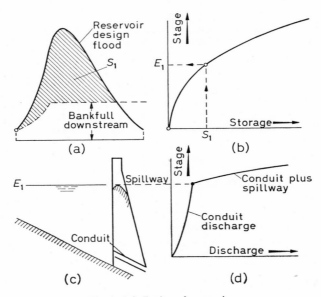

Figure 6.4. Design of reservoir

The results of this first trial may not be quite satisfactory. It is not unlikely that small changes in conduit size, spillway crest elevation and spillway width will result in substantial savings in cost without reducing the effectiveness of the reservoir. In this fashion several possibilities are explored, until the most satisfactory design has been found.

Multi-purpose Reservoirs

It is not very often that the construction of a large reservoir is justified on the basis of flood benefits alone. In most cases flood control is only one of the functions that a reservoir has to perform in addition to such other functions as power development and water supply. When the reservoir serves more than one purpose the operation of the reservoir becomes more complicated.

Let us discuss first the combination of flood control and water conservation for downstream interests. The reservoir content is allocated to these two purposes, as shown in *Figure 6.5(a)*. During the normal river discharges the reservoir is maintained at full supply level, as shown. Roughly speaking, the reservoir content above F.S.L. is available for flood control while the reservoir

216

content below F.S.L. is available for water conservation. However, with an adequate flow forecasting system, all or part of the reservoir storage below F.S.L. can also be made available for flood control. As soon as flood flows are predicted, the F.S.L. can be drawn down to such an extent that the

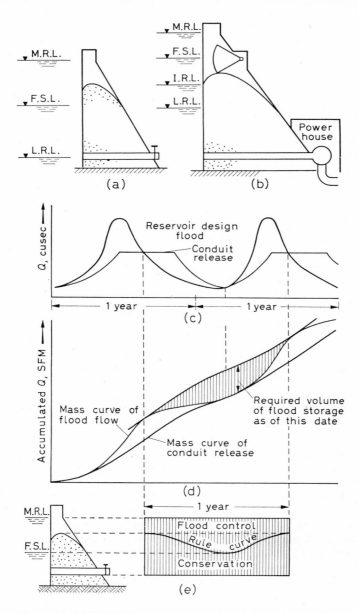

Figure 6.5. (a),(b) Multi-purpose reservoirs; (c),(d),(e) Flood control rule curve

217

minimum predicted run-off will bring the reservoir level back to F.S.L. If the floods are caused by snow melt, the volume of run-off may be forecast well in advance of the flood, so that there is ample time to empty the whole reservoir, if needed. Since the only two purposes of the reservoir are flood control and downstream water supply, there is no objection to emptying the reservoir. Hence the lowest reservoir level coincides with the invert of the conduit. The volume of water available for water conservation is the reservoir storage between the lowest reservoir level and the full supply level. With this storage volume and some recorded low periods, it is possible to determine the maximum dependable flow that can be released from the reservoir. Whenever the inflow of the reservoir is larger than this figure (but below flood proportions) the F.S.L. is maintained. Whenever the inflow to the reservoir is less than the maximum dependable flow, water is released from storage to make up the deficit.

If the floods in the drainage basin occur at a certain time of the year, such as the spring and early summer floods on the prairies and in the mountains, or the monsoon floods in the tropics, it might be possible to extend the use of the reservoir for conservation purposes to above the F.S.L. line shown in *Figure 6.5(a)*. If we plot a mass curve of a typical reservoir design flood, as shown in *Figure 6.5(d)*, and we plot also the mass curve of the release of water from the conduits, then the ordinates between the two curves indicate the amount of water that has gone into storage. If we lift up the mass curve of conduit release until it is above and tangent to the mass curve of the flood flows, we obtain the ordinates that indicate the required volume of flood storage as of any date. If we plot these ordinates, converted to the proper scale, in *Figure 6.5(e)*, and down from the spillway crest level, we obtain a rule curve for flood control operation that enables us to have, at all times of the year, adequate storage facility for the reservoir design flood of *Figure 6.5(c)*. It is likely that the reservoir design flood may occur somewhat sooner or later in the year, and that it might have different shapes. We would then repeat the above procedure and produce say half a dozen different curves in *Figure 6.5(e)*. We would then draw an enveloping curve below these different curves and regard it as the rule curve. The obvious advantage of the rule curve is that without harming the flood control interests, we enhance the operation of the reservoir for conservation purposes.

Operation of the reservoir with the rule curve would now be as follows. During normal flow conditions, the reservoir level is kept at the rule curve elevation as of any date of the year. During small floods, the level will rise above the rule curve. If exceptional floods occur, which can probably be predicted some time in advance, the reservoir is drawn down below the rule curve in advance of the flood, as discussed before. If low flow conditions prevail in the drainage basin, the reservoir is drawn down, below the rule curve, to release the dependable flow to downstream interests. If the driest flow period on record would be repeated, the reservoir would be completely emptied.

When the reservoir serves three or more purposes, including power, the situation becomes somewhat more complicated. Let us assume that the reservoir content is allocated to flood control, water power, and water conservation, as shown in *Figure 6.5(b)*. During normal flows the reservoir will be maintained at F.S.L., and all the water released from the reservoir will pass through the turbines. It is obvious that the F.S.L. cannot be maintained precisely since the inflow to the reservoir will not exactly match the energy demand on the power plant. However, let us assume that the power plant is part of a large power system so that its energy production can be readily absorbed, so that the reservoir level can normally be maintained in the vicinity of F.S.L. Whenever a flood is forecast, the reservoir is drawn down to the extent that the minimum forecast run-off will bring the reservoir back to F.S.L., allowing for the maximum amount of energy to be produced by the power plant during this whole operation. In other words there is a certain amount of reservoir storage, between the full supply level and some intermediate reservoir level, available for flood control and multi-purpose regulation. The reservoir storage above F.S.L. serves the purpose of flood control only. It may be noted that the reservoir spillway is equipped with a gate. This gate serves the purpose of conserving instead of spilling the small and medium floods that bring the reservoir level above F.S.L. The sill of the gate is well below F.S.L. in order to reduce the total length of the spillway for the passage of the spillway design flood.

During times of low river flows, the reservoir level will soon reach I.R.L. due to the power releases being larger than the inflow to the reservoir. However, when I.R.L. is reached, the release from the reservoir for power and water supply is reduced to its minimum value, namely the maximum dependable flow. This flow will have to maintain the firm capacity of the power plant and will have to satisfy minimum downstream water supply requirements. The maximum dependable flow is based on recorded low flow periods and the reservoir storage between I.R.L. and L.R.L. It should be noted that L.R.L. is well above the intake of the conduits. This is necessary in order to maintain a head on the power plant. If the reservoir were drawn down to the inlet of the conduits, the power plant could produce no power and would then lose its capacity benefits. It is not very likely that such a substantial loss in power benefits would be compensated by a gain in water supply or other benefits.

Small Dams Versus Big Dams

During the last decade some controversy has been aroused over the advantages of small dams on the headwaters of the river systems compared to large dams on the main stem of the river. The proponents of small-dam flood control claim that small reservoirs provide more efficient control, replenish the ground water table, provide local stock watering facilities, and serve recreational purposes. Granted the last three advantages let us investigate the claim of greater efficiency in controlling floods.

Figure 6.6(a) shows a hypothetical drainage basin, where the choice is between five small dams A or one large dam B, in order to provide flood protection for the city C. Let us assume that reservoir B has an effective flood storage capacity of 1,000,000 acre-ft., and that the five small reservoirs A also have a total storage capacity of 1,000,000 acre-ft. but control only 50 per cent of the drainage basin. The effect of the large reservoir B is shown qualitatively in *Figure 6.6(b)*, which resembles diagrams that have been discussed earlier. The effect of the small reservoirs A_1, A_2, A_3, A_4, and A_5 is shown in *Figure 6.6(c)*. Assuming that the snow melt or rainfall takes place at the same time

219

over the entire drainage basin, it may be expected that the floods from the various tributaries arrive at C in succession rather than at the same time. Hence the storage effect of one small reservoir will follow or precede the effect of another small reservoir and the cumulative result will be a spreading out of the storage over the entire hydrograph with the result that the peak of the flood at C is reduced to a lesser extent than with the one large reservoir at B.

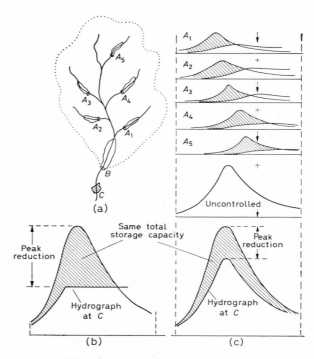

Figure 6.6. Small dams versus big dams

In addition to the above, it may be noted that there is more time lag between the rainfall (or snow melt) and the arrival of the flood at reservoir B than at the reservoir A—hence there is more time available for forecasting and hence more opportunity to manipulate reservoir B in a more efficient manner. This becomes particularly important when the reservoirs A or B are part of a larger flood control system. For example, reservoir B could reduce the river flow to zero for a short duration, if needed. The small reservoirs, controlling only 50 per cent of the drainage basin, could not do this.

From a viewpoint of cost, it may be noted that the average depth of small reservoirs will be smaller than the average depth of a large reservoir. Therefore, the small reservoirs will have a larger total surface area, and will flood more land. In areas where reservoir evaporation is a problem, more losses may be expected with the smaller reservoirs. Since the cost of a dam increases roughly with the second power of its height and the volume of the reservoir with the third power of the height of the dam, it may be expected that the cost of storing water in small reservoirs is greater per unit of volume than in large reservoirs. Even more so, because the river gradient at the site of a small head-water reservoir is normally steeper than at the site of a large main-stem reservoir; the steeper gradient calling for a higher and costlier dam to store a given amount of water.

The sum total of the above considerations is that small head-water reservoirs are several times as expensive for storing water than large main-stem reservoirs, and moreover they are less effective in controlling floods on the main stem. Nevertheless they may perform a useful function inasmuch as they provide local benefits. It may therefore be concluded that small reservoirs should be considered to supplement rather than replace the functions of large reservoirs, and that their construction should be justified largely by the local benefits derived.

DIKES

The oldest, most common, and often most economical means of flood protection is by constructing a system of dikes, leaving the river more or less in its existing course, and leaving the flood levels more or less in their natural condition. This chapter will discuss successively: (1) The danger of dikes; (2) The safety requirements; (3) The composition, slope and location of dikes; and (4) The required drainage works behind dikes.

Danger of Dikes

The great hazard involved in a diking system is that it provides full protection up to a certain flood stage and no protection at all for higher stages. A long period of absence of extreme flood stages will create a feeling of security for inhabitants of dike-protected areas, leaving them unprepared for an eventual failure of the dike. In order to illustrate this hazardous aspect of dikes, some historical accounts will be quoted of notorious dike failures.

In the early morning of Sunday, 1st February, 1953, Holland was hit by the most severe storm since the Middle Ages. The sea level rose one to two feet higher than the highest stage on record, and winds up to one hundred miles per hour, aiming at the sea coast, accompanied the surging tide. During the first part of the fatal night there was little concern among the population. A terrific storm was blowing, true enough, but terrific storms had been experienced before. The meteorological service had issued a warning for 'dangerous high water', true enough, but it had done so before and, so far as could be remembered, nothing had ever happened. Everybody felt safe and comfortable behind those big age-old dikes.

Even before the sea reached its highest level, the waves washed over the top of the dikes, sometimes to a depth of one to two feet. At these places the toe of the dike at the land side became thoroughly saturated, was damaged by the force of the down-rushing water, and gave way. Sections of dike, hundreds of feet long, failed almost instantly. The sea, being 10–15 ft. higher than the land behind the dikes, entered through the openings with a terrific flood wave. Houses close by were swept away and the polders (reclamation units) were quickly filled with water. The following is an eye-witness account, given by a doctor, of what happened at one of the many places where dikes failed:

'We went to bed without expecting anything unusual, although the noise of the wind was deafening. The tiles on the roof rattled constantly, and we heard nothing of the alarm bells in the church or the constant ringing of the telephone from four to five in the morning. After five my daughter, intending to go downstairs, discovered that water was standing half-way to the first floor. When daylight came we saw water as far as the eye could reach. We had no food and no drinking-water and stayed in our bedrooms. Looking out of the window, we saw other houses in our neighborhood. Some were rocking like ships on the waves, and occasionally one crumbled to pieces. People fled from one roof to another, or were drowned. A mother and two children came floating by upon a part of a roof. The van Hoogstraten family were hanging on to the eaves of the roof of their house that was still standing up—man, wife and four children. After an hour they dropped, one by one, and were swallowed by the water. Huge amounts of debris came slowly floating by, intermingled with drowned cattle, horses, sometimes a man or a child. The next day we were half crazy with thirst, cold, hunger and the horrible sights from our window. During the morning we were rescued by boat. On the way to higher ground we passed a barn with 14 people on top of the roof. The whole structure was swaying in the waves because the storm was still blowing with full force, accompanied by snow and hail. The people shouted to us: "Come back next trip to pick us up, we will manage for another half-hour." However, when the boat returned there was no barn and everybody had disappeared.'

In cases where the polder was large and the breaches in the dike relatively small, the water came up slowly, but the ultimate effect was in many cases the same. In general, people had the tendency to stay in their houses for a while, and after the roads were covered with water, they could not leave. Then the water got higher and higher and they fled to the roofs. Many walls could not stand the continuous pounding of the waves and then buildings completely disintegrated, affording no means of survival.

The toll of the disaster was 1,800 dead, 300,000 acres of agricultural land were flooded, and the total damage was estimated at one billion dollars.

During the spring of 1927 heavy rains in the lower reaches of the Missouri, Upper Mississippi and Ohio Rivers, brought all three streams into flood simultaneously, producing unprecedented flood conditions along the Lower Mississippi River. It took the flood wave almost two months to travel the length of the stream. On its way south it breached the existing levee system in 54 places, thus inundating 4,000,000 acres of cropland. It drove 700,000 people from their homes, and 246 lives were lost. Sometimes dike failures occurred close to villages, which were then quickly submerged to a depth of 10–15 ft. of water. Many homes were swept away and those that remained were sometimes buried in sand to a depth of 10 ft. In Arkansas City two dikes failed. As a result, the flood waters attained a height of 20 ft., sweeping from the foundations practically all the frame structures, with great damage inflicted upon the more substantial buildings. That there was not greater loss of life was due to prompt and efficient work of the Red Cross. This organization operated a rescue fleet including about 1,000 power boats. Marooned refugees would be immediately rescued before the rising water could drown them. A single mistake, or miscalculation as to the speed of the flood wave, could have meant the loss of thousands of lives, for when a levee broke the boats had to be there to save the people in the path of the water.

During the latter part of May, 1948, the Columbia Basin was ravaged by one of the severest floods on record. The greatest single disaster occurred at the community of Vanport at Portland, Oregon. This town is built on the flood plain of the Columbia River and protected by a system of dikes. One of the dikes, serving also as a railroad fill, broke on Sunday afternoon, 30th May. A wall of water poured through the breach with such rapidity that few of the 18,700 residents escaped with more than the clothing they were wearing. Within an hour the town was completely destroyed and 16 people had lost their lives. Because this area was surrounded by dikes, it became a placid lake after it filled with water. Houses that had not been demolished by the first onrushing wave floated in 10–15 ft. of water. It was a fortunate circumstance that the disaster occurred on a Sunday afternoon, the middle of a three-day holiday, when many people were absent or were using the holiday to move out of danger. It has been estimated that loss of life might have mounted to thousands had the disaster occurred at night.

The widespread floods of March, 1947, in Great Britain were unprecedented in their severity and duration in that country. The personal distress they caused was worse than anything recorded of similar events in the past. Nearly all the main rivers of the south, the midlands and the north-east of England were swollen into deep flood simultaneously, and persisted in flood for a great length of time. Included in the worst hit areas were the fertile lands reclaimed from fens and marshes, which are among the most valuable in England. Along the Trent one of the dikes gave way at 2 a.m., Sunday, 23rd March. A wide hole had been torn and through it the water was racing in a swift torrent out into the fenland beyond. The flood waters had a vast basin to fill—a long, low valley of farmsteads lying between the flood bank of the river on the one side and higher ground on the other side. Within a day or two the valley was a lake. Some farmers were still living in their upper storeys, others had escaped to higher ground. The floods were accompanied by gales with wind gusts up to 100 m.p.h. The waves battered against the inundated buildings and tore large holes through them, and swept away furniture, stacks, crops, out-buildings, all in one wild torrent of debris.

Many more cases could be cited, in Germany, Italy, China and other places all over the world, where the failure of dikes caused widespread disaster. All these examples have one thing in common: the complete confidence people had in the safety of the dikes. In most of the cases it is reported that up to the last minute people just would not believe the dikes could fail. They had stood there for as long as memory went back. They were built and maintained by government agencies and they were, in the eyes of the common man, a symbol of reliability. As a result he built his home behind the dike, whole villages were built right behind the dike, and so great was the confidence that in all but a few cases, emergency precautions were neglected.

This unwarranted feeling of safety is the greatest danger connected with the construction of dikes, because it actually increases the potential hazards of a diking system. Dikes are never built to provide protection from maximum possible water elevations, and hence there always remains the possibility that a dike will fail.

Safety Requirements

In view of these examples and utilizing hindsight it is evident that more thought should have been given to the aspects of safety before the dikes were constructed. Since that was not done, the present generation is paying dearly for the neglect. It is imperative to conclude that careful thinking is required in determining the elevation of a new diking system, or in reviewing an existing diking system.

Until a few decades ago it was standard practice all over the world to determine the elevation of a dike on the basis of the highest flood stage on record, and to this was added a certain amount of freeboard as a safety factor. This freeboard was usually about 3 ft. for river dikes and depended on the magnitude of the wave uprush for sea dikes. It is not uncommon for sea dikes that during severe storms the tip of an uprushing wave just reaches the top of the dike; nevertheless, these dikes still have a factor of safety because they are not really endangered until the waves go freely over the top.

The main reason for adding a certain amount of freeboard to the maximum recorded stage is to provide for the possibility that a still higher flood stage may occur at some time in the future. Since the application of probability theories to engineering problems is of fairly recent origin, it is no wonder that in early days practically no attempts were made to calculate the probability of occurrence of higher flood stages, and the choice of freeboard became more or less an intelligent guess.

A different attitude toward this problem has evolved during the last few decades, particularly in the United States, where it was instigated by the disasters of the Mississippi and the New England floods. Serious attempts are being made to determine hypothetical extreme flood stages at places where dikes are needed. In the design of river dikes extreme storms over the drainage basin and high coefficients of run-off are assumed in order to determine the run-off from the land into the river system. The resultant flood is routed down the river and stages are determined at desired points.

From a viewpoint of the security of people behind the dikes, there is no doubt that this approach is a vast improvement over the former method of merely guessing an extreme stage; however, the danger now exists of over-estimating the amount of protection which should be provided for the people. In determining hypothetical extreme floods it is left to the good judgement of the engineer to choose the size and location of the rainstorm and to calculate the proper run-off. As noted earlier, it is not possible to determine the probability of exceedence of such extreme conditions. Hence it can be expected that the extreme stage computed by one engineer may differ widely with one computed for the same case by another engineer, and since neither of the two reveals the actual degree of protection, a choice between them is difficult to make. It is true that a high degree of security is desirable, but there is a point at which a further increase in safety is not warranted with respect to the consequent increase in cost.

In essence, the problem of establishing water level criteria for a dike is nearly the same as determining a spillway design flood, which was discussed in the last chapter. However, there is a small but significant difference. The incremental cost of extra protection is relatively small for most dams and spill-ways, whereas it is relatively large for most diking systems. As a result, many spillway design floods are in the order of magnitude of the maximum possible flood or have return periods of some 10,000 years. The engineer can afford to stay on the safe side because it will only add a few per cent to the capital cost, which will not make the entire project uneconomic. Most diking systems, on the other hand, have been designed for flood conditions that have a return period of a few hundred years. Notable exceptions are the dikes in Holland and along the Mississippi River that are now redesigned for flood conditions with a return period in the order of 1,000 years. The engineer who has to design a diking system cannot afford to casually throw in a few more safety measures, because in doing so he may double or triple the cost of the whole project and make it uneconomical. Therefore, in the design of dikes, we must explore with greater prudence what safety criteria should be established.

It is important that we try to separate the policy decision on the degree of safety and the engineering problem of finding the corresponding flood condition, as was pointed out earlier. All people protected by a dike, and certainly their representatives, have a right to decide how much they want to spend on their security of living. Hence, it is the task of the engineer to carefully analyse the problem, to present it in an understandable form to the people or their representatives, and to request a decision on required degree of safety. Once this has been obtained, the engineer can go ahead and determine the corresponding flood conditions, and after that the corresponding elevation of the top of the dike. In these computations we must make due allowance for: (1) extremely high water levels that are caused by river floods or by sea tides, or by a combination of both; (2) wind set-up; (3) wave uprush; (4) possible deterioration of the top of the dike due to frost action or other causes; (5) expected settlement of the dike after construction.

Let us first discuss the policy decision on safety requirements. It will be recalled from the discussion on spillway design floods that problems of safety may be considered from three different viewpoints: engineering, economic, and social. It was suggested that an important engineering structure, such as a flood control or reclamation dike, should not fail under circumstances that have a probability of occurrence of greater than 0·1 per cent per year. Economic considerations may justify an increase in safety requirements; they may not be used to justify a decrease in safety. Social considerations should allow for the possible loss of human life, and may justify an increase in safety requirements.

In order to demonstrate the application of the above principles, the following example is given. In the delta area of a large river, reclamation of 1,000,000 acres of potential agricultural land is contemplated. The problem is to determine the most desirable top elevation of the diking system around the area to be reclaimed.

It is concluded from an appraisal of the situation that a dike failure would cause a major disaster; hence, the diking system is classified as a major engineering structure. Based on engineering considerations, it is decided that the probability of the dikes being overtopped should be less than 0·1 per cent per year. This figure cannot be increased because of other considerations and can only be decreased if other viewpoints prevail over the engineering viewpoint.

In order to appraise the situation from an economic viewpoint, estimates are made of the construction cost of the dikes and of the frequency of river stages, so that a means of comparing the capitalized average yearly flood losses with the capital construction cost of the dikes at various elevations of the river may be obtained. The result of these estimates are assembled in Table 6.1.

Table 6.1. *Economic Analysis of Dike Height**

Elevation Dike Above Bank (ft.)	Chance of Overtopping (% per yr.)	Annual Cost of Dikes ($)	Average Annual Damage ($)	Total Annual Cost ($)
4	5·00	100,000	1,250,000	1,350,000
6	1·00	200,000	250,000	450,000
8	0·50	300,000	125,000	425,000
10	0·10	500,000	25,000	525,000
15	0·01	1,000,000	2,500	1,002,500
20	0·001	2,000,000	250	2,000,250

* The figures in the columns of the table represent the following:
First column: The elevation of the dike, in feet, above the natural river bank.
Second column: The probability in per cent per year that the dikes are overtopped.
Third column: The estimated annual cost of the whole diking system.
Fourth column: It is estimated that the flood losses due to dike failure will amount to $25.00 per acre, or $25,000,000 in total. This figure is multiplied by the chance of overtopping, to yield the average annual damage.
Fifth column: Summation of the figures from the third and fourth columns.

The figures in Table 6.1 show that the most desirable dike elevation, from an economical point of view, would be 8 ft. above the river bank and would have a chance of being overtopped by 0·5 per cent per year. Since this is more than the 0·1 per cent per year based on engineering considerations, it can be disregarded as guidance for determining the elevation of the dikes.

In order to appraise the situation from a social viewpoint, the following estimates are made. The reclamation area will ultimately contain 200,000 people, who will live mostly on farms, but also in villages and towns throughout the region. A dike failure would cause the whole reclamation area to be covered with water from 5–15 ft. deep during a period of several weeks. Since there is no high ground in the area, it is assumed the death toll will be considerable. Table 6.2 is prepared in order to compare the incremental annual cost of the dikes with the safety provided.

Since the dikes will be built to an elevation of 10 ft. above the river bank in any case, on account of the previously discussed engineering considerations, the present problem is to decide whether or not additional expenditures for raising the dike still further would be justified in order to save the population from floods and the possibility of death.

It follows from Table 6.2 that in a range of 10–15 ft. the community would have to pay $2,800 per person, in order to save themselves from the experience of a flood. Would this expenditure be justified?

For the sake of the present discussion let us answer this question in the affirmative. However, the cost of a further increase in safety becomes prohibitive. For the present case, beyond the chance of an occurrence of a flood of 0·01 per cent per year, the cost of flood relief per capita is $55,000 per person per flood. Hence, it is tentatively concluded that the most desirable dike elevation will be 15 ft. above the river bank.

It must be pointed out that the economic and social aspects should be considered in conjunction with each other and not separately, since they are two different justifications for the same expenditure. This was not done in the present example for the sake of simplicity.

Table 6.2. *Social Analysis of Dike Height**

Elevation Dike Above Bank (ft.)	Chance of Overtopping (% per yr.)	Annual Cost of Dikes ($)	Annual Suffering (Capita)	Relief Cost Per Capita ($)
4	5·00	100,000	10,000	
6	1·00	200,000	2,000	12
8	0·50	300,000	1,000	100
10	0·10	500,000	200	250
15	0·01	1,000,000	20	2,800
20	0·001	2,000,000	2	55,000

* The figures in the table represent the following:
First–third columns: Same as in Table 6.1.
Fourth column: The average annual human suffering due to dike failures. These figures are obtained by multiplying the total number of the population of 200,000 by the yearly chance of dike failure.
Fifth column: The cost of flood relief per capita per flood for each interval. These figures are obtained by dividing the difference of two successive items of column three by the difference of the corresponding successive items of column four. For instance, when the dike is raised from 4–6 ft. above the river bank, the annual cost of the dike increases by $100,000 and the annual suffering decreases by 8,000 persons. This means a cost of $12 per person per flood.

The trend of thought in the preceding section can be summarized briefly as follows. The method of determining a dike elevation on the basis of the highest flood stage on record, and a more or less arbitrarily estimated freeboard, cannot properly take into account the flood potential of the sea or river which face the dike, nor can it evaluate the property and human life protected by the dike. For this reason it is highly unsatisfactory.

The method of determining a dike elevation on the basis of a 'maximum probable flood' or a 'standard project flood', is a great improvement, but leaves considerable room for different interpretation. It was pointed out that the application of this principle to the design of high dams is less sensitive to different interpretation because a considerable variation in estimated magnitude of the maximum probable flood may result in only a relatively small variation in the total cost of the dam. The cost of a diking system, however, increases rapidly with an increase in the design flood stage.

It would be more desirable to follow a method which would take into account the flood probability, the protected property and population, and the cost of the diking system, and which would evaluate these items in quantitative terms so that a deliberate choice can be made. The application of such a method has been attempted.

The requirement that an important dike should under no circumstances have a greater chance of being overtopped than 0·1 per cent per year, was chosen arbitrarily and is open to discussion. The requirement that a dike should have a minimum elevation based on economic considerations follows

227

established methods of computation. The requirement that a certain degree of protection should be based on social considerations is a universally accepted fact, but the method of evaluating it is somewhat new and also open to discussion.

If this method is to have any value it is essential that flood stages with very small frequencies of occurrence can be determined. To do this is not so much a matter of principle as a matter of technical ability, but it must be emphasized nevertheless that it is a prerequisite for the successful application of the method.

There is no doubt that the present methods of extrapolating flood frequency curves into regions of 0·1 and 0·01 per cent chance of exceedence per year are rather inaccurate, even when ample data are available. There is no reason, however, to believe that these methods cannot be improved upon, to the extent that they can provide results sufficiently accurate for application of the above principles. In appraising this aspect, it should be kept in mind that using the alternative design criterion of a 'standard project flood' does not reveal at all in what class of frequency the design flood belongs.

Even if it is assumed for a certain case that the frequency of extreme flood stages cannot be determined with sufficient accuracy, it may still be profitable to follow the above method, or one similar to it, in order to obtain insight into the relative importance of the various aspects of the problem. At least it can be found whether or not social and economic considerations prevail over engineering considerations and whether these considerations may lead to dikes which may be overtopped near the order of magnitude of 0·1 per cent per year, 0·01 per cent per year, or still smaller. Such insight will very likely prove beneficial even when a dike elevation is determined on account of other principles.

After having established the desired degree of protection, the next problem is to find the corresponding flood conditions. It will be recalled that the top of the dike is obtained by adding the freeboard to the design flood stage. The freeboard consists primarily of wind set-up and wave uprush. The design flood stage is the maximum water level resulting from the design flood, not allowing for any wind effect. The question now arises how probability analysis should be applied to the freeboard and the design flood stage.

Let us first discuss two relatively simple cases. Suppose the design flood level is for practical purposes a fixed elevation. For instance a reservoir or large lake that is controlled at one and the same elevation, regardless of the flow through the lake. In that case the probability analysis should be applied only to the freeboard, and since the freeboard is a function of the wind, this means a probability analysis of the winds, coming from the direction that causes the maximum set-up plus uprush against the dike. If the desired safety of the dike is a chance of failure less than 0·1 per cent per year, this would require the selection of a wind from the most unfavourable direction with a chance of exceedence of 0·1 per cent per year. With this wind velocity, the corresponding wind set-up and wave uprush are computed, as discussed in an earlier section. The total value is added to the fixed lake level, and the top of the dike is thus determined.

The second simple case would be a situation wherein wind set-up and wave uprush are negligible, such as in a narrow river channel with a dense growth of bush in front of the dikes. In that case the probability analysis should be applied only to the river discharge. If the desired safety of the dike is a chance of failure less than 0·1 per cent per year, this would require the selection of a river discharge and a corresponding river stage with a chance of exceedence of 0·1 per cent per year. To this stage we may add a token freeboard of say one foot to allow for small inaccuracies in the computations or uneven settlement of the dike, but apart from that, the river stage thus found becomes the elevation of the top of the dike.

There are many borderline cases, of course, that approach the above extremes. Without going too far astray, we may apply the following reasoning. When dealing with nearly constant water levels, apply the probability analysis to the wind only and add the resultant wind set-up plus wave uprush to a design flood level that can reasonably be expected to coincide with the extreme wind. When dealing with relatively small freeboard requirements, apply the probability analysis to the river discharge only, and add an amount of free-board that corresponds to a wind that can reasonably be expected to coincide with the extreme river discharge.

In the more general case, however, where the design flood stage as well as the freeboard are subject to substantial variation, the determination of the elevation of the top of the dike, with a predetermined degree of safety, becomes more complicated. In the following paragraphs a point by point outline will be given of the steps involved to arrive at the desired answer. In order to make the discussion specific, the following situation will be assumed. A large shallow lake, shown in *Figure 6.7(a)*, is part of a river system. The level of the lake is a function of the discharge from the lake. Adjacent to the lake are fertile low-lying farm lands that need protection from occasional inundation. The acceptable probability of overtopping of the dikes has been selected at 0·1 per cent per year. The problem is to find the elevation of the top of the dike.

(1) Prepare a rating curve showing the still lake level as a function of the river discharge, as in *Figure 6.7(b)*.

(2) Prepare a frequency curve of the daily discharges, as in *Figure 6.7(c)*. The horizontal scale shows the probability of exceedence and the vertical scale the magnitude of the average daily discharge. Since we are only interested in the relatively large discharges it will suffice to plot only the highest group of points and to draw the frequency curve through these points and the average discharge plotted at the 50 per cent mark.

(3) Convert the discharge frequency curve into a stage frequency curve, as is shown in *Figure 6.7(e)*, by using the rating curve of *Figure 6.7(b)*.

(4) Prepare a frequency curve of the south-to-north component of the maximum three hour winds during each day, as in *Figure 6.7(d)*, assuming that a three-hour wind will fully develop wind set-up and wave action.

(5) Prepare a diagram as in *Figure 6.7(f)*, showing the relationship between wind from the south and the required top of the projected dike, as a function of the mean lake level. It may be noted that the required top of the dike, to prevent overtopping due to wind set-up and wave uprush, coincides with the mean lake level when the magnitude of the wind is zero.

(6) Prepare a diagram as in *Figure 6.7 (g)*, showing the probability of exceed-ence of a certain dike elevation. This may be done as follows. Select 100 representative values from the frequency curve of south winds, and 100 repre-sentative values of the frequency curve of still lake levels. Pick one wind value

229

by chance and one lake level value by chance. Apply these two values to *Figure 6.7(f)* and obtain the required top of the dike to prevent overtopping from that particular combination. Repeat this 99 times, plot the 100 top of dike values, thus obtained, on frequency paper, and draw a curve.

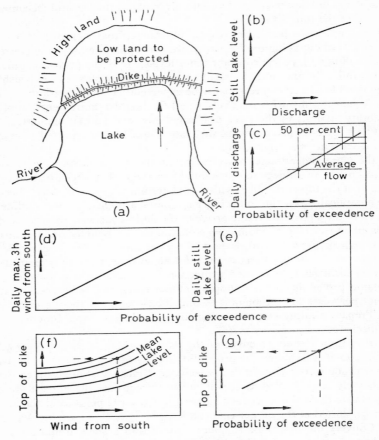

Figure 6.7. Determination of top of dike

(7) Take the selected probability of exceedence of 0·1 per cent per year, or 0·000275 per cent per day, enter *Figure 6.7(g)*, and find the corresponding elevation of the top of the dike.

Before we leave the subject of safety requirements of a dike, we should discuss how the above analysis should fit in, or can be reconciled with the over-all economic analysis of alternative flood control proposals. This issue is somewhat analogous to the dam and the spillway. We have a reservoir design flood that determines the capacity of the reservoir. This choice is only based on economic considerations. We have a spillway design flood that determines the capacity of the spillway. This choice is largely based on safety requirements.

When we design a dike, and we apply economic consideration only, we find a certain height. We must now apply engineering and social considerations and if we find a greater height, those considerations should prevail, regardless of the economics. This reasoning applies to all dikes: flood control, as well as reclamation dikes. We must realize that of the four flood control measures (reservoirs, dikes, diversions, and channel improvements) dikes have an unfortunate characteristic that the others do not have. When the design capacity of reservoirs, diversions and channel improvements is exceeded, those works keep on functioning and keep providing some degree of relief. They certainly do not make the flood situation worse than when there had been no such flood control works. Dikes, however, cease functioning when their design conditions are exceeded. After the dikes fail, the flood situation may be much worse than when there had been no dike at all. Without dikes, the flood levels might have risen so slowly that the people had a chance to escape. With dikes that fail, the disaster comes so suddenly that people are trapped. Therefore, we must apply to the design of dikes, certain safety criteria, over and above the economic analysis, like we apply spillway criteria, independent of the capacity of the reservoir.

Design of Dikes

After having determined the elevation of the top of the dike, the question arises as to how the dike should be constructed, with its given height, so that it will stand up under the design flood conditions, such as high water levels and breaking waves. The adequacy of the design of a dike should be appraised from three different viewpoints. First of all, the design of the dike should be such that failure due to seepage or piping is impossible. Second, the side slopes of the dike should be such that slope failures are impossible. Hereafter the dike slope on the sea or river side shall be called wet slope, while the slope on the land side shall be called dry slope. Third, the revetment on the wet side slope of the dike should be such that no damage to the dike will occur, even under sustained design conditions.

The composition of a dike should be, in principle, the same as that of an earth dam as discussed in Chapter 5: Hydraulic Structures. However, there are some differences. The dike may only be exposed to water during short and infrequent periods. The location of the dike may be dictated by flood protection requirements, regardless of foundation conditions. The fill for the dike, for reasons of economy, may be selected from nearby borrow pits, thus limiting the choice of materials. It may therefore be expected that established dike building practice in certain regions deviates considerably from practices in other regions. Therefore empirical methods may not be applicable to all places, and when the design engineer is faced with a new situation he is well advised to apply appropriate theoretical considerations.

It will be recalled from earlier discussions that seepage control is an important aspect in the design of dams and dikes. Seepage may be harmful for two reasons: (a) it makes the dry slope of the dam less stable, and (b) it can cause piping near the toe of the dike. Piping is the removal of earth particles by the seeping water as it comes to the surface near the toe of the dike. Piping at the exit of the flow lines is associated with an excessive upward hydraulic gradient, in the order of 0·5 to 0·8 which may be estimated from a flow net study. Piping may also begin with turbulent flow in gravel layers underneath the dike, dislodging particles in underlying or overlying strata of silts and fine sands. Once the dislodging of particles begins, channels may be formed that increase in size and length and that may finally result in the collapse of the dike. Such dike failures as a result of piping have been reported in the past, although post-mortem evidence is difficult to obtain because of the complete removal of the site where the phenomenon took place. An extensive discussion of seepage under dikes is presented in a Symposium by the ASCE (1961).

There are several means of controlling the seepage through and underneath a dike. First, it is important that the dike itself is composed partly or completely of impervious material. If ample clay or glacial till is available, the entire dike can be made of this material. In order to keep the small amount of seepage that would still come through the dike, away from the toe, one should design a horizontal filter of gravel underneath the toe. If impervious material is scarce, it is possible to make the dike of sand and design a central impervious core, or an impervious layer on the wet slope of the dike. In this last case it becomes important to select such impervious material that cracking is minimized and to place and tamp this material carefully so that no voids are left.

If the foundation of the dike is pervious or has pervious layers, measures must be taken to control seepage underneath the dike. It is not uncommon that the flood plain upon which the dike is being built has an impervious top layer of clay or silt and more pervious layers underneath. If that is the case, care should be taken not to disturb this layer on the wet side of the dike and to carefully key in the impervious core of the dike with the impervious flood plain surface. If there is no impervious top layer, it may be necessary to place an impervious blanket in front of the dike. The length of this blanket should be approximately 10 times the head on the dike. The next line of defence against seepage is a cut-off wall underneath the dike. This cut-off may consist of sheet piling or a clay trench. It should reach preferably to impervious substrata. The last line of defence is drainage or relief wells near the toe of the dike, as discussed under earth dams. If the head on the dike is high and the foundation conditions are poor, it may be necessary to apply all of the above measures.

In connection with seepage control, it is important where the borrow pits for the dike material are located. If the borrow pits are on the land side and close by, there is the danger that the seepage lines will flow towards these pits. Sand boils may occur in the pits and since these are likely filled with ground water, the boils will be difficult to detect and even more difficult to remedy. (If a sand boil is detected on the ground, a ring of sandbags is thrown up, around the boil, high enough so that it stops to flow, or at least until it stops to yield soil particles.) If the borrow pits are on the riverside and close by, they may facilitate seepage to enter the pervious layers underneath the dike. If the borrow pits are moreover in the form of a continuous deep, wide ditch, the river current will begin to flow in this ditch during high stages and thus threaten the heel of the dike by erosion. The best place for borrow pits is a good distance away from the dike, say 10 times the maximum head on the dike; to make them shallow, and if they are on the riverside, to leave plugs in between the pits so that no continuous current can flow through the pits.

232

We shall now consider the side slopes of the dike. We may observe in general that the wet slope depends largely on the severity of wave attack while the dry slope depends largely on soil stability. This depends again on the height of the dike, the type of soil, the foundation conditions, and the duration of high water levels against the dike. A relatively low sea dike, for instance, where water levels against the dike are infrequent and of a short duration, may have a dry slope of 2:1. A high river dike on the other hand, where high water against the dike may last for several weeks, may have a dry slope of 3:1. If the foundation conditions are poor, the dry slope may have to be increased substantially. In such cases a soil mechanics analysis should be made. The proper selection of the wet slope of the dike depends much on the type of wave onrush that the

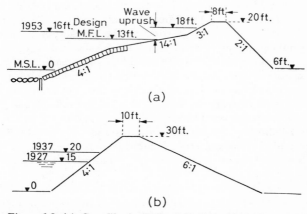

Figure 6.8. (a) Sea dike in Holland (built in 1616, did not fail in 1953); (b) Dike along Mississippi River

dike is exposed to. In fact, the slope of the dike, the top of the dike, and the revetment of the slope are interrelated. With the same type of revetment, a milder slope would reduce the wave uprush and would allow a lowering of the top of the dike. With the same slope, a more rugged type of revetment would reduce the wave uprush and would also allow a lowering of the top of the dike. It is therefore evident that in having established the top of the dike in advance, on considerations of safety, a certain slope and type of revetment must have been assumed. The wet slope of a river dike is quite often 3:1, while the wet slope of a sea dike may be as low as 10:1. It is not uncommon to find a berm incorporated in the slope of a dike. The purpose of the berm is to act as an energy dissipater by changing the direction of the onrushing water twice. Moreover, it may induce the waves to break at the site of the berm; or the berm may improve the timing of the recession of one wave so that it will cause maximum interference with the next onrushing wave. Finally, the berm may improve the stability of the slope, as was discussed in Chapter 5: Hydraulic Structures, *Figure 5.15*. It is evident that the selection of the proper slope and berm position calls for thorough experience under the given circumstances or extensive studies in a hydraulics laboratory. Two examples of sea and river dikes are given in *Figure 6.8*.

In connection with the stability of the wet and dry slopes of dikes it is of interest to review the experience with the failure of dikes in Holland during the 1953 disaster. A typical cross-section of one of the sea dikes is shown in *Figure 6.8(a)*. During the 1953 storm, the sea level came up more than two feet above the highest recorded level shown on *Figure 6.8(a)*. As a result many sections of dike failed. The sequence of failure was mostly as follows. During the initial stage of the storm the tip of some waves washed over the top of the dike. Nothing happened. During a later and higher stage of the flood the waves would break at the end of the berm, close to the top of the dike. Great quantities of water washed over the dike. The 'dry' slope of the dike became completely saturated with water. People that walked on the slope, reported that it felt like a soft, wet sponge. The grass cover stood up quite well, and no surface erosion would take place. However, at places along the dike a crack of a few inches wide and several (?) feet deep would appear at the top of the dike. Small slides on the dry slope, starting at the crack, began to take place. Water would pour in the cracks and aggravate the slides. Large masses of soil, between the top and the toe of the dike would now be steadily sliding down. In the mean time the sea side of the dike with its mild slope and its revetment would remain completely undamaged. The waves, rushing over the top of the dike, would now actively erode the soft and exposed interior of the dike, and at places breaches occurred through which the sea started pouring on to the low lying land behind the dikes. In some instances the local velocities were 10–20 ft./sec and the soil underneath the dike was so erodible that the dike breaches turned into scour holes of several hundred feet wide and 30–40 ft. deep.

It is believed that if the dry slopes of the dikes had been 3:1 or 4:1 instead of 2:1 and sometimes even steeper, many if not all of the dike failures would have been prevented. Practically all cases of failure or near failure were initiated by shallow slides on the upper part of the dry slope of the dikes. If these could have been prevented, most of the dikes would have withstood the overwash. It may therefore be concluded that all dikes should have a dry slope of at least 3:1, and that great care should be taken to maintain a solid, erosion resistant grass cover.

The selection of the revetment for a dike slope depends on various circumstances such as exposure to wave action and availability of materials. The part of the dike slope that is seldom exposed to water can be protected with a layer of clay and a well-maintained grass cover. A slope, thus protected, can resist a surprising amount of pounding by waves before any serious damage is done. The part of the dike slope that is frequently exposed to water, so that no firm grass cover can be maintained, needs artificial protection. This protection may consist of a layer of rip rap that resists the wave action and a sub-layer of gravel that prevents the soil of the dike from being washed away through the rip rap. When rip rap is costly to obtain, concrete or asphalt may be used instead. Depending on the severity of the wave action, the layer of loose rip rap may be from 1–4 ft. thick and the layer of gravel from 1–2 ft. thick. When the dike is exposed to ocean waves, the revetment may consist of the following layers. First a layer of 2–3 ft. of stiff clay; second, a layer of 1–2 ft. of gravel or bricks; and third, a layer of basalt stones, 2–3 ft. thick and carefully placed in a mosaic pattern so that they cannot be dislodged by the onrushing waves.

Before we leave the subject of the structures of dikes, it may be of interest to discuss briefly the construction of dikes in deep water, subject to tidal action. The cross-section of such a dike may look like *Figure 6.9(a)* and the sequence of construction would be as follows. The first step is to place sand from bottom-dump barges up to an elevation of about 6 ft. under water, as shown by *a–a*. After that, two clay dams *c–c* are placed by cranes that obtain the material from flush decked barges. Assuming that the left-hand side of the diagram represents the open sea, and that the right-hand side will be reclaimed, it would be advisable to give the left clay dam such proportions that it will

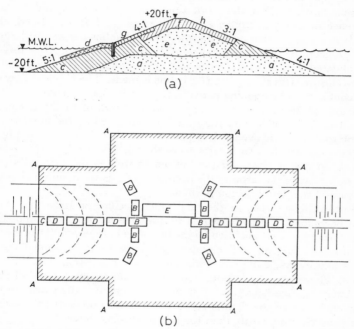

Figure 6.9. (a) Construction of dike in deep water; (b) Closure of dike in tidal bay

serve as the impervious 'core', preventing future seepage. The right clay dam, on the other hand, can have minimum proportions, as long as it serves its purpose of temporarily providing protection from damage by wave action to the core of the dike *e–e*, which may be composed of sand. The most convenient way to place this core is by the hydraulic fill method. After the core *e–e* has been placed, and trimmed off, cranes will place clay on top of the dike, as shown by *f–f*. The dike is now completed by placing the underwater revetment *d*, which may consist of willow mattresses and rip rap, or large sand–asphalt sheets. The above water revetment may consist of hand-placed basalt stones, as earlier discussed, or a continuous sand–asphalt layer. The borderline between the underwater and above water revetment may be located on a horizontal berm, and may consist of a continuous row of sheet piling to which both modes of revetment are carefully keyed in. The remainder of the dike is sodded and seeded to provide an adequate grass cover on top of the clay layer.

When the deep water dike closes off a sea arm or encloses a relatively large area that is subject to tidal action, it can be expected that a heavy current will see-saw in the last opening of the dike, just before closure. In such cases, special measures may have to be taken. When the bottom of the sea is erodible, the first measure would have to be to protect the sea bottom from the heavy currents. This may be done by placing willow mattresses and rip rap, as shown in *Figure 6.9(b)* by the outline *A–A*. On top of this bottom protection a series of concrete caissons is placed as shown by the rectangles *B*. The opening between the two rows of caissons will be the final closure opening. The measures, thus far described, are taken well in advance of the dikes approaching the closure area, so that the work can be done precisely and without the interference of heavy currents. After the caissons *B–B* are placed, the dike proceeds to *C*. From this point on, the tidal velocities become so large that a caisson *D* has to be placed before the dike can be advanced. After this caisson is consolidated with flanks of clay, the next caisson *D* is placed and so on, until the closure with the row of *B* caissons is made. The ebb and flow current is now confined to the opening between the *B* caissons, which may be 100–200 ft. wide and 20–30 ft. deep, with a maximum current of 15–20 ft./sec. A large caisson *E* is now hauled into the position shown, while the current reverses from ebb to flow or vice versa. It is placed such that the oncoming high water will force it against the two *B* caissons as shown. When *E* is in the right place, it is sunk, so that it closes off the bottom as well as the sides. Two or four cranes start placing clay around the perimeter of caisson *E*, and if all goes well, the closure of the dam is accomplished.

When the enclosed bay or the tidal range become so large that the above method becomes too risky, it would be possible to apply the following procedure. The entire area, enclosed by the line *A–A* in *Figure 6.9(b)*, is surrounded by an auxiliary ring dike. The water within the dike is pumped out, and a series of sluiceways, equipped with gates, is constructed. The ring dike is removed. The main dike is built towards the sluiceways, while all gates are open. After the dike is connected with the sluiceways, the gates are lowered and the enclosure of the dam is accomplished. It is obvious that such a dike closure is much more costly than the one described above.

Location of Dikes

When farmers settled the flood plains of rivers they often chose the river bank for the location of their farm buildings. This choice had several reasons. First, the river banks of alluvial rivers are usually higher than the adjacent flood plain and therefore less subject to flooding. Second, the strip of land along the river usually carries a stand of good timber, suitable for building purposes or for fuel. Third, being close to the river was convenient when the main traffic was still by boat. When a river valley became densely settled, and after the farmers had experienced one or more severe floods, they would get together and decide to build a dike along the river to protect themselves from flooding. It was only natural to locate the dike between the river and themselves; that is to say, on the very edge of the river bank! Moreover, those early dikes were of a rather primitive construction, and not very consistent in height with respect to the high water profile. It is therefore no surprise that such dikes would fail either due to undermining or due to overtopping. The next

236

step in development was usually to request technical and financial assistance from the local, the provincial, or the federal government. Quite often, such assistance would be provided, and since some sort of a dike existed already, the new and better dike would be located by engineers along the same route. This is the moment when the mistake is made! As soon as public funds are going to be invested in flood control measures, these funds should be applied on the basis of sound engineering practice.

Dikes should be kept a good distance away from river channels for two reasons. The first one is to avoid undermining. The second one is to provide ample discharge capacity. The very need for dikes indicates that we are probably dealing with an alluvial river flowing in a flood plain. Such rivers have a natural habit of continuously eroding their banks. The meanders of rivers

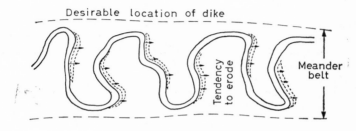

Figure 6.10. Location of river dikes

have a tendency to migrate in downstream direction, more or less within a strip of flood plain, called the meander belt. An ideal location of a diking system would be on the edge of this meander belt, as shown in *Figure 6.10*. Any attempt to intrude into the meander belt, let alone to follow closely the river bank, will result in frequent reconstruction of dikes or in costly revetment of the river channel to prevent erosion. It is of interest to note that most of the dikes that were built along the Mississippi River during the last century have disappeared in the river and had to be replaced by dikes farther inland.

As noted above, discharge capacity is another important consideration in locating the dikes away from the river channel. When the dikes would be on the edge of the river bank, only the channel provides discharge capacity and in order to accommodate the design flow, the dikes would have to be relatively high. However when the dikes are a good distance away from the channel and the flood plain between the river and the dikes is cleared of trees, bush and anything that may obstruct the flow, then the flood plain will carry a substantial portion of the design flow and consequently the top of the dike can be designed at a lower elevation than in the first case.

It may be concluded from the above that from an engineering viewpoint a diking system along an actively meandering alluvial river should be located such that undermining is unlikely. It is recognized that such a design may cause temporary hardship to farmers who have to move from the meander belt to a site behind the new dikes. However, there is no permanent economic loss of land involved because the flood plain between the dikes can be used for grazing and hayland. Since this land is periodically fertilized by river silt and has a high moisture content, it has proved to be very valuable and in great demand by the farming community.

Drainage

When floodwaters are excluded from flood plains by means of a diking system, it becomes necessary to pay particular attention to drainage problems which may arise in the reclaimed area. If outside flood levels are higher than the interior land levels and if they persist for any length of time, natural drainage becomes impeded and measures have to be taken either to store the water temporarily inside the dikes, or to pump it outside the dikes. In agricultural areas where precipitation is much greater than evaporation and transpiration, such measures become a necessity. In areas where precipitation is about equal to evapo-transpiration, the necessity of a drainage system may be questioned. There are large agricultural areas in North America where the soils are similar to those on flood plains, where the rainfall approximately equals evapo-transpiration, where virtually no drainage systems exist and where nevertheless good crops are grown!

There are several reasons, however, why a comparison between flood plains and other regions in North America is not justified. Primarily, most agricultural lands have a topography that provides natural drainage and, moreover, they usually have a ground-water table that is several feet or more below the ground surface. Therefore, they are differently affected by snow melt or heavy rains and observations at these places cannot be applied to flood plains where natural drainage is very poor or non-existent and where the ground-water table has a tendency to be extremely high due to the fact that the surrounding watercourses may have a summer water level higher than the interior lands.

It must be recognized that the existence of a proper ground-water table is very important. Too high a ground-water table, for instance, makes the soil difficult to plough and seed in the spring, retards the growth of the crops during the summer and causes difficulties during the harvest period in the autumn. Large fluctuations in the ground-water table within the region of the root zone are injurious to the plants, because the roots are alternately invited to reach low and are then killed off when submerged during a high stage. Too low a ground-water table is not desirable for the reason that the crops cannot obtain enough moisture during periods when the rainfall is less than plant requirements. When the average yearly precipitation approximately equals the estimated evaporation and transpiration, it can be expected that during years when the rainfall is near average and evenly spread over the season, a ground-water elevation will be established that is satisfactory for the proper growth of the crops. In those years, there will be no drainage problem. However, when the rainfall is above average or not evenly distributed, there is found to be a temporary surplus of water. The outstanding features of a wide flood plain are its flatness and its division into a number of isolated sub-basins. Therefore, it can be expected that surplus water will be ponded in the lower parts of these basins and will form lakes or marshes, since the ground-water table will always be relatively high. Consequently, there will be a definite drainage problem during such periods.

The simplest means of artificial drainage would consist of a number of small ditches, leading from the isolated sub-basins towards the natural main water-courses. These ditches would not be designed for a certain discharge capacity commensurate with the surface run-off during periods of snow melt or heavy summer showers, but they would be of small and uniform cross-section and run more or less continuously from break-up to freeze-up. There is no doubt that such a drainage system would be inexpensive and that it would render currently unproductive land capable of cultivation. However, on account of experience, some doubt as to the adequacy of such a system is justified. First, these ditches are not capable of lowering the ground-water table fast enough after the period of snow melt to permit quick access to the land for the purpose of ploughing and seeding. Secondly, the small size of the ditches would force the land to keep a high ground-water table for a long period of time after a heavy rainfall, thus preventing the land from producing the maximum yields of which it would be capable.

It would be desirable to try to overcome these disadvantages by constructing larger ditches that are capable of quickly removing the excess water from the land during and after snow melt or heavy rainstorms. These ditches have to be assembled in canals, ultimately leading towards the open river system. With such a drainage system, the desired ground water elevation can be restored after a period of one to two weeks, depending upon the capacity and number of ditches.

Now, a second important problem arises: 'How to determine the rate of run-off from the land for which the capacity and number of ditches and canals must be designed.' This problem has two components. The first pertains to the intensity of rainfall or snow melt that will be adopted for the design of the drainage system. The determination of this intensity could be made subject to economic analysis, but usually remains in the realm of choice, based on good judgement. The second component includes the problem of determining the maximum rate of run-off from the land when the intensity of the rainfall is given. Although this problem is rather difficult and its solution usually full of uncertainties, it is one of pure technical computation. In the following paragraphs, these two components will be discussed further.

The choice of a certain rainfall intensity which will form the basis of subsequent run-off computations can be based purely upon economic considerations. Loss of human life or severe property damage are not involved. The reputation of the engineering profession is not impaired when croplands or grasslands are allowed to be inundated occasionally. The proper choice of the magnitude and probability of occurrence of the design rainfall will be based upon a comparison of the increment of cost of a more elaborate drainage system versus the increment in benefits due to less frequent damage to the standing crops. It has been mentioned previously that such a comparison could actually be made in dollars and cents.

The run-off that will follow a period of heavy rainfall, will depend upon many factors, such as antecedent soil moisture conditions, elevation of ground-water table, number of ditches, soil permeability and evaporation during the run-off period. The only reliable way to obtain such information is by field observations. Lacking such guidance, it is necessary to adopt a certain unit-hydrograph which represents the distribution of the run-off and to assume certain losses due to evaporation and transpiration, in order to derive the volume of run-off. The result of this last phase of investigation will be that ditches and canals will have to be designed for a discharge of so many cubic feet per second per square mile of drainage area. The shape of the unit-hydrograph will depend partly upon the density and capacity of the ditch system. Hence, it will be a matter of trial and error to find the unit-hydrograph and to design the ditches. The criterion, in connection with the crop requirements, should be that the proper ground-water level is restored in, say, about 10 days.

If the rivers were always low enough to permit drainage by gravity, the layout of the drainage system would be rather simple and straightforward. However, the summer stages of the adjacent river may be higher than the desirable interior ground-water level, during periods that range from a few weeks to several months. This presents another problem: 'What must be done with the run-off from the land during periods when drainage by gravity is impossible?' If the drainage system continues to operate as described above, the run-off from the high lands would assemble on the lower lands which would quickly be drowned and cease to yield crops. This, of course, cannot be tolerated and means will have to be found to prevent it. There are, in general, two ways of accomplishing this purpose. The first is by temporarily storing the run-off within the reclaimed area in predesignated storage basins and the second is by pumping the run-off out of the reclaimed area. It is evident that a certain area can only store its run-off in a storage basin that has its water surface below the desired ground-water table of the area under consideration. This means that the lowest lying lands cannot be accommodated with storage facilities. To take care of the run-off of the whole reclaimed area by pumping alone, may require a tremendous pumping installation. The cost of a pumping plant with such a capacity may be prohibitive when compared to the benefits that can be derived from it. It would follow from the above discussion, that a logical solution could be to serve the lower lands by pumping and the higher lands by storage basins.

A necessary refinement of the system as sketched in the above, would be to have facilities at critical places to control the flow of run-off by check dams or culverts with gates. The purpose of such a control would be to store water as high as feasible in order to prevent the swamping of lower lands by the run-off from higher lands. Such a control would also be beneficial during dry years. If the rainfall is almost equal to the evaporation, it is not at all impossible that during dry years the ground-water table will become unfavourably low. Such a situation would be aggravated when the canals, which may have their bottom elevation from eight to ten feet below the average ground elevation, were permitted to continue draining the land. This situation could be relieved if, in due time, check dams and culverts were closed so as to preserve the available ground water. Such a control system would simply be an application on a small scale of the principles of water conservation.

In order to reduce the number of control points to a minimum, it would seem advantageous to divide the reclaimed area into polders which have approximately the same elevation within their boundaries. Each polder becomes a separate drainage unit with a more or less uniform ground-water table, with its own system of ditches and canals and its own drainage outlet upon exterior watercourses. In making this division into polders, attention must be paid, of course, to the existence of pronounced ridges and natural watercourses. The inclusion of such watercourses in the drainage system may substantially reduce the amount of required excavation.

When the division into polders is made, ditches and canals will have to be laid out within these polders. The spacing and layout of the ditches will likely be determined by the standard division of land in sections of one square mile. The location of the canals will be largely determined by the direction of the slope of the land and will follow as closely as possible the section lines,

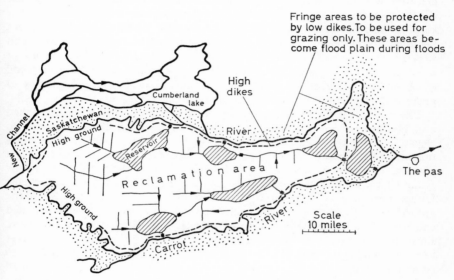

Figure 6.11. Reclamation of Saskatchewan Delta

in order to avoid unnecessary subdivisions of land. Once the layout of the ditches and canals is determined, the slope of these watercourses is practically fixed because of the topography of the land. Since the run-off from the land for which the ditches are to be designed is known, it will from then on be a routine matter to compute the size of the canals, the size of the storage basins, the size of the culverts and the capacity of the pumping plants.

To illustrate the above discussion *Figure 6.11* shows the reclamation plan for 1,000,000 acres of fertile land in the Saskatchewan Delta. In the early stages of planning, very little was known about the area, except that some isolated farmers had grown wonderful crops. While soil surveys were being conducted to determine the most attractive areas for reclamation, hydrometric surveys were conducted to explore the stream channel characteristics. Sedimentation studies were carried out to determine how the change in river regime, due to reclamation, would affect future flood levels. [See Kuiper (1960), in the bibliographic references of Chapter 4: River Morphology.] Hydrologic studies were carried out to determine the frequency of exceedence of flood peaks and flood volumes on the Saskatchewan River as shown in Chapter

2: Hydrology, *Figure 2.10*. Based on engineering, economic, and social considerations, it was decided that the dikes should be designed against flood conditions with a probability of exceedence of 0·1 per cent per year. Floods of corresponding magnitude were now routed through the delta area, assuming the future reclaimed conditions. The different rating curves that were required for the routing, were determined by backwater computation, starting well downstream of the reclamation area. The flood levels thus found, were the basis of determining the elevation of the top of the dikes. Where dikes were adjacent to lakes, so that wind set-up could be significant, the dike elevation was determined as described earlier in this chapter. In order to raise the flood stages not too much above their natural levels, with the objective of keeping the dikes as low as possible, the main river channels were at no point hemmed in too closely by the reclamation works. Wide bands of flood plain were reserved for passing the design flow. However, to still make some use of these areas, they were designed with low dikes with a probability of being exceeded of 5 per cent per year. These areas are to be used for hayland and grazing. Permanent buildings are not permitted.

The interior of the main reclamation area appears horizontal to the naked eye. However, there is a gentle slope from an elevation 880 in the west to elevation 850 in the east. The total area is divided into five polders or reclamation units. Each unit has its own storage reservoir for excess run-off during wet periods. The ground-water table will be maintained at two to three feet below ground level by means of numerous control gates in the ditches and canals. Under normal conditions all control gates will be closed. Only when the ground-water table rises due to rainfall or snow melt will the gates be opened and will the drainage system begin to operate. This is an important measure since the average annual precipitation is only 17 in. When the outside river levels are too high to permit drainage by gravity, all excess run-off will have to go temporarily into the storage basins. Since these basins have been designed with ample capacity, no pumping plants are needed. It is quite possible that in the future the reclaimed land will become so valuable that it will be economical to reduce the storage areas in size and provide pumping capacity instead. It may be of interest to note that the total cost of reclamation, including dikes, river diversions, dams, canals, main ditches and main roads, was estimated at $20 per acre.

FLOOD DIVERSIONS

The most direct and effective way to cope with a flood situation is to take water away from the river channel. A primitive way of accomplishing this is to breach a dike on purpose, in an area where the resultant damage is relatively small, in order to save the dikes that protect another area where the damage would be relatively large. In the olden days, one of the responsibilities of the appointed dike patrols, in time of flood danger, was to prevent 'friends' across the river from excavating a little pilot channel in their dike!

When this principle of emergency flood diversion is planned in advance, and when appropriate measures are taken, the outcome might be a very effective and relatively inexpensive means of flood control. Some requirements are that the area into which the floodwaters are diverted is free of habitation;

that it is bordered by secondary dikes to prevent the floodwaters from spreading in all directions; that there are reasonably good drainage facilities to get the water back into the river after the flood threat is over; and that there is reasonably good control over the diversion of water. With respect to this last point it is obvious that blowing up a section of dike with dynamite is a real emergency measure indeed. A better way of controlling the situation is to have a section of dike built as a fixed-crest spillway. A still better arrangement is to have spillway sluices with gates and energy dissipaters.

The function of such a floodway is twofold. First of all it provides storage capacity, and secondly it provides discharge capacity. It depends entirely on the local situation which of these two aspects is more important. In the early days, when the habitation of flood plains was not as dense as it is now, large tracts of land would be set aside for emergency floodways, and thus the storage effect was significant. However, with the increase in population density the old-time floodways were reduced in size or entirely eliminated, and replaced by diversion channels where discharge capacity is the main function and any storage effect incidental.

A diversion channel will be most effective in lowering water levels, if the diverted water can be taken away from the river, without returning it farther downstream, as shown in *Figure 6.12(a)*. Let it be assumed that the area to be protected is situated between B and C and that the diversion is located at A. In that case the water level downstream of A will become the normal depth for the reduced discharge; say 100,000 cusec natural flow, minus 40,000 cusec diversion = 60,000 cusec. It is not very often that such a permanent diversion is possible. Examples are the diversion from the Assiniboine River towards Lake Manitoba to protect Winnipeg, or the diversion from the Mississippi River via the Atchafalaya Floodway towards the Gulf of Mexico to protect New Orleans.

A less attractive but more common situation is that the diversion channel has to return the diverted water to the river channel, some distance downstream of the area to be protected, as shown in *Figure 6.12(b)*. If the return takes place shortly downstream of C, at D, the backwater effect of the return flow may be felt throughout the entire reach B–C, and therefore the diversion is only partially effective. In order to reduce or eliminate the backwater effect, point D will have to be moved farther downstream, as shown in *Figure 6.12(c)*. Needless to say, this will increase the cost of the diversion channel.

It may be expected that the cost of excavation of the diversion channel is a major portion of the total diversion project cost. If it so happens that the aim of the flood control project is to reduce stages between B and C and not upstream of A, it may be worth considering to construct a control dam in the river at A, just downstream of the intake of the diversion, as shown in *Figure 6.12(d)*, in order to restore the water levels at the intake to what they would have been under natural circumstances. This will increase the available hydraulic gradient on the diversion channel and therefore decrease the cost of excavation. If the flood damage upstream of A is negligible, one could even go so far as to increase the water surface at A above its natural level. It is then conceivable that part or all of the diversion channel could be designed as two dikes upon the flood plain, which brings us back to the old time floodway.

The above discussion may be illustrated by the example shown in *Figure 6.13*. The safe bankfull discharge in the city that must be protected, is 80,000 cusec. The design discharge of the diversion channel is 60,000 cusec. Hence the river flow at point *C* for design conditions is 140,000 cusec. The stipulation

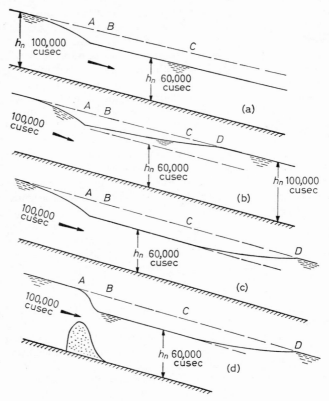

Figure 6.12. Profiles of river diversions

is made that the stage at *C* for design conditions must be the same as the stage under natural conditions (without flood control works). To achieve this, a control dam is designed in the river, downstream of the diversion intake. This dam may be any of the types that are discussed in Chapter 9: Navigation, page 376, and shown in *Figures 9.5* and *9.6*. In the present example we shall assume, for the sake of simplicity, a control dam with vertical lift gates, as shown in *Figure 6.13(b)*. The entrance to the diversion channel is designed with a sill, as shown in *Figure 6.13(d)*, so that no water will enter at low river stages. In *Figure 6.13(c)* are shown the natural rating curve 0–1–3–4, and the limits of controlling the stage at *C*, for any given discharge, by either completely closing or completely opening the gates of the control dam.

A likely operation of the control dam during the passage of a flood is as follows. During the beginning of the flood the gates are open and the stage at *C* follows the rating curve 0–1 (provided that the open control dam causes a negligible backwater effect). When the river stage exceeds the elevation of the sill to the diversion channel, the control gates are kept open till point 2 on the improved rating curve is reached. At this point, the discharge in the river, downstream of the control dam, has reached a magnitude of 80,000 cusec and should not be exceeded. From now on, the gates are lowered, so that the river flow in the city remains 80,000 cusec, while the stage at C and the discharge in the diversion channel increase gradually till point 3 is reached. This is the design condition of the diversion channel. The advantage of having followed the curve 1–2–3 instead of the natural rating curve 1–3, is twofold. First, by keeping the stages lower at C, it may provide some flood relief to the people

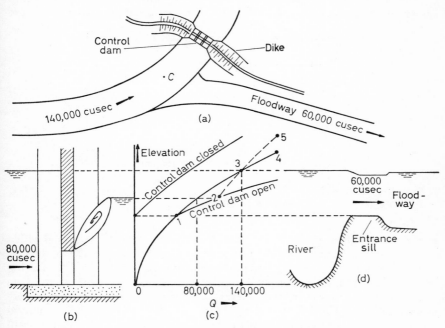

Figure 6.13. Rating curves at diversion site

that live upstream of C. Second, by keeping the flood stages as low as possible during the early part of the flood, we save natural flood plain storage capacity for the later part of the flood, and thus we may reduce the peak flow somewhat. When the river discharge exceeds 140,000 cusec we can either follow the curve 3–4, and make the flood situation for the people that live upstream not worse than it would have been under natural conditions, but exceed the bankfull stage in the city; or we can follow the curve 3–5, keeping the river flow in the city at 80,000 cusec, exceeding the design flow on the diversion channel and exceeding the natural flood stages at C. The choice between those two alternatives will depend on the alternative damages and political judgement.

If a diversion channel has to be excavated through relatively high lying terrain, as shown in *Figure 6.14(a)*, it may be advantageous to design the channel as narrow and deep as possible. It can be seen from *Figure 6.14(b)* that

245

a reduction in bottom width and enough deepening to keep the wetted area the same (area a = area b) will result in a substantial saving in excavation (area c). One has to watch out, however, for two possible limitations. The first one is slope stability. With an increase in channel depth the channel side slopes may have to become milder, thus nullifying the intended savings. The second limitation may be the permissible channel velocity. With a cross-section that becomes narrower and deeper, the hydraulic radius will increase, thus increasing the velocity of flow for a given hydraulic gradient.

If the water surface of a diversion channel is close to, or somewhat above, the natural ground surface, and if the longitudinal gradient of the diversion channel is such that the velocities would become too high with the design as shown in *Figure 6.14(c)*, it may be advantageous to design a channel that is

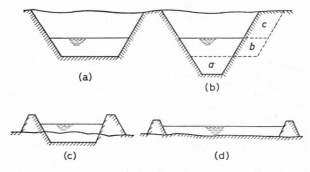

Figure 6.14. Diversion channels

wider and shallower as shown in *Figure 6.14(d)*. For instance, a channel with a depth of 10 ft., a slope of 0·0005, and a roughness coefficient of 0·025 would have an average velocity of 6·2 ft./sec. When the depth is reduced to 5 ft., leaving the slope and roughness the same, the velocity is reduced to 3·9 ft./sec. One has to be careful, however, not to design a too-shallow channel for a relatively large discharge. The slightest irregularity in depth would produce a local increase in velocity, which might produce some erosion, which in turn will result in higher velocities. Then, quickly, a deep channel will start meandering, attracting the entire diversion flow and endangering the safety of the dikes and bridges.

If the gradient of the diversion channel is so large that in spite of the above measure the velocities would still be too high, it may be necessary to introduce a drop structure, where the excess energy is locally dissipated. A drop structure may consist simply of a fixed crest weir and a stilling basin.

It is quite conceivable that the various aspects of channel design, as discussed in the above, are found in one diversion, as shown in *Figure 6.15*. Point A represents the diversion intake at the river channel and would correspond with point A in *Figure 6.12(d)*. Since the land adjacent to the river between B and C is relatively high, the water level at A will be raised as much as is commensurate with upstream interests. The channel between B and C will be designed so that the excavation is minimized. This may result in a very flat gradient. The gradient between C and D may now be too steep, so that a drop structure at the outlet of the floodway channel becomes necessary. Another likely cause for designing a drop structure at the end of the floodway is that the total length of the floodway may be substantially less than the total length of the meandering river channel between A and D. Hence the gradient on the floodway is steeper. Moreover, the roughness coefficient as well as the width ÷ depth ratio of the floodway may be less than that of the river channel. As a result, the velocities on the floodway, without drop structure, would become substantially larger than the river channel velocities. The latter ones may already be in the order of 6–10 ft./sec during extreme flood stages. It is obvious therefore that under such conditions a floodway without drop structure would produce velocities that are much too high.

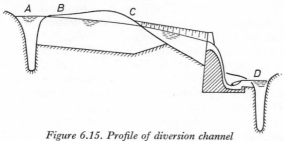

Figure 6.15. Profile of diversion channel

Frequent reference has been made to permissible velocities. This is an important aspect of channel design, since uncontrolled erosion during flood diversion may be dangerous and costly. Permissible velocity not only depends on the type of soil in which the channel is excavated but also on the depth of flow. For instance, it has been known for some time that the permissible velocity is higher in large canals than in small ones. It is therefore proposed by Lane (1955) to analyse the channel stability problem on the basis of the tractive force exerted by the water upon the stream bed. This tractive force is the component, in the direction of flow, of the weight of the water, and is expressed in pounds per square foot. Table 6.3 shows some data presented by Lane and compared to earlier permissible velocity values, prepared by Etcheverry and others.

Table 6.3. *Comparison of Permissible Velocities with Tractive Force Values*

Material	Size (mm)	Value of n	Permissible Velocity (ft./sec)	Limiting Tractive Force (lb./ft.²)
Fine loose sand	0·1	0·020	1·0	0·01
Coarse sand	1·0	0·020	2·0	0·04
Sandy loam	—	0·020	3·0	0·08
Stiff clay	—	0·025	4·0	0·30
Coarse gravel	10·0	0·030	6·0	0·70
Coarse pebbles	100·0	0·040	8·0	1·10

In the above table, the permissible velocity values are based on empirical findings with canals approximately three feet deep. These values have been converted to limiting tractive force values for a depth of 3 ft. and a roughness value as listed. The advantage of using the limiting tractive force figures, is that they allow for the experience that permissible velocities are higher in large canals than in small ones. For instance, for a canal, 3 ft. deep, with a bottom of coarse gravel, a roughness coefficient of 0·030 and a limiting tractive force of 0·70 we find, by using the Manning equation, a permissible velocity of 6·0 ft./sec, which confirms Table 6.3. When we increase the depth to 6, 10, 20, and 40 ft., leaving all other variables the same, we find permissible velocities of 6·8, 7·3, 8·2, and 9·2 ft./sec respectively. It should be pointed out that the above figures are very approximate and that they will substantially change when the soils are cohesive or when the water contains colloidal silts. In both cases the permissible velocity will increase. It is obvious that for actual channel design of any importance a research in the literature and field study of comparable channel conditions is required.

The selection of an appropriate roughness coefficient 'n', introduces another source of error in the design of the diversion channel. According to Ven Te Chow (1959), the roughness coefficient of an excavated straight earth channel may be as low as 0·016 when the channel is clean and recently completed and as high as 0·032 when the channel has short grass or a few weeds. Since the Manning formula contains the roughness factor to the first power, it follows that the discharge computed with the first value is 100 per cent larger than the discharge computed with the second value. For a given case it may be possible to narrow down the possible range of 'n' values. For instance, for the design of the Sea-level Panama Canal, as reported in a Symposium by the ASCE (1949), a study was made of the roughness coefficient of large comparable canals. The range of values from all sources was found generally between 0·024 and 0·031, including the present Panama Canal (Gaillard Cut) value of 0·026. It must be recognized, however, that although the possible range may be narrowed down, a substantial degree of uncertainty remains.

The question now arises, which 'n' value one should select from a possible range, and stay on the safe side. If a high 'n' value is chosen to design a channel that will carry the design flow with the permissible velocity, and it turns out that the real 'n' value is much lower, then the real velocity will be larger than permissible and the channel is in danger of being eroded. On the other hand, if a low 'n' value is chosen for the design, and it turns out that the real 'n' value is much higher, then the channel will not carry its design flow and its flood control function will not be adequately performed. To solve this dilemma, the channel may be designed such that the design capacity is carried with the high 'n' value (and a lower than permissible velocity) while the permissible velocity is obtained with the low 'n' value (and a higher than design flow).

CHANNEL IMPROVEMENTS

This last of the four basic flood control methods aims at lowering flood stages by improving the carrying capacity of the existing river channel. This may be accomplished by: (1) decreasing the roughness of the channel, (2) widening or deepening of the channel, (3) shortening the channel and thus steepening the hydraulic gradient, and (4) control of the sediment transport.

Before these methods are discussed, let us consider first over what length of river channel the improvements have to be applied in order to lower the flood stage in a given area. Let it be assumed that between A and B in *Figure 6.16(a)* the flood stages have to be lowered. If the channel improvements would be applied to the reach A–B only, the beneficial effect would not be very great. At B the stage would remain the same. From B to A the water surface will be a backwater curve that will ultimately reach the normal depth of flow that corresponds to the improved channel condition. Before this depth is reached, however, point A, where the improvement terminates, is reached, and from there upstream the water surface profile is a drawdown curve that will reach the normal depth of river flow again. In order to lower the levels between A and B more efficiently, we have to extend the channel improvements further downstream to C as shown in *Figure 6.16(b)*. The length of the

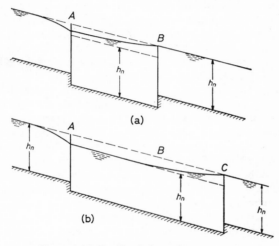

Figure 6.16. Profile of channel improvements

reach B–C that is required to make the lowering of the flood levels effective from A to B, depends on the slope of the river and the degree of improvement that is required. Let us now discuss the four methods of improvement in more detail.

Decreasing the roughness of the channel may be accomplished through the clearing of river banks of the brush, trees and other obstacles to smooth flow. It may readily be deduced from the Manning formula that a slight reduction in roughness coefficient, say from 0·033 to 0·030, could result in a stage reduction of a few feet, for a river channel of 30–40 ft. deep. However, it is rather difficult to estimate with sufficient accuracy the consequent reduction of the roughness coefficient as a result of a certain amount of clearing, so that benefits can be compared with costs. It would be advisable to obtain data on similar projects which have been carried out elsewhere or to proceed on an experimental scale. For instance, the Yazoo River whose banks were cleared of heavy brush, showed a reduction in roughness coefficient from 0·036 to 0·030. It should be realized that removal of natural vegetation may necessitate the placement of a revetment in order to prevent serious erosion due to the increased velocities near the banks.

By widening or deepening the channel—in other words, by enlarging the cross-section area—the water level will be lowered. Roughly speaking, if the river stage has to be lowered by 1 ft., the river bottom will have to be removed over the entire width of the channel and the entire length of the reach by 1 ft. In most cases it is more economical to build dikes or to raise existing dikes by a certain number of feet, than to dredge the bottom of the river by the same number of feet. Exception to this general rule may be made where the channel is very narrow, or where the construction or improvement of dikes is extremely expensive, or where local channel improvements reduce water levels for a long distance upstream.

Meander loop cut-offs have the effect of increasing the slope of the water surface, thus increasing the velocity of the water, and consequently reducing the flood stages. At first glance this method would appear to be very attractive, but the very means of the reduction in flood stage, namely, the increased velocities, may have the undesirable effect of producing uncontrolled erosion, thereby developing new bends and consequently bringing about severe damage to riparian property, adjacent roads and bridges.

Rectification of the Sangamon River, for instance, was undertaken in 1905 by making numerous cut-offs. The slope steepening effected in the improved 26 miles of river proved to be excessive. Increased velocities scoured out large volumes of riverbed sand and gravel, thereby causing excessive channel deepening. The eroded material came to rest in the unimproved lower seven-mile stretch, converting it into a large swamp, where an aggrading process of the river bottom began which caused the lower reaches of the river to lose their flood carrying capacity. This aggradation gradually extended upstream, with consequent return to higher flood stages.

A more successful application of the cut-off method is found along the Lower Mississippi River where a channel improvement programme was begun in 1932, resulting in a lowering of flood levels by 6–7 ft., without causing excessive instability to the river channel. The lessons to be learned from this experience as described by Matthes (1948) are the following:

(1) A cut-off must be aimed at ultimately establishing hydraulic gradients which the river can maintain without requiring corrective work at frequent intervals. In order to determine this, the first appraisal of the situation should consist of determining the bankfull river profile. This profile becomes a guide and indicates the hydraulic gradients which fit the river best in those stretches where curvature is not excessive.

(2) In order to counteract the tendency of the river to scour upstream and to deposit downstream of the cut-off, care should be taken to plan the pilot cut-off channel in such a way that it diverts only the swift top water of the river and that the sediment laden bottom water remains in the old river channel. This is accomplished by bringing the pilot channel in line with the upstream river channel and making the old river loop the branch channel of the new course.

(3) When navigation is an important factor to be considered it is advisable not to straighten the river channel over long distances, because this may result in meandering low water channels within the straight high water channel.

The fourth method to improve the carrying capacity of the river channel would be through control of the sediment transport. The cross-section of an alluvial river is more or less in balance with its sediment transport. If our aim is to enlarge or to deepen the cross-section, this may be accomplished by: (1) reducing the sediment load through upstream reservoirs, or sediment traps, or dredging; (2) increasing the sediment transporting capacity of the river channel by narrowing its cross-section through river regulation works or dikes. It is evident that such measures have no immediate effect but will only become effective after several years. A more thorough discussion of this topic is contained in Chapter 4: River Morphology.

ICE PROBLEMS

In the above discussions we have assumed that the flood stages were caused by extreme river discharge, by extreme sea levels or by a combination of both. However, flood stages may also be caused by ice conditions on the river channels. In the following paragraphs we shall discuss the ice formations on rivers and lakes; the nature of ice jams; and possible remedial measures.

When a river is exposed to freezing temperatures, there are three possible modes of ice formation: surface ice, anchor ice, and frazil ice. The first one is the most common and takes place in the same fashion as the formation of an ice cover over a lake or pond, namely by the growth of ice crystals that cling together and form a solid sheet of ice. When the average velocity of the river exceeds 3–4 ft./sec, the turbulence is so great that the formation of surface ice does not take place unless the temperature drops well below zero. With the absence of surface ice, anchor ice (on rocks and pebbles) and frazil ice (crystals in the water) may be formed, provided the water is clear. To understand its formation, let us first observe the flow of heat on a warm sunny day: the rays from the sun will penetrate through the water and warm the bottom of the river; small particles of sediment and organic matter, suspended in the water, will also be warmed up. Now let us reverse the situation: the night is clear and cold, and heat is flowing from the earth towards outer space. The bottom of the river and the impurities in the water will lose heat fastest. As a result, rocks and pebbles will receive a coating of ice and crystals of ice form throughout the body of water. The frazil ice crystals may cling together and form a slush-like substance that may clog up the trash racks of power plants. The most effective way to prevent the formation of anchor and frazil ice is to slow the river down by means of a dam so that a solid ice cover will form at the beginning of the winter. Anchor and frazil ice never form underneath an ice cover.

Let us see what happens to the river level when the river freezes over with a solid sheet of surface ice as shown in *Figure 6.17*. For the sake of this discussion we shall assume that the flow from the ground-water reservoir into the river system remains constant throughout the winter. Let us assume that the river freezes over instantaneously with a thin but continuous sheet of ice at point 1 in the diagram. Due to the added friction of the ice cover the water slows down and the discharge becomes momentarily less. However, the flow from ground water remains the same, and therefore the ice sheet is slowly lifted up to point 2. In this position the discharge is back to normal and the new depth of flow is in balance with the new friction. During the winter the sheet of surface ice grows thicker (in Central Canada river ice may be as thick as 5 ft.). Assuming the same discharge and the same friction, it follows that the surface of the ice must be lifted up, while the underside of the ice sheet, where the new ice crystals are being formed, remains in the same position. This continues until point 3, at the end of the winter. Let us now assume that the ice sheet suddenly breaks up into pieces and has no contact with the shore. The water would now suddenly speed up due to the decreased friction and the discharge would momentarily increase. Still assuming the same inflow into the river, the river level would gradually fall back to point 1, which is the same as in the previous autumn.

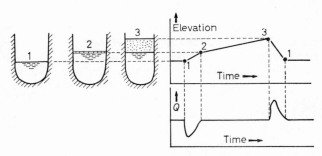

Figure 6.17. River levels during winter

The above discussion is highly academic since the discharge in a river does not remain the same throughout the winter. However, the purpose of the discussion was to highlight a few points that may easily become obscured when observing real river behaviour. To start with, the river discharge usually declines rather sharply during the freeze-up. As a result the river level remains the same or drops. During the winter the discharge keeps decreasing and hence the surface of the ice may remain the same or drop. During the spring thaw the discharge increases sharply. The ice sheet is lifted up by the swelling river. Since the sides of the ice sheet are frozen solid to the ground, they remain anchored down and are flooded with water. This situation may last anywhere from a few days to a few weeks. When the discharge has become large enough or when the ice has sufficiently deteriorated, the ice sheet will break up and move downstream with the current. In the beginning many cakes of ice are as large as the river is wide. When such ice floes come to bends or narrow sections or islands or sandbars or bridge piers they may get stuck and cause an ice jam. Before we discuss the nature of ice jams and its implications on flood levels, let us first see what has happened during the winter to the levels of a lake.

On a lake without inflow or outflow, it is obvious that the formation of the ice cover will cause no changes in level. During the winter, the ice sheet will grow thicker, with the surface of the ice remaining at the same level. Let us now assume that the lake has a constant inflow and outflow and that the outlet of the lake remains free of ice throughout the whole winter. This last phenomenon may be observed on many lakes, even during the severest winters. The explanation may be found in turbulence, or in the fact that slightly warmer bottom water is drawn from the lake, or because of heat generated by decaying vegetation in the outlet channels of the lake. Whatever the cause, if the outlet channel is open and the discharge is constant, the lake level must remain constant and hence the sheet of ice on the lake will keep its upper surface (which is practically identical with the water level when a hole is chopped in the ice) at the same elevation. Let us now assume that the lake has a constant inflow and outflow, but that the outlet channel is frozen over. In that case the rise of level in the outlet channel will determine the rise of level on the lake and we may expect the surface of the ice sheet to come up. During the spring thaw the ice sheet on the lake is subjected to less stress than on the river. As a result the break up may be several weeks to a month later. Once the ice sheet is broken, its disintegration depends very much on the force of the wind.

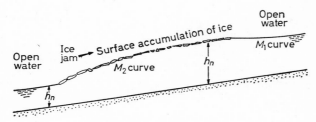

Figure 6.18. Profile over ice jam

Let us now come back to ice jams on river channels. We may distinguish between ice jams, whereby the cakes of ice remain in a horizontal or slightly slanted position at or near the surface of the river; and ice dams, whereby through the force of the current the cakes of ice are sucked underneath each other and plug up a good deal or nearly all of the cross-section of the river. In the case of ice jams, which usually occur when the velocities of the river are less than 3–4 ft./sec, we may consider the surface cover of broken and jagged cakes of ice, as providing an exceptionally large amount of friction. The profile of an ice jam is shown in *Figure 6.18*.

Downstream of the jam, the discharge is normal, the river is clear of ice, and hence the depth of flow must be normal. The downstream end of the jam must have that same elevation. In upstream direction the profile of the ice jam has the form of a drawdown curve reaching eventually its own normal depth of flow that corresponds to the given discharge and the increased roughness due to the surface accumulation of ice cakes. Some observations of ice jams on the Red River and the Saskatchewan River have verified this general shape of the ice jam profile. The normal depth of flow of the ice jam may be computed roughly by assuming a composite roughness coefficient that is substantially higher than for the open channel, and by allowing for the fact that the hydraulic radius is only half of the open channel value. Moreover, the upper layer of say 5–10 ft. of jumbled ice cakes does not contribute to the flow. As a result we may find that the new normal depth, under ice jam conditions, is more than twice the normal depth under open channel conditions. The normal depth of an ice

253

jam will only be reached, of course, if the jam is several miles long and if it is confined between high dikes or river banks. Upstream of the ice jam we may expect a backwater curve that brings the river back to its normal open flow conditions.

In the above paragraph we have assumed steady flow conditions with the normal discharge. It should be pointed out now that while the ice jam is being formed, a substantial amount of water goes into storage underneath the draw-down and backwater curves. Hence the stage at the downstream end of the jam will go down, immediately after its formation, but it will come up again as soon as steady flow conditions have set in. Due to the steep gradient near the jam, a tremendous pressure is exerted in horizontal downstream direction. This is often enough to break up the jam. Sometimes a well-placed discharge of dynamite will trigger off the break-up. Sometimes the jam is so solid that it will sit for weeks on end.

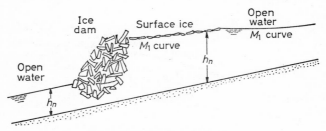

Figure 6.19. Profile over ice dam

If the river current is more than 3–4 ft./sec, the cakes of ice may not just set at the surface, but may be drawn under other ice cakes, thus forming a hanging ice dam, as shown in *Figure 6.19*. The hydraulic losses of the river flow underneath the ice dam are so large that a very steep gradient is built up. It is this steep gradient and its associated force that usually causes the destruction of the dam while it is being built up or soon afterwards. The forces involved are fantastic. Cakes of ice, 5 ft. thick and as large as tennis courts, may slide and crunch over the top of one another like pebbles and may knock over and push out of the way buildings and large trees.

Remedial measures to prevent damage from flood stages that are caused by ice jams are the following. First of all to prevent the ice jams or dams from occurring. To this end one should try to eliminate all the current causes of the jams. River bends may have to be straightened or cut off. Places where the river funnels from a wide section into a narrow section may have to be corrected by river training works. Islands and sand bars may have to be removed. Bridge piers should not be built too close. They may be equipped with a sharp steel knife edge on the upstream side to split any cakes that bounce into the pier. After the preventative measures have been exhausted or turn out to be too costly, it may be feasible to build adjacent dikes high enough to cope with the normal depth of flow associated with the ice jam. If this is done, it would be advisable to leave a fringe of solid trees in between the dike and the river for the purpose of protection. Another possibility may be to devise an ice-jam spillway into low lying land on one side of the river in order to protect

254

the other side. On rivers that have been converted by power dams into a series of reservoirs, it is found that the ice will simply remain in each reservoir until it is melted. This principle may be applied to other situations.

EFFECT OF FLOOD CONTROL WORKS

The foregoing paragraphs in this section on flood control have dealt with the various engineering measures which may be employed to reduce flood hazard. After having studied a particular flood problem it may be found that several alternative engineering measures are possible in order to cope with the situation and it then becomes necessary to select the most feasible measure, or combination of measures. In order to make this selection, the engineer needs to know to what extent the various measures will reduce the frequency of flooding in the area under consideration. This latter problem will form the subject of the present discussion. The problem of using the modified flood frequency in a cost–benefit analysis will be discussed later.

Let us first make some general observations with respect to the effectiveness of the four basic flood control methods. It will be assumed that all methods provide full protection up to their design flow conditions. However, we should not only be interested to what stage we have full protection but also what happens after the design flow conditions are exceeded.

Dikes, when properly designed and when exposed to flood levels that do not exceed the design level, will provide protection no matter how long the flood condition lasts. However, when dikes are overtopped they cease to function as a flood control measure and the damage from flooding will be as great or even greater than when no dikes had existed at all, as shown in *Figure 6.20(a)*.

In contrast to dikes, it may be noted that diversions increase their beneficial effect when the design flood is exceeded. The higher the flood stage, the larger the capacity of the diversion, as shown in *Figure 6.20(c)*. It may also be noted that diversions, like dikes, will perform their function regardless of the duration of the flood.

The effect of channel improvements is much the same as that of diversions. When the design flow is exceeded, the channel improvement remains effective, as shown in *Figure 6.20(d)*. Whether or not the stage reduction at higher flows is larger or smaller than the stage reduction for design flow conditions, depends on the type of channel improvement and on the channel characteristics. For instance, a uniform channel deepening over great length will result in approximately the same stage reduction for all stages. The removal of a local reef may only be effective at low stages and have an imperceptible effect at high stages. The widening of a canyon may have a small effect at low stages and a much larger effect at high stages.

The effectiveness of a flood control reservoir depends among other things on the duration of the flood. A reservoir may be designed for a typical flood hydrograph. However, when a flood occurs with the same peak flow, but a much longer base, the reservoir may be filled up by the time the peak arrives, with the result that hardly any storage capacity is left to reduce the peak! The same reasoning can be applied for a flood that has the shape of the typical hydrograph, but exceeds it in magnitude. It may be concluded that flood control reservoirs perform their function well, as long as the floods do not exceed the reservoir design flood in peak flow or in duration. When the design flood is exceeded, the beneficial effect of the reservoir is gradually reduced, until it may become practically nil for extremely large flood flows, as shown in *Figure 6.20(b)*.

An important aspect that influences the effectiveness of a flood control reservoir is the distance between the reservoir and the area to be protected. There are two reasons why distance reduces effectiveness. First, there is the possibility that floods will originate in parts of the drainage basin that are not

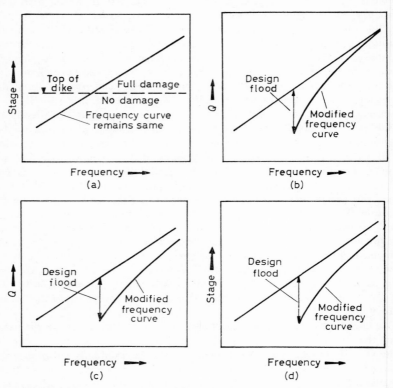

Figure 6.20. Effect of flood control measures: (a) dikes; (b) reservoirs; (c) diversions;
(d) channel improvements

controlled by the reservoir. Second, even if floods do originate upstream of the reservoir, the peak flow reduction at the reservoir tends to become less while the flood travels along the river system. Let us assume for instance that under natural conditions a flood with a peak flow of 100,000 cusec passes the prospective reservoir site. Assuming no inflow from tributaries, this flood peak will be reduced due to natural channel storage, to a peak of say 60,000 cusec, by the time it reaches the area to be protected. After having built the reservoir, the flood peak of 100,000 cusec is reduced to a safe bankfull discharge of say 20,000 cusec which is no further reduced by channel storage. In other words the effectiveness of the reservoir, as far as the area to be protected is concerned, is a peak flow reduction of 40,000 cusec and not 80,000 cusec! We may now add tributary inflow to the above example, but that will not change the principle of the matter.

The question now arises as to how the above aspects of the four flood control methods can be properly expressed in a quantitative analysis of alternative flood control measures. There are three methods that may be followed. All three have in common that they aim at establishing the frequency of exceedence of flood stages in the area under consideration, with and without the existence of a proposed flood control scheme. The three methods differ in the procedure of determining the modified frequency curve.

In the first method, all recorded floods represented on the original frequency curve are routed through the flood control works. The resultant maximum flood stages in the area under consideration are determined and a new frequency curve is prepared. This method is straight forward and one can easily visualize what is being done, but it may, in certain cases, have the serious disadvantage of not truly representing the beneficial effect of certain works. For example, a 40-year period of record is available, containing only one flood of extreme proportions. This flood may have originated in one part of the drainage basin, the remainder of the drainage basin having had only moderate rainfall, although the whole basin is equally susceptible to heavy storms. Let it be assumed that a number of flood control reservoirs are planned throughout the entire drainage basin. The application of this method will then indicate that only those reservoirs located in the area where the extreme flood originated are beneficial, whereas the others have hardly any effect. This, of course, would be an erroneous conclusion and some other approach would have to be made.

The second method aims at overcoming this disadvantage and utilizes a series of representative river basin floods, rather than recorded floods. The first step is to select a number of representative points on the frequency curve of flood peaks under natural conditions at the area to be protected. For instance, the flood magnitudes, corresponding to a frequency of exceedence of 0·10, 0·50, 1·0, 2·0, 5·0, and 10·0 per cent per year. The next step is to prepare representative hydrographs, upstream of any of the proposed flood control works, which, when routed down the river system under natural conditions, will yield approximately the above peak flows in the area under consideration. To accomplish this, all hydrological information of the drainage basin will have to be utilized. The above representative hydrographs (one set of sub-basin hydrographs for each frequency of exceedence) are now routed through the different proposed flood control schemes and for each scheme a new frequency curve is prepared.

This second method has also some shortcomings, although to a lesser degree. When carefully selected, the representative floods are probably more representative than the floods on record, particularly the extreme floods. In fact, with the second method one can select representative floods (with the frequency of exceedence of 0·10 and 0·50 per cent per year in the above example) that are probably larger than any recorded floods. Still, these 'representative' floods do not yet fully express the characteristics of the drainage basin. Instead of the flood originating in the most likely part of the drainage basin, as probably has been assumed, the flood could originate in some other part. Instead of the peaks of two representative tributary floods being separated by a few days, as they usually are, they may coincide, etc. There are numerous ways whereby the real situation may deviate from the 'representative' situation.

The third method aims at overcoming these disadvantages in the following way. Frequency curves are prepared of the flood flows, upstream of any of the proposed flood control works. Selected floods from these curves are routed downstream, with and without the proposed flood control works. At any desired point, a pertinent frequency curve can be prepared from the known frequencies of the selected floods and their peak flows, as changed by the stream flow routing. At the junction of tributaries, the flood frequency curve of one stream will be combined with the frequency curve of the other stream, making due allowance for coincidence and correlation of flood peaks. After such a combination of frequency curves, into one new frequency curve, new representative floods have to be composed and routed further down the river system. When finally the area under consideration is reached, the check on the accuracy of the computations is that the computed frequency curve under natural conditions must check with the frequency curve of the flood flows, recorded at the terminal point. When this is the case, a good deal of confidence may be placed in the computed modified frequency curves for each of the proposed flood control schemes.

It will be appreciated that this third method is somewhat complicated, particularly when it comes to combining frequency curves of flood flows that are correlated. It would only be justified to attempt this method when adequate hydrologic data throughout the basin are available. Otherwise the precision of this method will be invalidated by the numerous assumptions that would have to be made.

Before we discuss the combination of frequency curves, let us first make some general observations with respect to the probability of two events happening at the same time. For instance, to throw a five with one die may be considered as having a probability of occurrence of 1 in 6; while throwing two fives with two dice has a chance of 1 in 36. Because of this simple and well-known example, it is sometimes believed that one could also make the following statement: when a peak flow of 10,000 cusec on one stream has a probability of exceedence of 10 per cent per year, and a peak flow of 10,000 cusec on another stream has also a probability of exceedence of 10 per cent per year, then a peak flow of 20,000 cusec below the confluence of the two streams, must have a probability of exceedence of 1 per cent per year. This statement is incorrect for three reasons.

First of all, it is likely that there is no perfect coincidence of peak flows on both streams. When the peak flow on the one stream occurs, the other stream flow may still be a few days away from the peak. The result of non-coincidence is that the peak flow of 20,000 becomes a much rarer event than indicated.

Secondly, it is quite possible that there is some correlation between the peak flows on the two tributaries. When one peak is high the meteorological conditions over the other drainage basin are probably also above normal, producing an above-average peak flow. The result of such correlation is that the 20,000 cusec becomes a more common event than indicated. In fact, when we assume perfect coincidence and perfect correlation of peak flows, the 20,000 cusec peak below the confluence would have a probability of exceedence of 10 per cent per year.

Let us now assume that there is perfect coincidence of peak flows on the tributaries and no correlation between the magnitudes of the peaks at all. Even in that hypothetical case, the earlier statement of 20,000 cusec having a probability of exceedence of 1 per cent per year is incorrect. Such a statement

Figure 6.21

2	3	4	5	6	7	1
3	4	5	6	7	8	2
4	5	6	7	8	9	3
5	6	7	8	9	10	4
6	7	8	9	10	11	5
7	8	9	10	11	12	6
1	2	3	4	5	6	

would be equivalent to saying the following: to throw a five or more with one die has a chance of 1 in 3. Hence to throw five plus five or ten points or more with two dice must have a probability of occurrence of 1 in 9. Upon inspection this is found to be incorrect. There are 6 possibilities out of 36 to obtain ten points or more. Not only the combination of fives and sixes have to be counted, but also the fours and sixes! It may be concluded that this aspect of joint

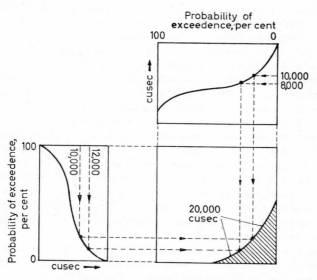

Figure 6.22. Combination of frequency curves

probability has the tendency to make the earlier quoted 20,000 cusec a more common event than indicated.

Determining the chance of throwing more than ten points with two dice can be represented numerically, as is shown in *Figure 6.21*. One would count the squares with a value of ten or larger, divide this number into the total number of squares and find the probability of equalling or exceeding ten points. In a

259

similar fashion we could represent the problem of river flows as is shown in *Figure 6.22*. One duration curve represents the probability of exceedence of the peak flows on one stream, while the other duration curve, placed at right-angles, represents the other stream. We now take the value of 10,000 cusec, enter one diagram till the duration curve is intersected, and follow the corresponding probability of exceedence value into the large square, formed by the limiting values of the two duration curve diagrams. The same thing is done with the 10,000 cusec of the other duration curve, and where the two lines meet in the large square a point is plotted. We will now take an arbitrary value in one diagram, say 12,000 cusec, use the complementary value of 8,000 cusec in the other diagram and find again the point of intersection in the square. When this is repeated a number of times, a line can be drawn as shown in the figure. This line represents all combinations that lead to a total of 20,000 cusec. All points to the right of that line must have higher values, and all points to the left, lower values. Therefore, if we determine the area to the right and divide it into the total area, we have obtained the probability of exceedence of 20,000 cusec. It should be noted that when the probability scales of the duration curves are not linear, the different areas in the square have no equal values. They must then be counted and converted to the proper value before they can be divided into one another.

This technique of combining frequency curves was developed in 1952 by C. Booy, hydraulic engineer. He also introduced the possibility of allowing for correlation and non-coincidence between the variables. The following paragraphs contain an abstract of his method:

'In the analysis of the effect of the flood control projects the problem was encountered of how to obtain the frequency curve of a function of two or more random variables when the frequency curves of the variables themselves and their correlation are known. Two special cases will be considered. The first case is one of a function of two or more independent variables. The second case is one of a function of two variables that are correlated, and it will be assumed that they are jointly normally distributed, or that simple functions of each variable are jointly normally distributed. With regard to the latter restriction, it may be noted that for meteorological data that follow a skewed frequency curve, a simple function of the original variable that approximately follows a normal distribution can generally be found. In this study the logarithm has been used for this purpose. The procedure for obtaining the frequency curve of a function of independent variables will be explained while considering an example.

'Suppose the variable Z is given as a function of two other variables X and Y such that $Z = X + Y$. The variables X and Y are assumed to be independent and a frequency curve is available for each. The problem is how to obtain the frequency curve of Z from this information. Before discussing this problem it will be necessary to explain a few terms that will be used.

'On *Figure 6.23* the frequency curve of X is shown on a lineal scale. It will be recalled that the probability that X belongs to a given interval $X_2 - X_1$ on the X axis is represented by a corresponding interval on the probability axis P_x. This interval will be indicated as $P_x(X_2 - X_1)$. If the interval on the X axis is taken as infinitely small it will be indicated briefly, for instance as $X = X_1$. The corresponding probability that X belongs to that interval is also infinitely small. This probability will then be denoted as $P_x(X_1)$ which quantity may be represented as a point on the probability axis P_x.

'If two independent variables are considered at the same time the one-dimensional probability axis may be replaced by a two-dimensional probability space. This may be accomplished by plotting the two frequency curves at right-angles, as was done in *Figure 6.23*. The probability axis P_x and P_y will then define a rectangular coordinate system in the probability space $P_x.P_y$. Every point in this space corresponds to one value of X and one value of Y, and every rectangle with sides parallel to the axis corresponds to one interval on the X axis and to one interval on the Y axis. Such a rectangle will be called an interval of the P_xP_y space.

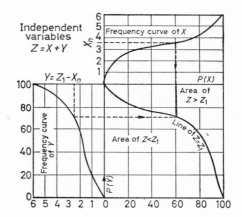

Figure 6.23. Combination of frequency curves

'From all possible combinations of X and Y, those may be selected that correspond to one infinitely small interval on the X axis, for instance, X_1. This set of combinations is evidently indicated on the P_xP_y space by all points lying on a straight line corresponding to $P(X_1)$. This line indicates a set of values for Y, for which set a frequency curve can be determined. This frequency curve will be called the conditional frequency curve of Y for $X=X_1$. In order to distinguish this frequency curve from the plotted frequency curve of Y, the latter will be called the marginal frequency curve of Y.

'When combining the variables X and Y it should be noted that these variables are assumed to be independent. This means that when X belongs to the interval X_2-X_1 and at the same time Y belongs to Y_2-Y_1 the probability of the combined event is $P_x(X_2-X_1)$ multiplied by $P_y(Y_2-Y_1)$. It follows then that the conditional frequency curves of one variable are the same as its marginal frequency curve. It also follows that the area of each interval on the P_xP_y space is a measure of the probability of the combined event indicated by that interval, since the sides of that interval correspond to the probability intervals on the axis.

'The probability will now be considered that Z exceeds a given value Z_1. For every fixed value of X the set of values of Y can be determined that would make Z greater than Z_1. For the infinitely small interval $X=X_n$ this is evidently the set of values of Y belonging to the interval from infinity to (Z_1-X_n) on the Y axis. The probability that X belongs to the infinitely small interval X_n and that at the same time Y belongs to the interval from infinity to $(Z_1 - X_n)$ may be written as: $P_x(X_1)$ times $P_y(\text{infinity} - (Z_1 - Z_n))$. On the two dimensional probability space this probability will be represented by a line segment. This line segment is shown heavy on *Figure 6.23*. If this line segment is determined for all possible values of X a surface will be defined on the probability plane, which represents the probability that Z exceeds the given value Z_1. In order to find this surface it is obviously sufficient to determine the line $Z = Z_1$ which line corresponds to all sets of values X and Y that would make $Z = Z_1$. By determining this line for several values of Z and computing the areas that are marked by these lines the frequency curve of Z can be obtained.

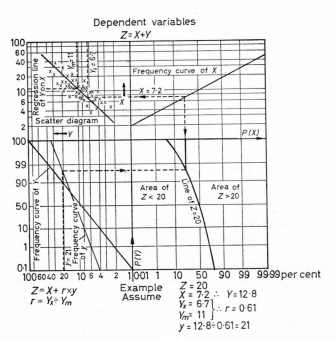

Figure 6.24

'This procedure can be extended to any number of independent variables by combining the resultant of two variables with the next. For the actual computations it was often more convenient to use a probability scale instead of a linear scale on the probability axis. This makes the evaluation of the areas on the XY plane more difficult, but it does not interfere with the principle of the method.

'The second case that will be considered is the frequency curve of a function of two variables, of which the logarithms are jointly normally distributed. I will again be assumed that the function is given by the equations $Z = X + Y$. On *Figure 6.24* the marginal frequency curves are plotted and a scatter diagram is also shown on a logarithmic scale.

262

'The same procedure as was followed for independent variables cannot be used here since for this case the multiplication rule for computing the probability of a combined event does not apply. In other words, the probability that X belongs to the interval $X_2 - X_1$ and that at the same time Y belongs to the interval $Y_2 - Y_1$ is not $P(X_2 - X_1)$ times $P(Y_2 - Y_1)$.

'A relatively simple solution of this problem can yet be found by splitting the variable Y into two factors r and y such that r is a definite function of X and that y varies independently of X. When that is accomplished the marginal frequency curve of Y can be replaced by the frequency curve of y. Intervals on the probability space are then a measure of the probability that the variables X and y simultaneously belong to given intervals on the X and Y axis. The line of $Z = Z_1$ may then be found on the probability space by computing r for various values of X and determining for those values of X and r the corresponding values of y that would make $Z = Z_1$.

'The problem thus remains of how to split Y into two factors r and y such that y is independent of X and that r is a definite function of X. From the theory of jointly normally distributed variables it follows that the regression curve of $\log Y$ and $\log X$ is a straight line. This regression curve indicates the means of all the conditional frequency distributions of $\log Y$ for fixed values of $\log X$. It also follows from this theory that all these conditional frequency distributions have the same standard deviation. Since all conditional distributions are normal they are completely defined by their means and their standard deviation. Thus, all the conditional frequency distributions can be made equal by subtracting from every item: $\log Y$, belonging to one conditional distribution, the difference between the mean of that distribution and the mean of all the values: $\log Y$. Then all standard deviations remain the same and all means of the conditional distributions are the same as the mean of all the values: $\log Y$. Since this case deals with the logarithms of the original variables this transformation may also be expressed as dividing every item Y in the scatter diagram by the ratio between the logarithmic mean of the conditional distribution to which it belongs and the logarithmic mean of all the items Y. The latter ratio will be called r. The logarithmic means of all the conditional distributions are given by the regression line of $\log Y$ on $\log X$, so r is uniquely determined as a function of X by this regression line.'

CHAPTER 7

WATER POWER

INTRODUCTION

The subject of water power development has many aspects. In fact, there are so many engineering problems involved that there is a danger of 'losing sight of the forest because of the many trees'. Let us therefore outline first of all the sequence of the studies that are made between the initial suggestion that a potential hydro site may be developed and the final construction of the project.

(1) Estimate of the future need for electricity.
    (*a*) Analysis of daily, weekly, annual load demand.
    (*b*) Extrapolation of past trends of growth into the future.
    (*c*) Comparison of existing resources with estimated loads.

(2) Appraisal of possible means of power production.
    (*a*) Steam and nuclear power.
    (*b*) Water power (provided by head × flow).
        Study of possible head developments.
        Study of available flows.
            Run-of-the-river.
            Improvement by pondage.
            Improvement by storage.

(3) Study of the least-costly alternative.
    (*a*) Estimates of plant cost.
        Over-all lay-out of power project.
        Preliminary design of components.
    (*b*) Estimates of energy production.
        Most economic regulation of reservoirs.
        Operation of entire power system.
        Sequential analysis of alternative projects.

(4) Development of the project.
    (*a*) The responsible agency reaches a decision.
    (*b*) The project is financed.
    (*c*) The detailed design is made.
    (*d*) Bids are invited—Contracts are let.
    (*e*) The project is constructed.

In the following paragraphs, the above sequence will be discussed in more detail.

When we discuss present and future electricity requirements we should realize first of all that such requirements are not constant in time. For instance, the electricity requirements of a residential area fluctuate not only during the day, but also from season to season; industrial requirements may be more constant from season to season but are reduced to a minimum during week-ends; whereas mining requirements may be nearly constant. In addition to the above changes, there may be a gradual change from year to year, due to a growth in population, or industrial development. For the sake of convenience all such changes may be expressed in approximate graphs and parameters such as daily, weekly, and annual load demand curves, load duration curves, load factors, etc.

In most areas, at least on the North American continent, we may assume that the present needs for energy are fulfilled, and our problem becomes one of estimating what the future needs for electricity may be. A widely accepted means of predicting future energy demands is to take the past record of energy production and to extrapolate the trend in growth, more or less corrected by economic considerations of the moment. The very people who make such predictions, are the first ones to admit the large amount of error that such estimates are subject to. However, as long as we have no better means of crystal-ball gazing, this method will retain its position for the simple reason that we need some sort of an estimate in order to be prepared for future conditions. A period from three to four years may elapse between the decision to go ahead with a certain power project, and the energy from this plant becoming available. Since it may also take a few years in advance of the decision to plan the project, and since the planners like to look a few years beyond the completion of the project to appraise how it will tie in with sub-sequent power development, it will be appreciated that load forecasts of one decade are minimum requirements and that forecasts of several decades are frequently made, although one sometimes wonders how realistic they are.

The result of this study of future energy needs is that we have specifications, from year to year, how much energy we need, in what localities, and in what shape. This prediction, adjusted from time to time when new information becomes available, forms the basis of forward planning of power development.

After having established what the electricity requirements are in the near future, and knowing of course what the capacity of the presently available system is, we are in a position to state how much additional capacity should be installed and at what time such capacity must be available. We do not know, as yet, if this capacity should come from steam, nuclear or hydro plants. This depends largely on the local circumstances. There are vast areas, on the North American continent, where coal is abundantly available and where water power sites are few or non-existent; in such areas the provision of steam capacity is the logical choice, without much deliberation. There are also areas, like the Pacific Coast, where good water power sites and ample stream flow is available; in such cases the provision of hydro-capacity is the obvious solution. Then we may quote Great Britain as an example where coal supplies are limited and further hydro development of no significance; as a result the power engineers of that country are concentrating on the development of nuclear power production.

In many areas, however, the choice of future capacity development is not obvious, and consequently we have to weigh carefully one alternative against the other. In this chapter we shall discuss how to determine the cost of water power development. We will assume that the cost of alternative steam and nuclear power development is made available by others.

265

After having decided to investigate the possibilities of future hydro development, we should make first of all a complete inventory of all potential hydro sites that are within transmission distance of the power market under consideration. Since hydro energy is the product of the available head at the plant times the available flow at the plant (times a certain factor), the investigation will be concentrated upon the sites with a good combination of head and flow. Since the transmission of energy from the hydro plant to the load centre is costly and involves the loss of energy, the investigations will largely be concentrated upon sites within reasonable distance.

The study of possible head development is a matter of carrying out topographic surveys to determine the gradient of the river and the shape of the river valley; conducting geological surveys to determine foundation conditions at prospective dam sites; and applying common sense in laying out dams, intakes, penstocks, spillways, power houses, etc. There is usually such an infinite variety of possibilities in locating these works, that it becomes impractical to prescribe hard and fast rules. The art of laying out a power development may be greatly improved, of course, by knowing how such problems have been solved elsewhere.

The study of available flows starts out by collecting available stream flow data, possibly supplemented by special hydrometric surveys. Of particular importance are the minimum recorded stream flows, from the viewpoint of dependable plant capacity; and the average stream flows from the viewpoint of energy output. If we realize that nearly all power companies are committed to satisfy, under all circumstances, the power demands of their customers, we appreciate how important the low flow conditions are, which determine to a large extent the dependable plant capacity of a power system. Upon inspection of the drainage basin, we may find it possible to create storage reservoirs that could increase the dependable flow as well as the energy output by storing water during times of flood flow, and to release this water during times of low flow. Such reservoirs may be created directly behind the power dam or they may be situated further upstream. If the reservoirs only serve to smooth out the weekly differences between stream flow and load demand, they are said to have pondage capacity; if they serve to store water from the wet season to the dry season they are said to have storage capacity.

After having appraised the water power potential of the drainage basin, we can lay out a number of tentative projects that are in accordance with ultimate full development of the river system, and prepare preliminary designs and cost estimates of the different components of the projects. This provisional design is carried to the point that the cost of the alternative projects are estimated with an accuracy of say 10–20 per cent. The result of our studies up to this point is that we have now available a list of water power projects (some of them alternatives) that could satisfy part or all of the increase in power demand that we expect in the near future. We shall assume that we have also available cost figures of producing power and energy by means of conventional thermal and nuclear thermal plants.

The next problem that we are faced with is to determine which of the many possible sequences of development over the next five or ten years, is the most economical. It is seldom possible to reduce this problem to a mere comparison of the cost of two single alternative additions to the system. The reason is that we cannot study each alternative by itself, but that we have to study it in conjunction with the rest of the power system to which it is going to be connected. The existing power system, under future load conditions, will be operated differently, whether the addition to the system consists of a steam plant, a nuclear power plant, or a hydro plant. Moreover, we may have found that a certain amount of hydro energy is available at one of the potential hydro sites, but unless we make a study of the whole power system, we do not know if this hydro energy is usable! We are therefore faced with the necessity of making a complete system study for each alternative addition to the system. If we have reservoirs in the system, their integrated regulation must be studied until maximum hydro energy output is obtained.

There is a further complication. A hydro project, which for practical purposes has to be built all at once at full size, is likely to require a relatively large capital investment within a short period of time. A steam plant, on the other hand, which may be built unit after unit, usually requires less capital investment, which is spread over a longer period of time. As a result, the interest charges weigh heavily against the early years of a hydro project when its available hydro energy is not yet fully usable in the system. Consequently, we have to study the various alternatives not for only one year, but for at least a decade into the future, in order to get a realistic comparison. After such a comparison of the various alternatives has been made, we are in a position to recommend the most economical addition to the existing power system.

After the engineers have made their recommendation it is up to the appropriate agency to decide upon a course of action. Although it may normally be expected that the engineers' recommendation is followed, there could be several reasons for not doing so. The engineers may have found that the most economical addition is a very costly hydro project. The economic position of the agency may make it impossible, however, to finance such a project! On the other hand, let us assume that a steam plant addition was found to be the most economical solution. In spite of that, it could be decided to go ahead with the hydro alternative since that makes use of renewable resources, while a steam plant uses non-renewable resources.

After the decision has been made, the selected project has to be financed. If a government agency is involved, the project may be financed by government bonds or a special series of bonds may be issued that are guaranteed by the government. If a private agency is involved, the project may be financed by a combination of bonds, debentures, and possibly stocks.

Immediately after the green light has been given on a project, and sometimes already before that, the detailed design will be started. This will be guided largely by the preliminary designs that were made during the planning stage. Field explorations will be completed, hydraulic computations will be finalized, and gradually the project takes shape, although it usually remains in a state of flux till the very last moment.

As soon as funds and designs are available, contracts can be opened for bids and let to the most attractive bidder. The sequence of construction may be as follows: access roads—construction of camp site—clearing—cofferdamming for a spillway—excavation for spillway—construction of spillway—diversion of river through spillway—cofferdamming for power house—excavation for power house—construction of power house—installation of turbines, generators

267

and hydraulic and electrical equipment—final testing—connection with the power system.

BASIC CONCEPTS

The main emphasis in this chapter will be placed on aspects of water resources development such as river flow utilization, reservoir regulation and system planning. However, before such aspects can be discussed, we should first of all define and try to understand the basic concept of water power engineering.

Plant Lay-out

Hydro-electric plants are generally divided into low-head plants, medium-head plants, and high-head plants. This division is rather arbitrary and no sharp line can be drawn between the three categories. A typical example of each group is shown in *Figures 7.1, 7.2,* and *7.3.*

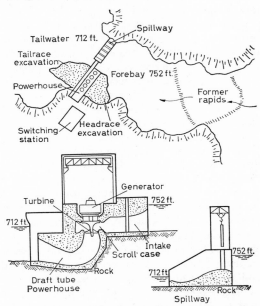

Figure 7.1. Low-head water power development at Pine Falls, Manitoba

Low-head power developments usually range in head from a few feet to about 50 feet. Since the head is relatively low, the power house, including intake, scroll case, and draft tube, performs also the function of dam. Since a low-head development is likely to be found on the lower reaches of a river, it can be expected that it uses a relatively large discharge. This results in relatively large dimensions of the power house. Besides the power house we usually find the spillway, passing all river flow that cannot be used through the turbines. Because of the large flow and low available head, the spillway also requires ample dimensions. It is not unusual to find low-head power developments where the entire width of the river channel is occupied by power house

268

and spillway, and where a regular dam section is non-existent. In fact, it is not unusual that the normal width of the river is insufficient to accommodate power house and spillway, and that one has to locate the plant at an unusual wide section, or excavate the channel, or place the structures at an angle. Low-head plants are equipped with propeller type turbines.

Medium-head power plants may range in head from 50–200 ft. Since it would be uneconomical to build a power house that could, at the same time, act as a high dam we usually find the dam and the power house detached. If there is no outspoken drop of the river bottom at the dam site, the height of the dam will be of the same order of magnitude as the available head of the plant.

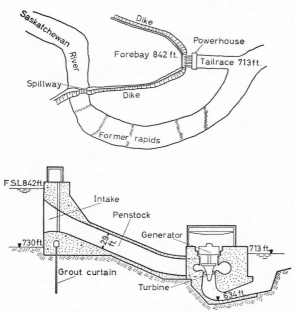

Figure 7.2. Medium-head water power development at Grand Rapids, Manitoba

If there is a significant, sharp drop in river bottom, it may be possible to build the dam on the edge of the falls, and the power house at the bottom, with a short conduit in between. If there is a significant but gradual drop in the river bottom, the situation may lend itself to a variety of solutions. The dam may be built at the head of the rapids and the power house at the foot, with any combination of open canals, tunnels or penstocks in between. Dam and power house may be built half way, requiring extensive excavation at the tailrace. Dam and power house may be built at the foot of the rapids, requiring a relatively high dam. Two dams may be built, each with their own power house. All such possibilities have to be investigated upon their costs, before a choice can be made. Medium-head plants may have propeller turbines (up to 150 ft. in head) or Francis turbines (over 50 ft. in head).

High-head power developments may range in head from 200–5,000 ft. Although some of these developments may resemble the medium-head type (high dam with power house and spillway), the usual arrangement is to intercept the river flow at a suitable place, to divert the water via an open canal with small gradient, until a point is reached where the head between the canal and river is as large as possible and the horizontal distance between the canal and the river as small as possible. The water is conveyed from the canal to the river by means of penstocks to the power house which is located on the river bank. It is understood, of course, that there are infinite variations on this theme. The open canal, or the penstocks, or both, may be replaced by tunnels; or the open canal may be replaced by a low pressure closed conduit. High-head plants may have Francis turbines (up to about 800 ft. in head) or impulse turbines (over 400 ft. in head). Since high-head hydro plants are likely to be

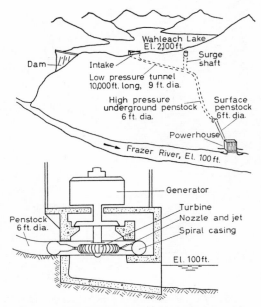

Figure 7.3. High-head water power development at Wahleach Lake, B.C.

found in the headwaters of a river basin where the gradients are steep, it may be expected that the discharges are low. The combination of high-head and low discharge will reduce the relative size of the turbines, and their cost. Although part of this saving may be offset by the cost of the penstocks and canals or tunnels, it is not unusual to find high-head plants among the least costly developments per kW of installed capacity.

Plant Capacity

It was discussed in Chapter 3: Hydraulics, that an element of water, flowing in a water course, may have three types of interchangeable energy: potential energy (due to its elevation above an assumed datum); pressure energy (due to the local pressure); and kinetic energy (due to its velocity).

The Bernoulli equation states that the sum of these forms of energy remains

270

constant, provided that there are no energy losses. The terms of the equation are usually expressed in energy per unit of weight: $(F.L)/F = L$, as follows:

$$h + \frac{p}{\gamma} + \frac{V^2}{2g} = \text{constant}$$

or, in words:

$$\text{elevation} + \text{pressure head} + \text{velocity head} = \text{constant}$$

When we observe an element of water flowing in an open water course, we may assume for practical purposes that its pressure energy and kinetic energy do not change, and we notice that its potential energy gradually decreases. This loss of energy is due to internal friction and friction between water and the channel boundary. Actually, the difference in potential energy is not lost, but converted into heat. When we compute the rise in temperature of the river, due to a lowering of its elevation, we find that it is relatively small. To raise the temperature of 1 lb. of water by 1°F requires 1 B.Th.U. of energy. The lowering of 1 lb. of water over 1 ft. is equivalent to a loss in potential energy of 1 ft. Lb. The conversion between these two forms of energy is: 1 B.Th.U. = 778 ft. Lb. Therefore, water would have to fall over 778 ft. in order to raise its temperature by 1°F (assuming that the loss in potential energy is completely converted into heat, and that all of this heat is contained in the water).

The above principle of energy conversion is sometimes used to measure the efficiency of a hydro plant. If the plant would operate at 100 per cent efficiency, no potential energy would be converted into heat, and the water at the intake and tail race would have the same temperature. Any deviation is therefore a measure of inefficiency. Since water temperatures cannot be measured with greater accuracies than about 0·003°F, which is equivalent to about 2 ft. of potential energy loss, it follows that this method will only give useful results for plants with a head well over 200 ft., in order to determine the plant efficiency accurate to within one per cent.

The principle of a hydro-electric development is to reduce to a minimum the natural conversion of potential energy into heat. Instead, the difference in potential energy between two points is converted into electrical energy. In order to minimize the loss of energy due to friction, we have to convey the water in channels and conduits with minimum hydraulic resistance (low velocities, smooth boundaries). In order to generate electricity we need turbines and generators.

Let us follow an element of water on its journey through a hydro development and let us see where energy losses may occur, how the Bernoulli equation may be applied and how the energy conversion takes place. *Figure 7.4* shows a typical high-head development. The total developed head between maximum reservoir level and tail water level is 1,000 ft. The reservoir, fluctuating between the levels of 980 and 1,000, has the function to modify the irregular river flows so that they can meet the regular load pattern. The first problem arises in conveying the water from the outlet of the reservoir at A to the intake of the penstocks at B. This may be done in an open channel. In that case, the next problem is whether we are going to have an open connection between the reservoir and the canal (resulting in a possible range of water levels at B from 970 to 1,000), or if we should design an outlet with energy dissipater at A that lowers the water to an elevation of 980 regardless of the reservoir level. In the first case the energy losses are minimized but the cost of the canal dikes may become prohibitive. In the second case we lose an average of 10 ft. of head at point A. A more economic solution may be to design a low pressure wood-stave conduit between A and B. Another solution that may be considered is to have an open canal from elevation 1,000 to 990 and a pumping plant at A to lift the water from the reservoir into the canal, whenever the reservoir is below elevation 1,000. We only have to lift over an average head of 10 ft., and we gain 20 ft. at B. This may offset the cost of the pumping plant and the required energy.

In *Figure 7.4*, an energy dissipater at A, and an open channel from A to B has been assumed. An element of water on this reach would undergo an energy loss of 30 ft. At point B, assuming that the element is near the surface of the canal, its elevation is 970 ft., while its pressure head and velocity head are

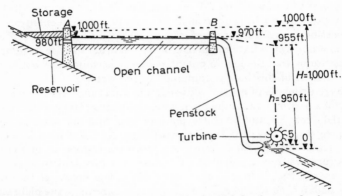

Figure 7.4. Gross head H and net head h of a high-head power development

practically zero. As soon as the element of water has entered the penstock, its velocity is, say, 18 ft./sec and hence its velocity head about 5 ft. This remains the same while the element descends towards the power plant. In the meantime an exchange between potential and pressure energy takes place. Between B and C the energy loss is assumed at 15 ft. Hence, at point C, just before the element of water enters the nozzle, its elevation is 5 ft., its velocity head 5 ft., and its pressure head 945 ft. At point C, immediately after the element has passed through the nozzle, and assuming no losses in the nozzle, the elevation is still 5 ft., the pressure head is zero (atmospheric), and the velocity head is 950 ft., which corresponds to a velocity of 248 ft./sec. After impact with the turbine, the velocity of the element of water is reduced to a small value, say 25 ft./sec, corresponding to a velocity head of 10 ft. This kinetic energy plus the last potential energy of 5 ft. will be lost when the element of water falls back in the river. In this example, the gross head H on the power development is 1,000 ft., while the net head h on the turbine is 950 ft. About 90 per cent of this energy will be converted into electrical energy.

For medium- and low-head developments with Francis or propeller turbines, the interchange of energy of an element of water during the last part of its journey from headwater to tailwater is somewhat different, as shown in

272

Figure 7.5. An element of water coming out of the penstock and entering the scroll case may have an elevation of 15 ft.; a velocity of 25 ft./sec. and hence a velocity head of 10 ft.; and since the total head is 97 ft., this leaves a pressure head of 72 ft. Immediately after leaving the turbine, and having converted most of its energy into electrical energy, the element of water has an elevation of 10 ft. The velocity may still be around 25 ft./sec and its velocity head 10 ft. The total head at this point is 3 ft. above the datum of tailwater. Hence the pressure head is 17 ft. negative. From here to the outlet of the draft tube, the element loses 3 ft. of energy and then flows back into the river, with practically no potential energy (with respect to the datum of the tailwater level), no kinetic energy and no pressure energy. In this example the energy conversion was mostly one of pressure energy into electrical energy. The gross head H on the power development is 100 ft., while the net head h on the turbine is 94 ft. About 90 per cent of this is converted, while the remainder is lost in heat.

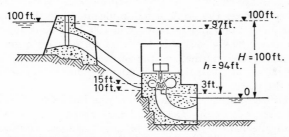

Figure 7.5. Gross head H and net head h of a medium-head power development

The net head h, acting on the turbine, is considered to be the total head from the entrance of the scroll case to the tailrace level at the exit of the draft tube. The difference between the gross head and the net head is equal to the sum of the hydraulic losses at the intake, the conduit, and the penstock. For low-head developments, the net head may be practically the same as the gross head. For high-head developments, the net head may be 90 per cent or even less, of the gross head. In view of the above definition of the net head, it is evident that the following hydraulic losses are charged against the inefficiency of the turbine: scroll case losses, guide vane losses, runner losses (surface resistance, form resistance, leakage), draft tube losses, and exit losses. In addition to these losses we have the bearing losses of the turbine-generator.

Power is the rate of producing energy. In the present case it would be more correct to say that power is the rate of converting energy, because the energy of the falling water is converted into electrical energy. To raise 1 lb. over h ft. required $1 . h$ ft. Lb. of energy. Lowering Q ft.³ over h ft. could release $Q \times 62 \cdot 4 \times h$ ft. Lb. of energy. If we lower Q ft.³ per second we could release energy at the rate of

$$Q \times 62 \cdot 4 \times h \, \frac{\text{ft. Lb.}}{\text{sec}}$$

Since

$$550 \, \frac{\text{ft. Lb.}}{\text{sec}} = 1 \text{ h.p.}$$

273

we may say that we produce a rate of energy (power) of

$$\frac{Q \times 62 \cdot 4 \times h}{550} = \frac{Q \times h}{8 \cdot 8} \text{ h.p.} = \frac{Q \times h}{11 \cdot 8} \text{ kW}$$

allowing for the loss of energy in turbine and generator.

$$P_{\text{kW}} = \frac{Q.h.e}{11 \cdot 8}$$

in which P_{kW} is the capacity of the plant in kilowatts, Q the discharge through the turbines, h the net head on the turbines, and e the efficiency of turbines and generators.

Energy is the amount of work done. It is most conveniently expressed in kilowatt-hours. If a power plant for instance produces a power of 10,000 kW for a duration of 5 h, the amount of energy produced is 50,000 kWh. In this chapter the amount of energy will also be expressed in 'MW continuous'. This is to be interpreted as the MW listed multiplied by the time listed. For instance: 'the energy demand in January is 100 MW continuous' means the energy demand is $100,000 \times 31 \times 24$ kWh. If we prepare a graph where the horizontal scale represents time, and the vertical scale the instantaneous power produced by the plant, then the area under the curve represents the energy produced by the plant.

It was noted earlier that the efficiency of the intake and conduits depends largely on the lay-out of the plant, and may range from 0·90 to 1·00. The hydraulic and mechanical efficiency of the turbines, combined with the mechanical and electrical efficiency of the generator, may range from 0·80 at older plants to 0·90 at new installations. The efficiency of transmission lines depends entirely on the voltage being used, the size of the conductors, and the transmission distance. It may range from 0·90 to 0·99.

Load Factor

Figure 7.6(a) is an example of how the residential industrial load of a fairly large power system may fluctuate during a week. Since the original weekly

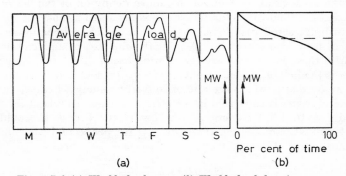

Figure 7.6. (a) Weekly load curve; (b) Weekly load-duration curve

load curve is not the most convenient curve to work with, we may prepare a so-called weekly load-duration curve, as shown in *Figure 7.6(b)*, where certain

274

characteristics of the load condition are more readily understood. The ratio of the average load to the peak load over any given period of time is known as the load factor. Normally, a daily load factor is larger than a weekly load factor, which in turn is larger than the monthly load factor, which in turn is larger than the annual load factor. *Figure 7.7* shows how the annual load duration curve is built up from a series of 12 monthly load duration curves. It may be

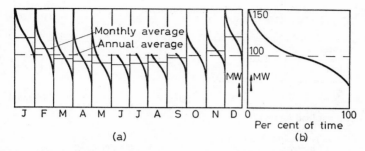

Figure 7.7. (a) *Monthly load duration curves;* (b) *Annual load duration curve*

seen from *Figure 7.7(b)* that the annual load factor is $100/150 = 0.66$. The installed capacity of the power system is 150 MW plus, say, 10 per cent spare capacity $= 165$ MW. The annual amount of energy produced is $100,000 \times 8,760 = 876,000,000$ kWh.

During the past 40 years the average load growth of power systems on the North American continent has been approximately 7 per cent per year, which comes very close to a doubling of the load every 10 years. Another experience

Figure 7.8. *Load forecast*

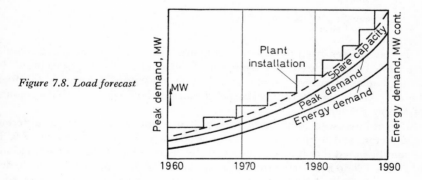

has been that with the growth of the system, the load factor also tends to increase somewhat. In a small system, an annual load factor of 0.5 is not unusual. In a large system, particularly when the cost of energy is low so that large industries are attracted, annual load factors up to 0.7 may be found. A future load forecast and system development may therefore look somewhat like the one shown in *Figure 7.8.*

275

Hydro and Thermal Costs

It was noted earlier that a power utility may extend its system by hydro as well as thermal capacity. In fact, it happens often that integration of these two types yields the greatest economy. In order to appreciate this, the difference in cost structure between hydro and thermal power is illustrated with the following examples.

			(dollars)
Capital cost of hydro plant including transmission: 200,000 kW × $350 =			70,000,000
Annual fixed charges (per cent):			
Interest	5		
Insurance and Taxes	1		
Depreciation	1		
Operation and Maintenance	1		
	8		5,600,000
Annual load factor:	0·6		
Spare capacity:	10%		
Annually produced energy:	945,000,000 kWh		
Cost of hydro:	5·9 mills per kWh		
Capital cost of steam plant: 200,000 kW × $150			30,000,000
Annual fixed charges (per cent):			
Interest	5		
Insurance and Taxes	2		
Depreciation	2		
Operation and Maintenance	2		
	11		3,300,000
Fuel cost: 945,000,000 × $0·005 =			4,700,000
			8,000,000
Annual load factor:	0·6		
Spare capacity:	10%		
Annually produced energy:	945,000,000 kWh		
Cost of steam:	8·5 mills per kWh		

It may be seen from the above cost figures that once a hydro plant is built, the annual charges are fixed, regardless of the amount of energy produced. In a steam plant, however, more than half of the annual charge is for fuel. In other words, the total annual cost of the steam plant goes up with the amount of energy produced.

It is of interest to compute the cost of hydro and steam energy for different load factors and to plot the results on a diagram, as shown in *Figure 7.9.* It should be noted that it is not quite realistic, as was done for the computations of *Figure 7.9*, to keep using the same capital cost of $350 per kW for the hydro plant, regardless of its load factor. Usually the cost of adding capacity to a hydro plant does not involve extra costs for the dam and spillway, and therefore the incremental cost per kW becomes somewhat less than the initial cost per kW. This would have the effect of somewhat lowering the cost per kWh figures for low load-factor hydro energy.

It can be seen from *Figure 7.9* that for high load factors, hydro energy is considerably cheaper than steam energy, in spite of the fact that the capital cost of the hydro plant is more than twice as large as the capital cost of the steam plant. The explanation is to be found partly in the fact that steam plants carry higher annual charges and partly because steam energy requires the

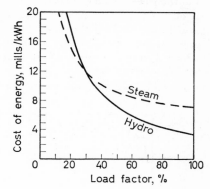

Figure 7.9. Cost of hydro and steam energy

burning of fuel. In this example the point of equal cost between hydro and steam energy is somewhere near a load factor of 0·3. Below this load factor, steam energy is less costly than hydro energy.

It appears from this analysis that an addition of hydro plants to carry the base of the load, supplemented by steam plants to carry the peak of the load, may be more economical than the addition of hydro plants only or steam plants only.

If we have steam plants and hydro plants in a system, it is obvious that we try to use the steam plants as little as possible since that requires extra expenses, while we can run the hydro plants at no extra expense. In other words, we put hydro on the base of the load, while we assign the peaks to steam. It is understood that the available hydro capacity plus the available steam capacity should equal the system peak demand plus 10 per cent for spare. Such an operation of hydro and steam is quite feasible when there is ample flow in the river system. However, when the river flows are low, we may not have enough water to run all hydro plants at full capacity, and of necessity we may have to place the hydro plants in the peak and steam on the base, as illustrated in *Figure 7.10(a)* and (*b*).

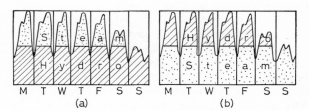

Figure 7.10. (a) Ample river flow; (b) Low river flow

Peak Percentage Curve

One of the important features in the planning of hydro development is to be reasonably sure that the full capacity of all hydro plants can be used, even

277

under the most adverse river flow conditions. A convenient tool in the analysis of the situation is the so-called peak percentage curve.

Let the weekly load curve be represented by *Figure 7.11(a)* and let us assume that one of the hydro plants in the system has a capacity of 100 MW and is placed in the peak of the load. The shaded area in the weekly load diagram represents then the amount of energy that the 100 MW installation has to produce during the week. If we planimeter this area and apply the proper conversion factor, we can determine the energy in kWh. Instead of doing this in the weekly load diagram, we may arrive at the same result by using the more

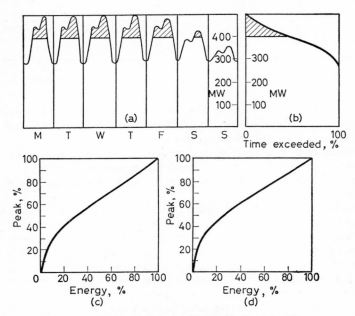

Figure 7.11. (a) *Weekly load curve;* (b) *Weekly load duration curve;*
(c) *Weekly peak-percentage curve;* (d) *Annual peak-percentage curve*

convenient weekly load duration curve, only we have to use a different conversion factor. Let us assume that the shaded area in the weekly load duration curve is 4 per cent of the total area under the curve. We may then conclude that any installation, placed in the peak of the load, and having a capacity equal to 20 per cent of the total system peak demand, only requires 4 per cent of the total system energy, to be fully operative. In order to have such information readily available, we can make such computations for the entire load duration curve and plot the results in a so-called weekly peak-percentage curve, shown in *Figure 7.11(c)*. If we repeat similar computations on an annual basis, we obtain an annual peak-percentage curve as shown in *Figure 7.11(d)*. It may be seen that the latter has a steeper inclination at the beginning of the curve.

To illustrate the above discussion let us assume the following: the hydro plant of 100 MW has a head of 148 ft. and an efficiency of 0·8. After having planimetered the shaded areas in *Figure 7.11(a)* we find they have a total area of say 1 in.[2] Assuming a horizontal scale of the diagram of 1 in. equals one day, and a vertical scale of the diagram of 1 in. equals 100 MW, we find that 1 in.[2] equals 2,400,000 kWh. To produce this much energy in one week we need an average flow at the hydro plant of 1,430 cusec. Assuming that the plant has pondage capacity (as will be explained in the following section) we may say that required dependable river flow is 1,430 cusec. Now let us use the weekly load duration curve of *Figure 7.11(b)* to make the same computations. We planimeter again the shaded peak and find it to be 0·43 in.[2] Assuming a vertical scale of the diagram of 1 in. equals 100 MW and a horizontal scale of 3 in. equals 100 per cent of the time which equals 7×24 h, this means that one square inch equals 5,600,000 kWh and hence the shaded area represents 2,400,000 kWh, which is the same as was found earlier. Finally, we can use the peak-percentage curve of *Figure 7.11(c)* to make the same computation. 100 MW represents 20 per cent of the total peak. According to the peak-percentage curve 20 per cent of the peak corresponds to 4·3 per cent of the total energy. Since the monthly load curve has a load factor of 0·66 (which is either known or can be found from planimetering), the energy requirement becomes: 4·3 per cent $\times 0·66 \times 500,000 \times 7 \times 24 = 2,400,000$ kWh. This again leads to a required dependable flow of 1,430 cusec, since $\dfrac{1,430 \times 148 \times 0·8}{11·8} \times 24 \times 7 = 2,400,000$ kWh.

Firm Capacity

A peak-percentage curve is convenient when appraising the firm capacity (dependable capacity) of a hydro installation. Let us assume that we have made an analysis of river flow conditions and storage capacity at a particular hydro installation, and that the dependable river flow was found to be 10,000 cusec. If we further assume a net head of 148 ft. and a plant efficiency of 0·80, we may say that we have available a dependable hydro energy (sometimes called prime power) of 100 MW continuous. During one week this would amount to $100,000 \times 24 \times 7$ kWh. Let us further assume that the total amount of required system energy is 330 MW continuous, and that the total system peak demand is 500 MW. In other words, the above hydro plant has available 30 per cent of the total system energy, and when this plant is placed in the peak of the load, it could, according to the weekly peak-percentage curve, operate with a capacity equal to 50 per cent of the total system peak demand, which comes to 250 MW. Therefore, the firm capacity of the hydro plant is 250 MW.

We may define firm hydro capacity as the capacity of a hydro plant that can be made fully effective, for given load conditions, with a given dependable stream flow, and with a given position in the generating system.

To amplify the above definition, let us assume that the dependable stream flow remains the same (10,000 cusec), that the position in the generating system remains the same (in the peak of the load during times of low stream flow), but that the system load is doubled in energy demand and peak demand. The plant has now available 15 per cent of the system energy, and according to the weekly peak-percentage curve it can make effective 35 per cent of the peak demand of 1,000 MW, which comes to 350 MW. In other words the firm capacity of a hydro plant will increase with a growth in load conditions, all other circumstances remaining the same.

This same principle is graphically illustrated in *Figure 7.12(a)*, (*b*), and (*c*). In each of the three load duration curves, the dependable stream flow, and hence the amount of prime power, and hence the area under the load duration curve, is the same. It may be seen that the vertical ordinate of the shaded area, representing the corresponding firm plant capacity, becomes larger when the same amount of prime power is fitted into the sharper peak of a larger system load.

Let us now assume that the load condition and the position of the plant remain the same, but that the dependable flow is increased, due to the provision of upstream storage. It is obvious that this will increase the dependable hydro energy, which in turn will increase the firm capacity of the plant. It is of interest to point out that there is a rather strong curvature in the graphical relation between storage capacity and dependable flow. Doubling the first will not

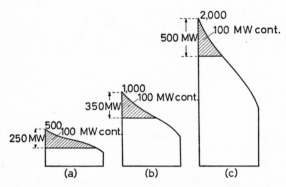

Figure 7.12. Firm capacity and load condition

result in doubling the latter, as may be seen from *Figure 2.4(b)*, page 28. Moreover, it may be seen from the shape of the peak-percentage curve, that a doubling of the dependable flow will result in less than doubling of the firm capacity. In fact, we may increase the dependable flow to the point where our hydro plant under consideration may produce all the required system energy. The firm capacity of the plant is then equal to the system peak demand. Any further increase in dependable flow has no effect, since there is no demand for more capacity!

For the sake of simplicity, we have considered in the above only one hydro plant, placed in the peak of the load during times of low stream flow, and we have disregarded what happened to the rest of the generating system. Let us now assume that we have two hydro plants, each with a prime power of 100 MW, and let us determine the firm capacity of each plant, using the same example as above. Only one plant can be placed in the peak, and may be assigned a firm capacity of 250 MW. The second hydro plant has to operate under the first plant, and it can be seen from the peak-percentage curve that its available 30 per cent of system energy (going from 30 to 60) will only make effective another 20 per cent of system peak demand (going from 50 to 70). Therefore its firm capacity is only 100 MW, although its prime power is the same as the first hydro plant! In other words the firm capacity of a hydro plant depends, among other things, upon its assigned position in the generating system.

It may be of interest to pursue this subject a little further and discuss what consideration may play a role in assigning a hydro plant to the first or second place in the peak of the load. One consideration may be the incremental cost of plant capacity. This is the cost required for installing additional units at a project, after the dam, reservoir and spillway have been provided. If all conditions at two hydro plants are the same, except that the incremental capacity cost of one plant (for instance a high-head plant) is lower than at the other plant (for instance a low-head plant), then it would be more economical to develop the first plant for operating in the peak of the load during times of low flow. In other words, the first plant will be developed to a larger capacity than the second plant, although they may both have the same prime power.

Another consideration may be the available head at the plants. Let us assume that we have one plant with a high head and small flow, and another plant with a low head and a large flow, while the two plants have the same pondage capacity in terms of prime power (flow × head × time). If each plant, in turn, would be operated in the peak of the load, the high-head plant would likely have a smaller percentage of fluctuation of head, and would therefore lose less energy. Hence the high-head plant would be more suitable for peaking development.

The above problem of assigning a hydro plant to a certain position in the peak of the load, becomes complicated if we are dealing with an all-hydro, or dominant hydro system, and we have to determine which plants should operate in the peak of the load and which plants in the base of the load. It is understood that we have first of all ascertained that the total amount of prime power at all plants, plus the available amount of steam energy, equals the system energy demand. With the amount of prime power (which represents area under the load-duration curve) approximately given at each plant, the problem is one of determining at which plant the least amount of energy is lost when that plant is operated in the peak of the load, rather than on the base of the load. Operation in the peak means installing a higher capacity and being idle part of the time. This requires pondage, which in turn means fluctuating forebay levels below the full supply level, with a consequent loss of energy. This loss of energy will be least when we select plants with a large forebay area (which reduces the fluctuations) or plants with a high head (which require a lesser volume of water for incremental capacity).

We may therefore conclude that in developing hydro plants for peaking purposes, preference will be given to those plants that have either a relatively high head, or a relatively large forebay, or a relatively low incremental capacity cost.

Before we leave the subject of firm hydro capacity, let us discuss how this is influenced by the addition of steam capacity to the system. We will assume that we have a dominant hydro system and that the load growth is met by steam additions. The amount of prime power at the hydro plants remains the same, but in a larger system the hydro plants can operate during times of low flow relatively higher up in the peaks, therefore operate at a lower load factor, and therefore have a larger firm capacity with the given amount of prime power. In other words, the addition of steam capacity increases the firm capacity of hydro plants. Since the addition of storage capacity also increases the firm hydro capacity, it is sometimes said that steam and storage have the same effect upon a hydro system, namely one of firming up its capacity.

REGULATION OF RESERVOIRS

It has become evident from the above discussions that the value of a water power development is enhanced when its dependable flow is increased or when the fluctuating river flows can be modified to agree with the energy demands of the power system. Such a function is performed by storage reservoirs. When this is done on a weekly basis it is called pondage. When it is done on a seasonal or cyclical basis it is called storage. The various aspects of pondage and storage will be discussed in the following paragraphs.

Pondage

If the energy that can be generated by a hydro plant from a steady stream flow equals the average weekly load demand, one may think that the plant will have no difficulty in meeting the load. However, the stream flow provides a steady supply of energy, whereas the system load is continuously fluctuating. The only possibility to let the steady supply meet the fluctuating demand is to store water in the reservoir above the plant whenever there is less demand than supply and to draw water from the reservoir when there is more demand than supply. If a hydro plant is able to do this over weekly periods, it is said to have sufficient pondage capacity. The fluctuation of the pondage reservoir may look

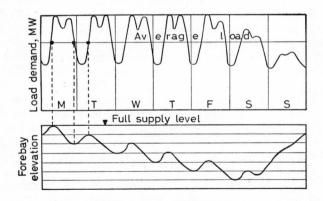

Figure 7.13. Pondage operation

somewhat as shown in *Figure 7.13*. It may be seen from this figure that due to the pondage operation, the average reservoir level is less than the full supply level. This is a disadvantage since it means that the plant operates at a less than maximum head and therefore produces less energy than possible.

If there are several hydro plants in the system on different rivers, some with large and some with small reservoirs, it would be logical to place the plants with small reservoirs on the base of the load, so that they can operate with a constant maximum head, and to place the plants with large forebays in the peaks of the load since their head will hardly be affected by pondage operation. If several plants are on one river, a good arrangement would be if the plants are served by one upstream reservoir of ample capacity. It is understood, of course, that these plants should be close together, or else the reservoir loses its effectiveness.

In the majority of cases, hydro plants have sufficiently large forebays to allow for pondage operation, although some of them may suffer from consequent energy loss. If a plant with its forebay can only be operated for pondage and not for seasonal storage, it is sometimes referred to as a run-of-the-river plant.

It is of interest to note that the possibility of pondage increases the firm capacity of a hydro plant for given flow conditions. Let us assume that we have a hydro plant with a net head of 148 ft. and an efficiency of 0·80. Now let us assume that the lowest recorded winter flow of 11,500 cusec is considered to be

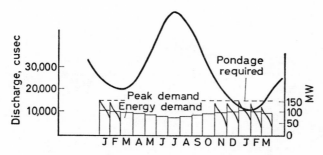

Figure 7.14. Effect of pondage on firm capacity

the natural dependable flow of the river. It is evident that without pondage, the dependable capacity of the hydro plant is 115 MW. With pondage, however, the dependable capacity becomes 115 MW divided by the weekly load factor of, say, 0·77 which is 150 MW, an increase of about 30 per cent! This situation is illustrated in *Figure 7.14.*

Storage

When a power system is served by a number of run-of-the-river hydro plants, and the load demand keeps increasing, there are basically two possibilities for meeting the extra demand. One possibility is to build more run-of-the-river plants in the river system. Another possibility is to provide storage capacity in the river basin, thus increasing the dependable flow, which will make it possible to increase the dependable hydro capacity of the existing plants. In the first stage of hydro development in a river basin, it may be found more economical to develop another good site. However, there comes a time when all good sites are developed and then it becomes more economical to increase the dependable flow rather than develop more head. This could be called the second stage in hydro development.

Let us illustrate the function of storage with the previous example of hydro plant and river flow. We will assume that the peak demand of the load system has risen to 450 MW, and the average demand to 300 MW. The existing run-of-the-river plant cannot supply this demand. The average flow in the

283

river, however, is 40,000 cusec and only 30,000 cusec are needed to produce the required 300 MW. Therefore, if sufficient storage capacity is provided at the plant, or somewhere upstream of the plant, the hydro development could still meet the entire power demand. In order to find out what volume of storage is required to accomplish this, two mass curves are drawn; one of the river flow over the period of record, and one of the power demand over the same period of years, both expressed either in cubic feet per second or in megawatts. The volume of storage needed may be found from these two mass curves as shown in *Figure 7.15*. A subsequent engineering study will reveal at what cost it may be provided and an economic study will indicate whether or not the cost of incremental plant capacity plus the cost of storage is less expensive than the cost of providing the same capacity by alternate means, like

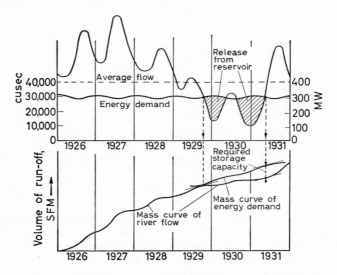

Figure 7.15. Mass curves of river flow and energy demand

a steam power plant or another hydro plant. The second stage of adding more storage capacity and plant capacity to existing hydro developments is a phase which many water power plants have gone through.

Let us now assume that the load demand of the power system has increased to a magnitude where it cannot be satisfied any more by the available and potential hydro plants in the area. The third stage in hydro development is then reached, whereby hydro power is supplemented by steam-generated power. This is a phase of power development in which many power producing companies, which started out at one time as hydro power companies, find themselves at the present time.

Regulation of Reservoirs

It was discussed in the above that the firm capacity of a hydro plant is a function, on the one hand of the nature of the power system in which it operates, and on the other hand of the dependable river flow.

284

In an all-hydro system the trend of thought in planning for sufficient firm capacity to meet a given load condition would be as follows. We have a certain number of sites with a given head and a natural dependable flow. We could produce the desired firm capacity by developing say five sites with the natural dependable flow or by developing say three sites with a larger-than-natural dependable flow, which could be provided by creating storage capacity at or upstream of the plants. A cost comparison of the various alternatives will reveal the most economical solution. Let it be assumed that we decide upon the alternative with storage capacity. The function of the reservoir is, then, to provide a certain dependable flow, which is determined by the load requirements and the total developed head. The capacity of the reservoir is found by a mass-curve study of recorded flows. In order to ensure this dependable flow the reservoir is operated as shown in the following example.

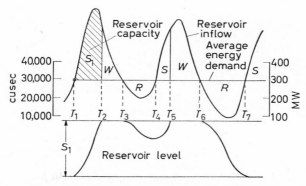

Figure 7.16. Reservoir regulation for dependable flow

One hydro plant with an upstream storage reservoir serves a load system with an average annual demand of 300 MW and an annual load factor of 0·66. The plant has a net head of 148 ft., and efficiency of 0·80 and an installed capacity of 450 MW (plus spare capacity which will not be considered in this example). The reservoir can provide a dependable flow of 30,000 cusec.

Let us assume in *Figure 7.16* that at time T_1 the reservoir is empty. Our first concern will be to get the reservoir full. During the next few months 30,000 cusec of the river flow will be used for power generation and the remainder for filling the reservoir, which has a storage capacity of S_1, represented by the shaded area in the figure. At T_2 the reservoir is full, and the volume of water W has to be wasted. From T_3 to T_4 the volume R has to be released from the reservoir to make up for the deficient river flow. From T_4 to T_5 the reservoir is filled again. From T_5 to T_6 water is wasted, and finally from T_6 to T_7 the reservoir is completely drained (the last R being the same as the area S_1) in order to provide the dependable flow of 30,000 cusec. It may be seen that a considerable amount of water is wasted in this operation. However, this is not so regrettable because there is no demand for more power than what is being generated. The provision of more storage capacity than S_1 would have served no purpose for the assumed load conditions, because a larger dependable flow is not needed.

285

It should be pointed out that in this example a few simplifications have been made. First, the average annual energy demand, shown in *Figure 7.16* as a horizontal line at 300 MW, is really a line that weaves up and down with annual cycles as shown in *Figure 7.7(a)*. Second, the possibility of producing surplus or 'dump' energy has been given no credit in this analysis. Third, in view of future system growth, it may be economic to build now a reservoir that has more capacity than is presently needed. This has been ignored in this example. However, such considerations will be given full attention when we discuss 'system planning' in a subsequent section.

Let us now assume that we have a system that is twice as large, consisting of the hydro plant of the previous example plus a steam plant with the same capacity of 450 MW. The average annual load demand of the system is 600 MW.

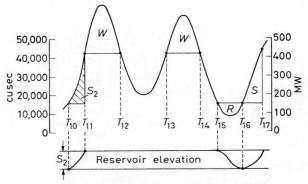

Figure 7.17. Reservoir regulation for dependable flow

The load factor of the system is again 0·66, resulting in a peak demand of 900 MW. Spare capacities shall not be considered in this example. First of all we should determine how much storage capacity we need to firm up the hydro plant. The capacity of the hydro plant is 50 per cent of the total system peak demand. Since we are dealing with average annual figures in this example we should use the annual peak-percentage curve of *Figure 7.11(d)*, to determine the relation between prime power and firm capacity. It may be seen that the upper 50 per cent of the system peak requires 26 per cent of the total system energy. In other words, the hydro plant requires 160 MW continuous of firm energy. This, in turn, calls for a dependable flow of 16,000 cusec, which can be provided by a reservoir capacity S_2 shown in *Figure 7.17*. This reservoir capacity is considerably smaller than the reservoir capacity S_1 of the previous example. The regulation of the reservoir with a capacity S_2 and the operation of the hydro plant would be somewhat as follows:

Assuming that the reservoir is empty at T_{10}, we will use all flow in excess of the minimum required flow of 16,000 cusec, to fill up the reservoir. At T_{11} the reservoir is full and the river flow is ample so that the hydro plant can be placed on the base of the load and operate at say 92 per cent load factor from T_{11} to T_{12} as can be seen from the load duration curve in *Figure 7.18(a)*. The average amount of usable river flow becomes: $0·92 \times 45,000 = 42,000$ cusec. It should be recalled, that for the sake of simplicity the annual load has been assumed constant. This is not correct, but it makes the present discussion simpler. Since the reservoir was full at T_{11}, no more water can be stored, and since the hydro plant operates already at nearly full capacity, the extra river flow between T_{11} and T_{12} has to be wasted. From T_{12} to T_{13} the hydro plant operates in a continuously changing position on the load curve, somewhere in between the base and the peak, so that the amount of generated energy corresponds to the available river flow as shown in *Figure 7.18(b)*. No water is drawn from the reservoir since the available river flow has not yet fallen below the dependable flow of 16,000 cusec.

This last statement may be explained as follows. From a mass-curve study as shown in *Figure 2.3*, page 27, or in *Figure 7.15* we find that we can maintain a dependable flow of 16,000 cusec throughout the driest period on record, provided we have a full reservoir at the beginning of the dry period. From

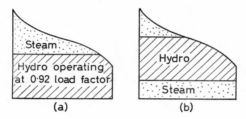

*Figure 7.18. (a) Hydro operating on the base;
(b) Hydro in intermediate position*

inspection of the mass curves in any of the two figures mentioned above, it may be seen that the dry period begins as soon as the natural flow (inflow to the reservoir) becomes less than the dependable flow. At that moment we may begin to release water from the reservoir. If we had done it already at an earlier date we would not have had a full reservoir at the beginning of the critical dry period and hence we could not maintain the dependable flow. In real operating conditions we can make allowance for the fact that certain times of the year (e.g. before the spring floods) are less critical than others. This leads to the development of rule curves as will be discussed in a subsequent section. Moreover, stream flow forecasting may somewhat improve the situation.

From T_{13} to T_{14} the plant is at the base again, the reservoir is still full and all extra flow has to be wasted. From T_{14} to T_{15} the hydro plant moves higher and higher in the load curve until, at T_{15}, it reaches the peak. The natural river flow falls below the dependable flow and water is drawn from storage between T_{15} and T_{16}. Assuming that this is the worst year on record, the area R would be equal to the volume S_2 From T_{16} to T_{17} the reservoir is replenished. All the time, from T_{15} to T_{17}, the hydro plant operates in the peak of the load at a load factor of about 35 per cent.

In comparing this type of regulation for a hydro-steam system with the previous example of regulation for an all-hydro system some points of interest may be noted. First of all we operate with a much smaller storage capacity since the required dependable flow is much smaller. In spite of the smaller storage capacity not much more water is wasted. This is because we can operate at a high load factor during times of ample river flow, whereas formerly we could

287

only operate at system load factor, even if we had ample river flow. However, we are still wasting a considerable amount of water in the hydro-steam example. It was noted in the all-hydro example that there was no purpose in providing extra storage capacity, because the wasted water could not be used at any time, since there was no demand for more power than was being generated. However in the hydro-steam example it would have been very useful to have had more storage capacity so that we could have stored the wasted water, to be released during periods that the steam plant was operating at the base of the load. It should be emphasized that such regulation with extra storage capacity would not affect the firm capacity of the existing hydro plant because that is firm already, and for the given load conditions we do not need to have the hydro capacity 'more than firm'. However, this extra storage

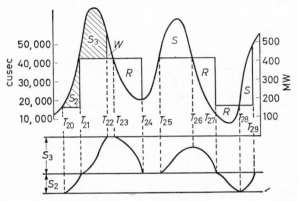

Figure 7.19. Reservoir regulation for maximum energy output

capacity will make water, and therefore hydro energy that can replace costly thermal energy, available. In order to illustrate the above discussion, let us repeat the last example, assuming that we add a volume S_3 to the existing storage capacity S_2, as shown in *Figure 7.19*.

Assuming again that the reservoir is empty at T_{20}, we start out with the same sort of regulation, namely releasing the dependable flow of 16,000 cusec to the hydro plant and using the remaining river flow to fill up the reservoir to the capacity S_2 which is needed to ensure dependable flow. At T_{21}, we place the hydro plant on the base of the load, and instead of wasting the remaining flow, we store it in the remainder of the reservoir which has now an additional capacity S_3. At T_{22} the reservoir is full and we have to waste water until T_{23}. From T_{23} to T_{24}, instead of using only the natural river flow, as we did before, we now keep the hydro plant at the base of the load and release water from the upper portion of the storage reservoir S_3. At T_{24} the volume S_3 is exhausted and we have to fall back on the natural river flow. We are not permitted to release any water from S_2 since that is reserved for conditions where the river flow falls below 16,000 cusec. From T_{24} to T_{25} we operate as we did in the previous example from T_{12} to T_{13}. From T_{25} to T_{26} we partly replenish the storage volume S_3 and release this water again from T_{26} to T_{27}. At T_{27}, when S_3 is empty, we are faced with a natural river flow less than 16,000 cusec, and therefore we draw from the storage volume S_2 to provide the dependable flow of 16,000 cusec. Assuming again that this is the worst condition on record, the storage S_2 will be completely drained at T_{28}.

It can be seen that in the above example the amount of wasted water has been reduced to practically nil, due to the provision of the storage capacity S_3. In order to find out whether this would be economical, we would have to compare the annual cost of providing the extra storage capacity S_3 with the annual benefits, consisting of the savings in thermal energy production. An economic analysis of such a problem may look in principle somewhat like *Figure 7.20*. The cost of storage may increase more rapidly than in straight proportion to the amount of storage. However, the usable flow and therefore

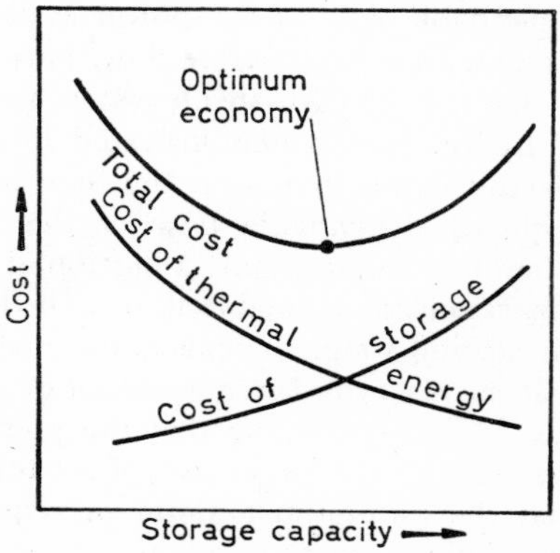

Figure 7.20. *Most economic storage capacity*

the thermal energy replacement will increase less rapidly than in straight proportion to the amount of storage, and will have the average river flow as a limit. Therefore it is not unlikely that the sum of the annual cost of storage capacity and the annual cost of thermal energy has an outspoken minimum which would indicate the most economical size of reservoir. We shall discuss more fully, in the section on system planning, page 309, how such problems must be analysed.

If we define river utilization of a hydro plant as the percentage of the long term average river flow that is converted into electrical energy we may observe that river utilization is a function of: (1) the installed generating capacity of the plant; (2) the size of the power system; (3) the capacity of upstream storage reservoirs. If two of the three variables remain constant, the river utilization will increase with an increase of the third variable. If all three are large: for instance, if we have an installed plant capacity that is well above the average river flow; if the power system is very large and predominantly thermal; and if we have a storage capacity that is several times the volume of the average annual run-off; then we are likely to have a 100 per cent utilization of the river flow. Such cases are very rare. However, a river utilization in the order of 80–90 per cent is not unusual. It must be understood that the above percentages of river utilization apply to long term averages. In one single above-average-river-flow year it would be possible, of course, to reach a river utilization that is more than 100 per cent of the long term average river flow.

Storage Benefits

In the above section we have considered a given power system and discussed the most desirable amount of storage capacity. Let us now look at the same problem of reservoir regulation from some other viewpoints. First, let us consider the function of a reservoir of certain size in a growing power system and then we shall discuss the benefits of adding storage capacity to all-hydro and hydro-steam power systems.

Let us assume that a power system has started out as an all-hydro system. Let us further assume that in the last part of this all-hydro phase, the total storage capacity is S_1 (comparable to the S_1 in the first example of the foregoing section). This storage capacity is needed to provide the required dependable flow of the power system, and we may say that the reservoir is regulated for dependable flow. Now let us assume that the power system has grown into a hydro-steam system with the larger part of the total capacity still in hydro. It was demonstrated in the above that, due to the possibility of placing steam plants on the base of the load during times of low flow, the required dependable flow of the hydro plants is considerably decreased. Therefore we need only a portion of the available reservoir capacity to provide this smaller dependable flow, and the remainder of the reservoir can be used for storing water to replace thermal energy at a later date. We may now say that the reservoir is regulated for dependable flow and increased energy output. Let us finally assume that the power system has grown into a hydro-steam system with the larger part of the total capacity in steam. It may easily happen that the minimum amount of required hydro energy to firm up all hydro installations, corresponds to a river flow which is less than the lowest natural river flow on record. In other words, we need no storage capacity at all to provide for a dependable flow, and consequently all storage capacity can be used for maximizing hydro energy output. We may now speak of reservoir regulation for maximum energy output.

In publications on the potential benefits of storage capacity, one may note a difference of opinion whether the benefits will decline or increase with advance in time. A discussion of this topic will form the subject of the following paragraphs. We will assume a river basin with extensive possibilities of hydro development on the middle and lower reaches, and the possibility of creating a large amount of storage capacity in the upper reaches of the river system. The various phases of power development, with the advance in time, are defined as follows: 'early-all-hydro' is the phase where we can still choose between more or less attractive hydro sites; 'late-all-hydro' is the phase where all potential hydro sites are developed and where we have to choose between storage with incremental hydro or steam; 'hydro-steam' is the phase where we have more hydro than steam in the composite system; 'steam-hydro' is the final phase where steam begins to dominate over hydro. It goes without saying that these four phases cannot be precisely separated, but that they gradually merge into one another. It may also be pointed out that where we speak of steam, we should refer to steam and possibly nuclear power.

The virtue of storage in the early-all-hydro phase is that it provides a certain dependable flow, which in turn provides a certain firm capacity at the various hydro plants. When the generating capacity of the system has to be increased, because of the increase in load demand, the alternatives are to develop other sites with the existing dependable flow, or to increase the dependable flow by creating more storage and to increase the capacity at the existing sites. We may therefore conclude that the benefit of incremental storage to an early-all-hydro system is that it enables us to extend capacity at existing plants (which is relatively inexpensive), rather than develop plants at new sites (which is usually more costly). It should be noted at this point in the discussion that the provision of incremental storage is of no benefit to an existing early-all-hydro system, assuming that it was adequately planned. The incremental capacity is only of benefit with respect to the extension of the system. Let us now illustrate this problem with the following example.

An existing power system has a peak demand of 150 MW in the year 1960; a load factor during January of 0·66; and therefore an energy demand in

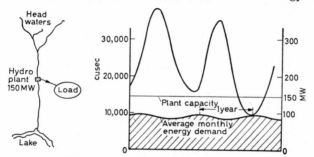

Figure 7.21. Existing situation in 1960

January of 100 MW continuous. The power system is served by one hydro plant with a head of 148 ft. and a dependable flow of 10,000 cusec as shown in Figure 7.21. The plant has sufficient pondage capacity.

We assume now that the load will double from 1960 to 1970. One possibility to meet a peak demand of 300 MW is to build a second hydro plant on the river. Assuming that the second site also has a head of 148 ft., the second plant would be identical to the first plant shown in Figure 7.21. Assume a capital cost of this second plant of $300 per kW, or a total of: $150,000 \times \$300 = \$45,000,000$. Annual cost: $8\% \times \$45,000,000 = \$3,600,000$. An alternative possibility to meet the 1970 load condition would be to build a reservoir with sufficient capacity to increase the dependable flow to 20,000 cusec, and to extend the first hydro plant to a capacity of 300 MW, as shown in Figure 7.22. Assuming an incremental capacity cost of $150 per kW, or a total of $22,500,000, the annual charge becomes $1,800,000. Assuming an annual cost of the reservoir of $1,000,000 the net benefits achieved are: $3,600,000 - $1,800,000 - $1,000,000 = $800,000. It should be noted that we produce the same amount of hydro energy and the same amount of hydro capacity in both cases. In other words we have not intensified the use of the water power resources by constructing the reservoir. The benefit of the reservoir is only a dollar benefit, inasmuch as it made possible a cheap extension of an existing plant instead of the costly construction of a new plant.

In the above example we have looked at the problem from the viewpoint of an agency that has control over the hydro sites as well as the reservoir sites.

291

If we look at this example, as if one agency would have control over the reservoir and another agency control over the hydro sites, the problem arises who pays for what, and how any benefits are to be shared. Actually such a problem has two components. First, one has to define the costs and benefits and, second, one has to agree how to share them. One reasonable solution to such a problem would be to determine first the gross benefits to both partners (these are the expenditures that they do not have to make as a result of the storage

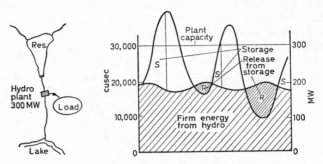

Figure 7.22. Possible development for 1970

reservoir), then to deduct from these gross benefits the costs of both partners (these are the expenditures that they do have to make), and thus arrive at the net benefits that will be shared equally between the two partners. This principle is illustrated graphically in *Figure 7.23*, and would apply as follows to the above example.

The gross benefits of the downstream agency are $3,600,000 (the annual cost of the hydro plants that do not have to be built if the storage reservoir is provided). The gross benefits of the upstream agency are nil assuming that it

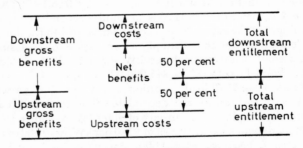

Figure 7.23. Possible division of benefits

only contemplates to build the reservoir, upon request, and not to add any hydro plants of its own. Hence the total gross benefits are $3,600,000. Out of this total fund, both agencies can now pay their respective costs: the downstream agency $1,800,000 for the annual cost of incremental capacity at the existing site, and the upstream agency $1,000,000 for the annual cost of the reservoir. The remainder of $800,000 can be considered as the net benefits of the joint venture and may be divided equally between the two partners.

Hence, the upstream agency would receive in total from the downstream agency: $1,000,000 (compensation for the cost of the storage reservoir) + $400,000 (half of the net benefits) = $1,400,000 per year, for as long as the assumed conditions prevail.

Let us assume that the upstream agency would request the downstream agency to make the payment of $1,400,000 per year in power rather than in dollars. The downstream agency would then have to produce more power than its own requirement of 300 MW. This extra power would cost $300 per kW or $24 per kW per year (assuming that the cheap incremental capacity is exhausted). Hence the payment of power could consist of $1,400,000 divided by $24 = 58,000 kW capacity with the associated energy at the system load factor of the downstream agency.

Let us now assume that we move several years into the future, that all hydro sites in the river basin have been developed, that all existing hydro plants have firm capacity, and that we need to plan for future extension of the generating system. The alternatives are to provide incremental storage capacity which will enable us to extend the existing hydro plants, or to build steam plants. Since a few steam plants in a large hydro system will operate nearly all the time in the very peak of the load, or serve as stand-by capacity, it is obvious that the steam capacity that would be added in this phase is not the type with high efficiency, designed for minimum-cost energy production, but rather the type where efficiency is sacrificed for low capital cost. Therefore, the annual cost of steam capacity may not be much larger than the annual cost of incremental hydro capacity in a late-all-hydro system. Moreover, small additions of steam to the all-hydro system would immediately be reflected in a lowering of the required dependable flow and hence a lowering of the required storage capacity. However, the storage capacity that would thus become available for increased energy output cannot be made effective yet because there is so little thermal energy to replace. It thus follows that the benefits of incremental storage capacity in a late-all-hydro system may be less than the benefits in an early-all-hydro system.

Proceeding still farther into the future let us assume now that a substantial portion of the total generating capacity is steam capacity. Part of the existing storage capacity serves to provide dependable flow and thus to ensure firm hydro capacity. The remainder of the storage capacity is regulated to replace thermal energy. The benefits of incremental storage capacity would therefore have two components, namely capacity benefits and energy benefits. Net capacity benefits may be evaluated as the difference in cost between incremental hydro capacity and steam capacity. Energy benefits may be evaluated as the saving in thermal energy production. This last item will gain in importance with the addition of more and more steam to the system. As a result, the combined benefits from incremental storage capacity in a hydro-steam system are generally larger than the storage benefits in a late-all-hydro system.

If we look at the problem of providing optimum storage facilities from the viewpoint of one agency, responsible for the entire river basin development, we would compare the cost of incremental storage with the resultant saving of capacity development (hydro instead of steam) plus the saving of thermal energy. Let us try to illustrate the main elements of the above discussion in another numerical example.

We shall assume that the peak demand of the earlier load system of *Figures 7.21* and *7.22*, has increased to a magnitude of 600 MW. Let us further assume that we had installed the two hydro plants without storage in 1970, and that there are no more hydro sites available on the river. We have the following alternatives for 1980: two hydro plants without storage, plus one steam plant of 300 MW, as shown in *Figure 7.24*, or two hydro plants of 300 MW plus a reservoir.

Assume a capital cost of the steam plant of $150 per kW or a total cost of 300,000 × $150 = $45,000,000. Annual cost: 10% × $45,000,000 = $4,500,000. From *Figure 7.24* it may be estimated that the average thermal requirements are about 125 MW continuous. Assume cost of incremental thermal energy 4·5 mills/kWh which is equivalent to $40 per kW per year. Total thermal energy cost: 125,000 × $40 = $5,000,000. Assume incremental cost of the two hydro

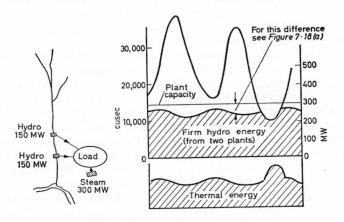

Figure 7.24. Possible development for 1980

plants: 300,000 × $150 = $45,000,000. Annual cost: 8% × $45,000,000 = $3,600,000. The net benefits achieved by the construction of the reservoir are now: $4,500,000 + $5,000,000 − $3,600,000 − $1,000,000 = $4,900,000 per year. It should be noted that the reservoir now has capacity benefits and energy benefits. Without the reservoir we need steam capacity and thermal energy to meet the load demand. With the reservoir we can meet the situation with hydro only, and we have therefore intensified the water power resources development.

If we look once more at this problem from the viewpoint of two separate entities, as discussed earlier, we could make the following observations. The gross benefits of the downstream agency are $4,500,000 for steam capacity and $5,000,000 for steam energy. The gross benefits of the upstream agency are still nil. The cost of the downstream agency is $3,600,000 for incremental hydro capacity. The cost of the upstream agency is $1,000,000 for the reservoir. Hence the net benefits, to be shared equally, are $4,900,000. The upstream agency would therefore receive: $1,000,000 (for storage costs) + $2,450,000 (its share in the net benefits) = $3,450,000. If payments are to be made in power, they should be such that the maximum capacity received × $15 (annual cost of steam capacity) plus the amount of energy received × 4·5 mills equals $3,450,000.

It may be of interest in this discussion of the hydro-steam phase, to pursue for a moment the thought that incremental hydro capacity will be provided as long as the cost of incremental hydro capacity plus the cost of associated incremental storage capacity minus the saving in thermal energy is less than the alternative cost of steam capacity. This may be true in general, but there are other factors that affect the ultimate capacity of a hydro development such as available pondage capacity, navigation on the river, and ice conditions. Development of a hydro plant for peaking purposes, means that the plant will operate only during a few hours per day at very high discharges and be idle for the rest of the time. This may result in significant fluctuations of forebay levels with a consequent loss in energy; it also means surges on the river channel which may not be tolerable from a viewpoint of navigation; or it may result in ice jams during winter time, particularly on northern rivers where successful winter operation depends on a solid ice cover. If such limiting circumstances exist, it may very well be that incremental storage capacity loses its capacity benefits, which brings us to the last phase in hydro development.

Let us now assume that we have advanced another decade into the future and that we have a system where the amount of steam capacity is substantially larger than the amount of hydro. It is likely that hydro has been developed to its practicable limit from a viewpoint of incremental cost, pondage capacity and hydraulic conditions. Therefore with the further growth of the system, we need less and less flow to firm up the hydro plants and hence less and less storage capacity to provide the decreasing dependable flow. As noted before, the time may soon arrive that the required dependable flow falls below the natural minimum flow, and consequently that incremental storage capacity has only energy benefits. Let us now consider for a moment the function of a reservoir from a viewpoint of maximum energy output. The reservoir stores the flood flows that are beyond the capacity of the hydro plants, and it stores also the flows that are within the capacity of the plants but that are not momentarily needed because of insufficient energy demand.

When we compare an advanced steam-hydro system with earlier systems we may note first of all that the total amount of hydro capacity has been developed to the practical limit, and hence that there are now fewer flood flows beyond the capacity of the plants; and secondly that the hydro plants, being the smaller part of the total system, can operate at all times on the base of the load if there is sufficient river flow. In other words, nearly all natural river energy can be absorbed in the power system, no matter when and how the water comes, and consequently the benefits of incremental storage capacity become insignificant.

Pursuing our earlier numerical examples, let us assume that for load conditions of the year 2000, the total peak demand has grown to 2,400 MW. We assume that the practical limit of hydro development at each of the two available sites is 300 MW. This limit is dictated by considerations of incremental cost, pondage capacity and hydraulic conditions. When the two hydro plants of 300 MW are placed in the peak of the load, they require, according to the annual peak-percentage curve, 100 MW continuous in hydro energy, which corresponds to 5,000 cusec. Since this is less than the natural dependable river flow, it follows that no reservoir capacity is needed to firm up the 300 MW capacity at each hydro plant, as is shown in *Figure 7.25*. However, a reservoir could store the water that is wasted in *Figure 7.25*, and release it during times of less-than-full-capacity flow, as is shown in *Figure 7.26*.

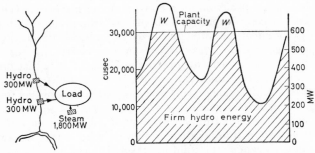

Figure 7.25. Possible development for 2000

It may be estimated from *Figure 7.26* that the average saving in thermal energy is about 15 MW continuous, or $15,000 \times \$40 = \$600,000$. Hence the net annual benefits of the reservoir now become negative to the amount of: $\$600,000 - \$1,000,000 = -\$400,000$. We may conclude that under these conditions the reservoir would not be constructed. If it existed already from earlier days the small amount of energy benefits would not compensate the annual cost.

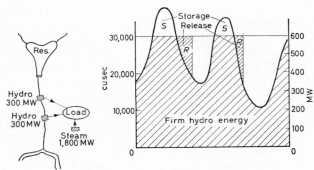

Figure 7.26. Alternative development for 2000

Before we leave this subject of storage benefits, it should be pointed out that much of the above discussion, particularly in the examples, has been presented in a simplified form. Energy benefits and capacity benefits of storage are difficult to separate and to define. For a given incremental amount of storage, one has the choice to allocate a certain portion to producing prime power and the remainder to maximize energy output. It requires extensive studies to determine how this allocation should take place. Moreover, this allocation may change from year to year. After one has set aside a certain amount of storage for maximum energy utilization, it requires again extensive studies to determine exactly how much thermal energy can be saved. Such studies will be discussed under the following section on rule curves.

We must also realize that the net benefits of a storage reservoir or an increment in storage, are different from year to year. Therefore, it is not sufficient to make an analysis of a certain 'point' in time and to conclude whether or not the reservoir is economical. It becomes necessary to analyse year by year the entire life of the project (say 50 years), to reduce all benefits and costs to present values, and then to compare the total benefits with the total costs. Such studies will be discussed in the section on system planning, page 309.

Rule Curves

A rule curve may be defined as a diagram showing storage requirements during the year. As such, a rule curve provides guidance for the operation of a reservoir from day to day. In fact, rule curves, supplemented with appropriate tables and notes, may be worked into a complete set of instructions for reservoir regulation. We will discuss in the following paragraphs how rule curves are constructed.

Let us take first the simple case of a reservoir with seasonal storage, operated for an all-hydro system. (It was noted earlier that seasonal storage only requires a carry-over of water from the wet season to the dry season, whereas year-to-year storage requires the carry-over of water from a wet period to a subsequent dry period, maybe several years later.) It follows that a reservoir with seasonal storage does not need to be full at all times of the year. In fact, if we take the driest year on record as the criterion for making up the rule curve, we find that the reservoir only needs to be full during a very short period, as shown in *Figure 7.27*.

In *Figure 7.27(a)* are shown the stream flow of the river during the driest year on record and the energy requirements of the all-hydro system (both converted to the same scale). At time A the reservoir needs to be full and at time B, the reservoir will be empty. The shaded area between the supply and demand curve between A and B represents the volume of the reservoir release. If we translate this figure into a mass curve, we obtain *Figure 7.27(b)*. From point A, to the right, the supply and demand curves diverge, and the ordinate represents the required reservoir capacity. If we pursue the two curves to the right we may note that at time D, the reservoir will be full again. Now let us make a slightly different analysis of the same situation. At point C in *Figure 7.27(b)*, the reservoir is empty. From this point we plot backwards in time the energy demand and obtain curve CE which is simply curve AB lowered and extended to the left. The vertical ordinates between the supply curve EAC and the demand curve EC represent the volume of water that is in storage in the reservoir during the period from E to C. Since we have assumed that this is the driest year on record, to be considered as the criterion for preparing the rule curve, we may now draw the conclusion that whenever there is more water in the reservoir at certain dates, than indicated by the ordinates in *Figure 7.27(b)*, there is no danger of subsequent emptying of the reservoir.

To make *Figure 7.27(b)* more convenient for ready use, we can take the vertical ordinates and plot them on a horizontal base, as shown in *Figure 7.27(c)*.

We have now obtained a rule curve, showing the storage requirements of the reservoir as of any date, assuming that a repetition of the most adverse stream flow conditions on record is possible. This rule curve represents the accumulation, in reverse order of time, of the deficiency between energy requirements and available stream flow during the critical period.

It was assumed in the above that one critical year would determine the rule curve. It is not unlikely, however, that there are other, near-critical years. Such years should also be analysed and the corresponding deficiency curves prepared.

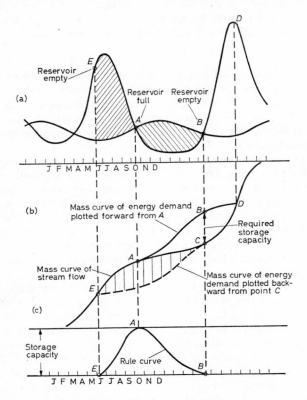

Figure 7.27. Preparation of rule curve

It is possible that these curves cross one another at places. It would then be advisable to draw the final rule curve as an enveloping curve of the various deficiency curves. We may even draw the rule curve some distance above the deficiency curves to guard against stream flow conditions that would be worse than recorded.

It may be pointed out that a rule curve for a reservoir that is exclusively operated for an all-hydro system is only of academic interest. Since there is no alternative use for the water there is no choice of operation: the required flow will be released when the reservoir is not full, and the inflow into the reservoir will be released when the full supply level is reached. From a mass curve study it has been ascertained that the reservoir capacity is adequate to guarantee the required flow during the critical flow period.

When there is alternative use for the water (e.g. irrigation or water supply), or alternative use for the storage capacity (e.g. flood control), it is evident that the rule curve becomes of practical interest. When the all-hydro system becomes a hydro-thermal system, as will be discussed shortly, the rule curve becomes a most important tool for economic operation of the whole power system.

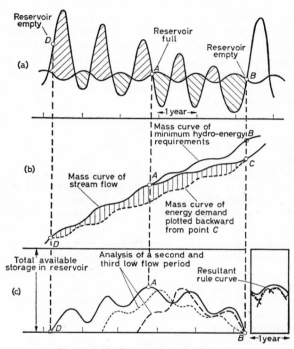

Figure 7.28. Preparation of rule curve

It should finally be noted that any rule curve, being the deficiency between energy requirements and dependable stream flow, is only valid for one particular load condition. If we go to another year, with different load conditions, we have to revise the rule curve.

Now let us consider a somewhat more complicated example of a reservoir with year-to-year storage, operated for a hydro-steam system, as shown in *Figure 7.28.* It was noted earlier that in a hydro-steam system part of the reservoir may be operated for dependable flow (place steam on the base, hydro in the peak, find the required minimum hydro energy and translate into stream flow) while the remainder of the reservoir is operated for maximum energy output (release all flow that can be converted into usable hydro energy). In view of this dual use of a storage reservoir it becomes now particularly important to what level we can draw the reservoir down as of any date during the year.

299

Figure 7.28(a) shows the stream flow during the most critical period on record and also the minimum hydro energy requirements for one given load condition. It may be seen that water has to be stored over a period of several years, the shaded areas above the energy demand line indicating water going into storage, and the shaded areas under this line indicating water released from storage. At time *B*, the end of the critical period, the reservoir is empty. From *Figure 7.28(b)*, it may be seen that the reservoir must be full at time *A*, and that the required volume of the reservoir for dependable flow is represented by ordinate *BC*. However, this volume does not have to be available at all times. If we plot the minimum hydro energy demand, starting at point *C* in *Figure 7.28(b)*, and working to the left, we find the required storage capacities, at any time, antecedent to the end of the critical period. When the ordinates of deficiency are plotted on a horizontal base, *Figure 7.28(c)* is obtained. Since this curve is spread over several years, we have to take the year with the highest ordinates in order to find the appropriate rule curve. If this procedure is repeated for other critical flow periods, other deficiency curves will be obtained. We have to scan *Figure 7.28(c)* month by month and plot the highest deficiency ordinates that are found on a one-year diagram shown in *Figure 7.28(d)*. After all the high points are plotted, an enveloping curve is drawn, representing the resultant rule curve for reservoir operation.

Operation of the reservoir will now take place as follows. When the reservoir as of any date is above the elevation of the rule curve as of that same date, all water that can be converted into usable electrical energy is released from the reservoir. When the available storage is only slightly above, or equals the critical storage capacity indicated by the rule curve, the release of water should be restricted to such a quantity that the reservoir storage does not fall below the rule curve. The lower limit of the release is, of course, the minimum hydro requirement as of that date. When, for some reason, the storage in the reservoir should fall below the rule curve, the release should be limited to the minimum hydro requirements, with the object of returning as soon as possible to the rule curve. It follows, that during such times, all steam plants in the system operate on the base of the load.

To illustrate the above discussion let us consider the same power system that was used for *Figure 7.19*, namely one hydro plant with a head of 148 ft., an efficiency of 0·80 and an installed capacity of 450 MW, plus one steam plant with a capacity of 450 MW. The load system has a peak demand of 900 MW and an annual load factor of 0·66. Spare capacity will not be considered. We shall assume that the steam plant can operate continuously on the base of the load, if needed. A reservoir with a capacity of 100,000 SFM (1 SFM = 1 second ft. month = 61 acre-ft.) is situated above the hydro plant. Part of the reservoir is used for maximum energy production, part is used to produce the required dependable flow. The 'rule curve' of the reservoir is shown as col. (7) in Table 7.1. The operation of the reservoir would be as shown in Table 7.1.

It should be noted that in regulating future flow conditions, there is always the possibility of encountering lower-than-recorded reservoir inflows. If we would operate, nevertheless, on the basis of a rule curve derived from recorded flow conditions, the result could be an empty reservoir with consequent power shortage! In other words, whenever we approach the basic rule curve we run the risk of capacity deficiency.

Table 7.1. *Rule Curve Regulation of One Reservoir*

Year Month (1)	Peak (2)	Ener. (3)	H. Cap (4)	Max. H (5)	Min. H (6)	Rule Curve (7)	Inflow (8)	Outflow (9)	Δ Stor. (10)	Stor. (11)	Th. En. (12)
1940 April	730	530	450	430	100	10,000	45,000	43,000	+2,000	20,000	100
May	690	500	400	380	75	12,000	80,000	38,000	+42,000	62,000	120
June	680	500	350	340	75	15,000	135,000	97,000	+38,000	100,000	160
July	670	490	350	340	70	25,000	80,000	80,000	—	100,000	150
Aug.	680	500	350	340	75	37,000	20,000	34,000	-14,000	86,000	160
Sept.	720	530	350	350	100	44,000	18,000	35,000	-17,000	69,000	180
Oct.	800	580	400	400	145	47,000	16,000	38,000	-22,000	47,000	200
Nov.	850	620	450	440	180	41,000	18,000	24,000	-6,000	41,000	380
Dec.	880	640	450	450	200	32,000	12,000	21,000	-9,000	32,000	430
1941 Jan.	900	660	450	450	220	22,000	12,000	22,000	-10,000	22,000	440
Feb.	850	620	450	440	180	15,000	11,000	18,000	-7,000	15,000	440
Mar.	810	590	450	440	150	11,000	14,000	18,000	-4,000	11,000	410
April	730	530	450	430	100	10,000	18,000	19,000	-1,000	10,000	340
May	690	500	400	380	75	12,000	60,000	38,000	+22,000	32,000	120

Col. (1). The period of April 1940 to May 1941 is an arbitrary selection out of a much longer regulation study. The purpose of the study could have been to determine thermal energy requirements for future load conditions using past flow records as a criterion.
Col. (2). represents the peak demand in MW for every month.
Col. (3). represents the monthly energy demand in MW continuous.
Col. (4). represents the available hydro capacity, in MW. From May to November, deductions have been made to allow for maintenance outage.
Col. (5). represents the maximum hydro energy in MW continuous that can be used in the system, if the hydro capacity of col. (4) is placed on the base of the load.
Col. (6). represents the minimum hydro energy, in MW continuous, that is required if the steam plant is placed in the base of the load. These figures are obtained by deducting 450 MW from the figures of col. (2). The remainder is expressed in per cent of total peak. Using the monthly peak-percentage curve, the percentage of required energy is found. This figure is multiplied with the figures of col. (3).
Col. (7). represents the amount of storage required in the reservoir, in SFM. The figures are obtained from a study as shown in *Figures 7.27* and *7.28*, or in *Table 7.2*.
Col. (8). represents the average monthly inflow into the reservoir in cusec.
Col. (9). The outflow from the reservoir in cusec is found by applying the following rules: (a) The outflow should never be less than the minimum requirements indicated by col. (6); (b) The outflow should not be more than the maximum usable flow indicated by col. (5), unless the reservoir is full. In that case, the excess flow must be spilled; (c) The total storage in the reservoir should not fall below the level indicated by col. (7); (d) Staying within these limits, as much water should be released as possible.
Col. (10). The difference between cols. (8) and (9), in SFM.
Col. (11). The total storage in the reservoir in SFM. This figure cannot be more than its total capacity of 100,000 SFM, and should not be less than the figures of col. (7). Only in the driest year on record will this figure fall down to zero.
Col. (12) represents the thermal energy requirements in MW continuous. This figure is obtained by deducting the usable hydro energy of col. (9) from col. (3).

On the other hand, if we would steer a safe distance away from the rule curve, we run the risk of subsequently filling our reservoirs too early and having to spill more water than otherwise. It appears that there is what we could call a 'risk zone' above the rule curve, where we have to take the risk of either capacity deficiency or energy waste. It may have merit to determine within the risk zone the probability of energy waste and of capacity deficiency. Such statistical information would provide further guidance to reservoir operation.

Another means to extend the usefulness of the rule curve for actual reservoir operation would be to devise a system of long range stream flow forecasting. In drainage basins where the spring run-off is mostly derived from snow melt, it may be feasible to devise a relationship between the volume of spring run-off on the one hand and the amount of snowfall and antecedent ground conditions on the other hand. Since the most critical flow conditions in such drainage basins usually occur at the end of the winter, the above relationship may greatly assist the selection of the proper reservoir release. If, for instance, a larger-than-normal spring run-off is expected, it would be permissible to draw the reservoir down below the normal rule curve, and thus increase the energy utilization of the stream.

The problem of regulating reservoirs by means of rule curves becomes somewhat complicated when there are several reservoirs in the system. We may distinguish the following situations:

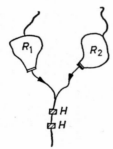

Figure 7.29. Parallel reservoirs

(*a*) Two or more reservoirs, parallel and above the hydro plants as shown in *Figure 7.29*. We may prepare one rule curve for the two reservoirs combined. This rule curve is found by analysing the combined inflow into the reservoirs and the combined minimum flow requirement of the hydro plants. When operating the two reservoirs with the one rule curve, we determine first the total outflow from the reservoirs and then distribute this outflow such that the resultant storage in one reservoir has approximately the same proportion to its full capacity as in the other reservoir.

(*b*) Two or more reservoirs, in series and above the hydro plants, as shown in *Figure 7.30*. We may also prepare one rule curve for the two reservoirs combined. If there is no appreciable inflow between the reservoirs, the problem may be reduced to one reservoir with the combined capacity. If there is significant inflow, the rule curve is found by analysing the inflow to the downstream reservoir and the minimum hydro requirements. Operation of the reservoirs can take place as described under (a). In both (a) and (b) it would be advisable to check the validity of the rule curve by regulating the reservoirs for future load conditions and past stream flow records, in order to verify that all reservoirs have sufficient inflow at all times to perform their assigned function. If this is not so, a more conservative rule curve should be adopted.

Figure 7.30. Reservoirs in series

(c) Two or more reservoirs, serving parallel groups of hydro plants as shown in *Figure 7.31*. In this case the release of a certain flow from one reservoir does not produce the same amount of energy as the release of an identical amount of flow from the other reservoir, since the amount of developed head below each reservoir may be different. For this reason it will be found more convenient to perform the book-keeping of reservoir releases and hydro energy requirements in terms of kWh instead of volumes of water. The storage in a reservoir thus becomes equal to the amount of kWh that would be produced when this volume of water would flow through the downstream power plants. The minimum hydro requirements, instead of being converted to stream flow, remain in terms of kWh. In this case of parallel plants, one rule curve will suffice for the two reservoirs. This rule curve is found by accumulating, in

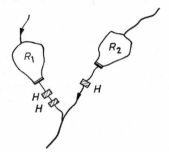

Figure 7.31. Parallel reservoirs

reverse order of time, the deficiency between the minimum hydro require-ments in kWh, and the inflow into the reservoirs in kWh. Operation of the reservoirs may be such that in each reservoir the ratio of available storage, in kWh, to storage capacity, in kWh, is nearly the same. In addition to the requirement that the total release, in kWh, from all reservoirs, should be equal or greater than the total minimum hydro requirements, we must also make sure that each of the parallel groups receives enough energy to be firm when placed in the peak of the load curve.

(d) Two or more reservoirs, serving groups of hydro plants, placed in series as shown in *Figure 7.32*. We may again prepare one rule curve for all

303

reservoirs. This rule curve is found by accumulating, in reverse order of time, the deficiency between the total minimum hydro requirements and the total amount of energy that would be produced by the natural, unregulated river discharge, flowing through the respective developed heads. Operation of the reservoirs may be performed again with the balancing principle, described above. The difference with case (c) is that it becomes somewhat more complicated to establish the appropriate additional requirements to ensure that each

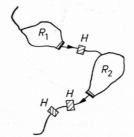

Figure 7.32. Reservoirs in series

hydro plant or group of hydro plants receives enough energy to be firm under the load curve. In fact, it may be desirable to establish supplementary rule curves for individual reservoirs in addition to the main rule curve for all reservoirs. Before a set of rule curves is applied to actual reservoir operation, it would be desirable to try it out first on the recorded flow conditions, so that any shortcomings may be corrected.

Table 7.2. *Preparation of Rule Curve for Two Parallel Reservoirs*

Year	Month (1)	Nat. Flow (2)	Min. Req. (3)	Draw (4)	Acc. Storage (5)
1940	April	25,000	12,000	−13,000	—
	May	28,000	11,000	−17,000	1,000
	June	18,000	10,000	−8,000	18,000
	July	24,000	9,000	−15,000	26,000
	Aug.	13,000	10,000	−3,000	41,000
	Sept.	9,000	12,000	+3,000	44,000
	Oct.	6,000	13,000	+7,000	41,000
	Nov.	12,000	15,000	+3,000	34,000
	Dec.	12,000	17,000	+5,000	31,000
1941	Jan.	10,000	20,000	+10,000	26,000
	Feb.	9,000	19,000	+10,000	16,000
	Mar.	10,000	16,000	+6,000	6,000
	April	20,000	12,000	−8,000	—

Col. (1). The period shown in the table is the most critical stream flow period on record.
Col. (2) lists the natural flow in cusec at the hydro sites. In other words, these figures equal the inflow into the two reservoirs plus the unregulated run-off between the reservoirs and the hydro plants.
Col. (3) shows the minimum hydro requirements expressed in SFM (cusec-months). These figures are obtained by assuming a certain load condition for the power system, placing the steam plants on the base and the hydro plants in the peak, determining the amount of hydro energy under the load curve, and converting this to average monthly stream flow.
Col. (4) shows the difference between cols. (2) and (3). Plus signs indicate flow deficiency and therefore draw from storage. Negative signs indicate flow surplus which may be stored in the reservoir.
Col. (5) represents the 'rule curve', in SFM, in storage. This column is computed backwards in time, starting at the last month with a flow deficiency: March 1941. The deficiencies are then accumulated until September 1940. From this month till May 1940 the accumulated storage is reduced to zero because of the flow surpluses. It must be noted that the storage volumes listed for every month are the minimum requirements at the beginning of every month. If we prepare the rule curve in the form of a graph, the values listed must be plotted on the first of the month and a smooth line drawn through the 12 points of the year. This curve can then be used for day-to-day regulation.

The following example will illustrate the preparation and use of rule curves, as discussed in the above. We shall assume that we have two parallel reservoirs, as shown in *Figure 7.29*. The hydro plants are assumed to be close together so that for practical purposes, they utilize the same river flow. The two hydro plants plus a number of steam plants form the entire power system. The preparation of the rule curve for regulating the two reservoirs is shown in the preceding Table 7.2.

It may be seen from Table 7.2 that the maximum storage requirement for dependable flow is 44,000 SFM in September. During the other months, the storage requirement is less. The actual storage capacity of the two reservoirs may be greater than 44,000 SFM, say 75,000 SFM. This would mean that a good deal of the reservoir capacity may be regulated for maximum energy output. An example of such regulation for two parallel reservoirs with downstream hydro plants, as shown in *Figure 7.29*, is presented in Table 7.3.

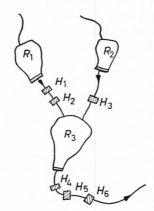

Figure 7.33. Reservoirs in series and parallel

We shall now discuss in principle how a complicated system of reservoirs and hydro plants would have to be operated as shown in *Figure 7.33*. First of all, we prepare a rule curve for the three reservoirs combined. The mechanics of preparing this curve is similar to the arrangement in Table 7.2. Col. (1) represents again the most critical stream flow period of the river system. In view of the larger storage capacity, this critical period may extend over several years and may be different for different load conditions. Col. (2) represents now the total amount of energy, in kWh, that would be available, if the natural unregulated river flow was converted at every hydro plant into electricity. Col. (3) represents the minimum hydro requirements, in kWh, of the total system (steam on the base). Col. (4) represents the surpluses and deficiencies of hydro energy, in kWh. Col. (5), computed backwards in time, represents the rule curve of the three reservoirs, in kWh.

However, this one rule curve is not sufficient to regulate the reservoirs. Let us assume, for instance, that the available storage, in kWh, is larger than the rule curve indicates, but all concentrated in the downstream reservoir, with the upstream reservoirs being empty. This would place the upstream hydro plants in a precarious position, particularly when the natural river flows fall below the minimum requirements! It is obvious that additional regulation rules are required. These rules may be listed as follows:

(1) The total hydro energy production of all plants should not exceed the maximum usable hydro energy. This figure is obtained by deducting the minimum thermal energy requirements (steam in the peak) from the total system energy requirements. This rule is similar to col. (2) in Table 7.3.

(2) The total hydro energy production of all plants should not be less than the minimum hydro requirements (all hydro in the peak). This rule is similar to col. (3) in Table 7.3.

(3) The energy production of plants H_1, H_2, and H_3 should not be less than what is required when these plants are placed in the peak of the load curve. If it is found that the energy provided by the natural river flow is at times less than this minimum requirement, a second rule curve has to be prepared for the regulation of R_1 and R_2 combined.

(4) The energy production of plants H_1 and H_2 should not be less than what is required when these plants are placed in the peak of the load curve. If it is found that the energy provided by the natural river flow is at times less than required, a third rule curve has to be prepared for the regulation of R_1. Similarly, a fourth rule curve may be required for R_2.

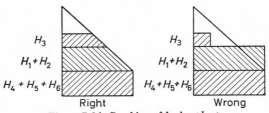

Figure 7.34. Stacking of hydro plants

(5) The energy production of all hydro plants, within the above described upper and lower limits, should moreover be such that the plants can be properly stacked under the load curve of the month, as shown in *Figure 7.34*.

(6) When a reservoir is full to capacity and its net inflow is greater than what may be used by the downstream plants, the reservoir outflow should equal the inflow, unless the outflow capacity is restricted. In that case the inflow has to be routed through the reservoir, using its storage curve and outflow rating curve. A separate account must be kept of spillage, since the usable hydro energy may never exceed what is set out under rules (1) and (5).

After having established the above rules and rule curves in quantitative terms, the actual operation of the system may be started. A table is prepared, in principle similar to Table 7.3, but somewhat more extensive, and all figures expressed in kWh instead of cusec. The regulation of the reservoirs becomes a trial and error procedure. The aim is to produce as much usable hydro energy as possible, while staying above the rule curves and within the rules. At the end of every month the reservoir storages should be in balance towards the rule curves, if possible. That is, for every reservoir or group of reservoirs the ratio of actual depletion to permissible depletion (rule curve value) should be the same. This has the advantage that all reservoirs reach the rule curves at the same time so that the whole system can simultaneously change from operation for maximum energy output to operation for dependable flow.

Table 7.3. *Rule Curve Regulation of Two Parallel Reservoirs*

Year	Month (1)	Max. Hydro. (2)	Min. Hydro. (3)	Rule Curve (4)	Unreg. R.O. (5)	Reservoir 1 (Capacity: 50,000 SFM)				Reservoir 2 (Capacity: 25,000 SFM)				Total Flow (14)
						Inflow (6)	Outflow (7)	Change in Storage (8)	Total Storage (9)	Inflow (10)	Outflow (11)	Change in Storage (12)	Total Storage (13)	
1930	April								20,000				10,000	
	May	25,000	11,000	1,000	10,000	23,000	15,000	+8,000	28,000	4,000	0	+4,000	14,000	25,000
	June	24,000	10,000	18,000	9,000	9,000	6,000	+3,000	31,000	10,000	9,000	+1,000	15,000	24,000
	July	23,000	9,000	26,000	7,000	6,000	7,000	-1,000	30,000	9,000	9,000	0	15,000	23,000
	Aug.	24,000	10,000	41,000	5,000	-1,000	2,000	-3,000	27,000	8,000	9,000	-1,000	14,000	16,000
	Sept.	25,000	12,000	44,000	4,000	3,000	2,000	+1,000	28,000	6,000	6,000	0	14,000	12,000
	Oct.	26,000	13,000	41,000	6,000	3,000	3,000	0	28,000	3,000	4,000	-1,000	13,000	13,000

Col. (1). The seven months shown in the table are an arbitrary selection out of a regulation study that may extend over several decades. The purpose of such a study may be to find the hydro energy output for future load conditions, using the past flow records to represent the most likely future conditions.

Col. (2) represents the maximum usable hydro energy in the system, expressed in cusec. This figure is found by placing all hydro plants on the base of the load and converting the average energy into river flow.

Col. (3) represents the minimum hydro requirements in cusec as explained under Table 7.1, col. (6).

Col. (4) shows the minimum storage requirements in SFM obtained from Table 7.2, col. (5).

Col. (5) shows the unregulated run-off, in cusec, between the two storage reservoirs and the power plants.

Col. (6), Col. (10) show the net inflow into the reservoirs in cusec. This net inflow represents the actual inflow plus the precipitation, minus the evaporation over the reservoir. If such figures are not available in published records, they must be prepared by a preliminary study. Note that the inflow in Reservoir 1, during August 1930, has a negative value of -1,000 cusec, due to excess evaporation.

Cols (7)-(9) and Cols. (11)-(13) are now manipulated simultaneously and by trial and error. First we see how much energy we can use in May: 25,000 cusec. From unregulated run-off we receive 10,000 cusec. Hence we only have to draw from the reservoirs 15,000 cusec. We decide to draw all of this from Reservoir 1, and nothing from Reservoir 2, in order to conclude the month with a total storage in Reservoir 1 (28,000 SFM) which is twice the total storage in Reservoir 2 (14,000) SFM, this being the same ratio as the total storage capacities of the reservoirs.

During June and July, we repeat the same procedure. In August, however, we cannot afford to draw the maximum usable flow, or the total storage in the two reservoirs would fall below the rule curve. Therefore we only draw 16,000 cusec, which makes the total reservoir content equal to the rule curve with a value of 41,000 SFM. In September, we find that we have to release at least 8,000 cusec from the reservoirs, in order to provide, with the uncontrolled flow of 4,000 cusec, the minimum hydro requirement of 12,000 cusec. However, the release of 8,000 cusec from the reservoirs results in a total storage of 42,000 SFM which is less than the rule curve value of 44,000 SFM. During the following month, September, our first concern is to get back to the rule curve. Therefore, we only release enough water to provide for the minimum hydro requirements, and as a result we are at the end of the month indeed back on the rule curve. It may be noted that consequently, during the months of August and September, the steam plants in the system have continuously operated on the base of the load.

Col. (14) represents the sum of cols. (5), (7) and (11) and therefore represents the total flow of water through the hydro plants. After having allowed for possible spillage during times of high river flows and full reservoirs, the figures of col. (14) represent the total amount of hydro energy production. When these figures are deducted from the total system energy requirements, the thermal energy requirements are obtained.

There may be situations where such a balancing regulation is impossible or undesirable. For instance, the flow conditions may be such that the reservoirs cannot be kept in balance for some time; or one of the reservoirs may be the forebay of a power plant, thus making it desirable to keep its level high as long as possible. Another exception to the general rule would be if the reservoirs have drainage areas that are much different in size or water yield. In such circumstances it would be advisable to draft heavier from the easy-to-fill reservoirs so as to avoid subsequent spilling of these reservoirs.

Now that we have discussed in principle how a reservoir rule curve can be prepared and applied, we may introduce a few refinements. First of all we may question if the past stream flow record is the best criterion for future dependable flow operation. Although this principle is widely accepted there may be good reasons, for certain power utilities, to desire a greater or smaller degree of risk than is indicated by 30, 40, or 50 years of record. If this is the case, it becomes necessary to conduct probability studies of low flow periods, before the rule curve can be determined.

Another refinement may have to be made in connection with changing load patterns. In the derivation of the rule curve in *Figure 7.28*, the mass curve of energy demand was based on one particular load year; and the resultant rule curve was only valid for that particular load year. This method is quite satisfactory for system planning studies where thermal energy requirements must be determined for one particular load year, for one particular sequence. However, for actual reservoir operation we must allow for the growing load demand. Hence the mass curve of energy demand in *Figure 7.28* must reflect the change in load pattern from year to year.

Another refinement in reservoir regulation is to establish so-called 'no-spill rule curves'. Their preparation is similar to what we have discussed earlier, but instead of using low flow conditions we use flood conditions, and instead of specifying that the reservoir must contain at least a given storage on a given date, we specify that the reservoir should contain no more than a given storage on a given date. The purpose of no-spill rule curves is to prevent flood damage around the reservoir, to prevent spilling of water, and thus to increase the energy utilization of the river system. The preparation of no-spill rule curves is only meaningful when the reservoir capacity is relatively large and when the installed hydro capacity is relatively large. Otherwise the no-spill rule curves may fall well below the dependable flow rule curves, which would be meaningless.

It was noted earlier that the area immediately above the rule curve may be looked upon as a risk zone, where we run the risk of capacity deficiency if we keep releasing maximum usable river flows, till we reach the rule curve, and where we run the risk of more spilling during subsequent flood flows if we would be more cautious and start cutting back flows before we reach the rule curve. It may be feasible to establish within this risk zone a bundle of so-called 'economy guide lines', more or less parallel to, and slightly above, the rule curve. When we reach the first line, we stop the generation of all surplus hydro energy; when the second line is reached, the least costly thermal energy producers are put on the line; when the third line is reached, the next best thermal plants are placed in operation, and so on. When finally the rule curve is reached, all thermal capacity must be placed in operation.

It will be readily appreciated that a great deal of tedious work is involved in reservoir regulation studies. For this reason, it would have merit to consider the use of a computer for the performance of the routine computations.

SYSTEM PLANNING

Nearly all power utilities are continuously faced with the necessity of having to decide upon increases in their generating capacity to meet the growing load demand. In this section we shall first discuss what general considerations may be applied to the most economic composition of a power system. After that, we shall discuss how a numerical analysis of alternative sequences of system development can be conducted.

Future System Composition

In general, we may observe that a power system may be composed of: gas turbines, steam plants, hydro plants and nuclear plants. Since the most economic combination, for given load conditions, depends very much on the costs of these generating sources, let us discuss cost aspects first.

Gas turbines are relatively new in the field of generating electricity. Presently, their cost per kilowatt of installed capacity ranges from $100 to $200. It is believed, however, that with the advance of technology in this field, the capital cost may be reduced to $100 or somewhat less. The cost of fuel depends much on local conditions. Even if the local cost of fuel is high, gas turbines may become attractive because of their low capital cost. For the purpose of the present discussion we shall assume the cost of figures quoted in Table 7.4.

Table 7.4. *Gas Turbines*

Capital cost	$100 per kW
Interest	6%
Depreciation (25-year life) . .	2%
Taxes and insurance . . .	2%
Operation and maintenance . .	$3·00 per kW per year
Fuel	7 mills per kWh

The cost structure of steam plants has been discussed earlier. Assuming some further advance in the technology of steam plants, aimed at lower capital costs, higher efficiencies and bigger units, we shall use for the present purpose the figures quoted in Table 7.5.

Table 7.5. *Steam Plants*

Capital cost	$150 per kW
Interest	6%
Depreciation (25-year life) . .	2%
Taxes and insurance . . .	2%
Operation and maintenance . .	$3·00 per kW per year
Fuel	4 mills per kWh

The cost of hydro development varies tremendously, depending on the available stream flow, head, site topography, and distance from the load centre. We may assume that by now nearly all feasible hydro sites close to load centres have been developed and that the problem of future hydro development is mostly a matter of developing new sites that are a considerable distance away from the load centres, or developing more capacity at existing sites. There is a third possibility that may be resorted to, when the conditions are favourable, namely the development of hydro pump-storage schemes. Before we quote the cost of these three alternatives, let us discuss for a moment the virtue of pump-storage schemes.

The function of a pump-storage scheme is to absorb energy during times when there is surplus energy in the system, and to release this energy again when there is a demand for energy. Let us assume, for example, the following situation, which is not unlike some real situations. A system is served by a hydro plant and a steam plant, as shown in *Figure 7.35*. New generating capacity is needed. There happens to be a site close to the load centre with a small lake in the mountains and another lake or river down in the valley, with a

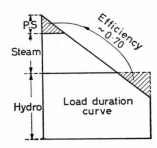

Figure 7.35. Source of energy for a pump-storage plant

head of some thousand feet in between. The main hydro plant is a run-of-river development on a stream with ample river flow. As a result we have surplus energy available during part of the time, indicated by the shaded area to the right. This energy may be used by a pump-storage scheme to pump water from the lower lake to the higher lake. Some time later, during the peak of the load this water flows from the high lake to the low lake to generate power. During a low flow period the steam plant could generate energy for pumping during off-peak hours, since its capacity exceeds base load.

It is obvious that several conditions have to be fulfilled, to make such a scheme not only possible but also economical. First of all we need favourable site conditions; secondly, we need surplus energy or very cheap energy; and thirdly, the capital cost of the pump-storage scheme should be relatively small. This last requirement points towards high heads.

The ultimate criterion of a pump-storage scheme is whether or not its addition to an existing power system will lower the overall cost of the system, as compared to the least-costly alternative addition to the system. A thorough analysis requires a study, as is described in the next section. Roughly speaking, we could make the following remarks. The energy that goes into the pump-storage plant is surplus energy and may have little or no value. The energy that comes out of the plant may displace thermal energy and will then have a certain value. (If we are dealing with an all-hydro system, the energy aspects, apart from prime power, may be disregarded.) The capacity of the plant will substitute for either thermal or hydro capacity. Therefore the annual fixed charges of the pump-storage plant plus the value of the annual energy that goes into it (if any), should be less than the annual fixed charges of the least-costly alternative equivalent capacity, plus the cost of its associated energy (if any). This is a rather severe criterion for a pump-storage scheme, since, roughly speaking, its capital cost has to be less than the capital cost of incremental hydro, or steam capacity!

Let us now assume some pertinent cost figures of hydro development, for the purpose of the present discussion:

Table 7.6. *Hydro Plants*

New capacity (incl. transm.).	$400 per kW
Incremental capacity (incl. transm.)	$300 per kW
Pump-storage scheme (incl. transm.)	$200 per kW
Interest	6%
Depreciation (40-year life)	1%
Taxes and insurance	1%
Operation and Maintenance.	$2·00 per kW per year

We come now to the last and most talked-about source of generation: nuclear power. It is superfluous to say that the cost of nuclear power generation is still in the realm of uncertainty, because its technology is still in the process of being developed. The best we can do for the present purpose is to assume figures that may reasonably be expected, say 10–20 years from now (Table 7.7). This will at least reveal to some extent how nuclear power may fit in the future generating system.

Table 7.7. *Nuclear Power*

Capital cost (incl. initial fuel)	$250 per kW
Interest	6%
Depreciation (25-year life)	2%
Taxes and insurance	3%
Operation and maintenance	$5.00 per kW per year
Fuel replacement and waste disposal	$10.00 per kW per year

It will be noted that the cost of fuel has been assumed as a fixed annual charge. This is because a nuclear power plant operates most efficiently when it

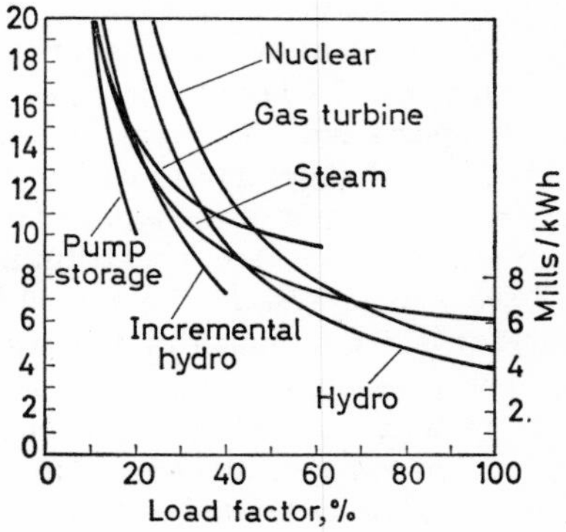

Figure 7.36. Cost of energy from different sources

produces continuous heat. In fact, it seems to be objectional to shut the plant down for an appreciable length of time. It may be pointed out here that the

311

problem of large-scale waste disposal has not yet been solved. It may therefore be that the estimated $10.00 per kW per year for fuel replacement and waste disposal is on the low side.

In order to illustrate the cost composition of the above sources of generation, let us review the cost of energy as a function of the load factor at which every plant could operate, as shown in *Figure 7.36*.

It may be seen that the hydro plant, although having the highest capital cost, produces the lowest-cost energy, when operating above a load factor of about 0·50. Below that load factor, steam becomes more economical than the new hydro plant. If there is. enough dependable flow to increase the capacity of the existing hydro plants, that might be still cheaper. Under the assumed cost conditions, the addition of a pumped-storage scheme might be cheapest

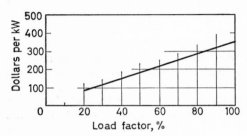

Figure 7.37. Capacity cost of nuclear plants

of all. However, it was noted earlier that such conditions are rather unusual. The addition of a gas turbine would only be economical if it would operate at a very low load factor, say below 0·10.

Nuclear power, in this example, would not be economical, until all hydro resources are exhausted. After that, nuclear power would be less costly than thermal power in the region of high load factors only. The location of the 'cross-over' point between steam and nuclear, which in this example is 0·70 LF, depends to a great extent on the capital cost of the nuclear plant. To illustrate this, *Figure 7.37* shows the capital cost of a nuclear plant required to make the cost of nuclear energy break even with the cost of thermal energy, operating at a given load factor. It may be seen that at a load factor of 1·00, the nuclear plant may cost as much as $350 per kW to be competitive, but at a load factor of 0·20 the nuclear plant would have to cost less than $100 per kW.

It appears that there may be a significant role for nuclear power in future system development, but that it will be supplemented for some time by thermal capacity, the ratio of the two depending on the advance of technology in both fields. Let us now see what a future generating system may be composed of. *Figure 7.38* represents the annual load curve of a power system that has, at present, say one-fourth of this magnitude. First of all, let us determine the proper position of the hydro plants. We know approximately how much energy is available. Say the average river flow past all available hydro sites, converted to kWh, times a utilization factor of 90 per cent. This energy gives us the area of hydro under the load curve. We can now shift it higher or lower, depending on the cost of incremental hydro capacity and incremental hydro energy gain versus the cost of alternative capacity and energy requirements. Let us assume that we arrive at the hydro position shown. (This is for average

312

flow conditions! For low flow conditions, hydro would be in the peak and we must verify that all hydro plants are firm!) Assuming the cost structure of *Figure 7.36*, it is obvious that nuclear capacity should be placed under the hydro position, and thermal capacity above. Since the nuclear plants will have

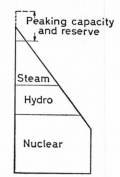

Figure 7.38. System composition

some surplus energy that can be produced at little or no incremental cost, it may be considered if this surplus energy can be utilized in a pump-storage scheme for peaking purposes. If such a possibility does not exist, one may consider the addition of gas turbines to the system for operation in the peak of the load.

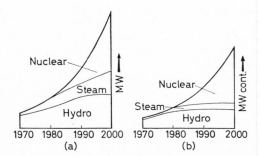

Figure 7.39. (a) Possible future capacity development; (b) Corresponding approximate energy supply

By such considerations as above, one may design an ultimate system that could produce electricity at the lowest overall cost. Reviewing the situation from decade to decade, and tying in the present with the future, one may arrive at system development diagrams as shown in *Figure 7.39*.

Sequential Analysis

Now that we have gained some insight in how to make a reasonable estimate of a likely future system composition, the question arises how we can precisely determine what plants should be built and in what sequence they should be built. In brief, we must find a method to compare the economic merits of alternative sequences of power development. Let us see what is involved in such a study.

313

One of the first efforts is the preparation of a load forecast for the next 10 or 20 years. It is important to keep in mind that this gradually increasing future load demand is considered as being fixed. This load demand has to be met in one way or another. If the load cannot be met, even under the most adverse river flow conditions, the development plan has to be rejected. If there is more capacity in the system than needed, the surplus capacity may be considered of little or no value.

When there are different possibilities to meet the future load demand (for instance by hydro, steam, or nuclear power), forward planning becomes a comparative cost analysis of the various alternatives. Each proposed power development will be called upon to produce the specified amount of electrical energy, determined by the load forecast; and the cost of doing so will be determined. After having analysed each proposal, the most economical development will be selected. At first glance, this may seem a simple procedure: we determine the cost of each single alternative, compare them, and there is the answer! Unfortunately, there are some complicating circumstances.

First of all, we cannot simply compare the single alternative additions with one another but we have to compare the entire alternative power systems with one another. This is because the various additions will affect the functioning of the existing power system in different ways. For instance, if we add steam capacity to the existing hydro-steam system, then the existing steam capacity is bound to operate at a lower load factor, than when we would add hydro capacity to the existing system. This change in the functioning of the existing steam plants has to be evaluated or else the comparison of the costs of the alternative additions has little value. It may therefore be concluded that for every alternative addition we have to analyse the cost of producing the total amount of required electricity with the total available power system (present system plus alternative additions).

A second complication arises from the fact that hydro and steam have a different cost structure. A hydro plant has a high capital cost (say $400 per kW installed capacity); a relatively low fixed annual charge (say 8 per cent); no variable charge; and usually provides a relatively large block of capacity. A steam plant, on the other hand, has a low capital cost (say $150 per kW); a higher fixed annual charge (say 10 per cent); a variable charge dependent on the cost of fuel (say 4 mills per kWh); and may be built in small convenient units. It is obvious that when we were to compare the cost of a series of small steam plant additions with the cost of one large hydro plant addition on the basis of the first year of operation only, we would give the steam plants a considerable advantage over the hydro plant. The first steam unit with its low capital charges may produce electricity at much lower cost than the large hydro plant, which has a good deal of its capacity potential idle, since the load demand has not yet developed. However, a few years later, when the load has fully developed, it may very well be that the hydro plant can produce electricity at much lower cost than the steam plants, which by that time require a large amount of fuel to meet the full load. It may therefore be concluded that, in order to make a fair cost comparison between alternative steam or hydro additions, we should base the comparison, not on one year only, but on a period of years, long enough to reflect the true merits of the hydro proposals.

314

This conclusion has a consequence that is going to make the cost comparison much more complicated than it originally seemed. Comparison of cost over a period of years, for instance, means that we have to take into consideration the further capacity additions during those years. In other words, we have no more the simple comparison of a few alternative power system extensions, but we have to compare alternative sequences of power development, extending over several years. It will be readily appreciated that the number of possible sequences of power development increases rapidly with the length of the period under consideration. Therefore, we have to apply plain common sense here and there, in order to prevent the problem from becoming unmanageable.

Let us assume now that we have decided upon investigating a period of a few decades, and that we have selected a dozen alternative sequences of power development to be analysed. It was noted earlier that every sequence is associated with the same amount of power generation. Therefore, we only have to compare the annual costs of each sequence during the period of analysis. One way to make this comparison would be to simply add the annual costs and compare the totals. However, let us assume that two sequences have the same total costs, but in one case (hydro) the annual costs are relatively high in the early years, while in the other case (steam) the costs are relatively high in the later years. We would then select the last alternative, because it is economically advantageous to postpone the payment of a certain amount of money. It would therefore be more correct to compare the annual costs on a common time basis. This could be done by taking, not simply the arithmetic sum of the annual costs, but the sum of the present values of the respective annual costs.

There is one more complication that may add to the amount of work that has to be done. We have seen already that we need to study a number of alternative sequences during a period of years. For every sequence and for every year, we have to determine what it will cost to produce the amount of required energy. Now the question arises, what river flow conditions we should assume for those annual cost computations. This is quite important because the available amount of hydro energy determines the required amount of thermal energy which has a large bearing on the total annual cost. If time, funds, and computer services are available, the most desirable procedure to find the annual cost for one sequence for one year would be to apply the entire stream flow period on record and take the average yearly cost. However, if time and funds are limited, we may have to resort to the use of a dry, an average, and a wet year and take the weighted mean of the results. Still simpler would be to use only an 'average-water-year'. This sounds quite attractive, but has its pitfalls. First of all, the average-water-year for one stream may not be the average-water-year for other streams under consideration. Secondly, the average-water-year may not produce the average amount of hydro energy. Lower-than-average years will obviously produce less hydro energy because of a lack of water. Higher-than-average years, however, may also produce less hydro energy because of an increase in tail water elevation at the hydro plants. It may therefore be concluded that we have to be careful in the selection of representative flow conditions for the computation of annual costs of proposed sequences.

We have discussed in the above paragraphs that, in order to properly evaluate the merits of alternative system additions, we should compare the accumulated present values of annual costs of alternative sequences of power development over an adequate period of time. Each alternative sequence consists of a series of system additions like: steam capacity of 400 MW, followed by a 250 MW addition to hydro plant A, followed by the construction of a new hydro plant D with a capacity of 600 MW, etc. Every addition is specified for its date of completion and the whole sequence adds up to a system that can satisfy the projected power demand at all times. We shall discuss in the following paragraphs the components that make up the total annual costs of one particular sequence of development for one particular year.

The cost of the hydro plants is composed of fixed charges. These consist of: interest, depreciation, operation, maintenance, taxes and insurance. The cost of the steam plants in the system is composed of fixed charges (the same as listed for hydro) plus variable charges which consist of incremental operation and maintenance plus the cost of fuel. In order to determine this last item, we have to determine the total amount of thermal energy that is required for the particular sequence and year under consideration. These computations of thermal energy requirements are the most time-consuming part of the whole sequential analysis.

The procedure to determine this figure is roughly as follows: after a sequence of power development has been laid out, we determine first of all if the planned hydro capacity is firm under the most adverse river flow conditions. This is a study in itself, involving the establishment of rule curves and the operation of reservoirs, as discussed in an earlier section. Then we select a period of flow records, extending over a few decades or, if time and funds are limited, we select an 'average-water-year' that is believed to be representative for long term river flow conditions. This flow period, or 'average-water-year', will be applied to every sequence and to every load year contained in the period of time under consideration.

Let us now restrict the discussion to an example of one particular sequence of development and one particular year to which certain flow conditions (the average-water-year) are going to be applied. The year is broken down in 12 months. Average river flows, average demands, and peak demands are determined for each month. The computation of thermal energy requirements is then performed as shown in Table 7.8 and is discussed in the notes beneath Table 7.8.

It may be of interest to dwell for a moment on the possible error that may be introduced by having used a representative water year instead of an extended period of flow records. It was noted that redistribution or river flow during the year was assumed to suit the load conditions. It was also noted that in two months there was a surplus of river flow, which was considered as lost due to spillage. It should also be pointed out that thermal energy requirements in dry years may be very much larger than in average years, the difference being more than will be offset by the wet years. If such possible errors are deemed too large, one must resort to a representative period of stream flow extending over a few decades and requiring a regulation of the available reservoirs as shown in Tables 7.1 and 7.3. Such a study requires a tremendous amount of work, since it must be completely repeated for every sequence and for every load year.

However, if one requires an accurate answer, such regulation studies become a necessary part of the sequential analysis.

Table 7.8. *Thermal Energy Requirements for Sequence 3, for Average Flow Conditions and for 1968 Load Conditions*

Month (1)	Peak (2)	Ener. (3)	H. Cap. (4)	H. En. (5)	Def. (6)	% Peak (7)	% En. (8)	Min. St. (9)	Tot. St. (10)
Jan.	1,073	576	792	514	281	26·2	7·2	42	62
Feb.	1,045	514	792	581	253	24·2	7·0	36	36
March	961	512	772	439	189	19·7	4·0	20	73
April	919	461	752	430	167	18·2	2·5	12	31
May	859	463	712	420	147	17·1	3·1	14	43
June	848	435	712	408	136	16·0	2·3	10	27
July	840	430	635	408	205	24·4	5·0	22	22
Aug.	850	437	635	414	215	25·3	6·2	27	27
Sept.	915	461	712	435	203	22·2	5·5	25	26
Oct.	937	500	752	475	185	19·7	4·5	23	25
Nov.	1,080	518	792	478	288	26·7	5·4	28	40
Dec.	1,132	603	870	549	262	23·1	5·3	32	54
									466

Col. (1) lists the months of the year 1968.
Col. (2) lists the peak demand for every month, in MW. The monthly figures are obtained by applying a certain ratio to the December figure, which is determined by the load forecast study.
Col. (3) lists the average monthly energy demand, in 1,000,000 kWh. These figures are obtained by multiplying the peak demand by the appropriate monthly load factor.
Col. (4) lists the total available hydro capacity at all plants, in MW. During seven months the capacity is reduced because of maintenance outage. In December 1968 the capacity is increased by one extra unit.
Col. (5) lists the total available hydro energy at all plants, in 1,000,000 kWh, based on the selected 'average-water-year' of 1922. The natural river flows have been somewhat redistributed over the year to fit the load demand better. This was considered permissible in view of the large storage capacities available. The total amount of river flow for the whole year 1922 was kept the same. We will not know until col. (9) if all this available hydro energy can be used in the system.
Col. (6) lists the hydro capacity deficiency in MW. These figures are obtained by deducting col. (4) from col. (2). This hydro capacity deficiency has to be made up by steam capacity (the total amount of available steam capacity in this sequence, in this year, is 356 MW and therefore ample).
Col. (7) lists the minimum required steam capacity of col. (6) as a percentage of the total peak demand of col. (2).
Col. (8) lists the minimum required thermal energy as a percentage of the total energy demand of col. (3). These figures are obtained from the appropriate monthly peak-percentage curves.
Col. (9) lists the minimum required thermal energy in 1,000,000 kWh. These figures are obtained by multiplying the percentage figure of col. (8) with the total energy requirement of col. (3). As noted earlier, these figures represent the minimum thermal energy requirements, assuming that the steam plants operate in the peaks of the load, assuming that we have ample river flow.
Col. (10). We now total the available hydro energy, listed in col. (5), and deduct it from the required energy, listed in col. (3). If the difference is less than the figure listed in col. (9) we have evidently ample river flow (in fact we may have to spill water). Hence the steam plants can operate in the peak, and the energy requirement is governed by the figures in col. (9). This happened in the months of February and August. If the difference is more than the figure listed in col. (9), we cannot operate the steam plants in the very peak, but have to go somewhat lower on the load curve. The thermal energy requirement is now equal to the hydro energy deficiency: col. (3) minus col. (5); as is shown in the remaining 10 months. The total of the figures in col. (10) represents the total thermal energy requirement, in 1,000,000 kWh, for the year 1968.

After the thermal energy requirement has been obtained in terms of kWh, we have to translate this figure in dollars. It may be recalled that we are dealing here with incremental cost of energy, since the fixed annual charges of steam have already been taken care of. The incremental cost of thermal energy consists primarily of fuel cost and to a much smaller extent on incremental operation and maintenance. Both items combined have been evaluated for the present example at 4·5 mills per kWh. In other words, the cost of thermal energy would have been in 1968: 466,000,000 kWh × 4·5 mills = $2,100,000.

Table 7.9 shows the components of the total annual cost. The notes beneath the table explain the details.

After having computed the total annual costs for all sequences, for all load years, we can now compare the merits of the various sequences as is illustrated by another example, shown in Table 7.10.

It is of interest to note that the annual costs of sequence NF are larger than the annual costs of sequence TO, till the year 1980. Thereafter the annual costs of the first sequence become smaller. The heavy capital investment of the hydro development of nearly one billion dollars now begins to pay off, and the cost of energy becomes less than that of the alternative thermal stations. However, it is not until 1990 that the total of the present values of the annual

Table 7.9. *Annual Cost of Sequence 3 for 1968 Load Conditions*

(1) Fixed annual charges on existing system	.	—
(2) Fixed annual charges of steam .	.	.	$ 1,400,000
(3) Variable annual charges of steam.	.	.	2,100,000
(4) Annual charges of hydro .	.	.	.	9,700,000
(5) Total annual cost .	.	.	.	.	$13,200,000
(6) Present (1962) value .	.	.	.	.	10,600,000

Col. (1). The fixed annual charges of the existing hydro and steam generating system have not been entered, since they can be considered as being the same for all sequences; therefore, they have no bearing on the sequential analysis.
Col. (2). Fixed annual charges of steam represent interest, depreciation, fixed operation and maintenance, taxes and insurance, on the total capital investment of the new steam capacity.
Col. (3). Variable annual charges of steam represent variable operation and maintenance plus the cost of fuel of all the steam plants in the system (new plus existing). This figure is based on the above computed thermal energy requirements and selected unit costs.
Col. (4) represents the annual charges, including operation and maintenance, on the total capital investment of the new hydro plants.
Col. (5) is the total of the above items.
Col. (6) is the 'present value' of the figure in col. (5), computed at a discount rate of 4 per cent per year.

costs of sequence NF equals the same figure of sequence TO. It is only thereafter that we can truly say that the hydro plants begin to show benefits, compared with the thermal. Admittedly, the annual benefits, by that time, are quite large: in the order of $50,000,000, with a present value in the order of $15,000,000.

The total difference in present values, over the full 30 years of the two sequences, is $145,000,000, which is significant enough to conclude that sequence NF is the more economic.

It is obvious that such long-term studies are very sensitive to the assumptions that are being made. One important item is the discount rate to determine the present values of the annual costs. This was assumed in this example to be equal to the interest rate of 5 per cent, which is conventional practice. However, one could make a plea for applying a higher discount rate to reflect the uncertainties of our estimates of load growth and construction costs, 20 and 30 years hence. A higher discount rate, in the present example, would have been to the disadvantage of sequence NF.

The assumed flow conditions, to determine the thermal energy requirements, are also important. It was noted earlier that the correct method is to route a representative flow period of a few decades through the reservoirs and hydro plants for every sequence and every load year. However, for preliminary studies, this entails too much work. A typical average flow year requires much less work. However, one should try, for at least one particular sequence and load year, at least a few 'typical' water years, and perhaps a low flow year and a high flow year, to see what difference these assumptions make on the computed thermal energy requirements.

318

The annual load growth in the present example was assumed at 6 per cent. The assumption of a larger figure would have been to the advantage of sequence NF, since it would have reduced the initial 'part load' period. The assumption of a smaller annual load growth would, of course, have had the reverse effect. As further alternatives we could assume a 'lag' or a 'bump' in the load growth. It may be stated, in a qualitative sense, that it is very important how the load growth takes place between the date that the decision is made to go ahead with a large hydro project and the date that this

Table 7.10. *Cost Analysis of Sequences* NF *and* TO

Year (1)	Plant Req. (2)	Sequence NF			Sequence TO		
		Addition (3)	T.A.C. (4)	P.V. (5)	Addition (6)	T.A.C. (7)	P.V. (8)
1970	1,290	100 Cap. Pur.	6·9	6·9	100 Thermal	6·0	6·0
1971	1,370	250 White Mud	20·4	19·4	100 ,,	9·3	8·8
1972	1,450		20·4	18·5	110 Gr. Rap.	11·4	10·3
1973	1,540	240 Red Rock	25·6	22·1	50 Win. Riv.	14·1	12·2
1974	1,630		25·6	21·0	90 ,,	17·6	14·5
1975	1,730	100 Kelsey	27·0	21·2	100 Thermal	21·3	16·7
1976	1,830	370 Kettle	34·2	25·4	100 ,,	24·5	18·3
1977	1,940		34·2	24·3	100 ,,	28·6	20·4
1978	2,060		34·2	23·1	150 ,,	33·3	22·5
1979	2,180	650 Long Spr.	45·4	29·2	150 ,,	38·0	24·4
1980	2,310		45·4	27·8	150 ,,	43·1	26·4
1981	2,450		45·4	26·5	150 ,,	46·6	27·2
1982	2,600		45·4	25·3	150 ,,	53·1	29·5
1983	2,760		45·4	24·0	150 ,,	59·6	31·6
1984	2,920	110 Gr. Rapids	45·4	23·0	150 ,,	65·5	33·0
1985	3,100	140 Win. River	45·4	21·8	150 ,,	71·0	34·1
1986	3,280	210 Birthday	58·2	26·6	200 ,,	78·2	35·8
1987	3,480	550 Turtle	67·9	29·6	200 ,,	85·4	37·2
1988	3,680		67·9	28·2	200 ,,	92·5	38·5
1989	3,910	900 Limestone	97·3	38·4	200 ,,	99·1	39·2
1990	4,150		97·3	36·5	200 ,,	107·6	40·6
1991	4,400		97·3	33·9	300 ,,	116·1	41·7
1992	4,650		97·3	33·1	300 ,,	125·3	42·9
1993	4,940	300 Thermal	100·6	32·8	300 ,,	134·7	43·9
1994	5,240	300 ,,	103·9	32·1	300 ,,	143·9	44·7
1995	5,550	300 ,,	107·2	31·7	300 ,,	154·1	45·5
1996	5,860	300 ,,	110·9	31·1	300 ,,	164·5	46·2
1997	6,220	300 ,,	114·2	30·7	300 ,,	175·2	47·0
1998	6,600	600 ,,	120·8	30·8	600 ,,	189·7	48·3
1999	6,960	300 ,,	131·7	32·0	300 ,,	201·2	48·8
2000	7,400	300 ,,	144·7	33·4	300 ,,	213·4	49·3
				840·4			985·5

Col. (1) shows the load years. A period of 30 years was deemed necessary to get the true comparison of a long-term hydro development with an alternative all-thermal development.
Col. (2) shows the total plant requirement for every load year. The installed capacity in 1969 will be 1,180 MW. The figures in this column were obtained on the basis of a 6 per cent per year load increase. The figures are listed in MW.
Cols. (3), (6) show the proposed additions to the system for the two alternative sequences of development. The figure indicates the capacity of the plant, in MW. The name indicates the location of the plants on the Winnipeg River, the Saskatchewan River and the Nelson River.
Cols. (4), (7) show the total annual cost of each sequence for every load year, as discussed in Table 7.9. The figures are in $1,000,000.
Cols. (5), (8) show the present (1970) values of the figures of cols (4) and (7). At the bottom of the table, the totals of these figures are given.

hydro project is fully loaded. However, since it is very difficult to predict bumps or lags in future load growth, such speculations are of small relevance to the study.

Interest and depreciation are of great importance to the cost comparison. Since the capital cost of hydro may be two to three times as large as the capital cost of steam, it is obvious that any increase in interest or depreciation charges is disadvantageous to all sequences with hydro development. In fact, these charges may become so high that electricity produced from hydro is no longer competitive with electricity produced from steam. The interest rate to be assumed, depends on what is expected to happen to the economy during the forthcoming years. In fact, the decision to go ahead with one sequence or another could depend on such speculation. If interest rates are expected to come gradually down, the installation of hydro may be postponed, and vice versa. However, to predict the rate of change of interest, several years from now, is indeed little more than speculation. The depreciation charge to be applied is important. It may be noted that the straight line method means in effect a saving for future investment. When it comes to comparing hydro and steam, we are demanding hydro to carry a much larger 'savings burden' than we do from steam. It must therefore be emphasized that a cost analysis for planning purposes should be based on a sinking fund depreciation method.

The outcome of a planning study may be rather sensitive to the assumed construction and fuel costs. These costs are usually estimated on the basis of the latest and best available information, but it is recognized that in spite of this they may be substantially in error a few years later. As an example: a hundred million dollar hydro plant was recently let to the lowest bidder at more than 30 per cent below the engineers' estimate! With respect to the cost of fuel it may be noted that the cost of transporting coal may change substantially on short notice. An upward change in construction costs would be to the disadvantage of hydro as compared to steam. However, if we assume a continuing increase in construction costs, in other words steady inflation, it may be advantageous to construct hydro as early as possible.

PLANT AND EQUIPMENT

The purpose of this last section is to provide the hydraulic engineer with a working knowledge of the different hydraulic structures that may go into a water power development, so that at least he can prepare a preliminary design and cost estimate of the whole project. Some of the important components such as dams, spillways, stilling basins, and conduits were discussed in Chapter 5: Hydraulic Structures. In this chapter we shall discuss some aspects of spillways and sluice gates that are typical for hydro developments; penstocks and surge tanks; and the power house with its hydraulic equipment.

Spillways and Sluice Gates

The reservoir behind a power dam may perform two distinct functions. First, to create head for the power plant. Second, to provide pondage capacity and perhaps storage capacity. It was noted earlier that these two interests may sometimes conflict inasmuch as storage operation may reduce the head on the plant. We may also recall that every hydro plant must have spillway facilities to discharge all river flow that is not passed through the turbines and that cannot be stored in the reservoir. The problem of determining the appropriate spillway lay-out has two components. First of all we must determine the spillway design flood, which must be considered as an inflow hydrograph into the reservoir. The second problem becomes one of designing the spillway and apportioning part of the reservoir for passage of the design flood. These problems were discussed in Chapter 5.

In addition to a spillway, which may be called upon rather infrequently, the power dam may be equipped with sluice gates for day-to-day regulation of the forebay level. Let us assume for instance that we have a power plant with a certain full supply level (F.S.L.) and pondage capacity. When the river flow is nearly average, the reservoir level may be below F.S.L. during most of the

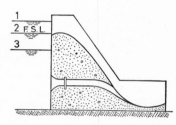

Figure 7.40. *Spillway without control, conduit with control*

week from Monday to Saturday as shown in *Figure 7.13*, page 282. The sluice gates, of course, remain closed. Over the week-end the reservoir is replenished and F.S.L. may be reached before the plant gets back to full operation. The sluice gates are then opened in order to maintain F.S.L. During a low flow period the hydro plant may operate in the peak of the load and the sluice gates remain continuously closed. During a high flow period, however, we may have a surplus of water for weeks in succession and the sluice gates are open all the time, and continuously being adjusted to maintain the proper F.S.L. Sluice gates may be selected from a large variety of possibilities such as: lift gates, radial gates, conduits through the dam equipped with sliding gates or butterfly valves, etc. It is quite common that one or more of the spillway gates is equipped with a permanent hoisting mechanism and is being used for sluice gates.

Let us now come back to the spillway requirements. We may distinguish two situations: (*a*) that we are free to exceed the F.S.L. if we desire; (*b*) that we may not exceed the F.S.L. under any circumstances, because of upstream interests. In the first case the situation may be as shown in *Figure 7.40*. The reservoir normally fluctuates between F.S.L. at elevation 2 and elevation 3. This range has been established from a study of pondage or storage requirements, taking into account the loss of head as a liability versus the gain in prime power as an asset. A sluice gate is not needed, unless the release of water from the reservoir for downstream interests, such as water supply or irrigation, over and above what goes through the turbines, is required. The crest of the spillway is located at F.S.L. When the reservoir level is at F.S.L. and the inflow into the reservoir exceeds the flow through the turbines, the excess discharge will raise the reservoir level above F.S.L. and cause overflow over the spillway. In order to determine the maximum reservoir flood level at elevation 1, we route the spillway design flood through the reservoir. Due to the storage capacity between elevation 1 and elevation 2, called 'surcharge', the maximum reservoir outflow will be somewhat smaller than the maximum reservoir inflow, the amount of reduction depending upon the volume and shape of the design flood and the amount of surcharge. Having determined the maximum reservoir level and the maximum reservoir outflow, we can now establish the elevation of the top of the dam by adding the required freeboard to the maximum reservoir level, and we can proceed to design the spillway and stilling basin. In order to find the most economical lay-out, we should prepare designs for different lengths of spillway crest and select the design that has the lowest total cost of dam plus spillway.

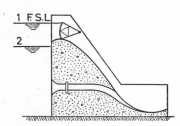

Figure 7.41. Spillway and conduit controlled

We shall now assume that the F.S.L. must not be exceeded. The situation may then be as shown in *Figure 7.41*. The normal range of operation is between elevation 1 and elevation 2. One of the spillway gates is being used as sluice gate. When the reservoir is at F.S.L. and the inflow exceeds the flow through the turbines, the sluice gate is lifted to release the excess flow. Assuming no advance warning of the spillway design flood, we must design the total spillway capacity such that the peak of the design flow can be discharged with the reservoir at elevation 1.

However, in many cases we may assume some advance warning of spillway design flood conditions, and we could then effect some economy in the design of the spillway because of advance release of water from the reservoir with consequent reduction of the flood peak. This is illustrated by the situation in *Figure 7.42*. Let it be assumed that we have a forecasting system that gives us a reliable estimate of the river flow three days in advance. At the beginning of the first day we know therefore that the river flow is going to be at least 40,000 cusec. From a study of experienced recession curves we have found the minimum volume of water that will come into the reservoir with a peak flow of 40,000 cusec. We can safely release that volume, without endangering our storage position. This is done during the first day by opening all spillway gates. The resultant release from the reservoir is about 60,000 cusec, while the average inflow during the first day is only 10,000 cusec. Hence we are drawing down the reservoir with a resultant decrease in spillway capacity. At the beginning of the second day we receive another flow forecast, confirming the arrival of still greater flows. Hence we keep the gates open. During the fourth day the inflow surpasses the outflow and therefore the reservoir level begins to rise again. At the end of the tenth day, the reservoir level is back to F.S.L. again and the peak of the outflow is reached. It can be seen that we have effected a reduction in maximum outflow from 80,000 cusec to 60,000 cusec.

In the above discussion we have treated the gated and ungated spillway as two distinct possibilities for the case where we are not allowed to exceed F.S.L. and the case where we have freedom to do so. However, there are many in-between cases for some reason or another. For instance, there may be restriction on the F.S.L. for normal operation, but we may have above F.S.L. a few extra feet of flood reserve for extraordinary circumstances. There is also the possibility that we select a gated spillway in favour of an ungated spillway purely for reasons of economy. Let us assume, for instance, that we have to design a spillway for a 400 ft. high earth dam with side slopes of 5:1. An

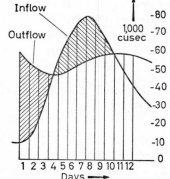

Figure 7.42. Flood operations in a power reservoir

ungated spillway, 500 ft. long, would raise the maximum reservoir level say 15 ft. above F.S.L., whereas a gated spillway, with gates of 20 ft. high, would keep the maximum reservoir level at F.S.L. Hence the dam would have to be 15 ft. higher with an ungated spillway. Assuming a length of the dam of 10,000 ft. and a cost of embankment of 20 cents per cubic yard, the extra cost of dam becomes $4,500,000, which may be much more than the cost of gates on the spillway which may amount to 20 ft. height times 500 ft. length times $100 per square foot equals $1,000,000.

Conduits and Penstocks

To convey water from the reservoir to the powerhouse in medium and high head developments, we may use open canals, tunnels, conduits (low pressure), and penstocks (high pressure), as shown in *Figures 7.2* and *7.3*. When we are dealing with low head developments, the water is usually conveyed from the reservoir to the turbine via an intake opening in the upstream face of the powerhouse, as shown in *Figure 7.1*. If water has to be conveyed over a large distance, as in a high-head development, one of the main problems is to design the open or closed conduits at the most economical dimensions. For a given plant discharge we can design the conduits at different dimensions. Larger dimensions will result in smaller velocities and hence less energy losses. However, the cost of the conduits becomes larger. The most economic solution is the one whereby the total system cost of energy becomes a minimum. Since loss in head may have to be made up by thermal capacity, it is obvious that the

323

final choice must be based on a sequential analysis of the whole power system, introducing alternative conduit sizes. For preliminary design, one could estimate the value of 1 ft. of head and then try to minimize the sum of the cost of the conduits and the lost head. Such a problem is similar to the one illustrated in *Figure 7.20*. Reasonable values for a first trial are the following design velocities: concrete lined open canals, 10 ft./sec; concrete lined conduits and tunnels, 20 ft./sec; steel lined conduits and penstocks, 30 ft./sec.

To estimate the value of 1 ft. of head of a potential or actual hydro development one could reason as follows. If we reduce the hydraulic losses in the conduits by 1 ft., we could develop that 1 ft. by installing more turbine and generating capacity at a cost of say $100 per kW. Let us further assume that the average annual flow through the plant is 10,000 cusec and that the plant operates at an annual plant factor of 0·5 (annual energy output in MW continuous divided by plant capacity). The annual cost of developing this 1 ft. of head with its associated energy, at 8 per cent annual charge, becomes:

$$\frac{10,000 \times 1 \times 0·80}{11·8} \times \$100 \times 8\% = \$5,400$$

The alternative way of producing the same capacity and energy by steam will cost:

$$
\begin{array}{ll}
680 \times \$150 \times 10\% & = \$10,000 \\
680 \times 0·5 \times 8,760 \times \$0·004 & = 12,000 \\
\hline
\text{Total} & \overline{\$22,200}
\end{array}
$$

Hence the benefit of gaining 1 ft. of head on the power development is $16,800 per year. This figure can now be compared to the increment in annual cost of the conduits to gain this 1 ft. of head. As long as the cost of gaining 1 ft. of head is less than $16,800, we can keep increasing the size of the conduits. In practice, it has been found that economical velocities for tunnels, conduits and penstocks are seldom less than 5 ft./sec, seldom more than 25 ft./sec, and mostly around 10–15 ft./sec.

Apart from economic considerations, there are certain limits to the velocities in conduits that must not be exceeded. In open canals, the velocities must not be so high as to cause scour. This subject was discussed in Chapter 6: Flood Control, section on diversion channels. In penstocks, high velocities would create excessive water hammer, as will be discussed in the following paragraphs.

If we want to stop a moving object, we must apply a force against the direction of movement. The simplest way to analyse problems of this kind is with the momentum equation:

$$\text{impulse} = \text{change in momentum}$$

or $$F.\Delta t = m.\Delta v$$

It may be seen from this equation that the force F required to stop the mass m, is inversely proportional to the time Δt that is available to slow the object down. If we would want to stop the motion instantaneously, we would need, theoretically, an infinitely large force F. Let us now assume that we have a power plant as shown in *Figure 7.43*, operating at full capacity. For some reason, there is a sudden load rejection on the turbine-generator unit. The first reaction would be a speeding up of the turbine. However, the automatic governing mechanism, which controls the speed of the turbine, senses this immediately, and begins to close the gates or the valve between the penstock and the turbine.

If we assumed that the closure of the gate was instantaneous, that the water was incompressible and that the penstock was inflexible, then we would build up an infinitely large pressure in the water, and the penstock would burst like an eggshell. Fortunately, this does not happen. To start with, water is not quite incompressible, and the penstock is not quite inflexible. Moreover, we can take a number of measures to further reduce the water hammer effect. First of all, we can set a lower limit to the time of closure, and design the hydraulic and electrical machinery accordingly. With impulse turbines, a simple solution is to deflect the jet first so that the water wheel loses its power, and to slowly close the valve afterwards. Secondly, we can design a relief valve

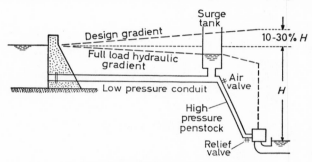

Figure 7.43. Design gradient for conduits

at the lower end of the penstock that will open and discharge water when the pressure becomes too high. Thirdly, we can try to keep the velocities in the penstock down to reasonable values, as mentioned earlier. Fourthly, we can install a surge tank, as shown in *Figure 7.43*. A surge tank is usually provided at the end of the long low pressure conduits. It is located as close as possible to the powerhouse in order to reduce the length of the costly penstock, and on high ground to reduce the height of the costly tower. The precise computation of pressures for given conditions (size and lengths of conduits, discharge, closure time, design of surge tank) is very complicated. In fact, some hydraulic engineers have devoted their entire career to the study of water hammer and the design of penstocks and surge tanks. For preliminary design one may assume a pressure gradient that is 10–30 per cent of the total head, above the reservoir level, as shown in *Figure 7.43*. The 10 per cent applies to the high heads; the 30 per cent applies to the lower heads. It should be noted that a sudden opening of the turbine gates causes a negative pressure wave in the penstock and conduit. If there is danger that the pressure could become less than atmospheric, an air valve should be installed.

Turbines

The hydraulic turbine converts the energy of the water into mechanical energy. This energy is passed on, via the turbine shaft, to the generator, where the mechanical energy is converted into electrical energy.

325

In order to have an efficient and economical generation of electricity, we must operate the turbine-generator combination within certain speed limits. Since there is a large range in possible heads on hydro plants, and therefore a large range in possible water velocity, it may be expected that we need different types of turbines in order to stay within the desired speed limits. During the past 100 years the following types of turbines have been developed:

I   Impulse
II Reaction..........(a) Francis
              (b) Propeller.........(1) Fixed blade
                                     (2) Adjustable blade

It has been found that the most suitable machine for very high heads is the impulse turbine; for medium and high heads the Francis turbine; and for

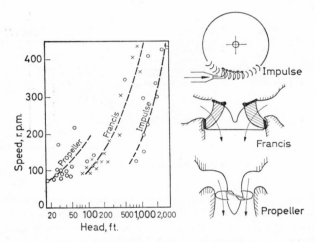

Figure 7.44. Angular velocities of turbines

low and medium heads the propeller turbine, as shown in *Figure 7.44*. Impulse turbines are driven by a free jet discharging from the nozzle of a penstock into the free air. When the jet hits the buckets of the turbine, the velocity of the water is greatly reduced and changes in direction, thus giving an impulse on to the buckets according to the equation:

$$F.dt = m.dv$$

or $$F = \rho.Q.dv$$

The change in direction and magnitude of velocity is shown graphically in *Figure 7.45* where V is the absolute velocity of water, v the velocity of water relative to turbine, and u the absolute velocity of turbine. In order to obtain the greatest force from the jet upon the turbine, the value of the exit velocity V_2 should be as small as possible and perpendicular to the jet velocity V_1. To achieve this, the buckets are designed in a certain fashion, and the peripheral speed u is selected at $\phi V_1 (\phi = 0.45 - 0.49)$. It may be noted that for a value $\phi = 1$ the speed of the bucket would equal the speed of the water. Hence there

326

would be no force on the bucket and therefore no power would be developed. For a value of $\phi=0$, the bucket would be standing still and again no power would be developed. It has been found from extensive testing that maximum power can be developed when the speed of the bucket is about half of the speed

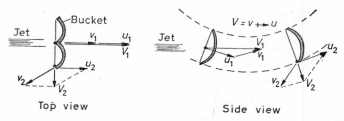

Top view   Side view

Figure 7.45. Velocity diagrams of impulse wheel: (a) view from above; (b) view from side

of the jet. Since the jet velocity V_1 is practically equal to (except for hydraulic losses) $\sqrt{2gh}$, we have now a relationship between the head at the plant, the angular velocity of the turbine N (in r.p.m.), and the diameter of the turbine D (in inches).

$$u = \phi\sqrt{2gh} \text{ ft./sec} \qquad \dots(7.1)$$

and since

$$u = \frac{\pi.D.N}{60 \cdot 12} \text{ r.p.m.}$$

$$D = \frac{1{,}840\phi\sqrt{h}}{N} \text{ in.} \qquad \dots(7.2)$$

Reaction turbines are completely immersed in water. The water, approaching the reaction turbine, has pressure energy and kinetic energy. While the water flows past the turbine, its pressure is reduced and its velocity is changed in direction and magnitude, thus transmitting energy to the turbine. The change of velocity is shown in graphical form in *Figure 7.46.*

In order to gain the greatest efficiency in energy transfer from the water to the turbine it has been found that the water should approach the turbine under a slight angle with the tangential direction, and leave the turbine in a radial direction as indicated by V_1 and V_2. In other words the water should make a spiral motion while it passes through the turbine. In order to achieve this, the water passages leading to the turbine are given a spiral shape and the stay vanes and guide vanes have a nearly tangential position.

The formulae (7.1) and (7.2) are also applied to reaction turbines, although they have lost much of their physical meaning. $\sqrt{2gh}$, sometimes called the spouting velocity, is no more the upper limit of the velocity of the water since we are now dealing with a closed conduit. Moreover, the direction of V_1 is no more the same as the direction of u and hence ϕ loses its meaning. However, the formulae have been found useful by changing the numerical values. From experience it has been found that:

$$\phi = 0 \cdot 053 N_s^{2/3} + 0 \cdot 09 \text{ for Francis turbines } (0 \cdot 5 - 1 \cdot 1)$$

$$\phi = 0 \cdot 065 N_s^{2/3} \qquad \text{for propeller turbines } (1 \cdot 3 - 1 \cdot 8)$$

327

In the early days of turbine design, the reaction turbines were mostly Francis turbines with vertical vanes and radial flow. As a result they had a large diameter, and hence required costly powerhouses. Because of their large diameter they also had a slow speed and hence required costly generators.

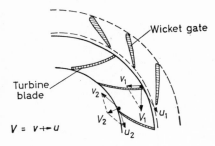

Figure 7.46. *Velocity diagrams of reaction wheel*

These disadvantages led to the design of mixed flow Francis turbines where the direction of flow is partly radial and partly axial, as shown in *Figure 7.47*. For low head turbines, with low water velocities, not only the diameter is reduced but also the position of the blades is changed in order to keep the turbine

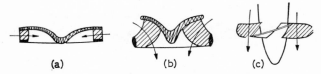

Figure 7.47. (*a*) *Radial flow turbine;* (*b*) *Mixed flow turbine;*
(*c*) *Axial flow turbine*

speed above a reasonable lower limit. The result is the axial flow propeller turbine. For plants with a large fluctuation of head or fluctuation of load, necessitating the frequent use of the wicket gates to regulate the flow, the propeller blades can be made adjustable, in order to keep the efficiency of the turbine at a high level. Examples of this type are the Kaplan and Deriaz turbines.

It was noted in eqn. (7.2) that the diameter of a turbine can be expressed as follows:

$$D = \frac{1{,}840\phi\sqrt{h}}{N}$$

and that the factor ϕ depends on the type of turbine. In other words, for a certain type of turbine we may say that:

$$D \propto \frac{\sqrt{h}}{N} \qquad \ldots.(7.3)$$

It was also noted earlier that:

$$P_{\text{h.p.}} = \frac{Q.h.e}{8{\cdot}8} \qquad \ldots.(7.4)$$

328

In this equation we may substitute for Q:

$$Q = K_1.A.\sqrt{2gh} = K_2 D^2 \sqrt{h} \qquad \ldots\ldots(7.5)$$

and substituting (7.5) in (7.4) we obtain:

$$P_{h.p.} \propto D^2.h^{3/2}$$

or

$$D \propto \frac{P^{1/2}}{h^{3/4}} \qquad \ldots\ldots(7.6)$$

In view of (7.3) and (7.6) we may now conclude that for a certain type of turbine, the value:

$$N_s = \frac{N.P^{1/2}}{h^{5/4}} = \text{constant} \qquad \ldots\ldots(7.7)$$

This parameter N_s is called the specific speed, and has been found to be useful in the design of turbines, although it must be admitted that it is an awkward figure inasmuch as it has an awkward dimension, and is difficult to visualize. At best we may say that for a value of $P=1$ and $h=1$, N_s becomes equal to N. In other words, the specific speed is the speed in revolutions per minute at which a turbine would run under a head of 1 ft., its dimensions having been adjusted to produce 1 h.p.

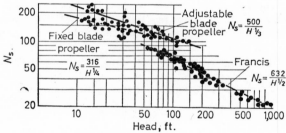

Figure 7.48. Specific speed of reaction turbines
[after Bureau of Reclamation (Monograph 2020)]

The most useful application of N_s is as follows: it was noted that N_s is constant for a certain type of turbine. It was also noted that we need a certain type of turbine in order to produce, with a given head, a desirable combination of speed and diameter. It may therefore be expected that a certain head will correspond to a certain N_s value. This has been found from an analysis of a great number of turbines as shown in *Figure 7.48*.

We may look upon *Figure 7.48* as an empirical diagram, giving the relationship between effective head on the turbine and N_s value of the turbine. For preliminary design purposes we can use this diagram as follows: for the given head, we look up the corresponding N_s value; from eqn. (7.4) we determine the available h.p. in the turbine; from the N_s value and eqn. (7.7) we determine the speed of the turbine; from this N value and eqn. (7.2) we determine the diameter of the turbine; from this diameter we prepare a preliminary design of the

329

powerhouse. It should be pointed out that the values in *Figure 7.48* are only approximate values for preliminary design. For final design, hydraulic model studies should be made. A higher specific speed for a given head and h.p output will result in a smaller size of turbine and generator and hence in a lower cost. However, too high a speed will result in too much energy loss and possibly cavitation. Too low a specific speed, on the other hand, will result in increased generator cost.

We shall now discuss an application of the above equations. Let it be assumed that a hydro site has to be developed, with an effective head on the turbines of 120 ft. From a system study, as discussed in an earlier chapter, it has been decided to develop the site to a total capacity of 450,000 h.p. From a study of practicability, flexibility, and spare capacity, it has been decided to install four units (of 112,500 h.p. each). From a viewpoint of generator and powerhouse economy it has been decided to select fixed blade propeller turbines. We can now determine the value of N_s from the equation of the experience curve of *Figure 7.48*.

$$N_s = \frac{316}{h^{1/4}} = \frac{316}{120^{1/4}} = 96$$

Hence:

$$96 = \frac{N.112,500^{1/2}}{120^{5/4}} \quad \text{or} \quad N = 114$$

We have to check if such a speed is feasible. The generator will have to produce electricity with 60 cycles per second, and will have an even number of poles. Hence the number of pole pairs times the speed in r.p.m. should equal 60×60. We must select such a speed, called the synchronous speed, close to the value of 114, that this is possible. We find that $N = 116$ with 31 pole pairs.

We can now compute the diameter of the turbine as follows:

$$\phi = 0 \cdot 065 N_s^{2/3} = 0 \cdot 065 \times 96^{2/3} = 1 \cdot 37$$

and

$$D = \frac{1,840\phi h}{N} = \frac{1,840 \times 1 \cdot 37 \times 120^{1/2}}{114} = 242 \text{ in.}$$

(Note that h is listed in feet and D in inches! This may be verified from the original derivation of the equation.) We may use the diameter of the turbine as the basis of a preliminary design of the powerhouse by using the diagrams shown in *Figure 7.54*, page 338.

After having determined the main elements of the substructure of the powerhouse, we should now establish its elevation with respect to the tailwater level. Before we try to establish this, let us discuss first of all the hydraulic function of the draft tube which connects the turbine exit with the tailwater.

Let us for simplicity's sake begin by assuming that there is no draft tube at all and that the turbine discharges into the free air as shown in *Figure 7.49(a)*. Since the pressure at the exit is atmospheric, the elevation of the energy line is a distance $\frac{V^2}{2g}$ above the turbine exit. Let us assume that the exit velocity is

25 ft./sec, and that the elevation of the turbine exit is 10 ft. above tailwater. The energy line at the exit is therefore 20 ft. above tailwater, and the effective head on the turbine is therefore the difference between headwater and tailwater minus 20 ft. Obviously an appreciable loss in head! The first improvement to this situation would be to connect the turbine exit and the tailwater with a straight piece of tube, as shown in *Figure 7.49(b)*. We may now observe that the pressure at the end of the tube, at the tailwater level, is atmospheric. Since the velocity is still 25 ft./sec, the energy line at this point is 10 ft. above tail-water level. Assuming negligible losses in the short tube, it follows that the energy line at the turbine exit is at the same elevation. In other words, we have

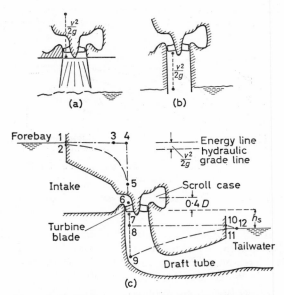

Figure 7.49. (a) Free exit of turbine; (b) Enclosed exit of turbine; (c) Pressure head at intake and draft tube

increased the head on the turbine by 10 ft. It may be noted that we have lowered the pressure at the turbine exit also by 10 ft. In other words, we have created a 'suction'. This explains the words 'draft tube'.

We can still further improve upon the situation, and reclaim nearly all of the remaining 10 ft. of head loss by gradually widening the draft tube from the small diameter at the turbine exit to a much larger cross-sectional area at the draft tube exit as shown in *Figure 7.49(c)*. It should be pointed out that in this figure the turbine exit is circular, while the draft tube exit is rectangular with a much larger width than height. As a result, the cross-sectional area of the latter may be 5–10 times as large as the former, and hence the velocity head at the draft tube exit has now become a fraction of the velocity head at the turbine exit. When inspecting *Figure 7.49(c)*, we can see that the elevation of the energy line for a point at the turbine exit is now very close to the tail-water elevation, and hence the effective head on the turbine is nearly the difference between headwater and tailwater levels. It may be noted that the pressure at the turbine exit has been further decreased. This brings us to the very essence of the present problem, namely to determine the proper elevation of the turbine so that the exit pressure does not become so low that cavitation will be avoided.

Let us now review *Figure 7.49(c)* and explain the significance of the points in the diagram. At the intake to the powerhouse, the energy line is at forebay level: point 1. Since the velocity at this point is small, the velocity head is small: 1–2. At the intake to the scroll case, the energy line has the elevation of point 3. The significance of this point is that the head on the turbine is considered to be the difference in elevation between point 3 and point 12. This, of course, is not quite correct, but it is common usage, since the turbine manufacturer is responsible not only for the design of the turbine, but also for the

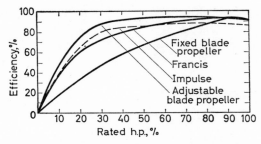

Figure 7.50. Turbine efficiencies as a function of their loading

design of the scroll case and draft tube, which are all combined in one efficiency figure. At point 6, just before the entrance to the turbine, the velocity is considerably increased, and the velocity head is represented by 4–5, while the pressure at 6 is represented by 5–6. Between 6 and 7, the energy transfer from the water to the turbine takes place. At point 7, just after the exit of the turbine, the energy line is at point 8 and therefore the effective head on the turbine in a strict sense is 4–8. The velocity head at point 7 is represented by 8–9, and it follows that the pressure at 7 is 7–9, below atmospheric. In the draft tube there is a deceleration of flow. As a result, there are energy losses, represented by the drop from 8 to 10, and a gradual diminishing of the negative pressure in the water. At the draft tube exit the pressure is nearly hydrostatic. Hence the hydraulic grade line reaches the water surface at point 11. The energy line is at 10, the distance 10–11 representing the exit velocity head. From 10 to 12 this velocity head is lost, and the energy line reaches tailwater level, assuming that the velocity head of the tailrace is negligible.

It may be of interest to mention here that the efficiency of the turbine for a given flow is defined as the amount of energy, delivered to the generator, divided by the amount of energy corresponding to the drop in energy line from point 3 to point 12. The total losses consist of: (1) scroll case losses; (2) guide vane losses; (3) wicket gate losses; (4) turbine runner losses: (*a*) surface resistance, (*b*) form resistance, (*c*) leakage; (5) draft tube losses; (6) exit losses; (7) mechanical friction losses. In modern turbine design, the overall efficiency, as defined above, is approximately 90 per cent, depending on the type of turbine and the load conditions, as shown in *Figure 7.50*.

Let us now come back to the subject of the present discussion: the setting of the turbine in connection with cavitation. Knowing the discharge through the turbine and the approximate dimensions of the draft tube, including the throat diameter of the turbine, we may compute the elevation of point 8 and point 9, in *Figure 7.49(c)*. Assuming a permissible negative pressure at the turbine exit, we would then be able to determine the maximum elevation of point 7. However, in most cases such considerations have limited value, since cavitation does not so much originate at the turbine exit as on the edges of the turbine blades where the velocities are highest. The water vapour bubbles thus formed are separated from the blades and swirled away. Due to the vortex motion of the water flowing through the turbine, the vapour bubbles tend to congregate near the centre of the draft tube inlet in the form of a 'swinging rope'. As soon as the bubbles come in a region of relatively higher pressure they will collapse. The unpleasant aspects of cavitation are mechanical damage and reduction of efficiency. In order to derive some criteria for avoiding cavitation we may look upon the problem as follows:

Applying the Bernoulli equation to point 7 in *Figure 7.49(c)* immediately underneath the turbine, and assuming tailwater elevation as datum, we can write:

$$Z + \frac{p}{\gamma} + \frac{V^2}{2g} = H$$

or

$$h_s + \frac{p}{\gamma} + \frac{V^2}{2g} = \frac{V_e^2}{2g} + hf$$

In this equation $\frac{p}{\gamma}$ has a negative value. We will make this positive by adding h_a = atmospheric pressure head (approximately 33 ft. of water) to both sides of the equation, and obtain:

$$h_s + \frac{p_a}{\gamma} + \frac{V^2}{2g} = \frac{V_e^2}{2g} + hf + h_a \quad (p_a = \text{absolute pressure})$$

We now consider the fact that $\frac{p_a}{\gamma}$ is the average absolute pressure under the turbine, but that, due to local velocity increases near the tips of the blades, the local pressures at the blade tips may be lower. This deviation from the average pressure is a function of the velocity V which in turn is a function of the head on the turbine h. Hence we call the deviation: $k_c.h$ and the minimum pressure at the blade tips becomes: $\frac{p_a}{\gamma} - k_c.h$. This pressure should not become less than the vapour pressure of water: h_v (about 2 ft. of water). Hence

$$\frac{p_a}{\gamma} - k_c.h = h_v$$

or

$$\frac{p_a}{\gamma} = h_v + k_c.h$$

We can look upon $\left(\frac{V_e^2}{2g} + hf\right)$, the total losses in the draft tube, as a function of

333

$\frac{V^2}{2g}$. Hence the three terms $\frac{V^2}{2g} - \left(\frac{V_e^2}{2g} + hf\right)$ may be considered as a function of $\frac{V^2}{2g}$. Since $\frac{V^2}{2g}$ in turn is a function of the head on the turbine h, we may write:

$$\frac{V^2}{2g} - \left(\frac{V_e^2}{2g} + hf\right) = k_d.h$$

and the original Bernoulli equation now becomes $h_s + h_v + k_c.h + k_d.h = h_a$ or:

$$h_s + h_v + \sigma_c.h = h_a$$

The k_c and k_d constants in the second and third term have been replaced by one constant σ_c, or: sigma critical value. Since we are mostly interested in h_s, the setting of the turbine with respect to the tailwater level, the equation is usually written in the form:

$$h_s = h_a - h_v - \sigma_c.h \text{ or approximately: } h_s = 31 - \sigma_c.h.$$

where: h_s is the setting of the turbine above tailwater (in feet), h_a the atmospheric pressure in feet of water, h_v the vapour pressure of water (in feet), σ_c the cavitation factor, h the head on the turbine.

It has been found from experience that the value of σ_c is a characteristic of a type of turbine. Since every type of turbine has its own N_s value we may expect a relationship between σ_c and N_s. This has been found indeed, and the relationship has been expressed in the following equation:

$$\sigma_c = \frac{N_s^{1\cdot64}}{4,325}$$

To illustrate the use of these equations, let us pursue the previous example of the hydro development with a head of 120 ft. and an N_s value of 96. Let it be further assumed that the plant is situated at 1,000 ft. above sea level, and that the water reaches a normal maximum temperature of 80 deg. We find:

$$\sigma_c = \frac{96^{1\cdot64}}{4,325} = 0\cdot41$$

and

$$h_s = 32\cdot7 - 1\cdot2 - 0\cdot41 \times 120 = -17\cdot5 \text{ ft.}$$

In other words, the turbine will have to be set 17·5 ft. below tailwater elevation. Since the centre line of the scroll case or distributor is about $0\cdot4D$ above the turbine, we find that the centre line distributor will have to be placed at $17\cdot5 - 0\cdot4 \times 20 = 12\cdot5$ ft. below tailwater.

In the above paragraphs we have discussed a procedure that may be followed to find preliminary dimensions of turbine and powerhouse. Let us now discuss some efficiency characteristics of turbines.

First of all it should be pointed out that a turbine must at all times run at its designated speed in order to produce electricity with the right frequency. That leaves as variables: the discharge, the head and the load. We may look upon the last two as the independent variables. For instance, if there is a change of the load on the turbine, then we have to change the discharge in order to keep the speed of the turbine the same. This is automatically performed by the governor, operating the wicket gates of the turbine. Now let us see what happens to the efficiency when the load or the head changes.

Let it be assumed that we have a turbine, operating under an average head of 100 ft., with a possible range in head from 80–120 ft., as shown in *Figure 7.51*. At the normal head of 100 ft., the discharge through the turbine is 9,600 cusec. The turbine has been designed so that it has its maximum efficiency of 0·92 at this average head, resulting in an output of 100,000 h.p. The position of the wicket gates, under these conditions, is 0·7 open. Now let us assume that the head is reduced to 90 ft. We want to maintain the designated speed and we would like to produce the same amount of power. Therefore we open the

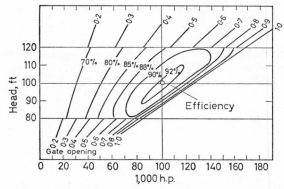

Figure 7.51. Turbine efficiencies as a function of head and gate opening

gates, to release more water. However, the efficiency drops to 0·88. An increase in head on the plant will necessitate the closure of the wicket gates, and will also be accompanied by a reduction in efficiency. Now let us assume that we have a constant head of 100 ft., but that the load on the turbine is reduced. In order to maintain the designated speed, we have to close the gates to reduce the discharge. Again, the efficiency decreases. An increase in load on the turbine would require a further opening of the gates and would also reduce the efficiency.

If we plot the efficiencies for all combinations of head and power output, and we draw lines of equal efficiency, we obtain the 'efficiency-hill' contours of *Figure 7.51*. We can also plot on the diagram the lines of equal gate opening. For every point on the diagram we can compute the discharge since we know the head, the efficiency, and the output. It is evident that we find the low discharges in the left lower corner and the high discharges in the right upper corner of the diagram.

When the plant operates at average head, on the base of the load, for maximum energy production, we should operate for 100,000 h.p. output, since that gives the highest efficiencies. When the plant operates in the peak of the load, and we wish to get the maximum capacity out of the turbine we may operate at full gate opening at so-called overload rating, at the expense of a decrease in efficiency.

In this example, the average operating head at which peak efficiency is developed, is called the design head; while the head of 87 ft., at which turbine full gate output equals the rated generator output, would be called the rated head.

It may be of interest to note that for a fixed gate opening an increase in head of say 10 per cent, corresponds to an increase in horsepower of about 15 per cent. This is because an increase in head has a twofold effect. First, it has a direct effect, to the first power, upon the power output. Then, it has a secondary effect upon the discharge which increases with the half power. Hence the power output, being the product of discharge times head, increases with the three halves power.

Since efficiency is largely a matter of under what direction the water meets the blades of the turbine, it follows that lines of equal efficiency are roughly parallel to the lines of equal gate opening, as may be seen from *Figure 7.51*.

Powerhouse

It will be recalled that the capacity of a hydro plant is directly proportional to the head as well as the discharge. Hence for the same power output, a high head development has a much smaller discharge than a low head development.

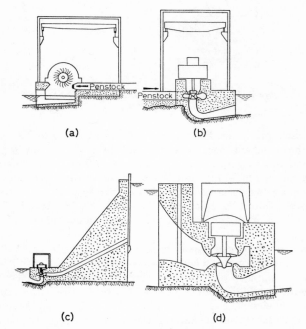

(a)   (b)

(c)   (d)

Figure 7.52. (a) Simple powerhouse for impulse turbine;
(b) Separate powerhouse for Francis turbine; (c) Gravity
dam and connected powerhouse; (d) Dam and powerhouse
integrated

Hence, the hydraulic machinery of the high head development is much smaller in size, and so is the powerhouse that accommodates the machinery. In its simplest form, the impulse turbine of a high head development rests on a concrete floor, a few feet above tailwater; and the powerhouse encloses the turbine, the nozzle of the penstock, the generator and the governor, as shown in *Figure 7.52(a)*. The cost of such a powerhouse is a relatively small portion of the total cost of the development.

For high head developments under 1,000 ft., a Francis turbine may be selected, and the hydraulic problem of conveying the water to and from the turbine becomes more complicated, as may be seen in *Figure 7.52(b)*. Between the penstock and the turbine we need a scroll case that enables the water to enter the turbine from all sides and under the proper angle. Between the turbine and the tailwater we need a draft tube to recapture the kinetic energy that is still in the water below the turbine, as discussed earlier. It is evident

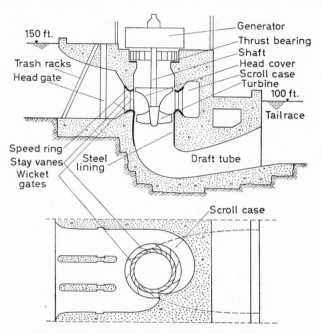

Figure 7.53. Elements of a low-head power plant

that all these measures result in a more costly powerhouse than was required for the impulse turbine. This extra cost, however, may be compensated by a lesser cost of the penstocks.

If the head of the power development is provided by a tall gravity dam, the powerhouse may be attached to the toe of the dam, as shown in *Figure 7.52(c)*. However, the design of such a powerhouse is basically the same as the design of *Figure 7.52(b)*, where the powerhouse is situated at the lower end of a penstock. When the gravity dam becomes lower, say below 100 ft., it becomes economical to unify the dam and the powerhouse, as shown in *Figure 7.52(d)*. When the head of the power development is still lower, say below 50 ft., the powerhouse dominates the design and the features of the gravity dam disappear, as shown in *Figure 7.53*.

It may be of interest to dwell for a moment on the functioning of the different components of the powerhouse, shown in *Figure 7.53*. The trash racks keep out all debris such as tree stumps, discarded lumber and cakes of ice in the winter from the river. The head gate is only used when the turbine is shut down for any length of time or when the unit is dewatered for inspection or repair. The wicket gates are used for regulating the discharge from minute to minute, to keep it in adjustment with the varying load, so that the speed of the turbine remains constant. The wicket gates are moved by the shifting ring, which is moved by two pistons, which are controlled by the governor, which is connected mechanically or electrically to the shaft of the turbine. Outside of the wicket gates are the stay vanes, which are streamlined columns that support the roof of the scroll case. The stay vanes are sometimes welded into a lower and upper steel ring. The whole assembly of the two rings and stay vanes is

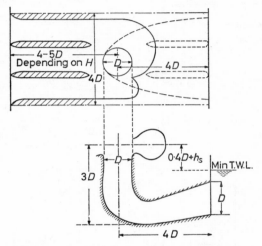

Figure 7.54. Approximate dimensions of a low-head power plant, in terms of D. (To be used for preliminary design only.)

called the speedring. The placing of the speedring during the construction of a power plant is one of the most precision demanding jobs in civil engineering. The turbine is situated in the turbine throat which may be lined with steel plating to prevent damage from cavitation. Above the rotating turbine is the stationary head cover that prevents the water from entering the powerhouse. The shaft of the turbine is directly connected with the generator. Underneath the generator is the thrust bearing, which is the only vertical support of the turbine-generator assembly. The thrust bearing must be designed for the weight of the generator, shaft and turbine, plus the hydraulic thrust on the turbine under running conditions. At the lower end of the draft tube there should be a gate or stoplog grooves so that the scroll case and draft tube can be dewatered.

For preliminary design of low and medium head powerhouses one may use empirical ratios between the dimensions of the powerhouse and the diameter of the turbine, as shown in *Figure 7.54*. To determine total volume of concrete, one may take the gross volume of the substructure multiplied by a factor 0·6.

CHAPTER 8

IRRIGATION

In the previous chapter, where we discussed the development of water power, the economic problem was roughly as follows. Electricity cannot be transported over distances of more than, say, 1,000 miles. Hence, electricity must be produced in or near by the regions where it is needed. A forecast of population increase and industrial activity indicates how much more power is needed at certain dates in the future. The problem is to provide that extra power at the lowest cost; but somehow it must be provided. The economic problem of irrigation development is different. Food can be transported over any distance, and therefore it is only attractive to develop an irrigation project if the cost of providing extra food through irrigation is competitive with importing that food from somewhere else. Hence, in irrigation development we must compare benefits to costs, whereas in power development we are only concerned with minimizing costs.

Whether or not irrigation is an attractive proposition in a certain region depends upon several circumstances, of which the following are the most important. First of all, the political, social and economic environment; second, the suitability of the land from a viewpoint of topography and soils; third, the availability of irrigation water in the region. If all circumstances are favourable, we can go ahead and prepare a preliminary plan of an irrigation project to determine if the benefits would justify the cost. These aspects of irrigation will be discussed in more detail in the following sections.

NEED FOR IRRIGATION

If we look at irrigation from the viewpoint of the private entrepreneur, we must apply the criterion whether or not the increase in yield from the land is more than the cost of irrigating the land plus the extra cost of planting and harvesting. In the early days of irrigation history this criterion was indeed applied, and often very successfully. However, with the advance of time, the simple and more profitable irrigation schemes were soon developed, and the difficult and dubious ones that were considered unattractive by private enterprise, were left undeveloped.

It was soon realized that the benefits of an irrigation project are more inclusive than just the direct benefits to the farmer who gets a greater yield from his land. Because of the added income and expenditures of the farmers, the entire region becomes more prosperous. Because of the elimination of the havoc formerly caused by drought periods, the economy of the region is stabilized. For these and other reasons the governments of nearly all countries where iirrgation is feasible, have adopted the policy of undertaking or subsidizing or otherwise encouraging all those irrigation projects that are beyond the means of private enterprise, but nevertheless of benefit to the country as a whole.

It is a most delicate matter, of course, to decide whether or not a proposed irrigation project is sufficiently 'to the benefit of the country as a whole' to justify its cost. An over-optimistic attitude in this regard may lead to disaster, as has been demonstrated by ancient civilizations as well as by recent irrigation projects in North America. Let us consider for example an irrigation project of 250,000 acres with a total capital cost of $100,000,000 for the reservoir, the main canals and the distribution system. Within 10 years an average of $60 per acre may be needed for surface drainage and land levelling. During the following decade it may be found that some 50 per cent of the total area requires sub-surface drainage at a cost of $200 per acre. In spite of these measures, it would not be uncommon that 10–20 per cent of the land becomes unproductive and will be abandoned. As a result, the average capital cost per acre has now become $700. Assuming annual charges of 5 per cent for interest, 1 per cent for depreciation, 1 per cent for taxes (real or foregone), and 1 per cent for operation and maintenance, the annual cost per acre becomes $56, which is probably much more than what the farmers can pay. In other words, the government must be prepared to subsidize the major part of the irrigation costs, probably in perpetuity. To this may be added the risk that a significant percentage of the farmers do not wish to change from dry land practices to irrigation farming, even after all the works have been installed (as happens frequently!). This would further increase the burden of the government. If the government would not be prepared to carry all of this burden, the farmers could not carry out such required corrective measures as sub-surface drainage and there would be the danger that the land would become salinized and unproductive. In the worst case, the entire irrigation project would be abandoned. Therefore, it cannot be over-emphasized that a government must recognize these risks in advance; that it must be prepared to subsidize in perpetuity if needed; and that it must consider the risk well worth taking in view of the potential benefits in terms of general prosperity of the nation.

The benefits of an irrigation project in North America are only appraised in terms of economic prosperity. The extra food that is produced through irrigation is not considered essential to living. In fact, during the last few decades, the production of more food products is looked upon as somewhat of a mixed blessing. This is an amazing state of affairs if one realizes that more than half of the world population is undernourished, according to U.N. statistics. Whatever the explanation, it seems reasonable to expect that in the foreseeable future the presently undernourished nations will gain access to the world food markets. If we also keep in mind that the world population is expected to double within the next 100 years, we may expect that the incentives to produce more food will be greatly increased. This will very likely be reflected in much higher prices for farm products; say two or three times what they are now, not allowing for inflation. It is evident that this will change completely the benefit–cost aspects of prospective irrigation developments. Projects that are at present considered unattractive may become very attractive when the value of the yield of the land is multiplied by a factor three.

Moreover, we must keep in mind, as was noted in Chapter 1: Planning, that in the foreseeable future the production of nuclear energy and sea-water conversion may become economical. If all three possibilities become reality (high food prices, low energy cost, cheap sea-water conversion), we may expect in countries with a good climate, good soils, but deficient rainfall, the development of irrigation projects that include a vast network of pump stations, pipelines and open canals that convey the fresh water from selected points along the coast towards agricultural lands on the coastal plains, as was pointed out by Kuiper (1964) in a report on irrigation development in Spain.

SUITABILITY OF LAND

One of the obvious prerequisites for a successful irrigation development is that the land be suitable to be irrigated. Suitability of the land has several aspects. First, the land should be situated so that it can be reached by irrigation canals. Second, the topography of the land should be such that the irrigation water can be readily distributed over the land. Third, the soil must be sufficiently fertile to justify its irrigation. Fourth, the topsoil and underlying soils must allow drainage of excess water or else salinization may set in.

There is, of course, no universal set of yardsticks, that can be applied in advance, to determine the suitability of land for irrigation. From a technical–agricultural viewpoint we could make crops grow anywhere. If the land is situated well above the supply of water, we can pump the water up. If the land surface is rolling, we could apply sprinkler irrigation. If the soil is sterile, we could treat it first with chemicals and then apply fertilizers. If the under-lying soils are too dense, we can install a system of sub-surface drainage. However, it is likely that the total cost of all these measures will become prohibitive. Therefore, it is not so much the physical circumstances but the economic aspects also that determine the suitability of the land for irrigation. These economic aspects are revealed by the benefit–cost analysis of the pro-posed irrigation project. Their acceptance depends very much upon the policy that the government decides to follow with respect to subsidizing irrigation projects. Hence we must conclude that defining the suitability of land for irrigation, and selecting the land for development, cannot be made until legislation has defined the extent of government subsidization of the capital cost and the annual cost of operation and maintenance of the main irrigation works, as well as the improvement of the land. Keeping these limitations in mind, let us discuss some of the suitability aspects in more detail.

Elevation

The ideal elevation of land to be irrigated is slightly below the source of the water, so that it can be reached by gravity flow. Much of the irrigated land in India and Pakistan, for instance, is situated on the alluvial plains of the rivers that originate in the Himalaya Mountains. Low dams back up the river till the water can be diverted into the main canals that distribute the flow throughout the irrigation districts. If the irrigated land is much below the place of origin, there will be added cost to the canals because of drop structures, canal linings and other measures to prevent erosion. Moreover, low land may require flood protection works and an elaborate drainage system. When land to be irrigated is situated above the place of origin, one must resort to pumping. During the

341

last two decades such pump-irrigation schemes have become rather common. Two outstanding examples are the Grand Coulee pumping plant on the Columbia Basin Project and the Tracy pumping plant on the Central Valley Project. In the first case, about 1,400,000 acre-ft. of water is pumped annually from the Grand Coulee storage reservoir, over a head of 300 ft. into an equalizing reservoir from where it is distributed over an irrigation district of nearly 1,000,000 acres. In the second case the pump capacity is 4,600 cusec at a head of about 200 ft.

Topography

The ideal topography of the land is a uniform gentle slope without local undulations. This enables a simple layout of the canals and it enables the water to flow evenly over the land from the irrigation ditches to the drainage ditches. The slope of the land determines how quick the water will run over the land and hence the percolation losses, and hence the method of irrigation and the kind of crops that can be grown. Some correction of the topography can be made by a land-levelling programme. However, this is costly, and it may remove precious topsoil and expose the unfertile clay or till subsoil. If the topography is too uneven, or the layer of topsoil too thin, so that a land-levelling programme is impractical, the best solution may be to irrigate by sprinkling rather than by surface flow.

Soils

The best soils, from a viewpoint of irrigation, are deep, well-drained alluvial deposits of medium to fine texture. Such soils permit the plants to develop an adequate root system and allow for proper circulation of air and water within the root zone. Sandy soils are less satisfactory because they allow rapid infiltration of irrigation water below the root zone. Heavy soils are difficult to cultivate and do not readily allow the irrigation water to penetrate to the root system. Shallow soils underlaid by impermeable strata do not provide much space for storing moisture for plant use between irrigations.

Accumulations of salts in soils are usually caused by upward capillary movement of soil moisture, and subsequent evaporation at the ground surface, leaving the formerly dissolved solids behind in the form of salt. Salinization of the root zone under irrigation conditions depends mostly on the salinity of the irrigation water and the drainage conditions of the irrigated lands. When adequate subsoil drainage is available or provided for, and sufficient quantities of water can be obtained to leach out the soil, the development of salt trouble can usually be avoided. It should be pointed out that the application of excessive amounts of irrigation water alone, without subsoil drainage, may be harmful instead of beneficial, inasmuch as it causes the ground water level to rise into the root zone, where it may rot the submerged roots, precipitate salt at the ground surface and waterlog the soil.

Notorious examples of irrigation land going to waste because of too high a salt concentration in the root zone are the Imperial Valley Irrigation System, where 18 years after initiation of the project more than 200,000 acres were out of production; and the ancient irrigation system in the delta of the Tigris and Euphrates in Iraq, where a great civilization declined because of too much salt accumulation on the land. In the first example, rehabilitation of the land through the construction of a drainage system was undertaken immediately afterwards. In the case of Iraq, rehabilitation is now slowly taking place, hundreds of years after the damage was done, by means of an extensive drainage system which aims at dissolving the salt concentration in the topsoil by a downward movement of water that comes from the irrigation ditches, over the land, and goes down to the water table and into the drainage ditches.

From the foregoing, it is evident that no irrigation project can prove successful if the soil conditions are not suitable, and that a thorough investigation of soil conditions is required before it can be decided whether or not irrigation is feasible and before the necessary plans and water requirements can be worked out.

For this reason, the U.S. Bureau of Reclamation has devised a system of classifying prospective irrigation land into the following six categories (see U.S. Bureau of Reclamation Manual; or Abbett (1956), Volume II, pages 17–33):

Class 1, arable; sandy loam to friable clay loam, 3–5 ft. thick, up to 4 per cent land surface slopes, and no artificial drainage required.

Class 2, arable; loamy sand to very permeable clay, 2–4 ft. thick, up to 8 per cent land surface slopes, inexpensive artificial drainage.

Class 3, arable; loamy sand to permeable clay, $1\frac{1}{2}$–$3\frac{1}{2}$ ft. thick, up to 12 per cent surface slopes, artificial drainage feasible.

Class 4, limited arable; includes lands having excessive deficiencies or restricted utility, but which special economic and engineering studies have shown to be irrigable.

Class 5, non-arable; includes lands requiring additional studies and temporarily non-productive lands.

Class 6, non-arable; includes lands which do not meet the minimum requirements.

The above land classification is performed in the field by appraising the soil characteristics, the topography, and the subsoil drainage features. The land classification data are recorded on topographic maps or aerial photographs, showing the boundaries between land classes. If the water supply in the area under consideration is a limiting factor, the land classification is used as the basis for selecting the location of the irrigable areas to be included in the irrigation projects. However, if there is ample water available, but good land is the limiting factor, the land classification is primarily used as the basis of designing the irrigation system.

Drainage

The need for good drainage of irrigated land has been noted in the earlier sections. A drainage system may have to perform three distinct functions in an irrigation system. First, it must remove the excess surface water that has been applied and that has not percolated into the ground. This function can be performed by open ditches at the lower end of an irrigation field, as shown in *Figure 8.1*. Second, it must induce sufficient percolation of water through the soil so that salts are leached out, or prevented from accumulating, to within permissible concentrations. This function may have to be performed by a system of subsurface drains. Third, it must keep the groundwater level suppressed below the root zone. This function may have to be performed by a combination of open and sub-surface drains.

Let us consider what happens to a certain amount of irrigation water that is diverted on to a farmer's field. A portion will keep flowing over the surface and be wasted into a drainage ditch. A portion will evaporate while the irrigation takes place. The remainder will percolate into the soil. A portion of this water will be absorbed by the crops and thus reach its intended destination. Another portion will keep percolating through the soil and finally drain away via groundwater flow, drains, and ditches back to the river system. Still another portion will remain in the soil, near the surface, and will evaporate during the days following the irrigation of the land. This last portion of water will leave, behind in the soil, the salts that it originally contained. Over the years, these

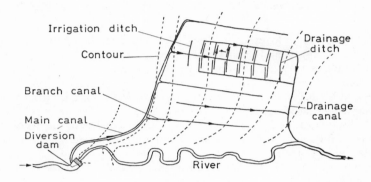

Figure 8.1. General lay-out of an irrigation project

salt accumulations may become harmful. This danger exists in particular when the irrigation water contains a high percentage of salts to start with, and when the climate is hot and dry so that much evaporation takes place. The most effective measures to prevent such salinization of the soil are: first, to use only water with a low concentration of salts; and second, to induce a liberal flow of water through the soil from the surface to the sub-drainage system, to continually dissolve any salts that may have precipitated in the surface layer. It is obvious that this measure requires a good deal of water, and may require a costly sub-surface drainage system.

The consistent application of irrigation water to a tract of land is bound to raise the ground water over and above its natural level. If the subsoil is rather impervious, the new ground water level may rise so high that it becomes injurious to the crops to be grown. Under such conditions it may become necessary to apply varying degrees of drainage works. Lands with heavy, slowly permeable soils from the surface down to considerable depth will waterlog quickly when irrigated in excess of surface evaporation and plant transpiration. Moreover, for reasons set out in the above paragraph they would quickly become salinized. Once this has taken place, it will be extremely difficult to apply corrective measures. Therefore, they should be provided right from the outset with adequate sub-surface drainage facilities.

WATER REQUIREMENTS

From a viewpoint of water resources development we are greatly interested in the water requirements of an actual or proposed irrigation project. This depends upon several factors. First of all, of course, the size of the project. Secondly, the method of irrigation. Thirdly, the type of crop to be grown. Fourthly, the prevailing climate, including rainfall, humidity and temperatures. These factors will be discussed below in more detail.

Methods of Irrigation

The most common method of irrigation in North America is to let water run over the surface of the ground from a supply ditch towards a drainage ditch (where preferably a minimum of water should arrive), as shown in *Figure 8.1*.

The overland flow of water may cover the entire surface, and is then called flooding. This is usually applied to such crops as hay, alfalfa, and grain. The duration of the flooding is relatively short, the purpose being to moisten the soil. In the case of rice cultivation it is necessary to have continuous submergence of the fields. From a viewpoint of water conservation it is more desirable to let the water flow through shallow furrows. This is usually applied to such crops as corn, potatoes, and orchard fruits. Water may be diverted from the supply ditch on to the land by making a temporary opening in the bank of the ditch or by small portable siphons. The spacing of the supply ditches, the size of the furrows and the rate of diverting water should be such that all water is just infiltrated in the soil, before the end of the furrow or field is reached.

Another method of irrigation is to sprinkle the water over the land. This is rather costly since it calls for a great amount of tubing, sprinkling equipment and pumps. Moreover, the evaporation losses are high. However, when the terrain is uneven, it may be the only practical method.

When the surface soils are very porous, it may be possible to apply the irrigation water below the ground surface by ditches that are spaced from 50–200 ft. apart. This is called sub-irrigation. The water percolates from the ditches into the soil and raises the ground water level high enough to moisten the root zone by capillary action. This method reduces evaporation losses and avoids the need for furrows or contour dikes.

Irrigation Requirements

Before we begin to discuss the water requirements for irrigation, let us define some of the terms that are normally used.

Water requirement, or consumptive use, or crop requirement, is the amount of water needed for crop growth.

Irrigation requirement is the consumptive use minus the precipitation available for plant consumption. In other words, it is the quantity of water that must be delivered by irrigation to the land in order to ensure crop production.

345

Farm delivery requirement is the irrigation requirement for the crops plus the losses due to evaporation, percolation, surface waste, and non-productive consumption.

Irrigation efficiency is the percentage of farm-delivered irrigation water that remains of the root zone and is available for crop growth. It may range from 20–50 per cent in regions where water supplies are ample, and from 40–70 per cent in regions where large expenditures are justified to conserve water.

Gross water requirement is the farm delivery requirement plus the seepage losses in the canals between the diversion dam and the farm unit (ranging from 15–45 per cent of diversion flow on unlined canals, and from 5–15 per cent on lined canals), plus the waste of water due to poor operation, breaks and overflows (ranging from 5–30 per cent of diversion flow on projects with ample water supplies, and from 1–10 per cent on projects with limited water supplies).

Project diversions are the gross amounts of water diverted from the reservoirs to the irrigation projects, expressed in feet of water over the irrigable area. On 24 major irrigation projects in the U.S., the average seasonal diversion ranged from 1·4 to 13·2 ft. with a mean of about 5·5 ft.

In determining the water requirements for irrigation, the first problem that should be dealt with is the anticipated consumptive use of the crops to ensure a maximum yield. Once this basic figure is established, it becomes a matter of subtracting the available effective precipitation and adding the various irrigation losses in order to arrive at desired diversion flows. The consumptive use of the crop depends on many circumstances, such as the type of crop, soils, climate and latitude. If an accurate estimate of consumptive use has to be made, say for final design of a project, it would be advisable to make a study of farming experience in similar areas. Some irrigators may be found who have carefully experimented with various crops and their optimum water needs. The establishment of an experimental irrigation farm may be warranted if sufficient time and funds are available. However, for a reconnaissance type of investigation, it would be more expedient to follow short-cut methods, like the one developed by Lowry and Johnson (1942), or the one developed by Blaney (1952). The following is an abstract of Blaney's method, illustrated by an example:

'The procedure is to correlate existing consumptive-use data with monthly temperature, percentage of daytime hours, precipitation, frost-free period, or irrigation season. The coefficients so developed for different crops are used to translocate or transpose consumptive-use data from one section to other areas in which climatological data alone are available.

'Consumptive use varies with the temperature and the extent of daytime hours, and with the available moisture (precipitation, irrigation, and ground water). By multiplying the mean monthly temperature in degrees Fahrenheit (t) by the monthly percentage of daytime hours of the year (p), by an empirical coefficient (k), and dividing by 100, the monthly consumptive use in inches of water is obtained.'

Table 8.3 illustrates the computations necessary to determine consumptive use.

After having established the consumptive use of the anticipated crop pattern in the proposed irrigation project, the next problem is to find the irrigation requirement. In other words, what is the effective precipitation that has to be subtracted from the consumptive use?

One should begin by collecting all available rainfall records of the region under consideration. The effective precipitation must now be estimated by considering the retention capacity of the soil. In other words all surface run-off,

Table 8.1. *For 'p' Values*
[from Blaney (1952), by courtesy of the ASCE]

Month	Latitudes in Degrees North of Equator						
	26	30	34	38	42	46	50
Jan.	7·49	7·30	7·10	6·87	6·62	6·33	5·98
Feb.	7·12	7·03	6·91	6·79	6·65	6·50	6·32
March	8·40	8·38	8·36	8·34	8·31	8·29	8·25
April	8·64	8·72	8·80	8·90	9·00	9·12	9·25
May	9·38	9·53	9·72	9·92	10·14	10·39	10·69
June	9·30	9·49	9·70	9·95	10·21	10·54	10·93
July	9·49	9·67	9·88	10·10	10·35	10·64	10·99
Aug.	9·10	9·22	9·33	9·47	9·62	9·79	10·00
Sept.	8·31	8·34	8·36	8·38	8·40	8·42	8·44
Oct.	8·06	7·99	7·90	7·80	7·70	7·58	7·43
Nov.	7·36	7·19	7·02	6·82	6·62	6·36	6·07
Dec.	7·35	7·14	6·92	6·66	6·38	6·04	5·65
Total	100·00	100·00	100·00	100·00	100·00	100·00	100·00

and all water lost to ground water flow away from the prospective drainage area, must be deducted from the actual precipitation. The U.S. Bureau of Reclamation proposes the approximate relations given in Table 8.4 for reducing total monthly precipitation to effective precipitation.

Table 8.2. *For 'k' Values*
[from Blaney (1952), by courtesy of the ASCE]

Item (a) Irrigated Land	Length of Growing Season or Period	Consumptive-use Coefficient (k)
Alfalfa	frost free	0·85
Beans	3 months	0·65
Citrus orchard	7 months	0·55
Corn	4 months	0·75
Cotton	7 months	0·62
Deciduous orchard	frost free	0·65
Pasture, grass, hay, annuals	frost free	0·75
Potatoes	3 months	0·70
Rice	3–4 months	1·00
Small grains	3 months	0·75
Sorghum	5 months	0·70
Sugar beets	5½ months	0·70
(b) Natural Vegetation		
Very dense (large cottonwoods, willows)	frost free	1·30
Dense (tamarisk, willows)	frost free	1·20
Medium (small willows, tamarisk)	frost free	1·00
Light (saltgrass, sacaton)	frost free	0·80

347

Let us now assume that we have determined the effective monthly precipitation figures for the entire period of record of, say, 50 years, and for all the months during the growing season of, say, from April to October inclusive. We could then determine the total effective precipitation during the growing season of seven months, for every year on record. From the resultant 50 figures we could prepare a frequency curve as discussed in Chapter 2: Hydrology. Let

Table 8.3. *Computed Normal Consumptive Use of Water by Grass Pasture*
(Victoria, Tamaulipas, Mexico)
[from Blaney (1952), by courtesy of the ASCE]

Month	t	p	k	Consumptive Use (in.)
April	77·0	8·60	0·60	4·0
May	80·2	9·29	0·65	4·8
June	82·0	9·18	0·70	5·3
July	83·3	9·39	0·75	5·9
Aug.	83·8	9·04	0·75	5·7
Sept.	79·9	8·31	0·65	4·3
Oct.	74·8	8·10	0·60	3·6
Total				33·6

us now assume that we wish to ensure full supply of water, required by the crops, during 9 out of 10 years. In other words, a mild shortage may occur on the average once in 10 years, while more serious shortages will occur at less frequent intervals. We should then determine from the frequency curve the 90 per cent value (exceeded during 90 per cent of the years) of the effective rainfall during the growing season. This figure may be distributed over the seven months of the growing season in the ratio, as indicated by the 10 driest

Table 8.4. *Relation of Monthly to Effective Precipitation*
(by courtesy of the Bureau of Reclamation)

Total Monthly Precipitation (in.)	Effective Monthly Precipitation (in.)
1·0	0·9
2·0	1·8
3·0	2·5
4·0	3·0
5·0	3·3
6·0 and over	3·4

years on record. We can now deduct these figures from the consumptive-use figures, as shown in the last column of Table 8.3, and thus obtain the irrigation requirement figures on a monthly basis.

After having established the irrigation requirement, the last problem is to determine all losses between the farm and the reservoir and thus establish the project diversion requirements. The best guidance for such determination would be a study of past experience in the project area or in similar areas. In case such information is not readily available, reference is made to the estimates of losses that were made earlier in this section. It may be of interest to note that in the Cauca Valley, Colombia, a loss of 50–60 per cent of the diverted water was assumed. In a study of a 22,000,000 acre irrigation project in Iraq, a loss of 35 per cent of the diverted water was assumed.

Return Flow

From a viewpoint of basin wide water resources development it is important to know how much of the gross project diversion from the reservoir returns to the river system. Return flow includes percolating water not retained in the root zone, surface run-off during irrigation, wasted water, and canal seepage. Part of the return flow reaches the original river channel as surface run-off in natural or artificial watercourses, and can easily be measured. The remainder, however, reaches the river as ground water flow and is not easily measured. It is learned from experience that return flow from a new irrigation project may begin within a few years after initiation of the project, but may not reach its full magnitude until after 10, 20, or even 30 years. On large irrigation projects, amply supplied with water, annual return flows may vary from 30–60 per cent of the diverted water, of which the invisible part may be one half or more. On many irrigation projects, about one half of the return flow reaches the stream during the irrigation season and thus becomes available for rediversion to other projects.

Quality of Water

It has been found that a high salt concentration in irrigation water is harmful to plants. In the beginning of an irrigation development the crops may not suffer too much. However, due to the continuous evaporation of water near the surface of the soil, as discussed earlier, the salt concentration in the soil will build up over the years and may become five to ten times as high as the salt concentration in the irrigation water. By now, the plants may begin to suffer and adequate measures to remove the salt in the soil must be taken or the project must be abandoned. It has been found from experience that when irrigation water contains more salts than 1,000 p.p.m. the situation becomes dubious and when the concentration exceeds 2,000 p.p.m. the water cannot be used. More precise limits are given in Table 8.5.

Table 8.5. *Criteria for Classification of Irrigation Waters*
[by courtsey of the California State Water Resources Board (Bull. 1, Sacramento, 1951)]

Classification of Water for Irrigation	Per Cent Sodium of Total	Total Salts (p.p.m.)	Boron (p.p.m.)	Chloride (p.p.m.)	Sulphate (p.p.m.)
Class I, excellent to good	0–60	0–700	0–0·5	0–177	0–960
Class II, good to injurious (harmful to some plants)	60–75	700–2,100	0·5–2·0	177–355	960–1,920
Class III, unsatisfactory	75	2,100	2·0	355	1,920

It should be pointed out that these limits apply to the irrigation water, but that it is the eventual salt concentration in the soil that counts. Hence, in applying the figures from this or similar tables it is important to consider also the conditions under which irrigation takes place. Are the drainage provisions adequate? Are frequent leaching operations possible? How intense is the rainfall or snowfall during the non-growing season? What are the prevailing temperatures during the growing season? What are the crops to be grown?

It is also important to note that the salt concentration in flood flows is much smaller than that in low water discharges, largely originating from ground water flow. Therefore it would be to the benefit of irrigation if it could be supplied with flood water, stored in reservoirs where provision is made to by-pass the low flows with the high salt concentrations.

It may be of interest to note here that artificial desalinization of water for irrigation purposes is generally considered too costly. With the presently available techniques it costs $1.25 to $2.00 per 1,000 U.S. gal. to convert sea-water into fresh water. It is expected that this figure may come down to $1.00 or somewhat less. The cost of desalinizing river water would depend on the concentration of salts, but would be substantially less than $1.00 per 1,000 gal., if undertaken on a large scale. Assuming a gross water requirement of 4 ft. of water per year, the cost of fresh water (at 50 cents per 1,000 gal.) would come to $4 \times 43,560 \times 7 \cdot 48 \times 0 \cdot 001 \times \$0.50 = \$650$ per acre per year. This is about 10 times as much as all other annual costs, and hence we may conclude that with the presently available techniques and under the present economic conditions, artificial desalinization of salt water for irrigation purposes is out of the question.

If irrigation water is derived from large reservoirs, it will be devoid of sediment. When it is derived from small reservoirs or from the pools behind river weirs, it may contain an appreciable amount of sediment. If this sediment originated from fertile topsoils it may constitute a desirable addition to the irrigated lands. If it originated from sterile subsoils it may be harmful, and measures should be applied to remove the sediment. If the soils of the irrigation area have a fine texture, there is a possibility that the suspended sediment will seal the pores and cause waterlogging of the surface.

AVAILABILITY OF WATER

After we have tentatively determined what lands are suitable for irrigation and how much water is required for irrigation, the next problem is to find an economic source for this water. It is also possible that our study proceeds in the reverse direction. We may have determined, first, how much water is available for irrigation and then we may search for the most suitable land to be irrigated. Whatever the sequence, let us see how the supply can be reconciled with the demand.

Natural Sources

The most nearby source of water for irrigation may be the ground water reservoir underneath. The availability of ground water for water supplies is discussed in Chapter 10: Water Supply. It is sufficient to note at this point, that usually the amounts of water that can be derived from ground water on a permanent basis, without harming other interests, are quite small. Moreover, the concentration of salts in ground water is often too high to make the water suitable for irrigation. Therefore, the scope of irrigation developments, based on ground water supplies, is usually very limited.

A better source for irrigation water may be provided by the natural lakes and streams in the vicinity of the project. To appraise the availability of water from these streams, we should collect all stream flow data that are available. If these data do not include at least a few decades, a hydrologic study should be made to extend the records over such a length of time. The total period considered should preferably include at least one cycle of dry years. From these flow figures we must deduct first of all downstream riparian rights, if any. Secondly, we must deduct the future downstream water requirements that should have priority over the irrigation requirements. Such priority requirements may include: water for domestic requirements and sewage dilution for downstream towns and cities; or minimum flow requirements for licensed hydro developments; or minimum lake levels for recreational purposes. The future period over which we consider the downstream requirements should have the same length as the useful life of the irrigation project; say 50 years.

Having deducted downstream priority requirements from the natural flows, we obtain the flows available for irrigation. We must now decide what degree of risk of water shortage for irrigation we wish to accept. Let us assume that we desire dependable flows during 90 per cent of the time. We could then prepare a duration curve of the flows available for irrigation (during the irrigation season, of course) and find the flow figure that is exceeded during 90 per cent of the time. This figure may then be looked upon as the natural river flow, available for irrigation. It is obvious that this figure may be considerably increased by means of storage reservoirs, as will be discussed in a subsequent section.

If water supplies in the vicinity of the irrigation project are inadequate, it may be possible and economic to import water via tunnels and canals, from other drainage basins. Notable examples are: the All-American canal that brings 15,000 cusec from the Colorado River to the Central Valley Project; the proposed California Water Plan that may bring water from northern California to southern California; and the proposed Prairie Irrigation Plan that may bring water from the Churchill River, via the Saskatchewan River to the southern prairies in Canada.

Reservoir Storage

It will be appreciated from *Figure 2.4(b)*, page 28, that the provision of storage capacity in a river system will greatly increase its dependable flow, and thus enhance the irrigation possibilities. If there are other interests to be served as well, such as water power and flood control, it may be economic to create reservoirs of substantial size. In such a multi-purpose reservoir, irrigation would be allocated a certain volume of storage.

Let us discuss how we can establish the relationship between irrigation potential and the storage volume of a reservoir. We prepare a mass curve of the inflow into the reservoir as shown in *Figure 8.2(a)*. We then plot reservoir demand lines that represent the total required flow from the reservoir for different assumed sizes of the irrigation project. The demand from the reservoir would consist of: downstream riparian water rights and other priority releases, the evaporation–precipitation excess over the lake, and the irrigation drafts. As a result of such a study, we find a relationship between size of irrigation project and required storage capacity. The curve representing this relationship would resemble *Figure 8.2(b)*, with the horizontal scale representing storage capacity and the vertical scale representing the size of the irrigation project. Having prepared such a figure and having assumed a certain storage volume we could then find the corresponding irrigation potential; or we may wish to use the figure for finding the storage capacity that is required for a given irrigation area.

Let us assume that we have an irrigation reservoir with a storage volume of 3,000,000 acre-ft. or 50,000 SFM. From a study as described above we have found that we can irrigate an area of 500,000 acres. The project diversion equals 3 ft. of water over the irrigated area and is distributed over the months of April to October inclusive, with most of the water being released in June and July. The long term average river flow at the reservoir site is about 10,000 cusec. The regulation of such a reservoir would be conducted as shown in Table 8.6.

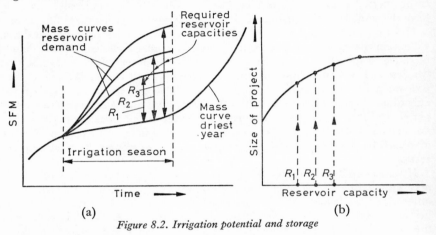

Figure 8.2. Irrigation potential and storage

It may be seen from Table 8.6 that during the three years listed, there is an appreciable waste of water. In fact, when the entire period of record is taken into consideration, about 16 per cent of the available river flow is wasted over the spillway. Unfortunately, with the given size of reservoir, and with the requirement that the irrigation diversion should be dependable, there is not much we can do to cut down on the waste of water. (We are ignoring, in this example, the possibility of multi-purpose or downstream use.)

However, if we would be willing to lessen the requirement on the dependability of the project diversion, we could make a more intense use of the available water supplies. This may be appreciated from reviewing *Figure 2.5(b)*, page 29, and the discussion that goes with it. This figure indicates that with a given reservoir capacity we can produce a 100 per cent dependable flow (assuming the period of flow record as the criterion of worst possible drought conditions) of say 16,000 cusec; with that same reservoir we can produce a 95 per cent dependable flow of 20,000 cusec, and so on. We could prepare a similar set of curves, but with the scales of *Figure 8.2(b)*, establishing the relationship between reservoir capacity, degree of dependability and size of irrigation project.

The operation of an irrigation reservoir under such circumstances would be somewhat like the regulation of a power reservoir for maximum energy output, as was shown in *Figure 7.19*. Part of the reservoir is used to maximize energy utilization, while the remainder is used to produce a small but dependable

352

flow. If we apply this thought to the irrigation reservoir, we would use the upper part of the reservoir to irrigate a larger-than-normal area, while we use the lower part of the reservoir to produce a small dependable flow for a smaller-than-normal area. This principle is illustrated in Table 8.7, which is a sequel to the example of Table 8.6. We have now assumed that the total

Table 8.6. *Reservoir Regulation for Dependable Project Diversions*

Month (1)	Inflow (2)	Riparian Reg. (3)	Evap. Precip. (4)	Irr. Draft (5)	Δ Storage (6)	Total Storage (7)	Waste (8)
1933							
Jan.	1,800	1,000	—	—	+ 800	800	
Feb.	4,600	1,000	200	—	+ 3,400	4,200	
March	16,400	1,000	300	—	+15,100	19,300	
April	34,800	1,000	400	1,000	+32,400	50,000	
May	11,400	1,200	600	2,000	+ 7,600	50,000	1,700
June	3,200	1,400	800	6,000	− 5,000	45,000	7,600
July	2,500	1,400	1,000	7,000	− 6,900	38,100	
Aug.	1,200	1,400	1,000	4,000	− 5,200	32,900	
Sept.	1,600	1,400	600	2,000	− 2,400	30,500	
Oct.	800	1,200	300	1,000	− 1,700	28,800	
Nov.	1,000	1,000	—	—	—	28,800	
Dec.	1,200	1,000	—	—	+ 200	29,000	
1934							
Jan.	2,000	1,000	—	—	+ 1,000	30,000	
Feb.	5,100	1,000	200	—	+ 3,900	33,900	
March	11,400	1,000	200	—	+10,200	44,100	
April	42,000	1,000	300	1,000	+39,700	50,000	33,800
May	18,600	1,200	500	2,000	+14,900	50,000	14,900
June	4,300	1,400	900	6,000	− 4,000	46,000	
July	2,100	1,400	1,200	7,000	− 7,500	38,500	
Aug.	1,200	1,400	800	4,000	− 5,000	33,500	
Sept.	800	1,400	700	2,000	− 3,300	30,200	
Oct.	1,000	1,200	200	1,000	− 1,400	28,800	
Nov.	900	1,000	—	—	− 100	28,700	
Dec.	800	1,000	—	—	− 200	28,500	
1935							
Jan.	600	1,000	—	—	− 400	28,100	
Feb.	900	1,000	200	—	− 300	27,800	
March	1,100	1,000	400	—	− 300	27,500	
April	3,100	1,000	500	1,000	+ 600	28,100	
May	1,900	1,200	800	2,000	− 2,100	26,000	
June	2,300	1,400	1,000	6,000	− 6,100	19,900	
July	1,600	1,400	1,400	7,000	− 8,200	11,700	
Aug.	1,400	1,400	1,800	4,000	− 5,800	5,900	
Sept.	600	1,400	1,100	2,000	− 3,900	2,000	
Oct.	900	1,200	400	1,000	− 1,700	300	
Nov.	800	1,000	100	—	− 300	0	
Dec.	1,200	1,000	—	—	+ 200	200	

Col. (1) shows a three year sample out of a longer regulation study. The sample includes the driest year on record: 1935. The years 1933 and 1934 are slightly below average.
Col. (2) represents the recorded natural river flow, in cusec, and may be regarded as the inflow into the reservoir.
Col. (3) represents the downstream riparian flow requirements, in cusec.
Col. (4) represents the evaporation–precipitation excess over the reservoir, expressed in cusec, and for the month of record shown.
Col. (5) represents the diversion for the irrigation project, in cusec.
Col. (6) is the sum of cols. (3), (4) and (5), deducted from col. (2), in SFM.
Col. (7) represents the total volume of water in storage, in SFM.
Col. (8) represents the water that has to be wasted over the spillway when the reservoir is full (50,000 SFM) and the inflow larger than the total demand.

irrigation project includes 750,000 acres, and therefore the project diversions are 50 per cent larger than they were in the first example. The total reservoir capacity is still 50,000 SFM, and we have found from an analysis of the mass curve that we can now provide dependable flow for only 300,000 acres. It may be seen from the reservoir regulation of Table 8.7, that in the normal years 1933 and 1934, we utilize more flow and reduce the waste over the spillway.

Table 8.7. *Reservoir Regulation for Maximum Project Diversion**

Month (1)	Inflow (2)	Riparian Reg. (3)	Evap. Precip. (4)	Irr. Draft (5)	Δ Storage (6)	Total Storage (7)	Waste (8)
1933							
Jan.	1,800	1,000	—	—	+ 800	800	
Feb.	4,600	1,000	200	—	+ 3,400	4,200	
March	16,400	1,000	300	—	+15,100	19,300	
April	34,800	1,000	400	1,500	+31,900	50,000	1,200
May	11,400	1,200	600	3,000	+ 6,600	50,000	6,600
June	3,200	1,400	800	9,000	− 8,000	42,000	
July	2,500	1,400	1,000	10,500	−10,400	31,600	
Aug.	1,200	1,400	1,000	6,000	− 7,200	24,400	
Sept.	1,600	1,400	600	3,000	− 3,400	21,000	
Oct.	800	1,200	300	1,500	− 2,200	18,800	
Nov.	1,000	1,000	—	—	—	18,800	
Dec.	1,200	1,000	—	—	+ 200	19,000	
1934							
Jan.	2,000	1,000	—	—	+ 1,000	20,000	
Feb.	5,100	1,000	200	—	+ 3,900	23,900	
March	11,400	1,000	200	—	+10,200	34,100	
April	42,000	1,000	300	1,500	+39,200	50,000	23,300
May	18,600	1,200	500	3,000	+13,900	50,000	13,900
June	4,300	1,400	900	9,000	− 7,000	43,000	
July	2,100	1,400	1,200	10,500	−11,000	32,000	
Aug.	1,200	1,400	800	6,000	− 7,000	25,000	
Sept.	800	1,400	700	3,000	− 4,300	20,700	
Oct.	1,000	1,200	200	1,500	− 1,900	18,800	
Nov.	900	1,000	—	—	− 100	18,700	
Dec.	800	1,000	—	—	− 200	18,500	
1935							
Jan.	600	1,000	—	—	− 400	18,100	
Feb.	900	1,000	200	—	− 300	17,800	
March	1,100	1,000	400	—	− 300	17,500	
April	3,100	1,000	500	600	+ 1,000	18,500	
May	1,900	1,200	800	1,100	− 1,200	17,300	
June	2,300	1,400	1,000	3,400	− 3,500	13,800	
July	1,600	1,400	1,400	4,000	− 5,200	8,600	
Aug.	1,400	1,400	1,800	2,200	− 4,000	4,600	
Sept.	600	1,400	1,100	1,100	− 2,900	1,700	
Oct.	900	1,200	400	600	− 1,300	400	
Nov.	800	1,000	100	—	− 400	0	
Dec.	1,200	1,000	—	—	+ 200	200	

*For key to table headings see bottom of Table 8.6.

When the dry year 1935 comes we know from our rule curve (which is not shown, but which is easily prepared as discussed in Chapter 7: Water Power, *Figure 7.27*) that we are only allowed to release the dependable flow for 300,000 acres. At the end of the driest year on record the reservoir is empty, like it was in Table 8.6.

The obvious advantage of the second method is the more intense utilization of the river flow. The long-term average use for irrigation in the first example is about 20 per cent of the available river flow. In the second case it is nearly 30 per cent. The disadvantage is that during exceptionally dry years we have to cut down our irrigated area from 750,000 acres to 300,000 acres. It depends largely on the local economic structure whether or not this can be tolerated. For instance, if livestock farms depend on the produce of the irrigated lands, and fodder cannot be imported, the second method would not be advisable. On the other hand, if the irrigated areas were used to produce vegetables and fruits that went to canning factories, the second method might be considered. In addition to such considerations we must, of course, also appraise the alternative use of the water that can be made for different purposes such as water power and recreation.

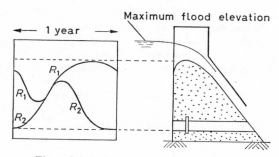

Figure 8.3. Rule curves for multi-purpose use

If the reservoir was jointly used for irrigation and flood control, the regulation of the reservoir would be facilitated by making use of rule curves. An example of such a rule curve is shown in *Figure 6.5(e)*, page 217. If the regulation of the irrigation storage was for maximum project diversion as discussed above, the rule curves may look like the ones shown in *Figure 8.3*.

Operation of the flood control irrigation reservoir would be as follows. Under normal flow conditions the reservoir level would be kept under R_1 and above R_2. Sufficient water would be released for the entire irrigation project and for downstream interests. No water would be spilled. Under high flow conditions, the level of rule curve 1 would be maintained, if possible, and water would be spilled. Under flood conditions the reservoir level would exceed R_1. When the design flood occurs, the spillway crest level is reached. When the spillway design flood occurs, the maximum reservoir level is reached. During low flow conditions, when the reservoir level falls below R_2, the reservoir release is restricted to downstream interests and the dependable irrigation draft (corresponding to the 300,000 acres in Table 8.7). When the lowest flow period on record is repeated the reservoir would be completely emptied.

IRRIGATION CANALS

The hydraulic structures associated with an irrigation project may include: a dam on the river to create a reservoir; a spillway to let flood flows bypass the dam; an intake to convey the water from the reservoir into the main supply canal; desilting works; a distribution system consisting of canals, tunnels, aqueducts, drop structures, siphons, conduits, and regulating works, to deliver water to all parts of the irrigation district; measuring devices and turnouts to bring the water on to the land; a drainage system and wasteways to convey the return flow back to the river system. The hydraulic features of reservoirs, dams, spillways, and conduits have been discussed in Chapter 5: Hydraulic Structures. This section is confined to a review of the different problems that may be encountered in planning and designing the main supply canal of an irrigation project. Such a canal may carry a discharge of several hundred or even several thousand cusec, while its length may be as much as 100 miles or more.

Intakes

When the land to be irrigated is near river level, and no storage reservoir is required, it is sufficient to build a low diversion dam in the river and have the intake to the irrigation canal adjacent to the dam, as shown in *Figure 8.4*. The diversion dam may be a fixed crest spillway without gates. As a result the full supply level will somewhat fluctuate with the discharge. To keep the diversion flow constant, a gate at the intake will provide for regulation. A sluiceway in front of the intake will remove occasionally the sediment that may have accumulated in front of the sill. If such provisions are inadequate to deal with the sediment problem, desilting works may be required.

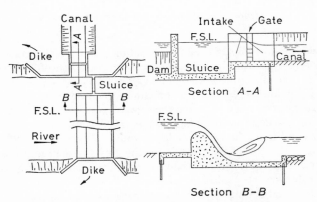

Figure 8.4. Low diversion dam and canal intake

When the land to be irrigated is substantially higher than the river and when reservoir capacity is desired, it may be economical to build a high dam, as shown in *Figure 8.5*. The outlet is situated at the bottom of the storage range and above the level of the main irrigation canal. When the reservoir level is high, a great deal of energy must be dissipated and a stilling basin at the outlet of the conduits becomes necessary. The gates of the conduits are at the downstream end. The intakes of the conduits are protected with trash racks.

356

Design Discharge

One of the first problems in designing a main supply canal for an irrigation project is to determine the maximum discharge for which the canal and its appurtenant structures must be designed. From the studies that have been described in the foregoing sections, we know the gross water requirement of the irrigation project in terms of average flow during the 12 months of the year. In Table 8.7, col. (5), for instance, the flow during Jan., Feb., March, Nov. and Dec. was zero; during April and Oct. 1,500 cusec; during May and Sept. 3,000 cusec; during Aug. 6,000 cusec; during June 9,000 cusec; and during July 10,500 cusec. On most irrigation projects such flows are more or less constant during 24 hours per day, 7 days per week, and 30 or 31 days per month. On other projects, particularly some of the older ones, the farmers do

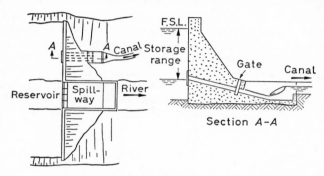

Figure 8.5. High irrigation dam and reservoir outlet

not irrigate during the night and during Sundays. Thus the actual daytime canal discharge becomes nearly twice the average discharge. It is obvious that such irrigation practice makes the project unduly costly. For this reason all irrigation authorities try to establish the 'continuous-use' policy on new projects.

Let us assume that in the above example the main supply canal is 100 miles long and that the project at the downstream end is irrigated 24 hours per day. At first glance, this would seem to require a canal capacity of 10,500 cusec. However, this capacity is only required during one month, while during the five winter months, the canal is empty and idle. This seems a waste of invest-ment! If there was a regulating reservoir at the lower end of the main canal, then the canal could be designed for the annual average flow of 2,900 cusec, while the reservoir would re-regulate the flow to suit the irrigation require-ments. It can easily be calculated that the storage requirements for a regulating reservoir in the above example, are in the order of 1 million acre-ft. To judge the merits of such a solution we must calculate the cost of 100 miles of a 10,500 cusec canal and compare it with the cost of a 2,900 cusec canal plus a 1 million acre-ft. reservoir. In case the last solution is more costly than the first, we may try an intermediate solution, such as a 6,000 cusec canal plus a 450,000 acre-ft. reservoir.

Another possibility to improve the utilization of the main supply canal is to use it during the winter for water power development. It is not unlikely that

357

the canal will run for some distance along the contour lines of river valley, and well elevated above the river bed. If there is enough water in the main reservoir to be used during the winter for power generation, if only for peaking, then it is likely that a small but profitable power station can be developed.

Alignment

Having established the intake elevation and the design discharge of the main canal, and knowing the location of the irrigation project, the lay-out of the canal seems a simple matter. However, there are still a few problems left. First of all, we must decide whether or not to line the canal. The advantages of lining the canal, be it with concrete or asphalt, are fourfold. First of all it reduces the seepage losses. If the canal is relatively long and constructed in porous soils, then lining becomes a necessity. Secondly, it prevents scour of the channel boundaries, and thus ensures a stable channel. This is particularly important when the gradient of the canal is relatively steep. Thirdly, a lined canal provides smooth boundaries, and thus, with a given gradient, it results in a smaller wetted cross-sectional area. This, in turn, means less excavation and less purchase of right of way. The fourth advantage is the maintenance of the canal. Without lining, a substantial amount of maintenance is required to prevent the growth of reeds on the banks and water plants on the bottom. With lining, only an occasional scraping of the boundary is required. These four advantages usually justify the cost of lining. As a result most large and new irrigation canals are designed with a lining. This lining may consist of un-reinforced or lightly reinforced Portland cement concrete, or asphalt concrete, 3–4 in. thick. The cost of such lining is in the order of $2–3/yd.^{2} The roughness coefficient 'n' is in the order of 0·012–0·015.

If it is decided not to line the canal, then we are faced with a problem of maximum permissible velocities. This problem was discussed in Chapter 6: Flood Control, under the section Flood Diversions. It was noted that the analysis of the limiting velocity by using the tractive force had more merits than the use of empirical velocity limits. The roughness coefficient of large well-maintained earth canals is in the order of 0·018–0·022.

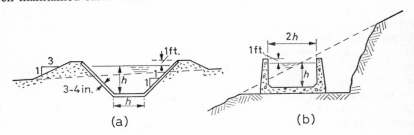

Figure 8.6. (a) Trapezoidal canal; (b) Rectangular canal

After having determined the design discharge, the approximate gradient, and the roughness of the canal boundary, we can compute approximately the wetted cross-sectional area. Now comes the problem of aligning a canal with these approximate dimensions, from the storage reservoir to the irrigation project. In the first instance, we try to follow contour lines, descending a few feet per mile. This may result in a very tortuous alignment, where the total distance can be two or three times the distance as the crow flies. We may then introduce 'short cuts' to straighten out the alignment. These short cuts may consist of tunnels, sections in fill, aqueducts, inverted siphons and drop structures. In each case it is necessary to make a cost comparison between alternative solutions. Examples of different canal and tunnel sections and special structures are shown in *Figures 8.6–8.10*, a discussion of which follows.

Canals in nearly horizontal terrain have usually a trapezoidal cross-section as shown in *Figure 8.6(a)*. When the canal follows a steep mountain slope, its cross-section is usually rectangular and relatively deep, in order to reduce the required excavation, as shown in *Figure 8.6(b)*. The ratio of depth to width may be 1 in 2, for maximum hydraulic efficiency. The boundaries of the canal may be a reinforced concrete 'U' section, or two gravity sections for the vertical walls with a horizontal slab in between for the bottom. Since such a canal is relatively costly, it may be economical to give it a relatively steep slope, to reduce its required dimensions. Thus, a long irrigation canal, composed of sections with different design and unit costs, would also have different slopes, if the over-all cost is to be minimized.

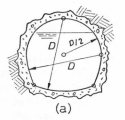 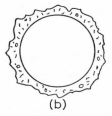

(a) (b)

Figure 8.7. (a) Gravity tunnel; (b) Pressure tunnel

A tunnel becomes economical when the alternative canal along the contour lines would be several times as long, particularly in difficult terrain. Tunnels may be designed as gravity tunnels, in which case they run partially full; or they may be designed as pressure tunnels, in which case they run full under pressure. In the first case the hydraulic flow conditions are analysed with open channel formulae; in the second case with pipe flow formulae.

The gravity tunnel has the advantage of minimizing the friction losses since there is no contact at the roof, without losing much area of flow. Moreover, the hydraulic design for different flow conditions is very simple. It is practically the same as for the adjacent open channel sections. The cross-section of the gravity tunnel is usually horse-shoe shaped, and is lined with concrete to reduce the resistance, as shown in *Figure 8.7(a)*. The gravity tunnel may be given a somewhat steeper slope than the adjacent canal, to reduce the over-all cost. However, care must be taken not to exceed critical velocities, to avoid hydraulic jumps and other unstable flow patterns.

If head loss is no problem, a pressure tunnel with high velocities and hence small cross-sectional area may be more economical. However, the following must be kept in mind: (1) the minimum economic diameter for machine excavated tunnels is about 8 ft.; (2) the hydraulic design for different discharges is more complicated—it is possible that a stilling basin at the exit and a weir at the entrance have to be included; (3) the minimum permissible depth of rock overburden shall not be less than 0·7 times the maximum hydrostatic head on the interior of the tunnel, unless the tunnel lining is reinforced.

If the canal has to cross a depression or river valley, and if it is not economical to detour around it, there are three different solutions: (1) when the depression

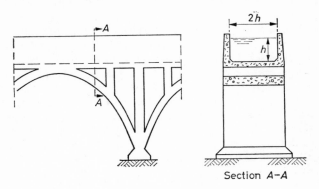

Section A-A

Figure 8.8. Aqueduct

is less than 10–20 ft. the most economic solution will probably be to continue the normal canal cross-section on top of an earth fill. If the depression contains a small creek, this may be accommodated in a culvert underneath the canal; (2) when the depression is between 10 and 100 ft. the most economic solution may be an aqueduct, as shown in *Figure 8.8*. The cross-section of the canal, on top of the aqueduct, becomes rectangular, with a depth–width ratio of 1 in 2,

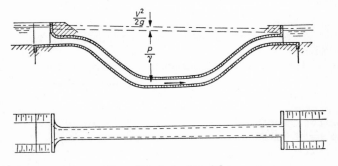

Figure 8.9. Irrigation siphon

and a slope that keeps the velocities just below critical; (3) when the depression is 100 ft. or more, the most economic solution will probably be a reinforced concrete inverted siphon, as shown in *Figure 8.9*; for very deep depressions, a steel penstock is the only solution.

At selected points along the main canal, it may be desirable to design gates and wasteways. The purpose of the gates is to be able to de-water any reach of canal, tunnel, aqueduct, or siphon, for inspection or repair; also, to be able to close down the canal on Sundays or for other special occasions, without losing the entire water content of the canal. The purpose of the wasteways, usually located directly upstream of the gates, is to discharge surplus water at designated points rather than have it spill at undesired places.

If the gradient of the canal has to be very steep, for some reason, or if a sudden drop in elevation has to be overcome, a so-called drop structure may have to be designed. This is simply an overflow spillway, with a stilling basin for energy dissipation, as shown in *Figure 8.10*. With such a structure the hydraulic energy line of the canal can be lowered by any desired amount. The design of a drop structure is based on the same principles as the design of spillways and stilling basins, discussed in Chapter 5: Hydraulic Structures.

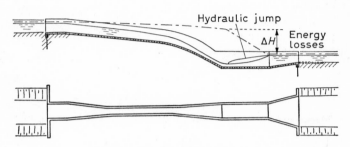

Figure 8.10. Irrigation drop structure

Pumping Schemes

If irrigation water has to be transported to a destination that is situated higher than the point of origin, the use of a pumping installation is obviously required. Even when the destination is below the point of origin, it is conceivable that a pumping station is economical. For instance, with a canal of 100 miles and a total drop of say only 10 ft., a gravity canal would require such large

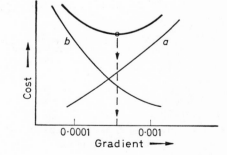

Figure 8.11. Most economical canal gradient

dimensions, that a smaller canal, plus a pumping installation, plus the cost of pumping would probably be more economical.

361

To find the optimum solution to such a problem, one could prepare a diagram as is shown in *Figure 8.11*. Curve '*a*' represents the annual cost of pumping. This consists of the annual charges (interest, depreciation, maintenance, taxes) on the capital cost of the pumping stations plus the annual cost of energy consumption. When the gradient of the canal is increased, the pumps must be designed for a higher head, and the cost becomes larger. Curve '*b*' represents the annual cost of the canal. This consists of the annual charges on the capital cost of the canal, including all appurtenant structures. When the gradient of the canal is increased, the velocities are increased, the required cross-sectional area becomes less, and hence the cost of the canal becomes less. Somewhere, the sum of curves *a* and *b* is a minimum. This point indicates the most economic canal gradient.

If the canal is long and the terrain is much different, the total distance may have to be broken up in reaches of similar terrain, with an economic analysis for each reach. For easy terrain, as is shown in *Figure 8.6(a)*, we may find an optimum gradient of say 0·0001, while for a difficult mountain slope as is shown in *Figure 8.6(b)* we may find 0·001.

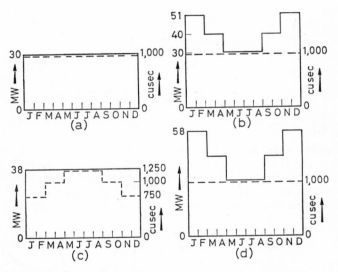

Figure 8.12. Different schemes of pumping: continuous line represents the maximum daily pump delivery (in MW), broken line the inflow into canal (in cusec)

There are three different ways in which the pumping plant can be designed: (1) continuous pumping; (2) discontinuous pumping; (3) pumping-generating. These three methods are discussed in the following paragraphs. To illustrate the discussion we shall assume that an average annual flow of 1,000 cusec must be pumped up over a height of 300 ft.

(1) With continuous pumping, the pumping station operates 24 hours per day, 365 days per year (with only minor outages for repair and maintenance), delivering a constant flow of 1,000 cusec to the main canal. This method is

graphically shown in *Figure 8.12(a)*. It is evident that this method requires a regulating reservoir at the end of the main canal, that receives the constant flow during the year and that releases the appropriate monthly irrigation requirements. From the viewpoint of electrical requirements, it will be necessary to install somewhere in the power system a generating capacity equal to the pump capacity of 30 MW. The cost of this generating capacity must be charged to the pump station, in addition to the energy that is consumed in pumping.

(2*a*) With discontinuous pumping we may avoid the cost of having to install generating capacity to drive the pump at all hours of the year. Let us assume that during the winter months the pump operates only during 14 off-peak hours of the day. Thus, we require a pump capacity of 51 MW, delivering a flow of 1,700 cusec during 14 hours into a small regulating reservoir at the head of the penstocks. From this reservoir flows a continuous 1,000 cusec into the main canal, as shown on *Figure 8.12(b)*.

(2*b*) It would be possible, with this principle of discontinuous pumping, to decrease the required pump capacity. This is illustrated in *Figure 8.12(c)*. The average annual water delivery is again 1,000 cusec. However, during the summer months, when no peak-hour outages have to be observed, the pump yield is 1,250 cusec, while in the winter it is only 750 cusec. Thus it is possible to reduce the pump capacity to 38 MW. From a total-cost viewpoint, we may observe that the size of the regulating reservoir at the end of the main canal can be somewhat reduced, but that the canal capacity must be increased from 1,000 to 1,250 cusec.

(3) With the pumping-generating principle we utilize the investment of the pumping station and penstock during the winter peak hours when it was sitting idle under the schemes (2*a*) and (2*b*). This is done by designing the pump as a pump-turbine, the motor as a motor-generator, and by enlarging the regulating reservoir at the head of the penstock. In *Figure 8.12(d)* it has been assumed that the pumping plant will be used in the winter for 2 hours per day for the generation of power. Thus, the pump requirements during the off-peak hours are increased from 51 MW (1,700 cusec) to 58 MW (1,940 cusec). The gain of this scheme is 41 MW of generating capacity. This may be sufficiently important to justify the extra costs.

Siltation

A last problem in canal design, to be discussed, is the possibility of siltation. If the irrigation water is obtained from a relatively large reservoir where all river sediment has already precipitated, there is of course no danger of siltation in the irrigation canals. There is only a problem of preventing scour, which was discussed before. If the irrigation water is obtained from a relatively small reservoir or from behind a diversion weir in the river, the water will probably contain silt. However, if the approach velocities to the canal intake are low enough over a sufficiently long reach, or if desilting works are installed, nearly all suspended bed material will be removed from the irrigation water. If, moreover, the velocities in the irrigation canals are reasonably large, the wash-load that may still be retained by the irrigation water, will have little chance to settle out. Only if a significant quantity of suspended bed material is introduced to the irrigation canals, is there danger of siltation. This was the situation in many irrigation districts in India where Lacey (1930) and others have made numerous observations to establish a relationship between pertinent parameters for stable, non-silting and non-eroding canals. Lacey's theory may be summed up as follows:

When we collect a great deal of observations on open channel flow in general (natural channels, artificial channels, fixed boundary flumes, etc.), we find an empirical relationship between the velocity, the hydraulic radius, the slope, and the roughness of the channel that is most conveniently expressed by the Manning formula:

$$Q = \frac{1\cdot49}{n}.R^{2/3}.S^{1/2} \qquad\qquad ....(8.1)$$

Lacey collected more than 100 observations on a special type of open channel, namely stable alluvial rivers and canals, and tried to establish additional relationships between the open channel parameters that would apply to those stable channels only. By trial and error he found that when V was plotted versus $f.R$ the points fell on a well-to-define curve. The parameter f is a so-called 'silt factor' that was first tied in with the roughness coefficient as follows:

$$n = 0\cdot022f^{0\cdot2} \qquad\qquad ....(8.2)$$

while in later publications it was tied in to the mean grain diameter, in inches, as follows:

$$f = 8.d^{0\cdot5} \qquad\qquad ....(8.3)$$

The graphical relationship between the velocity V and the parameter $f.R$ was expressed in the following formula:

$$V = 1\cdot17\sqrt{f.R} \qquad\qquad ....(8.4)$$

By further trial and error, Lacey also found a good empirical relationship between the velocity V and the parameter $A.f^2$, where A is the wetted cross-sectional area. This relationship was expressed in the following formula:

$$A.f^2 = 3\cdot8V^5 \qquad\qquad ....(8.5)$$

By combining eqns (8.4) and (8.5), the following formula is obtained:

$$P = 2\cdot67Q^{0\cdot5} \qquad\qquad ....(8.6)$$

where P is the wetted perimeter. This equation offers the advantage that by knowing the discharge Q, one can immediately find the stable width (wetted perimeter) of the channel. If eqns (8.3) and (8.4) are combined with (8.1), the following expression can be derived:

$$S = \frac{f^{1\cdot5}}{2,587.Q^{0\cdot111}} \qquad\qquad ....(8.7)$$

and we also know now the stable slope that belongs to the channel. By using eqn (8.1) we can find the missing dimensions of the channel.

The above is only a brief review of Lacey's regime theory. He has presented many more empirical formulae, more or less based on the above. Blench (1957) and others have continued his work and have presented still more empirical formulae. The regime theory of canal design has found many followers, although it is by no means universally accepted. The main objections are the following. The formulae only contain a parameter that indicates the size of the bed material. There is no parameter that allows for the quantity of the bed load that is introduced in the canal. It may be recalled from the discussions of river slope in Chapter 4: River Morphology, that the slope depends on discharge and sediment load as well as sediment diameter. Secondly, it must be pointed out that the eqns (8.3) and (8.4), upon which the whole theory is based, are strictly empirical and derived from one locality. Several critics have pointed out that some of the original data are unreliable. Finally, it must be noted that estimating a value for the silt factor 'f', before the canal is built and in operation, is open to considerable error.

CHAPTER 9

NAVIGATION

The main objective of a navigation development is to provide inexpensive transportation for bulk goods. This is an important objective since it may provide a great stimulus to the economy. New mining developments may be opened up. Industries may be established in locations that would otherwise be uneconomical. Existing industries may be able to compete more vigorously. One of the reasons that navigation may provide inexpensive transportation is the relatively low ratio of weight of ship to weight of cargo and the relatively small force that is required to move a ship. As a result, the force to move a ton of goods by rail is about 10 times as large as to move it by water. The disadvantage of transport by water is the slow movement and the interruption during the winter in northern climates. Moreover, if the waterways have to be improved by means of dams and shiplocks, much of the economic advantage may be lost.

There are three different methods to provide navigable waterways. The first is to improve a river channel by means of river training works, dredging, or increasing the low flows. This method is called river regulation. The second method is to make a river navigable by building a series of low dams and shiplocks to by-pass the dams. This method is called river canalization. The third method is to excavate artificial canals and to build shiplocks when differences in elevation have to be overcome. These three methods are discussed in subsequent sections of this chapter.

It may be possible and economical to combine a navigation development with other phases of water development. For instance, if a series of low dams is planned for river canalization, they may be slightly modified to incorporate power plants. Conversely, if a series of power dams is planned, they may be laid out in such a way that shiplocks can be built adjacent to the dams, and thus the entire reach of river may become navigable. If artificial navigation canals have to be excavated, they may possibly incorporate water supply or drainage functions.

RIVER REGULATION

This is by far the most attractive method to make a river navigable. No costly dams and shiplocks are needed, and the river traffic is not delayed at the locks. However, to make this method applicable, the river should meet certain requirements. Let us assume for instance that we want a navigable channel of 10 ft. deep and 200 ft. wide. Let us also assume that the river under consideration does not have that depth of flow under natural conditions. We may then apply three different methods to improve the situation. The first is to dredge the shallow places in the river, or to blast rock ledges or rapids. If this does not

366

accomplish our goal, we can force the river by means of spurs to flow in a narrower and hence deeper channel. If this is still not sufficient, we can increase the flow of the river by releases from upstream reservoirs. These three methods of river regulation will be discussed in the following paragraphs.

Dredging

Alluvial rivers are often meandering, and their bed configuration may resemble the one shown in *Figure 9.1*. It will be recalled from the discussion of river bends in Chapter 4: River Morphology that the main current of the river impinges on the concave bank, starting at or slightly upstream of the apex of the bend and extending well beyond the bend. In this region of the highest velocities the river has the greatest erosive and sediment transporting capacity and this is where we find the greatest channel depth. On the convex banks the

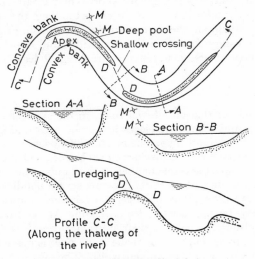

Figure 9.1. Dredging in alluvial river

velocities are low, sediment is being deposited and we find consequently sand bars and shallow depths. Ships that have little difficulty in navigating the deep troughs or pools in the bends may run into trouble when crossing the shallow bars in between the pools, particularly during low flows of the river.

In such a river, the situation could be considerably improved by dredging a channel from D to D. To guide the ships into this channel, four markers M may be placed on the shore as indicated in *Figure 9.1*. Since the formation of pools and crossings is a natural phenomenon in an alluvial river, it may be expected that the dredged channel will fill in with sediment after it has been dredged. The rapidity of this filling-in process depends on the amount of sediment transport in the river. As a result, the dredging operation becomes a perpetual event. One dredge may be assigned a certain reach of river, continuously cleaning out and maintaining navigation channels.

In straight reaches of an alluvial river, the depth of flow is not uniform. It is more than likely that pools, crossings and sandbars will alternate in an irregular and continuously changing pattern. This is very unpleasant since the dredging operations have to take place in a different location all the time; consequently beacons have to be replaced, and the skippers must get used to a new course every so often. To avoid this, it may be possible to force the river, by means of spurs, into a regular meander pattern. Not so outspoken that the total length of the navigable channel increases substantially, but just enough that the pools, sandbars and crossings remain in the same place. To devise such a pattern is mostly a matter of trial and error. It can be aided by a study of existing meander bends of the river, by studying successful training works on other rivers, and by hydraulic model studies.

To plan the location of a dredge cut in a crossing, between two pools, it is advisable to obtain first complete soundings of the river bottom. The axis of the cut should be located so that it has a smooth alignment with the pools, from a viewpoint of navigation. If the cut is too short it may create excessive velocities between the upper and lower pool. If the cut is too long, the velocities may not be high enough to maintain the channel during the season. Frequent inspection of the cuts is necessary to ensure that the width and depth of flow remains available. On most rivers, maintenance dredging at least once per year, is required.

Training Works

It was noted earlier that an alluvial river, in the state of nature, may have shifting sandbars and shoals. At places it may be wide and shallow. At other places it may be narrow and deep. The meander bends may be slowly migrating in a downstream direction. The discharge may be confined to one main channel, or it may be divided over several channels with islands in between. It is obvious that such a river, in its natural state, presents severe handicaps to navigation. For one thing, the river pilots must have a fabulous knowledge of the river channels (as may be appreciated from reading Mark Twain's *Life on the Mississippi*), or else they may lose their ships and cargo.

Substantial improvements can be made if the river is made to flow in a well-defined channel with a predetermined width. The two main advantages are: first, the maintenance of a minimum depth of flow; and second, the maintenance of one and the same channel from year to year. The means of establishing such a channel are spur dikes and longitudinal dikes as shown in *Figure 9.2*. Since we are primarily interested in creating sufficient depth of flow during low discharge conditions (the depth of flow during high discharges being large enough), and since we do not want to raise the flood levels on the river, it follows that the spur dikes should not be constructed much higher than the low water level of the river. We may determine approximately the width W of the navigable channel, between the front line of the spur dikes, by applying the Manning formula. We know the dependable flow in the river, we know the gradient of the river, we know approximately the roughness coefficient, and we know the desired depth of flow. From this follows the width of the channel. If this width is too small for the kind of navigation that we want to accommodate, we must resort to artificially increased flows, or to river canalization, as will be discussed in subsequent sections.

The above may be illustrated by the following example. The lower Rhine River has a dependable flow of 14,000 cusec. Its slope is 0·0001 and its roughness coefficient in the regulated reaches is about 0·025. The larger type of river boats that ply the river have a capacity of about 2,000 tons, a length of 300 ft., a width of 35 ft. and a depth of 10 ft. A minimum navigable depth of 12 ft. is desired. With the given slope and roughness, the unit discharge at 12 ft. depth of flow becomes 38 cusec/ft. width. Hence the average width of the river may be designed at 370 ft. Since the underwater slope of the spur dikes is about 1

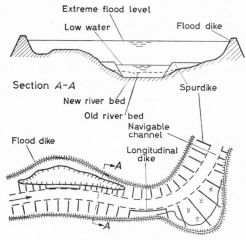

Figure 9.2. River training works

vertical in 4 horizontal, the bottom width of the channel becomes 320 ft. and the surface width 420 ft. This was considered sufficient for proper navigation, since a downstream traffic lane should have a width of at least five times the width of the largest ship, and an upstream traffic lane three times the width of the ship, while the total width of the channel should be preferably larger than the length of the largest ship.

In a preliminary lay-out of the spur dikes and longitudinal dikes, one may aim at a curvature of the axis of the navigable channel, that is about 10 times the surface width of the channel. Between the curves there should be transition sections with a length of about 3 times the width of the channel. It has been found from experience that much larger radii and much longer straight reaches may result in a wandering of thalweg (line of maximum depth) and the appearance of moving sandbars. A much smaller radius may cause local deepening of the channel at the outside of the bend and silting up of the channel at the inside of the bend, thus reducing the navigable width of the channel. When the transition sections are too short, the bottom topography of the upstream bend may begin to influence the downstream bend. The spacing of the spur dikes should be such that the distance between the riverside end of two successive dikes is less than the surface width of the navigable channel. When the spur dikes are short, their distance should also be smaller than one and a half times the length of the dikes. If the dikes would become very short, a longitudinal dike, or simply a revetment of the river bank, may be more economical and would provide a better guidance to the current. The spur dikes should be tied in solidly with high ground or with the flood dikes of the river, to prevent the formation of new channels during flood stages. The centre line of the spur dikes is usually designed perpendicular to the centre line of the river. Sometimes they are designed pointing upstream. This has the advantage that during flood stages the overflow is directed towards the centre of the river and thus avoids strong currents over the flood plains and near the flood dikes. Before a final lay-out of the training works is decided upon (which is for practical purposes irrevocable) a thorough study should be made of the river under natural conditions, of training works on similar rivers and of hydraulic models of the proposed training works.

Spur dikes may be constructed in different ways, depending on the functions they have to perform, on the available materials, and on the prevailing construction techniques. If a natural braided river channel is sprawled out over a

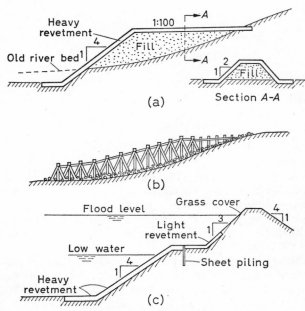

Figure 9.3. (a) Non-permeable spur dike; (b) Permeable spur dike; River bank protection

river valley of a mile wide and it has to be narrowed down to a few hundred feet, the major portion of the spur dikes may be of the permeable type, consisting of pile clumps or brush fences, to retard the flood plain currents, to close off all but the one major channel, and to promote sedimentation. Once this process has been set in motion, and the main channel begins to carry the entire low and medium river discharge, the velocities will increase, the bottom of the channel will be lowered, and the spur dikes may have to be built solidly with a heavily reinforced terminal end at the riverside to prevent undermining. Some examples of spur dikes are shown in *Figure 9.3*.

Sedimentation of the flood plain has a dual purpose. First, it forces the entire low and medium river discharge into the main channel, thus providing the required navigable depth of flow. Second, it makes available valuable farmland. The ideal way to treat the flood plain between the navigable channel and the flood dikes, as shown in *Figure 9.2*, is to induce by proper sedimentation measures the filling up of all old channels, creeks and gullies. In due time the flood plain will be levelled off and covered annually with a thin layer of fine silt. Those are ideal conditions for pasture and haylands, which provide an erosion resistant surface protection that will prevent the formation of new gullies during extreme flood stages. It is desirable that all bush, brush, buildings and fences be removed from the flood plain so that the flow resistance is minimized, so that the flood levels are kept as low as possible, so that the cost of the main flood dikes can be reduced.

In certain reaches the main channel may impinge upon the flood plain dike, and it becomes necessary to design a longitudinal guide dike or to protect the river bank itself with a suitable revetment as shown in *Figure 9.3*. Several design criteria must be applied to such a revetment. First of all the slope of the entire bank and dike must be such that no slide failures can occur, even with the most rapid river flow recession. Second, the revetment must be so tight that no soil particles can be removed by the current from underneath the revetment. Third, the revetment must start above the normal summer water level and must extend all the way down the slope on to the bottom of the river, so that no undermining of the toe of the slope is possible. Fourthly, the revetment must be so flexible that if, in spite of the above measures, some undermining takes place, the revetment will smoothly deform itself into the new bank contours, without breaking, cracking, or failing. Examples of revetments that have been successfully applied are: willow mattresses of about 1 ft. thick, sometimes with an inner core of reeds, and covered with 1 or 2 ft. of rip rap; articulated concrete mattresses of concrete blocks of a few square feet each, with a thickness of a few inches, and connected by a continuous wire mesh; or lightly reinforced asphalt cement mattresses a few inches thick with 15–20 per cent of asphalt, as described by Kuiper (1951)

It may be recalled from the discussion of river regulation in Chapter 4: River Morphology, that the initial result of the training works will be a rise in water levels over the regulated reach and deposition of sediment upstream of the reach. Only after some time will the river bottom of the regulated reach be lowered and the intended improvements become effective. It may be economic to accelerate this process by an initial dredging programme in the navigable channel during or shortly after the construction of the training works. It is not unlikely that afterwards some dredging maintenance is required to ensure that the navigable channel remains in its prescribed course.

Flow Release

It was noted above that if the natural dependable river flow is insufficient to fill the desired dimensions of a navigation channel, a possible improvement may be to release flows from upstream reservoirs. There are several reasons why this method has very limited applicability. First of all, it requires huge reservoir capacities. The annual flow deficiencies for navigation are likely to be in the order of several thousand cusec, for several months in succession. This amounts to several million acre-ft, of water. Even if reservoirs with such capacities can be built, it is likely that other resources development objectives, such as water power and irrigation, will receive priority of use. Secondly, the navigable reaches of a river are usually found in its lower course, while the reservoir sites are usually found in its upper course. Because of the substantial distance in between, it is difficult to release the right amount of water at the right time so that the desired navigation flow is maintained. In practice there will be much waste of water to guarantee the flow requirement. Thirdly, because of the large distance between the reservoirs and the regulated reach, it is likely that different countries or different jurisdictions are involved. This may pose practical difficulties, in addition to the technical and economic difficulties.

In spite of these drawbacks there are a few examples where reservoir release for navigation purposes is applied. In the upper Mississippi drainage basin, before 1900, about 1,500,000 acre-ft. of storage capacity was provided to increase the natural low flow by a few thousand cusec. As a result, the navigable depth was increased by about 1 ft. However, in 1930, the situation was deemed undesirable, and the river was canalized. On the Missouri River, five reservoirs with a total storage capacity of 75,000,000 acre-ft. have been built. This storage capacity is used for flood control, water power, irrigation, and navigation. The navigation releases are in the order of 20,000 cusec over and above the natural low flow of about 5,000 cusec, for a duration of about three months, to maintain a navigable channel on the lower Missouri River of 9 ft. deep and 300 ft. wide. This release of water is also of benefit to the navigation interests on the main stem of the Mississippi River, although to a lesser extent because of the thousands of miles of distance involved.

RIVER CANALIZATION

If the natural dependable flow of the river is insufficient to provide a navigable channel of desired depth and width, and if storage capacity in the drainage basin to increase the dependable flow to a desired magnitude is too costly, or insufficient or not available, then we may resort to the canalization of the river by means of dams and shiplocks. This method has been successfully applied to several rivers in Europe and North America. One example is the Ohio River where the natural dependable flow in the middle reach is in the order of 5,000 cusec. The slope of the river is in the order of 0·5 ft. per mile. A study of possible reservoir control indicated that the dependable flow could not be sufficiently increased to provide a navigable channel of 9 ft. minimum depth and 400 ft. width. The only alternative was to canalize the river. In spite of the drawbacks that are associated with the delays at the 46 locks on the Ohio River, the annual traffic has developed to 20,000,000 tons.

Profile

The principle of canalizing a river is that we convert the unsuitable low water profile in a series of backwater curves with sufficient depth of flow, as shown in *Figure 9.4(a)*. Let us assume that the natural dependable flow is 5,000 cusec, with a depth of flow of 5 ft. and hence unsuitable for navigation. The desired depth of flow is 12 ft.; to obtain this a series of dams is designed that will locally increase the depth of flow to say 22 ft. Upstream of the dams the river profile will be a backwater curve that is governed by the river channel characteristics and the dependable flow of 5,000 cusec. The distance between the dams is such that the backwater curve of any dam reaches a depth of flow of 12 ft. at the upstream dam. For instance, if the roughness coefficient of the river channel is 0·025 and the bottom slope 0·0001, it would require about 25 miles to reduce the depth of flow from 22 ft. to 12 ft., with a discharge of 5,000 cusec and a channel width of 600 ft.

Let us assume that the river under natural conditions will obtain a depth of flow of 12 ft. when it has a discharge of 25,000 cusec. One way of designing the dams on the river, would be to make them removable when the river discharge is 25,000 cusec or more, as is shown in *Figure 9.4(b)*. The ships can then move freely up and down the river without having to pass the locks. In

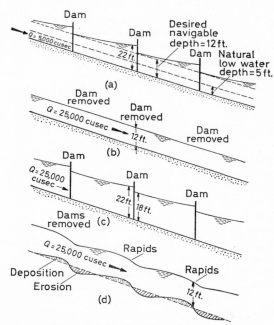

Figure 9.4. Profiles of canalized river

this way many river canalizations have been designed. However, there are some drawbacks. First of all, the river velocities in *Figure 9.4(b)* are much higher than in *Figure 9.4(c)* where the dams remain in place and where as a result the depth of flow ranges from 22 to say 18 ft., instead of 12 ft. all the way through. If the distance between dams is reasonably long, skippers may prefer low river velocities and locking, over high river velocities and no locking. Another drawback may arise from the disturbance of the sediment transport in the river channel. The construction of the dams will result in sedimentation upstream of the dam and consequently erosion downstream of the dam as shown in *Figure 9.4(d)*. If the dams are now removed, extra high velocities, perhaps even rapids, will occur in the vicinity of the dam, which will constitute a hazard to navigation. Finally, it should be pointed out that the so-called

373

navigable dams, which can be removed during high water, are limited to a difference in water level of about 10 ft., and that they are difficult to inspect and costly to maintain.

If the above drawbacks are serious enough to outweigh the advantages of the navigable dam, one may design instead a non-navigable dam. Such dams can be designed with a much greater difference in upstream and downstream water levels. Hence they can be located farther apart, which is of advantage to navigation because of less frequent locking and greater depth of flow, and hence less ship resistance. Let us discuss what considerations have to be applied to determine the difference in water level at the dam site. A very important consideration may be the interests of the flood plain occupants. If the flood plains are extensively used for agriculture, or if there are towns, cities and industries on the flood plains, we must keep the high level at the dam well below the flood plain level so that drainage and sewage disposal are not affected. Under such circumstances we must also be careful to design the dams in such a way that they offer practically no obstruction during times of flood. This requires a low sill, narrow piers, and gates that can be hauled up above the highest flood levels, as will be discussed in a subsequent section.

If the flood plain interests are of a minor nature or non-existent, we may create a substantial difference in water level at the dam and it may become economical to incorporate a water power plant. It is obvious that the power potential is enhanced with a reduction in the number of dams and a consequent increase of head at each dam. At a great number of small dams, all head would be drowned out at medium and high river flows. At one large dam, practically all head would remain available. Examples of such dual purpose developments are the St. Lawrence Seaway where navigation was of primary interest and the lower Columbia River where the power development was of primary interest.

Another variable that may affect the canalization plan is suitable sites for dams and locks. From the standpoint of navigation requirements there should be a good alignment of approach from the river to the locks. The magnitude and direction of the river current should allow an easy entrance and exit of traffic. From the standpoint of shiplock and dam construction there should be good foundation conditions with a minimum possibility of seepage underneath the structures. Such sites may not be plentiful along the river!

After all possible designs have been considered, it becomes a matter of economic analysis to select the best one. On the one hand is the cost of the scheme that may increase with fewer dams and larger heads. On the other hand is the saving in time for the ships and perhaps an increase in power output. Such an economic analysis should take into consideration anticipated future river traffic, future flood plain use, and future water use. In principle, we must strive towards the lowest total annual cost for the engineering works, the river freight, and the power output.

River Dams

It was noted in the above section that the dams of a canalization plan may be of three different types. First, the navigable type, that is completely removable so that ships can travel over them at high river stages. Second, the non-navigable type, where the gates can be lifted up above flood level, but where the river traffic cannot pass over or through the dam during high discharges. Third, the fixed dam (gravity dam, earth dam, rock-fill dam) that is incorporated with a power plant. In this section, only the first two types will be discussed. The third type was discussed in Chapter 5: Hydraulic Structures.

The lay-out of the dam and the shiplock is usually very simple and may resemble the one shown in *Figure 9.5*. In this particular example there is a combination of the navigable and non-navigable type of dam. The first has been installed to allow free river traffic past the dam during high river stages. The second has the function to keep the upper pool level constant during fluctuating low discharges of the river. The pool should not rise above its established level because of flood plain interests; and it should not drop below this level because navigation will benefit from the lowest possible velocities.

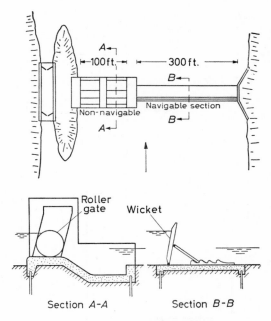

Figure 9.5. Canalization dam

The navigable section in *Figure 9.5* is equipped with so-called wickets, which are pivoted to the foundation and pivoted to a support that can slide back and forth and be locked in the position shown. The gates have a ratio of width to height of about 1 in 5. Raising and lowering of the individual gate sections may take place from an overhead bridge or from a barge in the upstream pool. Alternative designs for a navigable dam are the Poiree dam (collapsible trestles which support needles or planking or panels, operated from a barge or overhead bridge) and the Bear Trap dam (two flexible leafs that form the roof of a chamber wherein the hydrostatic pressure can be changed), as shown in *Figure 9.6*.

The non-navigable section in *Figure 9.5* is equipped with roller gates, which are simply hollow steel cylinders with closed ends, which are hung up between two piers. In its lowered position the gate closes off the opening. The gate is moved by means of a cable or chain which is wrapped around the cylinder. The gate rolls up under a slant towards the top of the piers, above flood level. The hydraulic efficiency of the gate can be improved by a curved segmental plate that makes contact with the sill. Water can be discharged either over the top of the gate, or underneath. Alternative designs for a non-navigable dam are the Taintor gate (a steel-plated section of a circle, hinged at its centre) or the Sydney gate (a Taintor gate assembly that can be hauled up like a roller gate), or a vertical lift gate, as shown in *Figure 9.6*.

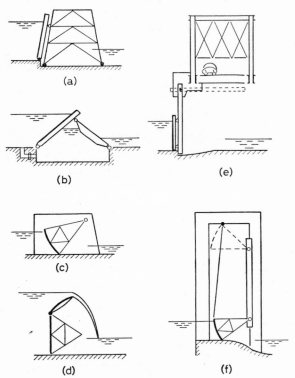

Figure 9.6. *Different types of river dams: (a) Poiree dam (navigable); (b) Beartrap dam (navigable); (c) Taintor gate (non-navigable); (d) vertical lift gate with movable flap (non-navigable); (e) panels and supports lowered from bridge (navigable, depending on height of bridge); (f) Sydney gate (non-navigable)*

The dam sill and downstream apron of the navigable and non-navigable dam are designed for different criteria. The navigable dam in *Figure 9.5* is either completely closed or completely opened. In the first case there is no water flowing over the sill. In the second case there is very little energy dissipation over the sill. From a viewpoint of hydraulic performance it is sufficient if the sill has approximately the elevation of the river bottom and is large enough to accommodate the gate equipment. From a viewpoint of seepage underneath, the sill must be sufficiently long and perhaps be designed with a row of sheet piling or an upstream impervious blanket. Downstream of the sill there should be some rip rap to prevent scour.

The sill and apron of the non-navigable section in *Figure 9.5* should be designed for energy dissipation. At times it may be necessary to pass substantial flows over or under the gates with a substantial difference in upstream and downstream water levels. The kinetic energy contained in this flow release must be dissipated before the water returns to the normal river channel. Hence it becomes necessary to design the downstream apron for the criteria of a stilling basin, as discussed in Chapter 5: Hydraulic Structures. The sill of the gate is usually placed near river bottom level, or a few feet lower, to increase the discharge capacity of the gate opening.

In deciding whether or not the dam should be designed with a navigable section, as shown in *Figure 9.5*, the following two aspects should be taken into consideration, in addition to what has been discussed earlier. First, the percentage of the time that open river navigation is possible. If this percentage is small, the disadvantages of the two-system navigation, as discussed above, may not outweigh the advantages. Second, the need for flushing ice and sediment through the river system. If there is such a need, and if the non-navigable dams cannot perform this function satisfactorily, the navigable type of dam may be chosen.

CANALS

We may distinguish three different types of canals. First, the canals that connect two points on one river. Their purpose is to by-pass a particularly difficult reach with rapids or falls. They may become part of a canalized river system. The second type is the canal that connects two different navigable waterways. The third type is the canal that connects a large city with a navigable waterway or with the sea. All three types have in common that they must be excavated and equipped with locks. This makes them rather costly and hence it becomes important to give due consideration to their most economic location, their most economic cross-section, and to ensure their water supply.

Location

In determining the most economic location for a canal to connect two given points we must consider the following four requirements.

First, the alignment of the canal should be as straight as possible. If curves are introduced, a rough rule of thumb is that the radius of the curve, in feet, should be about three times the tonnage of the ships for which it is designed. Thus a 2,000 ton ship would require a curve of about 6,000 ft.

Second, there should be as few shiplocks as possible, since every shiplock means a significant delay to the canal traffic.

Third, the water level in the canal should be approximately the same as the surrounding ground water level. If the canal level is significantly lower, it may lower the ground water level with consequent damage to agricultural interests. If it is significantly higher there may be undesirable water loss and water-logging of adjacent agricultural lands. This could be reduced by lining the canal with a layer of clay, but this is a rather costly measure.

Fourth, the total amount of earth movement should be a minimum. This includes the excavation as well as the embankments where the canal may be locally higher than the surrounding land. This requirement assumes, of course, a wetted cross-section of the canal of given dimensions.

It is obvious that, in planning the location of a canal, one must compromise in trying to meet these four requirements. If the terrain is uneven, one simply cannot keep the alignment straight and also keep the canal level the same as the ground water level, without introducing ship locks. The ultimate criterion to be applied to a number of alternative proposals must be again, as was noted earlier, that the total annual cost of the canal plus the total annual cost of traffic is a minimum. It is of interest to note that in canals that have been built and that have proved to be economical, the vertical distances overcome, and

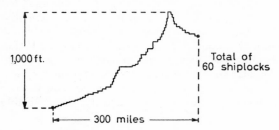

Figure 9.7. Profile of canal between the Rhine and the Danube

the number of shiplocks applied, is quite substantial, as is illustrated by *Figure 9.7.* In hilly terrain, cuts of 50–100 ft. deep are not unusual. The deepest cut in the Panama Canal is 250 ft. Canal embankments have been constructed up to 85 ft. high. However, if the length of such embankments becomes substantial, it may be more economic to lower the canal and introduce two shiplocks.

Cross-section

It is obvious that a larger cross-sectional area offers advantages to the canal traffic: there is less possibility of accidents and there is less resistance. However, a larger cross-sectional area is also more costly, and hence we must find again the most economic compromise. Let us first discuss ship resistance.

If a ship moves in wide-open still water, its total resistance has two components: the form resistance and the skin friction. Roughly speaking, the total resistance of the ship varies with the square of the velocity of the ship.

If a ship moves in a canal of limited dimensions, the water underneath and beside the ship will not remain still, but will move in the opposite direction of the ship. Hence, a ship will have more resistance in a canal than in wide-open water when it travels with the same speed with respect to the shore. The fact that water underneath the ship moves against the motion of the ship may be explained by making reference to *Figure 9.8(a)*. While the ship moves from position A to position B there is no movement of water in the canal well in front of the ship and well behind the ship. However, during the displacement of the ship, the volume of water V must be displaced from its position at B to the position at A. This movement of water must take place underneath and beside the ship. In other words, while the ship moves to the left, the water moves to the right. The smaller the opening through which the water must move (the cross-section of the canal minus the cross-section of the ship) the larger the velocities, and hence the greater the resistance exerted on the ship. We can quantitatively analyse this problem by pretending that the ship is at rest and that the water moves to the right with the absolute velocity of the ship (the velocity with respect to the shore) as shown in *Figure 9.8(b)*.

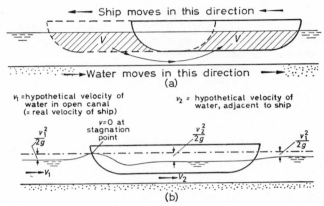

Figure 9.8. Motion of ship in canal

If one would wish to make a precise economic analysis of the optimum canal cross-section, one would have to determine: water velocities, resultant resistance, resultant power requirements, and resultant power costs for the average annual canal traffic, for different canal designs; add to each figure the annual costs of the canal and find the design with the lowest total cost. This is very seldom done. Instead, it is more common to apply such empirical criteria as shown in *Figure 9.9*. A satisfactory cross-section is obtained: when the shape

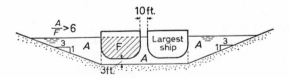

Figure 9.9. Standard canal cross-section

is trapezoidal; when the side slopes are 3 horizontal to 1 vertical; when the largest ships can pass one another with a clearance of 10 ft. while remaining above the horizontal bottom; when the clearance with the bottom is 2 ft. for ships up to 500 tons and 3 ft. for ships over 1,000 tons; and when the ratio of the wetted cross-sectional area of the canal to the wetted cross-sectional area of the largest ship is at least 6.

The above empirical rules apply to inland canals with still water and in sheltered reaches. If there is an appreciable current in the canal for drainage, water supply, or power purposes, navigation becomes more difficult and the horizontal dimensions must be increased. Exposure to strong winds also makes navigation more difficult and requires larger dimensions. For the design of sea ship canals the clearance requirements between ships, between ship and slope and between ship and bottom should be increased. The desirable ratio of channel area to ship area remains the same.

Water Requirements

In a canal system with shiplocks there are four different ways of losing water. These losses must be compensated by an adequate water supply at suitable locations. The first loss is the evaporation-precipitation deficit over the surface area of the canals. This loss depends very much, of course, upon the prevailing climate of the region in which the canal system is planned. Let us assume, for example, a monthly evaporation-precipitation deficit of 10 in. during the height of the summer, and a surface width of the canal of 150 ft. This would result in a water loss of 25 cusec per 100 miles of canal.

The second cause of water losses is seepage into the ground, assuming that the ground water level is lower than the canal level. This loss also depends on local circumstances. To estimate it, one should prepare maps showing the ground topography, the ground water levels, the canal levels and the types of soil traversed. Ground water flow computations must be made to estimate the rate of loss per section of canal. If the losses become too high (from the view-point of adjacent agricultural interests or from the viewpoint of available water supplies) one may resort to lining the entire canal perimeter or perhaps just the side slopes, with a layer of clay. It is good practice to cover this layer of clay with ordinary soil, to prevent the clay from being punctured by anchors or otherwise disturbed. Seepage losses ranging from 10–50 cusec per 100 miles have been observed on existing canals.

The third cause of water losses is from the locking operations. As will be discussed in the following section, every locking operation requires the release of water in downstream direction. Let us assume, for example, a shiplock that is 1,200 ft. long, 110 ft. wide, and has a difference in upstream and downstream water level of 20 ft., as shown in *Figure 9.10*. Let us consider one complete cycle of locking operations with ships of 1,000,000 ft.[3] of water displacement. The initial condition is: the water level in the lock is at downstream level, and two ships a and d waiting to be locked. We could observe the following: ship a moves to position b and 1,000,000 ft.[3] of water moves from the lock into the downstream canal; the downstream gates are closed, 2,640,000 ft.[3] of water is released from the upstream canal into the lock chamber and the ship moves from position b to c; the upstream gates are opened, the ship moves from position c to d and 1,000,000 ft.[3] of water moves from the canal into the lock chamber; now the other ship that was waiting in position d moves into the lock chamber and 1,000,000 ft.[3] of water moves from the chamber back into the upstream canal; the upstream gates are closed, 2,640,000 ft.[3] of water is released from the lock chamber into the downstream canal and the ship moves from position c to b; the downstream gates are opened, the ship moves from position b to a and 1,000,000 ft.[3] of water moves from the downstream canal into the lock chamber. One cycle of locking is now completed and if we make up the balance of water movement, we find that the volume of the lock differential V, equal to 2,640,000 ft.³, has been released from the upstream canal into the downstream canal. If the locks are designed for efficiency and the traffic is heavy, we could expect say 20 complete cycles per day. This would be equivalent to an average flow of 600 cusec.

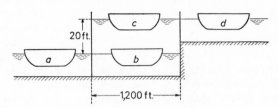

Figure 9.10. Locking operation

The fourth and last cause of water losses is leakage of the gates and valves at the shiplocks. This loss depends on the difference in upstream and downstream water level, and on how well the gates and valves are designed, constructed, and maintained. Under normal conditions, leakage may amount to about 1 cusec for 20 ft. height of lift. Compared to the water losses through locking operations, such leakage is negligible.

It should be pointed out that lockage losses on a canal are not accumulative. The water that is lost by one lock, can subsequently be used by the next lock downstream, and so on. It should also be noted that the water requirements for locking are directly proportional to the difference in water levels at the lock. Assuming uniform traffic throughout the length of the canal, it follows that the water requirements of the canal for lockage purposes, are determined by the lock with the greatest difference in water levels. Since canal levels are to be maintained within fairly narrow limits, it follows that at every lock, except the one with the highest head, provision must be made to by-pass the surplus discharge coming from upstream.

If the profile of a canal has a summit, as shown in *Figure 9.7*, it is obvious that the main water supply is required at the summit. From this point, water must be supplied in two directions, each of them being in the order of perhaps several hundred cusec. If there is no stream near the summit with a natural dependable flow equal to the water requirements of the canal system, measures must be taken to ensure a sufficient supply of water. Such measures may include: a local reservoir, pumping, or reducing the water requirements. The application of reservoirs to increase the dependable flow of a stream has been discussed in other chapters. The application of pumping is technically simple but economically unattractive. From the nearest dependable source, water is pumped via the canal (with a pumping station at every lock) to the summit. The possible reduction of the water requirements merits some more discussion.

It was noted above that the main water requirements are caused by locking operations. It was also noted that the lock with the largest difference in water levels governs the whole canal. Hence a first improvement could be made by avoiding one or a few locks with an exceptionally large head in a string of locks with lesser heads. The second improvement could be made by building adjacent twin locks and to integrate their operation. When the one lock is high,

381

the other is low. The high lock empties into the low lock, until they are equal. After that, the water is emptied in the downstream canal. This operation would reduce the water requirement by 50 per cent. Of course, it could only be applied when there is enough canal traffic to justify the construction of twin locks. The third improvement follows the same principle and adapts it to the single lock, as shown in *Figure 9.11*. On each side of the lock is a so-called thrift basin with a surface area equal to that of the lock chamber. While the level in the lock is lowered from 4 to 3, the left basin is filled. While it is lowered from 3 to 2, the right basin is filled. From 2 to 0, the lock is emptied in the downstream canal. During filling, the right basin is emptied first, bringing the level up to 1. Then the left basin is emptied, bringing the level up to 2. The remainder of the lock chamber is filled from the upstream canal. In this

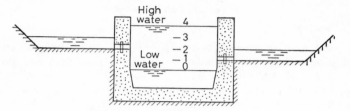

Figure 9.11. Lock with thrift basins

example, there is also a saving in water use of 50 per cent. By introducing more and larger basins, one could further increase the efficiency of operation. It should be pointed out that there are also disadvantages associated with the thrift basins: the total time of locking tends to increase, thus reducing the daily capacity of the lock; and the lock will become more costly, in construction cost as well as maintenance. The fourth improvement that can be made, is to install an extra set of gates halfway up the length of the lock. Whenever the traffic is light and the ships are small, only half of the chamber is used, thus reducing the water consumption for that occasion by 50 per cent.

SHIPLOCKS

The first shiplock was reputedly built in Holland, around the year 1200. Its purpose is to bring ships from one water level to a different water level. Its principle is to have a lock chamber, as shown in *Figure 9.12(a)*, with gates on each side. When the water level in the chamber is made equal with the high level, the upper gates can be opened and the ship can be moved in the chamber. The upper gates are closed, and water is released from the chamber into the lower canal. When the chamber level is equal with the low canal level, the lower gates are opened and the ship can proceed on its journey. In modern shiplocks this whole process does not take more time than about 20 minutes.

If there is heavy traffic on the canal, it may happen that the skippers lose more time with tying up and waiting for their turn than with the actual locking. It is therefore very important to give adequate consideration to the lay-out of the shiplock with respect to the canal or river; to design a good alignment from the traffic route to the mooring facilities, into the lock, and back to the traffic route. For final design of a shiplock one would be well advised to study existing facilities under similar conditions and talk to the skippers that use them to find out where improvements may be made.

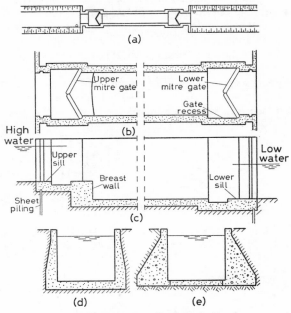

Figure 9.12. Details of shiplock

In designing the shiplocks, we must make a number of decisions, of which the following are the most important: the size of the shiplock; the type of walls; the type of gates; and the method of filling and emptying the chamber. These topics will be discussed in the following paragraphs.

Table 9.1. *Dimensions of Shiplocks (in ft.)*

Waterway	Lift	Length	Width	Draft
Ohio River	35	1,200	110	12
Upper Mississippi	49	400	56	9
St. Lawrence Seaway	40	766	80	30
Upper Rhine River	54	600	82	20
Rhone River	40	650	43	15
Dutch canals	10	450	45	12

The size of the shiplocks depends on the largest ship for which the canal is designed and upon the volume of traffic. If the traffic is light, the chamber of the shiplock could be designed so that it can just hold the largest ship with clearance of a few feet on all sides. If the traffic is heavy, it may be economic to design the lock larger so that a few ships can be locked at the same time. It saves a good deal of time if a tugboat with its string of barges can be locked in one unit, instead of in two or three parts. The new locks of the Ohio River, for instance, can accommodate a 24-barge 'river liner' at once, and thus will reduce the locking time for such a unit from 3 h to 30 min. If the traffic is very heavy, it may be economic to build twin locks. In Table 9.1 some dimensions of recently constructed shiplocks for major inland waterways are listed. The first column, 'lift', indicates the difference in upstream and downstream water level and in the last column 'draft' indicates the low water depth over the downstream sill.

Modern shiplocks are usually built of concrete. If the cost of materials is relatively high and the cost of labour relatively low, it may be economical to design the shiplock in reinforced concrete. A cross-section over the chamber

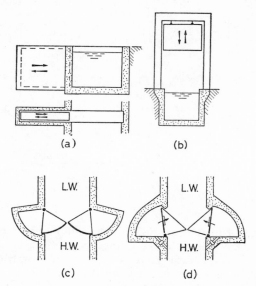

(a)   (b)

(c)   (d)

Figure 9.13. Types of shiplock gates: (a) rolling gate; (b) vertical lift gate; (c) sector gate (can be opened and closed against water pressure by mechanical operation); (d) swing gate (can be opened and closed against water pressure by hydraulic operation)

will then look like *Figure 9.12(d)*. If the cost of materials is relatively low and the cost of labour is high, the walls may be designed as gravity dams, with a floor slab in between, as shown in *Figure 9.12(e)*. If the foundation is good, the lock may simply rest on its foundation as shown. If the foundation is poor, which happens often in alluvial valleys, the lock may need a foundation of piles. In that case, the reinforced cross-section of *Figure 9.12(d)* may be preferred because it is lighter, and when it settles it will do so as a unit without cracking between the wall and the floor. Adequate measures must be taken to prevent the water from seeping behind the walls and underneath the floor. For this reason, several rows of sheet piling are embedded in the concrete and extend at least 20 ft. under the floor and at least as much sideways, as the soil is excavated for the construction of the lock. If there is much difference in water level, the analysis of the 'line of creep' shown in *Figure 5.12*, page 178, should be applied. For final design a soils analysis should be made and flow nets sketched, to ascertain that no piping can occur at the downstream end of the lock.

The upper sill of the lock should have approximately the same elevation as the bottom of the upstream canal, so that the ships can pass over it with sufficient clearance. Likewise, the lower sill should have the elevation of the bottom of the downstream canal. Hence, the difference in elevation between the two sills is approximately equal to the lift of the lock. This difference in level should be attained immediately downstream of the upper gates, so that the ships cannot bump into the breast wall when they are being lowered in the chamber. The top of the lock walls should be at least 1 ft. above the highest level of the upstream canal, including wind set-up. If the high level is a large

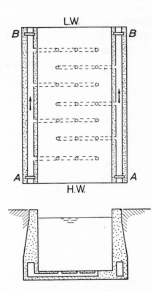

Figure 9.14. Culvert system for filling and emptying a shiplock

lake, reservoir or the sea (as it is in Holland), the highest level including wind set-up may be substantially above normal operating levels. To avoid building the whole lock to that extreme level, only the upper gates with adjacent walls may be built to that level. It is obvious that no shiplocking can take place while such flood levels prevail since they exceed the top of the walls of the lower end of the lock.

There are several types of gates that can be used in a shiplock. The simplest and most commonly used type is the mitre gate shown in *Figure 9.12(b)*. Other types are shown in *Figure 9.13*.

The simplest way to fill and empty a lock chamber is to have a few small vertical sliding gates near the bottom of the mitre gates and to regulate the in and outflow with these sliding gates. This is the way it was done for centuries, and some of the modern shiplocks are still designed in this fashion. The main disadvantage is that for high heads the jets coming out of the upper mitre gates are a great hindrance to the ships that are being locked. For this reason, the sliding gate in the mitre gate has been practically abandoned in favour of valve controlled culverts in the lock walls that connect the upstream and downstream canal. From the main longitudinal culverts, smaller ones take off into the floor of the lock, where they have several exits into the chamber, as shown in *Figure 9.14*. For filling of the lock the valves A are opened and the valves B are closed. For emptying of the lock, the valves A are closed and the valves B are opened. The capacity of such a culvert system becomes quite large if one considers such locks as listed in Table 9.1, where a volume of water of $35 \times 1,200 \times 110$ ft.[3] must be admitted in say 15 min. This is equivalent to a flow of 5,000 cusec. It is considered good practice not to immediately admit or release flows of such a magnitude (or even much smaller). This is for two reasons. First, it is beneficial to the ships in the lock if there is a transition period from zero to full flow of about 5 min. Second, the locking operations create surges on the canals, which would be very detrimental to navigation and moored ships if the flow releases were started instantaneously. For final design of the lock filling system it is recommended that hydraulic model studies of the locks as well as the surges on the canals be made.

CHAPTER 10

WATER SUPPLY

In the previous chapters we have discussed the water requirements of water power, irrigation and navigation. We have seen that water power developments may use nearly all of the available river flow. However, its use is non-consumptive. Irrigation developments may require substantial quantities of water. California, for instance, has 7,000,000 acres of irrigated land, using 20,000,000 acre-ft. of water per year, which is equivalent to an average flow of 27,000 cusec. More than half of this represents consumptive use. Navigation may require large flows when navigable depths have to be maintained through reservoir release. However, such developments are rare. Most navigation projects require relatively small flows, in the order of a few hundred cusec. Moreover, nearly all of this represents non-consumptive use.

In this chapter we shall discuss how much water is required by cities and how this water may be obtained. We shall see that, although the consumptive water use of cities is small, their total water requirements are relatively large. In fact, in some areas in North America, available water supplies may set the limit to future economic development. For the purpose of the present discussion we shall divide the water requirements of a city into three broad categories. First, the requirement for pure, clean water. This is the water that is supplied by the municipal water supply system. Its total flow is relatively small (the city of New York, for instance, has a metropolitan water supply system with a total capacity of about 2,000 cusec). The consumptive use of this water is in the order of 10–20 per cent. The remainder is collected by sewers, treated, and returned to the river system. The second category of water requirement is for industrial use. Large manufacturing plants and thermal power stations have their own water supply system. Since most of this water is used for cooling purposes, the quality does not have to be so high and it would be uneconomic to use water from the municipal supply system. Moreover, the discharges required may be several times larger than what the municipal system could deliver. The third category of water requirement is for diluting the treated municipal sewage and industrial waste, where they are returned to the river system. The quantities of water needed for this purpose are again several times larger than the municipal and industrial uses combined. We shall discuss in the following paragraphs a numerical example to illustrate the above.

WATER REQUIREMENTS

The numerical example shown in *Figure 10.1* and the corresponding description below have the purpose of illustrating the different kinds of water use, their order of magnitude, and their degree of consumption for a typical city of about 200,000 people in North America.

The municipal use is based on a rate of consumption of 100 U.S. gallons per capita per day (gpcd), resulting in an average flow of 30 cusec. This water may be obtained anywhere from a few to a few hundred miles away from the city where a clear lake or stream provides a desirable quality and quantity of water. Municipal use includes: domestic use (drinking, cooking, washing, lawn sprinkling, air conditioning, etc.); public use (public buildings, street cleaning, fire fighting, etc.); commercial use (commercial buildings, hotels, laundries, etc.); industrial use (small industries that cannot afford a private water supply); system losses. The consumptive use of municipal water has been assumed in this example at 18 per cent. The remaining 25 cusec passes through a sewage treatment plant, before it is discharged into the river.

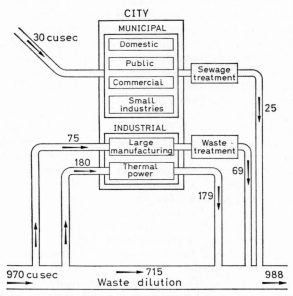

Figure 10.1. Typical water requirements for a city of 200,000 people

The total industrial water of this city is nearly 10 times as large as the municipal use. A flow of 75 cusec is used by large manufacturing establishments such as steel, chemical, or food industries. Most of this water is used for cooling processes. In this example the water is obtained from the nearby river. Another common source is ground water. Water that is to be used for product processing may have to be treated first, depending on the product and on the quality of the river water. The consumptive use of industrial water ranges widely with the type of industry. In this example it has been assumed near the average value of 8 per cent. Part of the remaining 69 cusec may have to be treated before it is returned to the river. The largest withdrawal from the river is made by the thermal power station, for cooling purposes. Practically none of this water is consumed. It is returned to the river unpolluted, but with a slightly higher temperature.

The total amount of impure water that is discharged into the river is about 50 cusec in this example. This includes 25 cusec treated municipal sewage, and about 25 cusec industrial waste water that was used for processing of the products of manufacturing. The remaining 44 cusec was used for cooling processes and cannot be considered as polluted waste flow. This 50 cusec of effluent, although treated, must be diluted in order to ensure sanitary river conditions downstream of the city. The ratio of river flow to waste flow depends, of course, on the degree of sewage and waste treatment. In this example a waste dilution requirement of 715 cusec has been assumed. This water, plus the return flow from the power plant, will prevent downstream river pollution.

We may sum up the entire water budget for the city as follows. The total water supply requirements are 1,000 cusec, divided into 30 cusec for municipal requirements, 255 cusec for industrial requirements, and 715 cusec for water dilution requirements. The consumptive use of the city is only 12 cusec.

It is evident that the total water supply requirements of the cities in a region are not necessarily the sum of the requirements for the individual cities. This would only be so if all cities were situated on different rivers and no two on the same river. On the other hand, if all cities were situated on the same river, with reaches of sufficient length for purification of the water in between, the total water requirements of all cities would only be a little larger than the water supply requirements of the largest city. In practice, we may have to do a good deal of book-keeping before we can establish the total water supply requirements of all towns and cities in a drainage basin. In the following sections we shall discuss the three categories of water requirements in more detail.

Municipal Requirements

The so-called water requirement of people is a flexible figure. In a warm dry climate, where water is not rationed and not metered (the customers pay a flat rate without having their actual use measured), and where it is used for lawn sprinkling, air conditioning, swimming pools and other such amenities of gracious living, a family may consume water at a rate of a few hundred gallons per capita per day. When that same family spends a vacation in a mountain cabin where they have to carry their water in pails from the nearest lake, they will probably reduce their water consumption to a few gallons per capita per day, for drinking, cooking and washing. In general, we may observe that water requirement depends mainly on the following circumstances. First of all the standard of living. If people become wealthy enough to buy an air-conditioning unit, they do not mind paying a few extra dollars for water. Second, the climate. In a warm, dry climate there is a much greater demand for water than in a cool, moist climate. Third, the cost of water. Although the monthly water bill is a relatively small portion of total living expenses, it has been found from experience that water consumption will go down when prices go up. In fact, when cities change over from an unmetered to a metered water distribution system, the total water consumption may decrease by 20–30 per cent.

A reasonable average figure for household use in North America is 50 gpcd. Since we are mostly interested in the water use of the whole city we must add to this figure the public use of water (public buildings, fountains, fire fighting,

389

street cleaning); the possible losses in the water system; the commercial and small manufacturing use in the city; and the agricultural use on the fringe of the city if it is also served by the municipal water system. If all these conditions are near average, the resultant water consumption figure may be in the order of 150 gpcd for large cities and 100 gpcd for smaller cities and towns.

It may be of interest to quote the following figures from the U.S. Senate Select Committee on National Water Resources (1959). The water consumption in large cities with substantial industries, like Philadelphia, Detroit, and Chicago is in the order of 200 gpcd. A large city in a warm, dry climate, like Salt Lake City may use as much as 225 gpcd. Cities that are not particularly industrialized and in the northern part of the U.S., like Boston and Minneapolis, consume in the order of 100 gpcd. The overall average for the U.S., dividing the total municipal water supplies by the total population, is 100 gpcd.

Industrial Requirements

The total yearly water intake for all manufacturing establishments (not counting the electrical industry) in the U.S. in the year 1959 was in the order of 15,000 billion gallons, or approximately 250 gpcd. The three major water-using industries were steel, chemicals and paper as may be seen from Table 10.1.

Table 10.1. *Industrial Water Use in the U.S.*

Industry	Water Use (gal. per ton prod.)	Water Used For			% of Total
		Product (%)	Cooling (%)	Misc. %	
Steel	30–50,000	24	72	4	28
Chemicals	10–400,000	7	88	5	24
Paper	50,000	68	21	11	16
Food	1–50,000	25	54	21	6
Other	—				26
					100

The total yearly water intake for thermal electric power stations in the U.S. for the year 1959 was in the order of 36,000 billion gallons, or approximately 600 gpcd. To amplify this figure it should be noted that power development in the U.S. is predominantly thermal. If this total water intake is divided by the total annual thermal electric energy output, we find a water requirement of about 50 gal./kWh. Nearly all of this water used in thermal plants is for condensing the steam.

It may be seen from the above that more than 90 per cent of all the water that is used for the manufacturing industries and for thermal power stations is for cooling purposes. Fortunately, there are several measures that can be applied, if needed, to reduce the water requirements for cooling purposes. One is to construct an artificial reservoir or pond to recirculate the cooling water. Another is to build cooling towers for the cooling water. By such measures, some industries have reduced their water requirements to less than one-tenth of what they used to be. Another possibility is to design the machinery for the use of saline cooling water and use sea water or brackish water or other water of poor quality for cooling purposes. Thus we would release water of good quality for other more important purposes. We may therefore conclude that the water requirement figures, listed in the above, may be applied when there is ample water in the drainage basin. However, when water is scarce, these figures can be substantially reduced.

To illustrate the above, it is of interest to quote from the California Department of Water Resources (1957), where it is estimated that a future California population of 40,000,000 will consume for municipal and industrial purposes a total of 8,500,000 acre-ft. per year. Assuming no re-use of water, this would come to a figure of 190 gpcd, which seems none too high, allowing for the warm dry climate and the tendency of increased water demands with increased living standards. However, on a region-wide basis a good deal of re-use may be possible, which would increase the available water per capita. Such re-use may take the form of treated waste water being pumped on to selected areas for percolation into underground aquifers from where it may be pumped out again at a later date when it is purified.

Waste Dilution Requirements

In the early days of city development it was common practice to dump the raw municipal sewage and industrial waste in the river, taking full advantage of the self-purifying capacity of the stream flow. As long as the river discharge was at least 20 times the combined waste flow, and if there was a sufficiently long reach of river to the next city, there would be sufficient natural treatment to prevent nuisance conditions, to maintain fish and aquatic life, and to protect downstream water supplies.

However, with the growth of cities and their industrial activity, and with the increased use of rivers for recreational purposes, the practice of dumping raw sewage into the rivers is rapidly disappearing from the North American scene. The problem is now to what extent sewage should be treated before it is discharged into the river.

Modern sewage treatment involves four different steps. Primary treatment removes all floating and coarse suspended solid matter by screening, skimming and sedimentation. Secondary treatment oxidizes all colloidal and dissolved organic matter. In the third step the water is disinfected by chlorine. In the fourth step all remaining solids are removed by processes such as lagooning. Sewage treatment can be applied with different degrees of effectiveness, and of course different degrees of cost. The minimum treatment that is applied will remove 50 per cent of the solids. The maximum that can be obtained with modern methods is 97 per cent removal. The approximate annual cost of such treatment for a sewage flow of 1 cusec is shown in *Figure 10.2*.

The treatment of industrial waste may be different from the treatment of municipal sewage. The industrial water that is used for cooling, needs, of course, no treatment at all. Only the water that is used in the processing of the product is considered industrial waste. This waste flow may contain organic as well as inorganic matter. In the last case, some chemical or mechanical rather than biological treatment may be required to clean the water. When discharged into streams, industrial waste flows may cause more nuisance than municipal waste flows. Hence, adequate control over treatment measures is required. The cost of treating industrial wastes may vary widely. For a rough approximation the values of *Figure 10.2* may be used.

391

The amount of river flow required for the satisfactory dilution of waste flow from municipal sewage and industry, is a function of the oxygen content, the dissolved minerals, and the temperature of the river water, and of the length of the downstream reach that is available for the purification action of the flowing stream. For average conditions it has been found that the ratio of river flow to waste flow should be approximately as shown in *Figure 10.2*. Raw sewage requires a ratio of 40, while thoroughly treated sewage requires a ratio of only 2.

Assuming that we have made a study of available storage capacity in the drainage basin and that we know the annual costs of providing a given dependable flow, we are now in a position to determine the most economic treatment of the city waste. For every degree of waste treatment we can

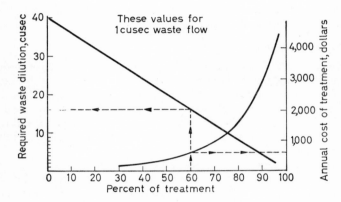

Figure 10.2. Waste dilution requirements

determine the waste dilution requirements. This figure is converted into annual cost of storage to produce that flow (after having allowed for other flow uses). To this figure is added the cost of treatment. This is repeated until the lowest total cost is found.

If the conservation measures, discussed in the foregoing, were applied to the example in *Figure 10.1*, the city water supply requirements would be as follows: municipal use—the same at 30 cusec; industrial use reduced from 255 to 30, because of the recirculation of cooling water for all purposes. The waste dilution in the river reduced from 715 to 120, because of maximum sewage and waste treatment. Thus, the total city water requirements are reduced from 1,000 cusec to 180 cusec. The consumptive use remains 12 cusec. Hence there is still 168 cusec available for downstream use.

The figures that have been used in the above diagrams and examples are very approximate. They should be applied only to establish the order of magnitude and relative importance of city water requirements. For more precise purposes one should consult the professional literature and make a more elaborate analysis. Among the more important recent documents on water requirements are the reports by the U.S. Senate Committee that were mentioned earlier and that are listed in the Bibliography.

Quality Requirements

Cities usually have a single system of pipes for their municipal water supply. From an economic viewpoint it might be advantageous to have a dual supply system since water for such purposes as lawn sprinkling, street cleaning and car washing does not have to be of the same high quality as water for household use. However, public health officials would strongly object because of the danger of cross connections and the resultant danger to public health. Therefore, all municipal water must be of drinking water quality. This means that the water must be free of bacteria, that it must be free of sediments, that is, should not contain an excessive amount of dissolved solids, and preferably that it has no odour, no colour and a pleasant taste. Many cities import their drinking water over distances of more than 100 miles to obtain such quality standards, rather than use nearby river water with a high mineral content and an unpleasant taste. It is evident that careful control of pollution must be exercised in the area of the water supply source. In spite of those measures, some form of water treatment may be required before the water can be consumed.

Industrial water for cooling purposes should preferably have a low temperature and be free of corrosive chemicals, or sediment. However, it was noted earlier that the machinery can be built to withstand the chemicals. Alternatively, it may be more economic to treat the water and recirculate it via cooling ponds. Suspended sediments can be removed in sedimentation tanks. Industrial water for processing may have to meet special requirements, depending on the product.

Water for diluting waste flows will perform its purification function more efficiently when it has a high oxygen content. For instance, if river water had a dissolved oxygen content of zero, its rate of purification would only be half of what it is with an oxygen content of 4 mg/l., which is normally available. Since warm water contains less oxygen than cold, the amount of diluting water must be greater in summer than in winter.

SURFACE WATER SUPPLIES

In humid climates with ample amounts of surface run-off, all or nearly all of the three categories of city water requirements may be obtained from surface streams. In arid climates, on the other hand, most of the municipal water is derived from ground water. A good deal of industrial water may also come from wells. Even some of the water used for waste dilution may have been indirectly obtained from ground water sources. In this section we shall discuss how city water may be obtained from surface streams, and in the next section we shall discuss ground water as a possible source.

Stream Flow

The most economic source of water for a city is a natural stream or river, close by. The two criteria that we must apply to judge the suitability of the stream are: the quality of the water, which was discussed earlier, and the dependability of the flow. Without storage capacity, the water requirement of the city plus downstream riparian requirements must be less than the lowest recorded flow of the river, including such periods as the dry thirties. Important factors in the dependability of stream flow are the size of the drainage area, the climate, and the characteristics of the drainage basin. The first two factors are obvious: the larger the drainage basin and the wetter the climate, the larger the dependable flow. As far as the characteristics of the drainage basin are concerned, the soils and geology play the most important role. Porous soils, like sands and gravels, will absorb precipitation like a sponge and release this water slowly. Impervious soils, like shales, hardly allow the rainfall to infiltrate, and most of it runs off during and shortly after the rainstorm. All other circumstances being the same, a drainage basin with sandy soils may yield a dependable flow that is 10 times as large as the dependable flow from a basin with impervious soils.

Reservoirs

If the dependable flow of a watershed is insufficient to satisfy the estimated demand, but the average flow is ample, the demand may be met by constructing a reservoir. In Chapter 2: Hydrology, we discussed how the storage capacity of a reservoir can be determined in order to produce a given dependable flow.

Another benefit of a reservoir to the water supply objective is the purification of the water. First of all, the suspended sediment will settle out. Second, there will be a reduction in the amount of bacteria in the water. The extent to which such improvements take place depends upon the length of the storage period; that is, the volume of the reservoir divided by the rate of flow.

It is obvious that recreational use of reservoirs that are to supply drinking water must be restricted. This may take the form of allowing boating and swimming only in areas that are several miles away from the waterworks intake. Moreover, it may be desirable to prohibit the use of outboard motors since they leave a residue of oil and stir up the bottom and shores.

Apart from small seasonal fluctuations, the demand for water (municipal, industrial and waste dilution) is fairly constant. Therefore, the operation of water supply reservoirs is relatively simple. If it is a single-purpose reservoir, we simply release the flows that are required. The level of the reservoir will fluctuate with the changing inflows. When the inflow is consistently large, the extra water will flow over the spillway. When the inflows are consistently small, the reservoir level will keep dropping. We have ascertained with a mass curve analysis that the storage capacity is large enough to maintain the required flows during the driest period on record. If such a period would occur again, the reservoir would be empty at the end of that period, but there would be no flow deficiencies. If we are dealing with a multi-purpose reservoir, a certain portion of that reservoir will be allocated to water supply, and the operation of that portion will be as described above.

Aqueducts

A municipal water supply system may consist of the following components: an intake structure from a reservoir, a natural river channel, or a ground water well; aqueducts to convey the water from its place of origin to the city; possibly a treatment plant for filtering, disinfecting and softening the water; a city distribution system including small reservoirs, pumps, water towers, water mains, valves, meters, and taps.

An industrial water supply system may consist of the same components as a municipal supply system. However, in most cases the water quality requirements are not so rigid and the required quantities are much larger. As a result we may find a simple water intake from the nearest river or ground water well, a short pipeline to the factory, probably no treatment plant, and a simple distribution system towards the points of use.

A waste dilution supply system is in most cases the river itself. If the minimum recorded flows do not meet the city requirements, it may be necessary to construct upstream reservoirs or to effect a diversion of water from another drainage area.

In the following paragraphs we shall discuss some aspects of conveying water from its place of origin to cities by means of aqueducts. In the broad sense of the word, an aqueduct is an artificial conduit to convey water from its place of origin to its place of destination; including open conduits as well as closed conduits. The open conduits may be earth or rock canals, or open flumes supported by trestles, bridges or arches. The closed conduits may be wood stave, masonry, concrete or steel conduits above or under ground, or tunnels excavated in rock. The use of aqueducts for conveying water from springs to places of habitation dates back to several centuries B.C. They were developed to near perfection during the Roman Empire, when a system of underground conduits plus a system of elevated aqueducts on arches brought into the city of Rome, over distances ranging from 10–60 miles, a total flow of 100 cusec. Ancient conduits were mostly free-flowing open channels with a mild gradient, or masonry conduits under low pressure. High pressure artificial conduits were not developed. Therefore, when it was necessary to cross a valley, the channel was continued on its predetermined level and gradient in a masonry flume, supported by masonry piers and arches, until the other side of the valley was reached, whereupon the aqueduct was continued as an open canal on the ground, following the contours of the land. It was not until the development of modern techniques that aqueducts could become a succession of open channels, high-pressure conduits and tunnels.

Some modern municipal water supply systems convey substantial quantities of water over respectable distances. Los Angeles, for instance, is served by two aqueducts. The Los Angeles Aqueduct imports 500 cusec from the Sierra Nevada Mountains over a distance of 240 miles. The Colorado River Aqueduct imports 1,500 cusec from the Parker Dam on the Colorado River over a distance of 242 miles. These aqueducts include open concrete-lined channels; nearly 100 miles of concrete-lined tunnels; pumping plants with heads of several hundred feet; and reinforced concrete siphons up to several miles in length. The City of San Francisco has an aqueduct with a capacity of 600 cusec that imports water from the Sierra Nevada Mountains over a distance of 250 miles. Included in this aqueduct are two welded steel siphons across the San Joaquin Valley, 47 miles long and with a maximum head of 500 ft. The city of New York imports its drinking water from the headwaters of the Delaware River, over a distance of 125 miles. The water is collected in five reservoirs and conveyed through aqueducts with a total capacity of 1,500 cusec. Included in these aqueducts are open channels, pressure conduits and tunnels.

The most grandiose water supply scheme will be the California Water Plan, which is scheduled to be built during the forthcoming decades. In this plan some 10,000 cusec will be transported from northern California, where a surplus of water is available, to southern California, where a deficiency of water is expected in the near future. This water plan includes a few hundred reservoirs, 5,000 miles of canals, 600 miles of tunnels, 100 hydro-electric plants to drive 75 pumping plants, including a lift of 3,300 ft. over the Tehachapis Mountains. The total cost of all works, to be built over a period of 50 years, is estimated at 12 billion dollars.

GROUND WATER

It was noted earlier that a substantial amount of municipal and industrial water is obtained from wells. In large areas in North America, particularly in the Middle West where dependable surface run-off is not very large, more than half of the municipal and industrial requirements are obtained from ground water. In this section we shall discuss qualitatively to what extent ground water supplies can supplement the surface water supplies of a drainage basin. First we shall review a few of the more elementary principles of ground water movement.

Hydraulic Principles

It has been known for a long time that the head loss in pipe flow is proportional to v^2 when the flow is turbulent; and proportional to v when the flow is laminar. This empirical knowledge has led to the Darcy–Weisbach formula for pipe flow:

$$hf = f \cdot \frac{L}{D} \cdot \frac{v^2}{2g}$$

Only when the flow is fully turbulent is the head loss proportional to v^2 and hence the friction factor f constant. For laminar flow conditions the friction factor f becomes:

$$f = \frac{64}{R_e} = \frac{64 \times \nu}{V.D}$$

which makes the original formula:

$$hf = \frac{64 \times \nu}{V.D} \cdot \frac{L}{D} \cdot \frac{v^2}{2g}$$

or:
$$\frac{hf}{L} = S_e = C.V$$

In other words: the gradient of the energy line is proportional to the velocity to the first power. It has been found that under normal ground water gradients the flow is always laminar. In nature, rates of flow of more than a few feet per day are exceptional. Hence it may be expected that the energy gradient of ground water flow is also proportional to the velocity to the first power. This law has been expressed for ground water by Darcy in the form:

$$V = K.S$$

and
$$Q = A.K.S$$

V is the hypothetical velocity that would take place if the discharge Q flowed through the cross-sectional area A, without the presence of soil particles, in feet per second. The true velocity of the ground water could be obtained by dividing V by the soil porosity, in per cent.

K is the coefficient of permeability, in the same units as V, hence in feet per second. The approximate value of K, observed from laboratory and field tests, is as follows:

| Clay | $10^{-8}-10^{-10}$ ft./sec | Coarse Sand | $10^{-2}-10^{-4}$ ft./sec |
| Fine Sand | $10^{-4}-10^{-8}$ ft./sec | Gravel | $1\ -10^{-2}$ ft./sec |

S is the energy gradient or hydraulic gradient, which is for practical purposes the same since the velocity head is negligible.

Q is the discharge, in cubic feet per second.

A is the cross-sectional area of the 'soil tube' under consideration, measured perpendicular to the direction of flow, in square feet.

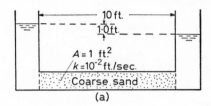

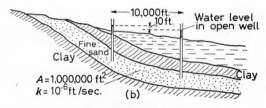

Figure 10.3. (a) Discharge through soil sample; (b) Discharge through aquifer

To illustrate the use of this equation, let us compute the discharge through an element of sand tested in a laboratory, as shown in *Figure 10.3(a)*. The length of the sample is 10 ft., its cross-sectional area is 1 ft.², the coefficient of permeability is 10^{-2} ft./sec, and the hydraulic gradient is 1 ft. in 10 ft. Hence the discharge is:

$$Q = A.K.S = 1 \times 10^{-2} \times 0 \cdot 1 = 0 \cdot 001 \text{ cusec}$$

In *Figure 10.3(b)* is shown a confined aquifer (a porous water-bearing geologic formation enclosed by relatively impervious formation) of fine sand 100 ft. deep and 10,000 ft. wide, with a hydraulic gradient of 5 ft. per mile, and a coefficient of permeability of 10^{-6} ft./sec. Hence the discharge is:

$$Q = A.K.S = 1,000,000 \times 10^{-6} \times 0 \cdot 001 = 0 \cdot 001 \text{ cusec}$$

It may be seen that the discharge through the laboratory example is the same as the discharge through the aquifer. The fact that the cross-sectional area is a million times as large is offset by the smaller coefficient of permeability, and the smaller gradient.

If water is pumped out of a well, as shown in *Figure 10.4(a)*, we can develop the following relationship. The discharge towards the well, passing a vertical cylindrical section with radius x, is:

$$Q = 2.\pi.x.y.\frac{dy}{dx}$$

or:

$$\frac{Q}{2.\pi.k}.\frac{dx}{x} = y.dy$$

When integrated:

$$\frac{Q}{2.\pi.k}.\ln x = \frac{y^2}{2}+C$$

Since $y=H$ for $x=R$ and $y=h$ for $x=2$:

$$Q = \frac{\pi.k.(H^2-h^2)}{\ln R-\ln r}$$

This formula is known as the Dupuit formula.

The corresponding formula for pumping a well in an aquifer, as shown in *Figure 10.4(b)*, is:

$$Q = \frac{2.\pi.k.d(H-h)}{\ln R-\ln r}$$

If, in the above formulae, the quantities k, H, h, R, r, and d, are known, the discharge Q can be computed. The radius R would be known if pumping took place in the centre of an island, surrounded by an open fixed water surface. However, wells are normally situated in places where the cone of depression is not limited by fixed boundaries. For such cases the following empirical formula for R may be used:

$$R = 1,700.S.\sqrt{K}$$

wherein S is the amount of drawdown in the well, as shown in *Figure 10.4*. If this drawdown was 10 ft., and the coefficient of permeability $K=10^{-2}$ ft./sec., the radius of the cone of depression would become 1,700 ft. It may be seen from the well discharge formulae that a substantial error in the empirical formula for R will only result in a small error in the computed discharge.

Another application of the Dupuit formulae is to determine the coefficient of permeability. The formulae are written in the following form:

$$Q = \frac{\pi.K.(y_2^2-y_1^2)}{\ln x_2-\ln x_1} \quad \text{and} \quad Q = \frac{2.\pi.K.d.(y_2-y_1)}{\ln x_2-\ln x_1}$$

The drawdown ground water level or piezometric level y_2 and y_1 are measured at the distances x_2 and x_1 from the well, the discharge from the well is measured, and then the coefficient of permeability K can be computed.

It should be pointed out that the derivation of Dupuit's formulae assumes parallel flow lines. If the drawdown S in *Figure 10.4(a)* is more than about half the value of H, the stream lines becomes so curved near the well, that the formula is no longer applicable. In fact, the open water level in the well may be drawn below the point where the flow lines enter into the well.

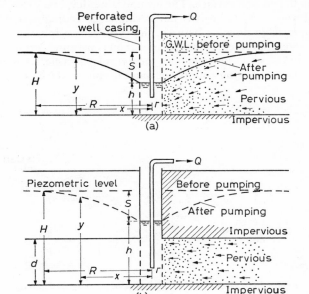

Figure 10.4. (a) Pumping from a well in an unconfined pervious formation; (b) Pumping from a well in a confined aquifer

Ground Water Budget

The occurrence of ground water may begin immediately under the ground surface and will extend as deep as pores in the soil are maintained. Thus, the thickness of the ground water may range from practically zero in rocky regions to several miles in geologically depressed areas. The majority of the soils consist of shales, silts and clays. The ground water contained in these soils is practically unavailable to man because of the extremely slow movement of water through these soils, even under large gradients. Hence we are primarily interested in the ground water contained in the porous formation, such as fine sand, coarse sand, and gravel. Such permeable formations are called aquifers. If they are exposed to the surface, they are called unconfined aquifers. If they are enclosed, above and below, by impervious layers, they are called confined aquifers, as shown in *Figure 10.5*.

Confined aquifers may find their origin in ancient river valleys that were subsequently covered with sedimentary deposits of silt and clay in lakes or seas. They may also be glacial stream gullies, filled with gravel and coarse sand and subsequently covered with glacial till. When confined aquifers are extremely porous and when the ground water is subject to a steep gradient, the aquifer may be popularly referred to as an 'underground river'. Nevertheless, the velocity of water in such an aquifer may only be a few feet per day, and its total discharge in the order of one cusec or less.

We may look upon natural ground water as being a huge underground reservoir that is filled to overflowing. The water content of the reservoir may be a hundred or a thousand times as large as the annual amount of precipitation that falls on the ground surface. The ground water level may slowly rise during a period of wet years, and slowly fall during a succeeding period of dry years with an appreciable time lag between the cycle of precipitation and the cycle of ground water movement. This phenomenon constitutes one of the beneficial aspects of ground water movement in the hydrologic cycle, inasmuch as creeks and streams keep receiving a supply of water in spite of the lack of precipitation.

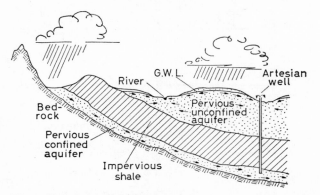

Figure 10.5. Ground water movement in confined and unconfined aquifers

The main supply of water to the ground water reservoir is from that part of the precipitation that percolates towards the ground water surface. This portion may range from practically nil during a light shower on a dense soil with a low ground water table, to practically 100 per cent during a prolonged rainfall on a porous soil with a relatively high ground water table. Other sources of supply (or recharge as it is often called in the ground water literature) may be seepage from streams, lakes and ponds. However, this is rather uncommon since the natural watercourses are normally the recipients of ground water flow. When considering a particular ground water basin, one may have to explore the possibility of sub-surface inflow from adjacent basins. When significant quantities of water are released on the ground surface through irrigation, reservoirs, or industrial waste disposal, they should be taken into account.

The main discharge from the ground water reservoir is towards springs, streams, lakes, marshes, or directly into the sea. Normally, the ground water table is relatively close to the ground surface and well above the level of the nearest stream. During times of relatively high precipitation, the ground water level rises, the gradient towards the streams increases, and the discharge into springs, or on to low-lying lands, or into the stream channels proper, increases. During times of relatively low precipitation, the reverse process takes place. In this fashion, there is a continuous flow of water from the highest points of the undulating ground water table towards the lowest points, where the water emerges to the open surface. An example of such a flow pattern is given in *Figure 10.6.*

400

Part of the ground water may get lost from the underground reservoir through evapo-transpiration rather than through discharge into open water courses. If the water table is close to the ground surface, ground water may be within reach of the roots of the plants and thus return to the atmosphere before it can flow away. If the water table is well below the root zone, the ground water may slowly flow towards places where it can be reached by the plant roots.

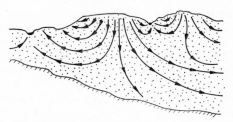

Figure 10.6. Ground water movement in uncon-fined aquifer

When considering a particular ground water basin, one may have to explore the possibility of underground discharge into an adjacent basin. When significant quantities of ground water are extracted by means of artesian wells or pumping, these should be taken into account in making up the ground water budget of the drainage basin.

Ground Water Availability

It is obvious from the above sketch of the water budget of a ground water reservoir, that if we start pumping water from this reservoir at any place at any time, we shall eventually reduce other forms of discharge from this reservoir. Over a long period of time the supply to the ground water reservoir must equal the discharge from that reservoir. If we add a new form of discharge, in the form of wells, we are bound to reduce the natural ground water discharge to streams, or the availability of water to plants, or we may lower the ground water table.

It is conceivable that the water pumped from a well has a great economic value, while the alternative natural discharge of that same water has no economic value. For instance, a deep aquifer may drain into the sea, or into a large lake, or into a large river with ample discharge. Another possibility is that ground water normally drains towards, and emerges into, saline creeks and lakes where it cannot be used for any purpose. If in such cases the ground water is intercepted and pumped to the surface and used for industrial or other purposes, no harm is done. However, such cases are the exception rather than the rule.

It happens more often that ground water is extracted without having given due prior consideration to the detrimental effect it may have on other interests. These detrimental effects may include: the reduction of stream flow, particularly the low water flows; a lowering of the water table and consequent damage to agricultural interests; the intrusion of salt water underneath the fresh water in coastal areas.

There are several reasons why such unfortunate situations have developed. First of all, the interests that desire to use ground water (industries, municipalities) are usually different from the interests that are affected (farmers, other municipalities). In many regions, no adequate legislation has been established to prevent or regulate the extraction of ground water. Therefore, the interests that benefit from ground water use can proceed with the pumping of wells, in spite of the fact that other interests suffer. Secondly, it should be pointed out that it is an extremely difficult problem to predict in advance and quantitatively what stream flows will be reduced by how much; and at what time and locations the ground water table will be lowered by how much. The only way to find a conclusive answer is by field observations! Now we run into another difficulty, namely that it requires a very long time before equilibrium conditions are established. It was pointed out earlier that the total water content of a ground water reservoir may be many times the annual rainfall or the annual surface run-off from that basin. The movement of the water in the underground reservoir is very slow. Over hundreds of years a delicate equilibrium between supply and discharge has been established. If we suddenly come along and begin to extract water somewhere, it may require decades, if not centuries, before a new equilibrium is established. By the time we find out that the detrimental effects are more harmful than the gain from the wells, it may require another few decades of corrective measures before the natural conditions are re-established.

The conclusion from this discussion is that all pumping from wells should be made subject to legislative control; that comprehensive studies should be made of the natural ground water budget in the region under consideration, and the possible changes in stream flow and ground water levels, resulting from different amounts of ground water use in different places. This study may include such measures as the recharge of the ground water table by dumping the waste flows back on to the soil for infiltration, rather than discharging it into surface channels; or the importation of water from outside the region, to be released on to selected areas in the drainage basin. The resultant ground water flow may be the least costly method of conveying the water to its place of destination. Moreover, the water will be purified while it flows through the ground.

Let us assume that we have completed, for a certain region, the studies discussed in the above paragraph. The problem then arises to what extent the pumping from wells can be permitted. We may deal with this problem from a viewpoint of ground water availability and from a viewpoint of economics.

From a viewpoint of ground water availability, we may look upon the problem as follows. There is a limit to the yield of water from one individual well, as is evident from inspection of the Dupuit formulae. The main limiting factor is the permeability of the ground. In the beginning of the operation of the well, the yield is larger (since the drawdown profile has a steeper gradient) than later on. After a long time (anywhere from several months to several years, depending on the geologic formation) a constant yield will be reached. From the viewpoint of the owner of the well, and ignoring what happens elsewhere in the basin, and assuming that no other wells will begin to operate in the vicinity, this constant discharge is the 'safe yield' of the well. There is a good deal of theoretical and empirical literature on this subject, mainly aimed at determining the safe yield from wells.

When a second well is sunk very close to the first one, it is obvious that the safe yield of the two is not much greater than that of the first one alone. When the wells are placed a good distance apart, the safe yield of the second may be nearly as large (or larger, if the geologic formation is more favourable) than the first one. Let us now assume that we have a great number of wells throughout the entire region under consideration. Provided that the carrying capacity of the aquifers is ample, it may very well be that the total of the 'safe yields' of the individual wells, exceeds the ground water availability in the region. It was noted earlier that the supply of the ground water reservoir in a region is from the precipitation that percolates down to the ground water table; from seepage, from streams, rivers and lakes; and from underground flow from outside the region. This total supply, or recharge, of the region constitutes an upper limit of 'safe yield', for the region as a whole. It also means that when the total discharge from all wells reaches that magnitude, and when the waste flow of this water is not artificially recharged into the ground, all other natural discharges from the ground water reservoir will terminate as soon as equilibrium conditions set in.

If the total pumping capacity of the wells exceeds the natural and artificial recharge of the region, the result will be a gradual lowering of the ground water table. This is referred to as 'mining' of ground water. Actually, only the difference between the total well discharge and the total recharge is the water that is being mined. This water can only be used once, and will never be replenished, as long as the well discharge is not decreased.

Let us now look at the problem of ground water availability from an economic viewpoint, and let us first assume the extreme case where the natural discharge from the ground water reservoir has absolutely no value, neither at the present, nor in the foreseeable future. In that exceptional case it would be justifiable to regard the ground water as a non-renewable resource, and to consider where and at what rate it should be extracted to yield the greatest economic value. An example of such a situation is the Llano Estacado basin in New Mexico where the natural supply to the ground water reservoir is in the order of 100,000 acre-ft. per year, and where the total discharge from wells is in the order of 5,000,000 acre-ft. per year. Since most of this water is used for irrigation, and therefore consumed, it is evident that there is a substantial amount of 'overdraft' (the difference between pumping capacity and ground water supply). It has been estimated that the total amount of ground water available in accessible aquifers will be exhausted in about 40 years.

Normally, the natural discharge from a ground water reservoir (stream flow and evapo-transpiration) does have value and the problem becomes one of maximizing the total benefits from natural discharge and well discharge. Moderate amounts of ground water, extracted in areas where surface streams are non-existent or unreliable, may yield substantial benefits, while the resultant reduction in stream flow and lowering of ground water level may cause damages of a much smaller magnitude. The next increment in well discharge will probably result in relatively smaller benefits and relatively larger damages, and soon the point is reached where the increment in benefits from increased pumping is smaller than the corresponding increment in damages. It is obvious that we should not exceed that limit. In view of the difficulties involved in analysing the situation, and in view of the long periods of time involved before equilibrium conditions are reached, it would be advisable to approach these problems with a great deal of caution and to maintain a close control over the operation of the wells.

From a viewpoint of basin-wide water resources development, the use of ground water is of limited interest. What we gain from wells, we lose somewhere else. Mining the underground reservoir yields only temporary benefits, and cannot become an integral part of a water development plan. Moreover, the total amount of water that can be extracted from the ground is relatively small, compared to the surface run-off. Normally, we do not want to harm agricultural interests or damage the natural vegetation in the drainage basin. In other words, we should not reduce the evapo-transpiration discharge of the ground water reservoir. This leaves only the discharge from the ground water reservoir into the open-stream system. The magnitude of this flow can be appraised by reviewing, in the stream flow records, the periods where all surface run-off had ceased to take place. Such periods include the latter part of dry spells, or freezing spells. Normally one will find that the residual flows in the river system during such times are in the order of say 10 per cent of the average flow. Assuming that we do not wish to completely dry up these low flows, we may conclude that the permissible ground water extraction in a drainage basin is only in the order of a few per cent of the average stream flow originating from that area.

CHAPTER 11

ECONOMIC ANALYSIS

The purpose of this chapter is to provide insight into the economic aspects of water resources development. It is quite likely that in a given drainage basin several water development schemes are technically possible and that all of them would provide great benefits to the people. However, the funds of society to launch large public undertakings are limited, and there are competitive uses for those funds. Therefore every project must be carefully analysed on its economic merits before the go-ahead signal is given; and when several alternative projects would serve more or less the same purpose a careful cost comparison must be made.

It would be fairly correct to state that the development of the civil technology, in contrast to the military technology, is primarily governed by economic considerations. If it had been profitable to do so, we might already have a traffic tunnel underneath the English Channel, connecting England with France; and a power dam between Spain and Africa. The Congo River might have had a hydro-electric development 10 times larger than on any other river in the world, and the Sahara might have been irrigated. Pleasant as it may be to indulge in such daydreams as a form of relaxation, as soon as the civil engineer is responsible for spending public funds, he must apply very strict reasoning to justify his decisions. This reasoning is not necessarily restricted to economic analysis alone. He must take social aspects into consideration. Although he does not bear the sole responsibility for decisions regarding the development of national resources, the engineer should feel accountable for having made the correct recommendations, based on an appraisal of the problem from all possible viewpoints. We could define his responsibility as follows. The aim of the engineer should be to use the available resources of brain-power, man-power and materials in the most efficient manner, in order to create the physical living conditions that are needed for the spiritual, intellectual and moral development of mankind.

The economic analysis of water development projects may be performed in several ways, depending on the nature of the problem. When the function of the project is predetermined, the problem is rather simple. For instance, when water power has to be developed on a certain reach of river with a given head and a given load factor, the technical problem merely becomes one of comparing the cost of alternative designs. However, when the function of the project is not predetermined, the problem becomes more complicated since not only the costs but also the benefits of the different alternatives have to be taken into consideration. To facilitate the discussion of cost comparison and benefit–cost analysis, we shall first review some of the highlights of the mathematics of finance.

MATHEMATICS OF FINANCE

Interest

Interest can be looked upon as a reward for having made capital available to someone who needs it. Some of us may have been introduced early to the meaning of this statement when mother, near the end of the month, pleaded for a loan from our piggy bank, to be returned on pay day with an appropriate reward for the services rendered. The rate of interest depends largely on the circumstances. When a father gives a loan to his son he may do so without charging any interest at all. Likewise, a wealthy country may extend a loan to a friendly but poor country, without charging interest. On the other extreme, a person without any securities, and in great need of money, may fall in the hands of a loan shark and may have to pay 10 per cent per month. Normally, the rate of interest depends on two main factors: the state of the economy, and the risk involved in the loan. During times of an expanding economy, when profits are high, there is great demand for capital to build new factories or to open up new mines, or to develop new products. This great demand for the always-limited amount of capital will likely result in high interest rates. On the other hand, during times of economic stagnation, there is little demand for capital and as a result, interest rates may be relatively low. For instance, during the depressed thirties, the interest rates on federal government bonds went below three per cent per year, while during the booming fifties the interest rate on government bonds went above five per cent.

The other reason for different interest rates, as noted above, is risk. The federal government, being the most dependable borrower, will be able to obtain funds at the lowest current interest rate, say 4 per cent. A private company, on the other hand, may have to pay 6 per cent interest, under the same economic conditions, while a private person, engaging a second mortgage on a house, may have to pay as much as 8 per cent.

When i represents the annual interest rate and P represents a present sum of money, while n represents a number of years, then the interest at the end of one year is $i.P$. When the investor chooses to withdraw his interest every year, he will have collected after n years a total amount of interest of $n.i.P$.

However, when the investor does not withdraw his interest and is able to invest his interest at the same interest rate as his original investment, then we may observe that the original P has increased to $P+i.P$ after one year and to $P(1+i)+i.P(1+i)=P(1+i)^2$ after two years and to $P(1+i)^3$ after three years. Hence we find that the amount F to which an original investment P has increased when subjected to compound interest is:

$$F = P(1+i)^n \qquad \ldots.(11.1)$$

Some numerical values of this formula may be found in Table 11.1.

Present Value

From Table 11.1 it may be seen that when $100 is invested now, on a compound interest basis, at an interest rate of 4 per cent, it will have obtained a value of $219 after 20 years. Therefore, if one were asked what he would rather have, $100 now or $219 twenty years from now, the answer would be indifferent (assuming, of course, that the person in question is in no great

406

immediate need of funds and is assured of a continuous 4 per cent interest). We may now turn the situation around and observe that the equivalent of receiving $219, twenty years from now, is to receive $100 now. The financial language of making this same observation is to say that the present value of $219 twenty years from now, at a discount rate of 4 per cent, is $100. The algebraic language of saying the same thing is the reverse of eqn. (11.1)

$$P = \frac{F}{(1+i)^n} \qquad \qquad \dots (11.2)$$

Some numerical values of this formula may be found in Table 11.1. It may be seen that the present value figures can be obtained by dividing the corresponding compound interest figures into 100.

In the economic analysis of engineering projects the concept of present value is often used to compare estimated costs that will occur at different times. For example, a hydro plant is at present under construction. The incremental cost of the sub-structure for a future power house extension is estimated at $1,000,000. The cost of constructing that same sub-structure 10 years from now, when the plant extension is needed, is estimated at $2,000,000. Which of the two possibilities is more attractive? If we use a discount rate of 5 per cent, we find that $2,000,000 ten years from now, is equivalent to $1,230,000 at the present time. Hence, the present extension at a cost of $1,000,000 would be more attractive. However, if it is likely that the plant extension is not needed until 20 years from now, the present value of the future expenditure becomes $755,000. Hence, it would be advisable to delay the construction. We could observe that the 'break-even' position is at 14 years. If there is more than an even chance that the plant extension is needed before that time, the sub-structure should be built now. If not, the construction should be postponed till the extension is needed.

When comparing engineering projects that have different annual costs from year to year, as will be discussed shortly, it becomes necessary to reduce all annual costs, for the entire useful life of the alternative projects, to a common time basis, such as the first year of the earliest project. To enable such computations, Table 11.2 presents discount values for different discount rates for a duration of 50 years. An example of the use of this table will follow in the discussion of Table 11.5.

Annuities

Let us now assume that an annual sum A is invested at the end of every year, on a compound interest basis. We would like to know to what sum this annual investment or annuity has grown, after n years. The very last deposit of A, at the end of the nth year, has not accumulated any interest yet, and has therefore a value of A. The second last deposit is worth $A(1+i)$. The third last is worth $A(1+i)^2$ and so on. We may therefore write:

$$F = A + A(1+i) + A(1+i)^2 \dots + A(1+i)^{n-1}$$

This expression can be simplified by the following manipulation. Multiply both sides of the equation by $(1+i)$ and subtract the original equation from the new equation: $F(1+i) = A(1+i) + A(1+i)^2 + A(1+i)^3 \dots + A(1+i)^n$.

The result:
$$Fi = -A + A(1+i)^n$$

or
$$F = A\frac{(1+i)^n - 1}{i} \qquad \qquad \dots(11.3)$$

Some numerical values of this formula may be found in Table 11.1. For example, a sum of \$1,000 is set aside, every year, at 5 per cent compound

Table 11.1. *Numerical Values of Interest Formulae* (i is the annual interest rate, n the number of years, P the present sum of money, F the future sum of money, and A the annual payment of money)

	Compound Interest (The sum to which \$1 will accumulate after n years) $F = P(1+i)^n$						Present Value (The present sum that is equivalent to \$100 after n years) $P = \dfrac{F}{(1+i)^n}$				
Years	3%	4%	5%	6%	7%	*Years*	3%	4%	5%	6%	7%
5	1·16	1·22	1·27	1·34	1·40	5	86·26	82·19	78·35	74·73	71·30
10	1·34	1·48	1·63	1·79	1·97	10	71·41	67·56	61·39	55·84	50·83
20	1·81	2·19	2·65	3·21	3·87	20	55·37	45·64	37·69	31·18	25·84
30	2·43	3·24	4·32	5·74	7·61	30	41·20	30·83	23·14	17·41	13·14
50	4·38	7·11	11·47	18·42	29·46	50	22·81	14·07	8·72	5·43	3·39
100	19·22	50·51	131·50	339·30	867·72	100	5·20	1·98	0·76	0·29	0·12

	Amount of Annuity (The sum to which \$1 per year will accumulate after n years) $F = \dfrac{A((1+i)^n - 1)}{i}$						Present Value of Annuity (The present sum that is equivalent to \$1 per year during n years) $P = \dfrac{A((1+i)^n - 1)}{i(1+i)^n}$				
Years	3%	4%	5%	6%	7%	*Years*	3%	4%	5%	6%	7%
5	5·3	5·4	5·5	5·6	5·8	5	4·58	4·45	4·33	4·21	4·10
10	11·5	12·0	12·6	13·2	13·8	10	8·53	8·11	7·72	7·36	7·02
20	26·9	29·8	33·1	36·8	41·0	20	14·88	13·59	12·46	11·47	10·59
30	47·6	56·1	66·4	79·1	94·5	30	19·60	17·29	15·37	13·77	12·41
50	112·8	152·7	209·3	290·3	406·5	50	25·73	21·48	18·26	15·76	12·80
100	607·3	1237·6	2610·0	5638·4	2381·7	100	31·60	24·51	19·85	16·62	14·27

	Sinking Fund Payment (What has to be paid every year to amount to \$100 after n years) $A = F\dfrac{i}{(1+i)^n - 1}$						Capital Recovery Factor (What has to be paid every year to be equivalent to \$100 now) $A = \dfrac{P.i(1+i)^n}{(1+i)^n - 1}$				
Years	3%	4%	5%	6%	7%	*Years*	3%	4%	5%	6%	7%
5	18·88	18·46	18·10	17·74	17·39	5	21·83	22·46	23·10	23·74	23·39
10	8·72	8·32	7·95	7·59	7·24	10	11·72	12·33	12·95	13·59	14·24
20	3·72	3·35	3·02	2·71	2·44	20	6·72	7·36	8·02	8·72	9·44
30	2·10	1·78	1·51	1·27	1·06	30	5·10	5·78	6·51	7·27	8·06
50	0·89	0·66	0·48	0·34	0·25	50	3·89	4·65	5·48	6·34	7·25
100	0·17	0·08	0·03	0·02	0·01	100	3·17	4·08	5·04	6·02	7·01

interest to provide for some future expenditure. This fund will have grown after 20 years to a total sum of $33,100.

Table 11.2. *Discount Values*. (The same as 'present values'. These figures can be used in determining the present value of the future annual costs of a project with a useful life of 50 years or less)

Years	3%	5%	7%	10%	Years	3%	5%	7%	10%
1	0·971	0·952	0·935	0·909	26	0·464	0·281	0·172	0·084
2	0·943	0·907	0·873	0·826	27	0·450	0·268	0·161	0·076
3	0·915	0·864	0·816	0·751	28	0·437	0.255	0·150	0·069
4	0·888	0·823	0·763	0·683	29	0·424	0·243	0·141	0·063
5	0·863	0·784	0·713	0·621	30	0·412	0·231	0·131	0·057
6	0·837	0·746	0·666	0·564	31	0·400	0·220	0·123	0·052
7	0·813	0·711	0·623	0·513	32	0·388	0·210	0·115	0·047
8	0·789	0·677	0·582	0·467	33	0·377	0·200	0·107	0·043
9	0.766	0·645	0·544	0·424	34	0·366	0·190	0·100	0·039
10	0.744	0·614	0·508	0·386	35	0·355	0·181	0·094	0·035
11	0.722	0·585	0·475	0·350	36	0·345	0·173	0·088	0·032
12	0·701	0·557	0·444	0·319	37	0·335	0·164	0·082	0·029
13	0·681	0.530	0·415	0·290	38	0·325	0·157	0·076	0·027
14	0·661	0.505	0·388	0·263	39	0·315	0·149	0·071	0·024
15	0·642	0·481	0·362	0·239	40	0·307	0·142	0·067	0·022
16	0·623	0·458	0·339	0·218	41	0·298	0·135	0·062	0·020
17	0·605	0·436	0·317	0·198	42	0·289	0·129	0·058	0·018
18	0·587	0·416	0·296	0·180	43	0·281	0·123	0·055	0·017
19	0·570	0·396	0·277	0·164	44	0·272	0·117	0·051	0·015
20	0.554	0·377	0·258	0·149	45	0·264	0·111	0·048	0·014
21	0.537	0·359	0·242	0·135	46	0·257	0·106	0·044	0·012
22	0.522	0·342	0·226	0·123	47	0·249	0·101	0·042	0·011
23	0.507	0·326	0·211	0·112	48	0·242	0·096	0·039	0·010
24	0.492	0·310	0·197	0·102	49	0·235	0·092	0·036	0·009
25	0.478	0·295	0·184	0·092	50	0·228	0·087	0·034	0·009

It is possible that we would be interested in the present value of an annuity, running for the duration of n years. Eqn. (11.3) gives us the future value, after n years. If we apply to this eqn. (11.2), we should get the present value at the beginning of the n year period. The new equation for the present value of the annuity becomes:

$$P = \frac{A((1+i)^n - 1)}{i(1+i)^n} \qquad \ldots (11.4)$$

Some numerical values of this formula may be found in Table 11.1. For example, the present value of $1,000 per year, during 20 years, at 5 per cent compound interest, is $12,460. In other words, if we placed $12,460 in the bank at the present time, at 5 per cent interest, we could draw $1,000 per year for 20 years, before the fund was depleted. It may be noted that the sum of $12,460 times the factor 2·65 from the compound interest table becomes the $33,100 of the above example. This inter-relationship is graphically illustrated in *Figure 11.1*.

In the above example, it was noted that $1,000, at 5 per cent compound interest, for 20 years, amounted to $33,100. In engineering problems we often have to provide for a certain fund after a certain period of years. In such cases we know the future sum that has to be obtained, and we wish to know the annual payment that will produce this sum. To arrive at an appropriate expression we can simply use eqn. (11.3) in reverse:

$$A = \frac{F.i}{(1+i)^{n-1}} \qquad \ldots.(11.5)$$

Some numerical values of this formula may be found in Table 11.1 under the heading of Sinking Fund Payment. For example, to provide for a sum of $100,000, 20 years from now, allowing for 5 per cent compound interest, we have to set aside at the end of every year, during the intervening 20 years, an amount of $3,020.

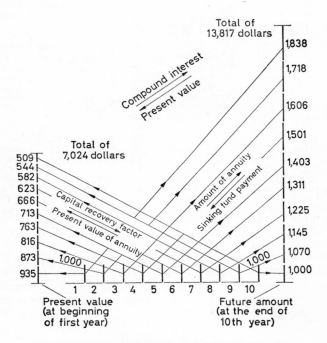

Figure 11.1. Graphical illustration of interest formulae.
Annuity of $1,000 during 10 years at 7 per cent compound interest

In case we are interested to know how much we ought to receive every year, for a duration of n years, in order to break even with a given investment at the present time, we may use eqn. (11.4) in reverse:

$$A = \frac{P.i(1+i)^n}{(1+i)^{n-1}} \qquad \ldots.(11.6)$$

Some numerical values of this formula may be found in Table 11.1 under the heading of Capital Recovery Factor. For example, to recover a present capital investment of $100,000, over a period of 20 years, allowing for 7 per cent compound interest, we should receive, at the end of every year, an amount of $9,440.

COST COMPARISON

It was noted earlier that when two alternative projects perform the same function, it will suffice to compare their costs to find the most desirable of the two. This cost comparison should be made on the basis of the annual costs, since the capital cost of a project is no real yardstick for comparison. Let us consider, for instance, a hydro plant and a steam plant of the same capacity, operating at the same load factor. The capital cost of the hydro plant may be three times as high as that of the steam plant, and still its annual cost, and hence its energy cost, may be lower ! The reason is that the annual depreciation and insurance on a hydro plant are lower; the operation and maintenance require less manpower; and no costly fuel is needed to turn the turbines.

In the following sections we shall discuss successively the components of the capital cost of a project and the nature of the annual charges, such as interest, depreciation, operation, maintenance, taxes and insurance. Thereafter we shall discuss cost comparison of different projects when the annual costs are constant, and when they are not.

Capital Cost

The capital cost of a project may be defined as the sum of all expenditures required to bring the project to completion. We may arbitrarily divide these expenditures into direct items and indirect items, as shown in Table 11.3.

The direct items in the table require little explanation. They are obtained from more detailed cost estimates, not shown. For these detailed estimates, preliminary designs have been made to obtain quantities, which were multiplied by appropriate unit prices, to obtain total costs of project components.

The indirect items, as may be seen from Table 11.3, form a substantial portion of the total capital cost and therefore they should receive careful consideration. The item 'Contingencies' represents expenditures that are possible but not certain; also expenditures that may come up but are as yet unforeseen; and finally, this item reflects the preliminary nature of the design and the uncertainty of future cost trends. It may therefore be expected that the percentage that is allowed for contingencies is much higher in preliminary than in final cost estimates.

The item 'Engineering' stands for all expenditures associated with engineering activities, beginning with the preliminary field surveys, proceeding with the office studies, the consulting services, the detailed design, and terminating with the supervision of the construction of the project. The item 'Interest during construction' allows for the necessity of having to acquire the capital expenditure of the project in advance of the date that the project begins to function. For instance, in the example given, the total duration of construction may be four years, with the peak of activities occurring during the third year. Contractors have to be paid while they proceed with the construction. Funds have to be made available in advance of payment dates. Hence it may be expected that money has to be borrowed as follows: $6,000,000

411

at the beginning of the first year, requiring $1,000,000 of interest till the end of the fourth year, assuming 4 per cent interest; $12,000,000 at the beginning of the second year, requiring $1,500,000 of interest till the end of the fourth year; $18,000,000 at the beginning of the third year, requiring $1,500,000 of interest till the end of the fourth year; and $11,000,000 at the beginning of the fourth year, requiring $400,000 of interest till the end of the fourth year; or a total of $4,400,000 for interest.

Table 11.3. *Estimated Capital Cost* (Whitemud Falls Power Development)

Direct Items	
Access roads	$3,800,000
Camps	2,000,000
Reservoir damages	3,000,000
Unwatering	1,300,000
Spillway.	2,700,000
Dam	1,000,000
Powerhouse	8,700,000
Generators and turbines	10,400,000
Transformation	1,700,000
	$34,600,000
Indirect Items	
Contingencies, 15 per cent of direct items .	5,200,000
Engineering, 8 per cent of direct items .	2,800,000
Interest during construction . . .	4,400,000
Capital Cost	$47,000,000

It should be pointed out that advance cost estimates of engineering projects are intelligent guesses at best. First of all, the design on which the estimate is based is preliminary. Secondly, the real cost of construction is difficult to estimate for engineers who have not been in the building business themselves. Thirdly, and this is by far the most important reason, it is nearly impossible to guess at the intensity of competition among contractors a few years, or even one year, in advance. When there is too little equipment and manpower to handle all the construction that is offered for bidding, the cost of projects may exceed the estimated cost by 10, 20, or even 50 per cent. On the other hand, when contractors are idle, and desperate to keep their key personnel on the payroll and their equipment rolling, the projects may be built for 10, 20, or even 50 per cent less than estimated. In view of such uncertainty, it serves no purpose to submit cost estimates with a great degree of precision. In fact, such an estimate suggests to the innocent layman a degree of accuracy that really does not exist. It is somewhat amusing to see, for instance, a cost estimate of a hydro project of $95,053,400 submitted by an engineering firm to its client, and to find this estimate a few years later revised to $140,000,000!

Annual Cost

Interest is the first, and often the largest, of the annual charges of a project. The fluctuation of the rate of interest with the state of the economy and the dependability of the borrower was discussed earlier. Whether interest is charged from year to year on the full capital cost, or on a steadily reducing figure, depends on how the depreciation of the project is computed, as will be discussed in the following paragraphs.

First of all, let us make distinction between depreciation and amortization. We make an annual allowance, called depreciation, because of the decline in value of the project, or its components, due to such causes as wear and tear, or obsolescence, or depletion of the basic resource, or an arbitrarily assumed limited life of the project. The purpose of these annual depreciation allowances is to build up a fund with which we can finance a new project when the existing one is worn out or obsolete.

At the same time that we are making depreciation payments, we may also be making annual amortization payments for the gradual extinction of the debt that was engaged to finance the project. The question now arises if both the depreciation payments and amortization payments should be looked upon as constituting annual cost components.

Let us assume that we have under construction a $100,000,000 navigation project with a useful life of 30 years. This project is financed by a 30 year 5 per cent bond issue. What would happen if we allowed, in our cost calculations, for amortization as well as depreciation? Over the forthcoming 30 years we would allow for $1,510 per year to amortize the bond issue and we would also allow for $1,510 per year to build up a fund to replace the navigation works. As a result, the second project, from 30–60 years, would be without interest charges, since no funds had to be borrowed to finance it. This would, of course, be very attractive for the future generation that will make use of the second project; but it must be realized that the present generation has paid for it.

When the subject of discussion is corporate financial policy, and the object to build up reserves for future expansion, it may be a sound decision to allow for depreciation payments as well as amortization payments, when calculating the cost of the product, or when determining the profits of the corporation. However, when the subject of discussion is water resources development, and the object to compare two alternative possibilities, either depreciation or amortization should be allowed for in determining the annual cost and not both. A convenient way of simplifying the problem for the present purpose is to consider that repayment of the bond issues at maturity is financed by issuing new bonds. Then we have a constant and perpetual interest charge. In addition, we build up a fund with which we can finance the replacement of the existing project, when needed. Thus we have introduced depreciation payments that are based on the estimated useful life of the project, and on average long-term interest rates.

There are several methods to calculate annual depreciation charges such as the straight line method, the sinking fund method, the fixed percentage of depreciated value method, the present value of future returns method, and the production units method. For the economic analysis of water development projects, only one method should be considered: the sinking fund method. Unfortunately, it is often believed that the straight line method is preferable. We shall therefore devote the following paragraphs to a discussion of these two methods, and we shall distinguish between the following possibilities:

1. Straight line depreciation
    (a) not allowing for interest on the depreciation fund
    (b) allowing for interest on the depreciation fund
2. Sinking fund depreciation.

413

Assuming for the purpose of illustration a hydro plant with a capital cost of $100,000,000, interest at 4 per cent, and a useful life of 50 years, the depreciation charge of method 1(*a*) would become a fixed sum of $2,000,000 per year, while the interest charge would remain during the life of the project at its full value of $4,000,000 per year. In other words, the total 'cost-of-money' is $6,000,000 per year. It is obvious that we are neglecting in this computation the interest that can be earned on the steadily increasing depreciation fund, either as a bank interest when the depreciation payments are placed on a bank account, or as indirect earnings when the depreciation 'payments' are partly or wholly absorbed in the financing of subsequent power developments. However, some reasons could be advanced for the neglect of the interest benefit. First of all, the economy of the past has been subject to a steady inflation of prices. It may therefore be expected that when the hydro plant needs to be replaced, 50 years from now, a capital well in excess of $100,000,000 will be needed. If we would place $2,000,000 annually in the bank, at an improvement rate of 4 per cent, we would obtain after 50 years a capital of $305,000,000, which appears more than ample to cope with inflation. However, before we accept this reasoning, let us ask the question: 'What is the purpose of annual depreciation allowance?' Is it to retire the original debt that has been made to build the plant, or is it to build up a fund to replace the plant 50 years hence? Let us not lose sight of the fact that it is the present generation that builds the plant and uses its product. Why then should we allow, in our cost computations, for anything over and above what is needed to amortize the presently incurred debt? We do not know what will happen after 50 years; maybe the future generation will not replace the existing plant at all, and may have found other and less costly means of energy production. But even if the plant will have to be replaced by a costlier plant, we may look upon that responsibility as belonging to the future generation, who will benefit from that future development. It may be noted at this place that in view of the above discussions, it would be more appropriate in the present case to use the word amortization instead of depreciation.

There is another reason that may be advanced to justify the neglect of interest benefits. In an expanding economy, funds are continuously needed, not only to replace old plants but also to build additional ones. The capital needed for expansion could be partly provided by the compounded interest of the depreciation fund. This would be a sound policy if we were dealing with private financing, particularly in view of the fact that the taxation structure allows such a policy. However, we are dealing with public financing and, as noted earlier, we would in effect be letting the present generation pay for future developments. It is therefore concluded that in determining depreciation charges, due allowance should be made for the interest that will accumulate on the depreciation fund. This brings us to methods 1(*b*) and 2.

In method 1(*b*), straight line depreciation, allowing for interest on the depreciation fund, the earlier quoted example would be modified as follows. The annual depreciation charge is again $2,000,000. The interest, however, is only taken over the depreciated value of the plant. Hence, the interest during the first year is the full $4,000,000, but then it is constantly reduced until a near-zero value is reached at the last year, when the plant has been written off. As a result, the 'cost of money' of the project is not a constant item, but declines from $6,000,000 in the first year to $2,000,000 in the last year. Such a method of cost computing would be logical if we had annual operation and maintenance costs that would steadily increase with the advance in time to such an extent that they would compensate the decrease of 'cost-of-money'. The total annual costs of the project would then be nearly constant, which is desirable from a viewpoint of determining the price of energy. However, the cost of operation and maintenance of a hydro plant has by no means such a sharp increase with time, as indicated above. It may therefore be concluded that a method of cost computing whereby the cost-of-money remains constant over the years would be more suitable for planning purposes. This brings us to method 2.

We may visualize the sinking fund depreciation method as follows: every year we set aside a certain amount of money that will yield, at compound interest, after 50 years, the original $100,000,000 capital investment so that we are in a position either to retire the original debt or to finance partly or completely a new plant. The interest rate at which the annual depreciation deposits will improve is not necessarily the same as the interest rate that we have to pay on the bond issue that has provided our original capital of $100,000,000. However, assuming this also at 4 per cent for the present discussion, the corresponding annual depreciation payments then become $660,000, and the total cost-of-money becomes $4,660,000.

It is concluded from the above discussion that the sinking fund method is the more suitable depreciation method to be used in planning studies. As compared to the straight line method without allowing for interest, there is a significant reduction of about 1·3 per cent in annual charges. As compared to the straight line method, with allowance for interest, the sinking fund method has the advantage of a constant annual cost so that the study of unreasonably long periods of time is avoided.

In an economic analysis of water development projects, it is recommended to assign a limited period of time to the useful life of the project. Suppose a flood-control plan envisages a diversion canal around the city. The canal, as a physical structure, when properly maintained, has an indefinite life. However, some time in the future, the whole flood control project may not be needed any more! Suppose a gravity dam is built for hydro development. Several hundred years may pass before the dam is deteriorated or before the reservoir is silted up. However, before that time, the hydro plant may have been replaced by a more economical source of energy!

In order to make a conservative comparison of the costs and benefits, the useful life of a project is usually taken as 50 years. In other words, the capital that has to be borrowed for building the project, has to be repaid over a period of 50 years.

It should be pointed out that not necessarily all components of a project have the same useful life, and therefore that not all components have the same depreciation charge. This may be illustrated by the following example.

Suppose a power project with a capital cost of $100,000,000 and an interest rate of 5 per cent is under consideration. The project consists of dams, bridges, powerhouses, transmission lines and steam plants. The useful life of the whole project is taken as 50 years. Some items of the project, namely the gates of the spillways, the bridges, and the powerhouse equipment, at a total cost of

$30,000,000, may have deteriorated after 30 years. Other items, like the boilers and condensers of the steam plants, at a total cost of $10,000,000, may become obsolete after 20 years. The cost-of-money of this project is then:

Interest .	. . 5·00% of $100,000,000 =	$5,000,000
Depreciation .	. . 0·48% of $ 60,000,000 =	$ 290,000
,, .	. . 1·51% of $ 30,000,000 =	$ 450,000
,, .	. . 3·02% of $ 10,000,000 =	$ 300,000
		$6,040,000

In the above example it was assumed that money was borrowed at 5 per cent interest and that the annual depreciation payments into the sinking fund would grow at the same 5 per cent. However, it is quite possible that the borrower may not be in a position to invest its annual depreciation allowances at the same interest rate, but may have to accept a lower rate, say 3 per cent. In that case the above example would become:

Interest .	. . 5·00% of $100,000,000 =	$5,000,000
Depreciation .	. . 0·89% of $ 60,000,000 =	$ 530,000
,, .	. . 2·10% of $ 30,000,000 =	$ 630,000
,, .	. . 3·72% of $ 10,000,000 =	$ 370,000
		$6,530,000

For an accurate analysis, the annual cost of operation and maintenance has to be determined by analysing the various services that have to be performed: so much for labour, so much for equipment, so much for materials, etc. For a rough analysis, however, it is permissible to express the cost of operation and maintenance as a percentage of the capital cost of the project, as shown in Table 11.4. Most of these figures are taken from 'Multiple purpose river basin development', U.N. (1955).

Table 11.4. *Annual Operation and Maintenance Costs in Percentage of Capital Cost*

Dams and reservoirs	0·1
Intake and outlet works. . . .	1·0
Hydro plants	1·0
Steam plants (excluding fuel). . .	2·5
Transmission lines	1·0
Unlined canals	2·0
Lined canals	1·0
Steel conduits	1·5
Concrete conduits	1·0
Wooden conduits	8·0
Irrigation distribution work . . .	3·0
Bridges, concrete and steel . . .	3·0
Bridges, timber	8·0
Gates, hoists, miscellaneous metal work .	1·5

On federal projects it is not customary to include taxes and insurance in the determination of annual costs, for the reason that the Federal Government does not pay taxes to itself, and does not insure its property. However, it may nevertheless be advisable to include such an item in an economic analysis of a federal project. The 1957 Report of the Engineers' Joint Council of the A.S.C.E. makes the following statement on this subject:

'The costs reported for a proposed federal project should be detailed and thoroughly inclusive. All items of first cost, including cost of investigation, lands and rights-of-way, construction, interest during construction, and the cost of measures needed to correct detrimental results of the project, should be presented. The costs of all components of the project required to achieve its intended purpose, including costs to others than the Federal Government, should be included in the cost summary.

'The estimates of annual costs should include those for operation and maintenance, interest on the unamortized amount of first cost, proper allowance for taxes (paid or foregone), amortization, replacements (to the extent required during the period of amortization), and allowance for insurance (or its equivalent) and contingencies.'

Comparison of Annual Costs

When two projects serve the same purpose to the same degree and they have constant annual costs, a selection between the projects may be based on a comparison of those costs. This will be illustrated by the following example. Somewhere in a remote wilderness a rich ore deposit has been discovered. It has been estimated that all ore can be mined during a period of 20 years. The most economical way to bring out the ore is by river. To make the river navigable there are two alternative projects:

(A) Rock ledges in the river have to be removed and a channel has to be dredged at a total initial cost of $10,000,000. Afterwards, the channel has to be maintained by dredging at an annual cost of $800,000.

(B) Two dams and shiplocks have to be built at a capital cost of $18,000,000. Operation and maintenance of these works require $50,000 per year.

In both cases it may be assumed that money is available at 7 per cent interest; that the sinking fund will improve at 4 per cent interest; that taxes and insurance are nil; and that the scrap value of both projects is nil. The cost comparison would then look as follows.

	Project (A)	Project (B)
Interest (7·0%)	$ 700,000	$1,260,000
Depreciation (3·35%)	340,000	600,000
Operation and maintenance	800,000	50,000
Total annual cost	$1,840,000	$1,910,000

The conclusion is that project (A) would be preferable to project (B).

Varying Annual Costs

When two projects serve the same purpose to the same degree, but they do not have constant annual costs, the problem becomes more complicated. Let it be assumed in the previous example that in alternative (A) the cost of dredging is not a constant $800,000 per year, but that it is $1,700,000 during the first year, $1,600,000 during the second year, and so on, declining by $100,000 per year, and nil during the last three years. Although the average annual cost of dredging comes to slightly less than $800,000, we cannot draw the same conclusion as before, because the heavy expenditures occur now in the early years, which is from an economic viewpoint less desirable. Therefore, we have to make a new appraisal of the situation.

A convenient and accurate way of comparing one expenditure, say 10 years from now, with another expenditure, say 20 years from now, would be to calculate the present value of each, with an applicable interest (actually discount) rate. The lowest present value would then indicate the most desirable expenditure. Applying this thought to the example of the alternative river improvement schemes, we could determine the annual cost for each project, for every year during the 20 years, calculate the present values of these annual expenditures, total up the present values, and compare the totals of alternatives (A) and (B). The applicable discount rate would be 7 per cent, since that is the rate at which the project is financed, and the annual costs determined. For alternative (B), with its constant annual cost, the total present value of all annual costs can be found by applying eqn. (11.4): $P = \$1,910,000 \times 10.59 = \$20,200,000$. For alternative (A) with its varying annual cost, the total present value of all annual costs can be found as follows:

Table 11.5. *Computation of Total Present Value of Annual Costs*

Year	Fixed Cost	Variable Cost	Total Cost	Discount	Present Value
1	$1,040,000	$1,700,000	$2,740,000	0·935	$2,560,000
2	1,040,000	1,600,000	2,640,000	0·873	2,310,000
3	1,040,000	1,500,000	2,540,000	0·816	2,080,000
4	1,040,000	1,400,000	2,440,000	0·763	1,860,000
—	—	—	—	—	—
17	1,040,000	100,000	1,140,000	0·317	340,000
18	1,040,000	0	1,040,000	0·296	310,000
19	1,040,000	0	1,040,000	0·277	290,000
20	1,040,000	0	1,040,000	0·258	270,000

Total present value: $P = \$21,350,000$

The conclusion is that project (B) $(P = \$20,400,000)$ would be preferable to project (A) $(P = \$21,350,000)$.

MEASUREMENT OF BENEFITS

It may be expected in the general case, that alternative water development projects will not perform exactly the same function to exactly the same degree. In all cases where the difference in benefits becomes significant, a mere comparison of the cost of the projects, as was discussed in the foregoing paragraphs, will not suffice, and an analysis of the benefits as well as the costs becomes necessary.

Benefits may be divided into tangible and intangible benefits, and also into direct and indirect benefits. Tangible benefits are those measurable in dollars. Intangible benefits are the ones that do not yield to dollar evaluation, such as greater security or enhancing the beauty of the landscape. Direct benefits are the immediate result of the project, such as the production of electricity, or the prevention of flood damages, or the increased farm production. Indirect benefits are of a subsequent nature, such as the stimulation of industries or the increase in general taxation level. It is considered conservative practice to judge the merits of a project on its tangible and direct benefits only.

In the following paragraphs we shall discuss first the nature of the benefits of the various means of water resources development, such as flood control, reclamation, water power, navigation, and water conservation. Thereafter will follow a discussion of the benefit–cost analysis and the selection of the most desirable project.

Flood Control Benefits

The elimination of flooding, or the reduction of the frequency of flooding, has a two-fold beneficial effect. First it prevents the occurrence of flood damages. Second, it will stimulate increased production in the protected areas. In most cases, flood control benefits fall primarily in the first category, with the increased production benefits being of a secondary nature. However, it is conceivable that flood control benefits fall entirely in the first, or entirely in the second group. For instance, in 1953 when the dikes in Holland failed, no one had expected the possibility of such a disaster. Therefore, the subsequent flood control measures only restored the existing situation, without inducing any more intensive use of the land. On the other hand, the construction of dikes in the Saskatchewan Delta, preventing flooding from some one million acres of land, will create a new situation where the benefits are exclusively the future use of the land, since the present use is nil. The first group of benefits, namely the prevention of flood losses, will be discussed in this section. The second group of benefits, namely the more intensive land use, is very similar to reclamation benefits and will be discussed in that section.

Before we discuss flood damages and flood control benefits in detail, let us briefly glance over the subject as a whole. It was noted in Chapter 6: Flood Control how the effectiveness of any combination of flood control works can be expressed in a modified frequency curve of flood stages at the location under consideration. We have, of course, also the frequency curve of flood stages under natural conditions. Since every flood stage is associated with a certain amount of damage, that can be expressed in dollars, it is possible to construct a stage-damage diagram that expresses the amount of damage as a function of the flood stage. From this stage-damage curve, and the stage-frequency curve, and applying some mathematical manipulation, it is possible to come up with a figure of average annual damage. Thus, we are in a position to compute the average annual flood damage under natural conditions, under the conditions of flood control scheme (A), scheme (B), etc. The annual benefits of the different flood control schemes are now by definition the difference between the average annual damage under natural conditions and the average annual damage associated with each scheme. If permissible, these benefits may be increased with annual net benefits, derived from more intensive land use. The total annual benefits for each flood control scheme, thus obtained, can now be compared to the total annual cost of that same scheme. By making an analysis of the different costs and benefits of all schemes, a selection can be made. In the following paragraphs we shall discuss successively the determination of flood damages as a function of stage, the computation of the average annual damage, and the determination of the average annual benefits. The selection of the most feasible project will be discussed at the end of this chapter. Flood damage may consist of the following items:

(1) Physical damage to buildings and their contents, bridges, highways, railways, etc. The amount of damage is to be appraised in terms of the cost of replacement, or repair of the affected property.

(2) Agricultural crop losses. These losses are to be appraised in terms of market value, less any costs that had not yet been incurred at the time of loss.

(3) Loss of income due to interruption of business. This loss should be appraised in terms of the goods or services that would have been produced if the flood had not occurred.

(4) Cost of flood fighting, and the evacuation, care and rehabilitation of flood victims. It must be emphasized that double counting between items (3) and (4) is to be avoided.

To illustrate the relative importance of the above items in a given situation, the following Table 11.6 lists the damages estimated to have occurred in Greater Winnipeg and the Red River Valley as a result of the 1950 flood.

Table 11.6. *Flood Damages in Greater Winnipeg*

Damage to residential property	$ 51,000,000
Damage to industrial property	47,000,000
Damage to bridges, highways, railways	6,000,000
Agricultural crop losses	3,000,000
Loss of income	14,000,000
Flood fighting and evacuation	5,000,000
Total flood damage	$126,000,000

Damage estimates as shown in Table 11.6 are obtained from a systematic survey of the area by qualified economists. The method of survey may be one of sampling representative areas. The initial survey will concern the floods on record. Subsequent studies will be made of what the damage would be if floods of greater magnitude would occur, up to the maximum possible flood. Thus a stage-damage curve can be prepared as shown in *Figure 11.4*. If different sections of the total area under consideration are affected to different degrees by certain flood control measures, it may be desirable to prepare separate stage-damage curves for each of those sections. For instance, removal of a river control (such as a rapids or a narrow stretch of river channel) downstream of a city, will lower the flood levels of the city. However, the lowering will be more pronounced in the downstream part of the city than in the upstream part. A diversion channel around the city, on the other hand, may lower the levels in the upstream part of the city to a greater degree than in the downstream part. In order to compare two such measures, the damage reduction in the city must be determined section by section.

After having determined the stage-damage curve, and having available the stage-frequency curves for every proposed combination of flood control measures, sufficient information is available to compute the average annual benefits of every flood control scheme. The stage-frequency curve is represented by *Figures 11.2* and *11.3*. *Figure 11.2* is the discharge-frequency curve, and *Figure 11.3* is the rating curve of the section of river for which the stage-damage curve has been prepared. Due to the proposed flood control measures, the discharge-frequency curve may change, while the rating curve remains the same (for instance due to an upstream reservoir); or the rating curve may change, while the discharge-frequency curve remains the same (for instance due to a downstream channel improvement); or both curves may

420

change. Whatever the change is, from the *Figures 11.2, 11.3,* and *11.4, Figure 11.5* can be prepared showing the damage in terms of its frequency of exceedence. For instance, let *Figures 11.2* and *11.3* represent the natural conditions. It may be seen from *Figure 11.2* that a peak flow of 200,000 cusec has a probability of being exceeded of 1 per cent per year. *Figure 11.3* shows that 200,000 cusec corresponds with a stage of 40 ft. above datum (the river bottom or some arbitrary city datum), while *Figure 11.4* shows that a stage of 40 ft. corresponds to a damage of $300,000,000. In other words, every year there is a chance of 1 per cent that, under natural conditions, a damage of $300,000,000 will be exceeded. This statement is translated in graphical language in *Figure 11.5.*

We may now make the following observations with respect to *Figure 11.5.* For instance, a damage of $100,000,000 has a chance of being exceeded of 8 per cent per year. A damage of $120,000,000 has a chance of being exceeded of 7 per cent per year. Hence there is a chance of $8 - 7 = 1$ per cent per year that a damage between $100,000,000 and $120,000,000 will occur. Likewise, there is also a chance of 1 per cent per year, that a damage between $120,000,000 and $135,000,000 will occur, and so on. When there is a chance of 1 per cent per year that a damage of $110,000,000 (the average of the first interval) will occur, we assume that this is equivalent to a damage of $1\% \times \$110,000,000 = \$1,100,000$ every year, year after year. It may be noted that we are now in effect multiplying an interval of vertical ordinate in *Figure 11.5*, with the corresponding horizontal ordinate. In other words, when we do this for all intervals, we have determined the area under the curve. Thus we have come to the conclusion that the total average annual damage is represented by the area under the damage-frequency curve, assuming, of course, that appropriate conversion factors are applied.

Let us dwell for a moment on the correctness of the statement that a chance of 1 per cent per year of $100,000,000 damage is equivalent to an average annual damage of $1,000,000. Assuming an interest rate of 4 per cent, we may say that the present value of an average annual damage of $1,000,000 perpetually, is $25,000,000. Let us try to convert the possible $100,000,000 damage also to its present value. This is somewhat more difficult. That damage may occur this year, and its present value would be $100,000,000. It may also occur after 100 years, and its present value would be $2,000,000. Let us assume that we have 10 such cases and that the $100,000,000 damages occur at regular intervals; the first one after 5 years, the second one after 15 years, and so on. The present value of those 10 damages then becomes:

$$P = \$100,000,000(0.82 + 0.55 + 0.38 \ldots + 0.02) = \$250,000,000$$

which is 10 times as large as the present value of the average annual damage of $1,000,000! We may therefore conclude that the above statement is sufficiently correct to be used in this type of economic analysis where other inaccuracies are much greater.

After having obtained the average annual damage for natural conditions, as shown in *Figure 11.5*, the same computations are repeated for the modified frequency curves and rating curves that correspond to the proposed flood control schemes. Each new computation yields a different damage-frequency

421

curve, and the area between the original and the new curve represents, by definition, the average annual benefits of the flood control project under consideration.

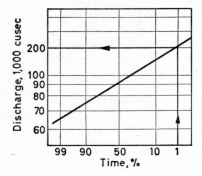

Figure 11.2. Frequency curve of maximum annual flood flows

Figure 11.3. Rating curve of river at centre of area subject to flooding

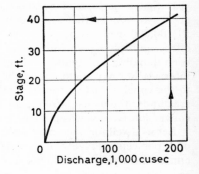

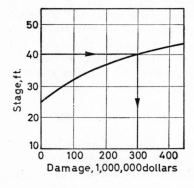

Figure 11.4. Stage–damage curve of area subject to flooding

It was noted in Chapter 6: Flood Control that different measures such as dikes, or reservoirs, or diversions, have a different effectiveness with respect to extreme flood flows. This finds its expression in the damage–frequency curves. In *Figure 11.6*, for instance, are shown the benefits of a diking system.

Let it be assumed that the dikes are designed for a flow of 150,000 cusec, corresponding to a frequency of exceedence of 7 per cent per year. It is obvious that benefits are only obtained for flows less than 150,000 cusec. When this

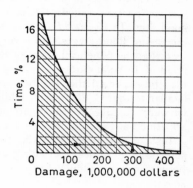

Figure 11.5. Average annual damage in area subject to flooding (shaded area)

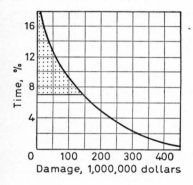

Figure 11.6. Average annual benefits as a result of the construction of a diking system (shaded area)

Figure 11.7. Average annual benefits as a result of a flood diversion or a reservoir (shaded area)

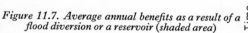
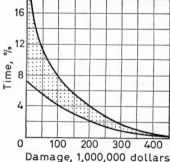

flow is exceeded, the dikes must be assumed to fail, and the damage is as large as if there had been no dikes. Let us now assume that a diversion is designed to cope with the same river flow of 150,000 cusec. Up to that discharge, no damage will occur. Above that discharge (and hence below the frequency of 7

423

per cent) the diversion will keep functioning and hence will continue to relieve the flood situation and hence reduce the damage below what it would have been without the diversion. Depending on the local circumstances (shape of rating curve, shape of stage-damage curve) the reduction of damage may further increase, remain the same, or decrease, with increasing magnitude of flood flow. In *Figure 11.7*, there is a slight decrease in damage reduction with increasing flood stages. However, in spite of this slight decrease, it may be seen that the beneficial effect of the diversion for flows, larger than the design flood of 150,000 cusec, is as large as the beneficial effect below that stage. It may be expected that the beneficial effect of channel improvements is very similar to that of diversions, while the effect of reservoirs will probably be in between the dike and the diversion, assuming, of course, that all four measures are aimed at the same design flood.

Instead of determining the average annual damage of each flood control scheme by graphical methods, as discussed above, the same computations can be performed in tabular form. The only difference is that where we can draw a smooth curve through the plotted points in the graph, we will be dealing with 'straight lines', when we compute in table form. The small advantage is that the preparation of cumbersome graphs and the planimetering of large areas is avoided. Table 11.7, as an example, shows the computation of the average annual damage of *Figures 11.2, 11.3,* and *11.4,* under natural conditions.

Table 11.7. *Computation of Average Annual Damage*

Peak Flow (cusec)	Frequency (%)	Damage ($1,000)	Frequency of Interval	Av. Damage of Interval	Av. Annual Damage ($1,000)
100,000	30	0			
			16	20,000	3,200
125,000	14	40,000			
			6	70,000	4,200
150,000	8	100,000			
			5	150,000	7,500
175,000	3	200,000			
			2	250,000	5,000
200,000	1·0	300,000			
			0·6	350,000	2,100
225,000	0·4	400,000			
			0·3	450,000	1,400
250,000	0·1	500,000			

Total average annual damage: $23,400,000

It may be of interest to point out that in the early stages of the planning of flood control schemes, it is of advantage to have available an approximate figure of the average annual damage under natural conditions. This is the maximum benefit that can be obtained, when the works would prevent all future flooding. Since this is never the case, one could take a figure somewhat smaller, and regard that as an upper limit of the annual cost of the flood control works that are going to be planned. Rough as such a guide may be, it may perform a useful function in that it will eliminate immediately all sorts of suggestions that are way beyond the realm of practicability.

Reclamation Benefits

In the above section we have discussed flood control benefits in the sense of benefits derived from the reduction of damages. It was noted that there were also more positive benefits to flood control, namely the increased production of the land, due to a different type of land use, which becomes feasible because of the flood protection measures. These last benefits could actually be classified as reclamation benefits. When we reclaim land, we bring this land from its raw, natural condition, to a desired state of cultivation. This can be accomplished in two ways, depending on the natural state of the land. If the land is subject to periodic flooding, the reclamation measures consist of building dikes to keep the floodwaters off the land, and excavating ditches and canals to maintain a desired ground water level. If the land suffers from a shortage of moisture, the reclamation measures consist of irrigating the land by means of storage reservoirs, canals and ditches. In both cases, however, the object is the same, namely to provide the land with optimum moisture conditions.

To illustrate the concept that reclamation benefits are the value of the increased production, after allowance for increased associated costs, reference is made to Table 11.8. It will be assumed that an area of 15,000 acres is converted from dry land farming to irrigation farming. The budget of the entire area may look somewhat as follows:

Table 11.8. *Example of Reclamation Benefits*

	Irrigated Land	Dry Land	Benefits
Number of farms	100	10	
Acres per farm	150	1,500	
Farm products sold	$1,500,000 −	150,000 =	$1,350,000 +
Production expenses	$900,000 −	75,000 =	$825,000 −
Family allowance	$250,000 −	25,000 =	$225,000 −
	Total farm benefit		$300,000
	Benefit per acre, per year		$20

If one tries hard, the above benefits may be inflated somewhat. It could be argued that the people moving into the irrigation area would otherwise have received an annual allowance of less than $2,500 per family. It could be argued that products consumed on the farm, and hence not sold, are of relatively greater quantity on the irrigated land than on the dry land. However, such arguments do not sound very convincing and their resultant adjustments would be less than the order of magnitude of the possible error in estimating the value of the future farm products!

From the viewpoint of the economic analysis of an irrigation project, the total farm benefits, such as the $300,000 listed in Table 11.8, are the annual benefits that have to be compared to the annual costs of the project. The annual cost will consist of the interest, amortization, operation and maintenance of the storage reservoir, the intake works, the main canal, and the irrigation ditches leading to the farmers' fields, and the drainage ditches and canals leading back to the river system.

From the viewpoint of the farmer, the farm benefits of Table 11.8 are the gross benefits. Out of these gross benefits he will have to make water payments in order to defray the annual costs of the irrigation project. Assuming that he would have to pay $10 per acre per year for water, his net benefits would then be $10 per acre per year. It may be expected that such net benefits will be reflected in the market value of the land. Hence, an approximation of these net benefits may be made by estimating the increase in market value of the land and converting it to an annual net benefit by applying an appropriate rate of return. When this annual figure is increased by the annual charge for water, the annual gross benefits of the project are obtained.

It must be pointed out that when reclamation benefits are compared to reclamation costs, this should be done on the basis of present values. It is not uncommon that the conversion of dry land farming to irrigation farming is spread over several decades, following the construction of the main (and costly!) irrigation works, such as the reservoir, outlet works, and main canals. It is obvious that such a lag in time between the capital expenditure and the reaping of the benefits could be a heavy liability against the project. This should be duly expressed by bringing all costs and benefits to a common time basis.

In the beginning of this section on benefit–cost analysis, the distinction was made between direct and indirect benefits. Direct benefits, such as discussed above, pertain to the immediate increase in productivity due to the project. Indirect benefits reflect the impact of the project on the rest of the economy. Although indirect benefits may be very real indeed, there are two serious objections against including such benefits in an economic analysis. First of all, they are very difficult to define, and any quantitative appraisal is little more than speculation. Secondly, not only irrigation projects but any water development project, any capital expenditure, will have an impact on the economy as a whole! It would therefore seem inadvisable to introduce indirect benefits as a measuring device for the economic evaluation of water development projects. It is of interest to note that the Presidential Advisory Committee on Water Resources Policy in the U.S. has recommended that economic analyses should be made on the basis of direct benefits alone.

Water Power Benefits

Before we make an attempt to define the nature and the magnitude of water power benefits, let us realize first of all that normally the scope of power development is governed by the increase of the load demand. Power utilities have to predict load demand five to ten years in advance and they have to take adequate measures with respect to increasing their generating capacity so that they can meet this power demand when it comes. One of the important activities in power engineering is to plan the extension of the system capacity such that the lowest annual cost of energy is obtained. To achieve this, comparative studies are made of different sequences of system development, with hydro plants, steam plants, gas turbines, or nuclear plants as alternative units of increment. From this comparison, the most feasible sequence will be

426

adopted and the power system extension will take place accordingly. Some of the units of this plan may be hydro plants. We could ask now: 'What are the water power benefits of these hydro plants?'

An accepted way of defining these benefits is to compute what the production of electrical energy in the entire system would cost, if the hydro plants under consideration did not exist, and were replaced by the least costly alternative source of power. The annual cost of this alternative generating capacity may be regarded as the benefit of the hydro plants. Does it make much sense to compute water power benefits in this fashion? Sometimes it does, sometimes it does not.

We must keep in mind that an economic analysis of water development projects is made for the purpose of deciding first what project or combination of projects is most feasible, and secondly, if such projects should be undertaken at all. In power engineering, both questions have been answered after the system planning study, as described above (and in more detail in Chapter 7: Water Power), is completed. We have established that power is needed, and that sequence X, including hydro plant A, is the least costly way to obtain it. Assuming that licences for the hydro sites can be obtained and that there is no alternative use for the water, it would serve no purpose to determine water power benefits by computing what it would cost to generate power by alternative means. Such information would be of no help in the decision-making process.

However, the situation changes when there are alternative and competing uses for the limited amount of available water. Let it be assumed, for instance, that in one proposed development the river flow is used to generate electricity in a series of hydro plants, while in the alternative scheme the river flow is used to irrigate land in the upstream part of the drainage basin so that the hydro development becomes unattractive. To compare the two schemes on a dollar basis, we would compute the annual cost of the hydro project and its associated benefits (the annual cost of producing the same amount of power, with the least costly alternative means); and also the annual cost of the irrigation project (all costs up to the farmers' fields) and its associated benefits (the total farm benefit, as defined in the previous section). These costs and benefits of the two alternative projects will then be compared as will be discussed shortly under the section Selection of Project.

There are other situations wherein the evaluation of water power benefits is desirable. For instance, when an upstream agency (private or public) provides storage capacity in the river basin, it may benefit the present and future hydro development of a downstream agency. Due to the greater dependable flow, more capacity may be installed, and more energy may be produced at the downstream plants. The quantitative treatment of such a problem is discussed in Chapter 7 under the section Storage Benefits.

When the cost of producing power by alternative means is computed, one has to apply the same interest and tax charges as were applied when computing the cost of the hydro plants, or else the benefit–cost analysis will be biased. Surprisingly enough, this is not always done, as may be illustrated by quoting from a report on St. Lawrence Power by the New York State Power Authority. According to this report, the St. Lawrence hydro development would result in a saving of $12,720,000 per year over alternative steam plants. Upon inspection of the computations, however, it is found that the annual charges on hydro include 3 per cent interest and no taxes, while the annual charges on steam include 6·0 per cent interest and 5·5 per cent taxes!

The discussions in the above paragraphs were based on two assumptions. First, that the load growth is determined by the economy of the country, and not by the availability of power. Second, that in the absence of hydro development, its place would be filled by alternative power development. These assumptions may not always be valid. In the Pacific Northwest of the United States, or along the St. Lawrence in Canada it is the availability of cheap hydro energy that has stimulated the remarkable industrial growth. Hence it would be unrealistic to evaluate water power benefits in terms of steam power costs. Instead, one would have to ask what would happen if a certain hydro development could not take place. Maybe it would be replaced by another, slightly more costly, hydro development. Maybe it would not be replaced at all, and, as a result, the dependent industrial development would not take place. In that case the water power benefits would be somewhere in between the hydro development under consideration and the least costly alternative power development.

Navigation Benefits

When a system of waterways and navigation facilities is included in a water resources development plan, the benefits of those navigation works will have to be estimated. We may distinguish again between direct and indirect benefits. The direct benefits may be determined in terms of the cost of the most likely alternative means of providing transportation. The indirect benefits are the impact of the project on the rest of the economy. It was noted earlier that it is conservative practice to take only the direct benefits into consideration. Nevertheless, it is not uncommon that navigation projects are considered justified because of their indirect benefits. Shortly after the Second World War, for instance, a thirty million dollar canal system was built in the northeastern part of Holland, to promote industrialization in order to relieve the surplus labour situation in that part of the country.

As an example of computing direct benefits, it is of interest to quote from a publication, *The Bonneville Project*, The Engineer School, Fort Belvoir, 1950:

'The Bonneville project was constructed during 1934–42 to improve navigation and develop hydro-electric power on the middle Columbia River. With a 66 ft. lift shiplock and a 518,000 kW hydro plant, this $84,000,000 development ranks as one of the largest achievements of the Corps of Engineers.

'The allocation of cost between the project functions of power and navigation was made as follows:

Navigation Lock.	.	.	$ 6,000,000
Power Plant	.	.	$38,000,000
Joint Facilities	.	.	$40,000,000

'Navigation is made possible over a distance of 47 miles at a cost of $26,000,000, which equals $55,000 per mile. (For comparison, waterways usually cost from $150–$500,000 per mile; a first-class single-track railway will cost in excess of $150,000 per mile.)

'Assuming an interest and amortization charge of 2·5 per cent on the navigation investment, amounting to $650,000 per year, plus lock operating and maintenance cost of $100,000 per year, the annual cost of $750,000 per year amounts to $1·09 per ton for the 695,000 tons per year of present traffic. River traffic is rapidly increasing and for a potential traffic of 10 million tons per year, the cost per ton will reduce to 13·4 cents. The cost per ton mile would be 2·8 mills. Assuming an average operating cost of 2 mills per ton mile, the total estimated transportation cost of 4·8 mills per ton mile is far below the average railroad haul cost of 1 cent per ton mile.'

In many instances the issue whether or not navigation facilities will pay for themselves is not so clear cut, as illustrated by an abstract from a paper by Curran (1955).

'The authorized Missouri River Navigation Project covers the lower 762 miles of the Missouri River. Nearly $300,000,000 has been spent on this project in efforts to stabilize a 9 ft. deep and 300 ft. wide channel, by channel regulation, dredging and flow regulation up to 25,000 cusec. The estimated additional amounts needed to complete construction are over $100,000,000.

'Despite these large expenditures of federal funds, traffic on the Missouri River has been almost negligible. In 1953, for both projects from the mouth to Kansas City and on to Sioux City, the total traffic was 2,544,026 tons, of which only 16 per cent was commercial traffic and the rest was waterway improvement material and sand, gravel and crushed rock. However, the Corps of Engineers is hopeful of greater traffic development in the future, noting a "marked increase of interest in Missouri River navigation even though the overall project is only about two-thirds complete and full project depths have not yet been obtained !"

'In a report dated September 1954, the Corps of Engineers presented the Task Force on Water Resources and Power with additional data bearing on the economic justification of the Missouri River navigation project. Included therein is an estimate derived by a "recent study" that "savings in transportation costs (to shippers) will average $14,650,000 of which $7,325,000 is assigned to the project Sioux City to the mouth and an equal amount to upstream reservoirs !"

'Thus, within the span of a few years, three different estimates of "transportation savings" to result from the Missouri River navigation project, increasing from $2,450,000 to $6,699,000 and then to $7,325,000 annually, have come from the same agency.

'A report by the Missouri Basin Inter-Agency Committee concludes that the water supply on the Missouri River is adequate to satisfy the requirements of the 9 ft. navigation project for a full 8 months' navigation season with developments as proposed to 1960. Under the same water supply situation, the developments by 1970 would anticipate reducing the navigation season by about 2 weeks for 25 per cent of the years. By 1980 the development would require further reductions in navigation. By the year 2000, with the proposed developments, the reduction in navigation season would amount to 2 weeks in 1 year and to 4 months in 13 years in the event of drought recurrence. In the event that recurrence of the severe prolonged drought conditions of the '30s are experienced too frequently under fully developed conditions, and the resulting curtailment of the navigation season is found to place too heavy a handicap on successful operation of the navigation project, canalization of the river could be resorted to, which would provide for 9 ft. navigation on much lower levels of flows.'

The above report was directed to hydrologic problems and it presented no evaluation of the economic impact on the project of the reduction of the navigation season by 50 per cent over 13 full years or of the cost of the channel maintenance by dredging which could be undertaken "without excessive difficulty", or the cost of the canalization that "could be resorted to". It might be desirable to obtain some estimate of the probable cost of such additional developments before any great amount of additional monies are added to the $300,000,000 used to develop the existing improvement.'

A last example of dealing with the problem of navigation versus other means of transport is provided by a report on Saskatchewan River Navigation by Kuiper (1958). The cost of adding navigation facilities to a series of proposed power dams on the Saskatchewan River, from Empress to Lake Winnipeg, was estimated at $230,000,000. The shiplocks would have a size of 86 by 540 ft. and an annual capacity of 10,000,000 tons. The total annual cost of all navigation facilities was estimated at $18,000,000. The total freight traffic by rail across the Prairie Provinces in 1957 was 50,000,000 tons. It was assumed that about 10 per cent or 5,000,000 tons per year would be diverted to transport by water. The cost of transport from Selkirk to Empress of bulk goods by rail is 8 mills per ton mile, and the total distance by rail is 725 miles! This results in a cost of $5.80 per ton. The cost of transport by water was calculated at 4 mills per ton mile, over a distance of 1,100 miles. Hence the cost per ton becomes $4.40. Therefore, the benefits of the navigation facilities would amount to $5,000,000 \times \$1.40 = \$7,000,000$ per year, which is less than half of the annual cost of the project. As a result of these preliminary findings, the proposal has been no further pursued for the time being.

The above examples were intended to illustrate some methods of approach, and some difficulties encountered in answering the question whether or not the provision of navigation facilities should be incorporated in a development plan. The trend of thought in dealing with this problem can be summarized as follows:

First, a rough study is made of the possibility of establishing or improving waterways in the drainage basin under consideration. If the possibility exists, another rough study is made of the potential traffic that could be shifted from rail and road to water transport, the direct savings that could accrue, and the indirect benefits that may ultimately be attained. If the situation looks promising, a detailed study of providing navigation facilities will be carried out, and the relationship with other water uses will have to be appraised carefully. After the navigation plan has been worked out satisfactorily from an engineering viewpoint, the annual cost of the project can be determined. At the same time, an economic study is carried out to determine direct annual benefits and long-term prospects for the instigation of industrial development. After having compared the results, a decision can be made and the study of the over-all water plan can proceed with or without the provision of waterways for navigation.

Water Supply Benefits

In the above sections we have discussed that flood control benefits are measured in terms of damages prevented, and reclamation benefits in terms of increased farm income. Both benefits are measured in terms of real goods that the community gains with the project. The benefits from water power and navigation, on the other hand, were measured in terms of the cost of producing the same goods by other means. This is also a reasonable and convenient yardstick. It may be noted that this principle could not be applied to flood control and reclamation, because there are no alternative means to accomplish those purposes. We can only prevent flood damage by controlling the floods; we can only increase the farm output by improving the moisture conditions on the land.

The problem of evaluating the benefits of a water supply project is somewhat different again. There is no alternative commodity for drinking water, and yet its value is very difficult to determine. We may recall from the previous chapter, Water Supply, that domestic water requirements are relatively small, that industrial water requirements may be somewhat larger, and that waste disposal requirements may be quite substantial. A city of one million people may only need an average flow of a few hundred cusec for domestic and industrial purposes. More than half of this amount will be returned to the river system. The waste disposal requirements of this city may amount to a few thousand cusec. However, none of this water is consumptively used.

By lack of any direct method to determine the real value of domestic and industrial water, it is common practice to express the benefits of providing such water in terms of the cost of the least expensive alternative source of water supply. However, if the nearest alternative is much more costly than the project under consideration it is possible that the beneficiaries would not be willing to pay the higher price, or that the consumption of water would be substantially reduced. In such cases we should temper our evaluation of water supply benefits accordingly.

Evaluating the benefits of providing waste disposal water may be done in terms of the cost of alternative sewage treatment plants, as was discussed in the previous chapter.

The difficulty of finding a proper evaluation of water supply benefits is not as regrettable as it may appear. The main purpose of the benefit–cost analysis is to have a tool for making decisions. Quite often, the decision to provide adequate water supplies for municipal purposes is made without a prior benefit–cost analysis. We may then look upon the water supply requirements as having priority in the overall basin development plan. The subsequent benefit–cost analysis applies then primarily to the remaining aspects of water development.

SELECTION OF PROJECT

In the above sections has been discussed how the annual costs and the annual benefits of alternative project proposals have to be determined. After this is done, the question arises what sort of an economic analysis we should apply to these figures, in order to select the most feasible project.

One of the most efficient ways to grasp the situation is to plot all project benefit–cost results in one diagram, as is shown in *Figure 11.8*. There are 12 different, more or less alternative, proposals of basin development. Each project has its own annual cost and annual benefit estimate. (If the annual costs or the annual benefits change during the life of the project, the analysis must be

431

made on the basis of the present values of the total costs and benefits during the life of the project.) It may be seen from *Figure 11.8* that, except for No. 8, all projects have a benefit–cost ratio larger than 1. At first glance it would seem therefore that all these projects would merit consideration. In fact, No. 12, having the largest benefits of all, and still a benefit–cost ratio larger than 1, would seem to be the most desirable. This conclusion, however, is incorrect, as will be discussed in the following paragraph.

Let us start at the left side of the diagram. It may be seen that project No. 2 has a larger ratio of benefits over costs than any other project. This also appears to be an attractive feature. If a private investor had to make a choice between the 12 alternatives, he might very well choose No. 2, because it would give him a greater return on his investment than any other project

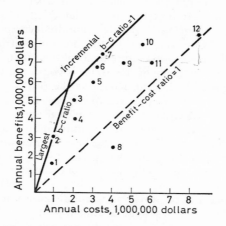

Figure 11.8. Selection of most economic project

(assuming that all projects have approximately the same annual charges). However, when discussing water resources development, we should analyse the situation from a public viewpoint and not from the viewpoint of the private investor. When we proceed from project No. 2 to No. 3, or 5, or 6, or 7, we may note that the increment in annual benefits is larger than the increment in annual costs. Assuming that the interest, included in the annual charges, represents a satisfactory return on the investment, we must therefore conclude that it is advantageous to proceed to a larger project as long as the increment in benefits is larger than the increment in costs.

It may be seen from *Figure 11.8* that when we take a line under 45 degrees (incremental benefit–cost ratio = 1) and we let it descend upon the cloud of points, the first point we touch, No. 7, represents the project with the greatest excess of annual benefits over annual costs. We may also observe that from that point, we can proceed to no other point and obtain an increase in benefits that is larger than the corresponding increase in costs. Hence we have now found the most feasible project of the entire group. If we would proceed from project No. 7 to No. 12, which seemed at first glance a desirable development, we would be adding a considerable amount of annual costs and only a small amount of annual benefits, which would be wrong from a viewpoint of public investment.

It may therefore be concluded that the most feasible water development project is the combination of elements that results in maximum net benefits. Net benefits are maximized when the scope of development is extended to the point where the benefits, added by the last increment in scope, are larger than, or equal to, the cost of adding that increment; and where the benefits, added by the next increment in scope, are smaller than the corresponding cost.

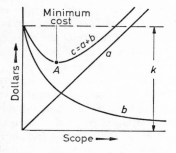

Figure 11.9. Minimizing costs

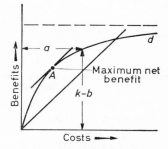

Figure 11.10. Maximizing benefits

It may be pointed out that the above analysis is essentially the same as the more familiar analysis of minimizing total costs, as shown in *Figure 11.9*. On the horizontal scale is shown the scope of the project, either in dollars, or in some other convenient unit (storage capacity, reservoir level, channel capacity, installed generating capacity, etc.). The unit of the vertical scale is dollars. Curve *a* in *Figure 11.9* represents the annual cost of the project. Curve *b* represents the associated 'costs'. For instance, in the case of a flood control project, curve *b* would represent the residual average annual damages for any project size, while the value *k* represents the average annual damage under natural conditions. In the case of a water power development, curve *b* would represent the annual cost of alternative steam power to satisfy a given load demand, with the value *k* being the maximum cost of steam, when no hydro is introduced at all. Curve *c* represents the sum of curves *a* and *b*. The lowest point on curve *c* indicates the most desirable scope of development.

By comparing *Figures 11.9* and *11.10*, the latter being a schematized presentation of *Figure 11.8*, it may be seen that point *A* in *Figure 11.9* has the same meaning as point *A* in *Figure 11.10*. In *Figure 11.9* point *A* is the minimum value of $a+b$; in *Figure 11.10* point *A* is the maximum value of $k-b-a$ which must indicate the same scope of project.

After having completed an economic analysis, as described above, one may now be tempted to make a definite choice with respect to the size of the project. However, before we do so, let us dwell for a moment on some aspects that may have been overlooked. First of all, let us ask with what degree of accuracy our computations have been performed. There are three areas where main errors can be made. First of all, in deciding upon the annual charges that have to be applied to the capital cost. It is very difficult to estimate several years in advance at what interest rate money can be obtained to finance the project. An error of

433

one or two per cent in the estimated rate of interest would not be impossible. When the total annual charge is say 8 per cent, a 2 per cent error would mean a 25 per cent error in the estimated annual costs! Secondly, the estimates of the benefits are also open to substantial error. In estimating future farm benefits, for instance, one group of economists may come up with one figure, while another independent group may come up with a figure that is 25 per cent smaller or 50 per cent larger! In the economic analysis of flood control projects, the third source of error could be the estimate of flood frequencies. A seemingly slight upturn or downturn of the frequency curve on log probability paper may make a large difference in the computed average annual damages and benefits, particularly when the damages keep increasing rapidly with increasing flood magnitudes. One does not have to stretch the imagination very far to obtain average annual damages that are either 50 per cent smaller or 100 per cent larger than the values obtained by someone else.

When such inaccuracies as described above, which cannot be avoided because of the limited amount of basic data and foresight, aggravate one another, it will be appreciated that the resultant benefit–cost ratios should not be shown in more than one figure behind the decimal point, and even then a benefit–cost ratio should not be considered more than merely indicating an order of magnitude. It may be noted here that such errors as described above, are not too harmful when the purpose of the economic analysis is to select the most feasible scheme. It may be assumed namely that any error in interest rate, or damage estimate, or flood frequency, will affect every project benefit–cost ratio to nearly the same extent. However, when the benefit–cost ratio is used to justify the construction of a project, we should review the whole procedure, and whenever there is doubt about the basic assumptions, we should stay on the conservative side; that is, we should make assumptions that keep the benefit–cost ratio on the low side.

We must realize that the economic analysis, discussed so far, has been based on direct and tangible benefits. It was noted earlier that it is conservative practice to base the analysis on these benefits only, but in the final analysis, when a selection between projects has to be made, it would be quite correct to acknowledge the existence of indirect and intangible benefits. It is not impossible that several alternative projects are so close to being the most feasible one that their differences in net benefits become insignificant in view of the probable errors, discussed above. In such a case an appraisal of the indirect and intangible benefits, associated with each project, would provide much better guidance for the final selection, than placing complete confidence in precisely quoted benefit and cost figures. For this reason it is advisable to include in the economic analysis of water development projects a clear qualitative description of expected indirect and intangible benefits. Such a description should not be used to justify the construction of a project, but to select the most feasible project from an array of projects that are already justified because of their net direct benefits.

Intangible benefits that should be acknowledged include the following: the prevention of loss of life as a result of flood control measures; the enhancement of scenic values due to stabilizing lake levels and the creation of reservoirs. (We must, of course, also take into account the possible loss of scenic values, which has to be added as an intangible cost to the project.) When comparing irrigation and water supply benefits with water power and navigation benefits, we must keep in mind that electricity and transportation can also be obtained by other means, but that crops and people must have water, or else they cannot live. The construction of reservoirs may be associated with recreation benefits. When describing such benefits it is important to note the alternative recreational facilities within a few hundred miles of the project.

Up to now we have applied our economic analysis to one particular river basin development or maybe only one phase of river basin development. From different, alternative project proposals we have selected the most feasible one, taking into account the magnitude of net direct benefits, and if necessary allowing for intangible benefits. Let us now look at the problem from the viewpoint of public investment, and let us recognize that there is not only this particular river development project, but that there may be many more across the country, all of them competing for the limited amount of available public

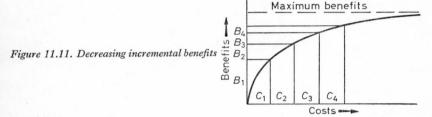

Figure 11.11. Decreasing incremental benefits

funds. The question now arises, what project should be started first. In order to assure the greatest flow of future economic values, as soon as possible, the government may develop first the projects with the highest benefit–cost ratios.

Anticipating such a policy, we may be tempted to somewhat reduce the size of our particular project under consideration in order to increase its benefit–cost ratio, so that its competitive position with respect to other projects is improved. However, in doing so, we must be very careful that we do not make a choice that would result in permanent underdevelopment of the natural resources. For instance, if the most feasible project, with the highest net benefit, consists of three separate reservoirs, and we can improve the benefit–cost ratio by eliminating one of the three reservoirs, then it would be quite permissible to submit the two reservoirs as a definite proposal, indicating that eventually the third reservoir should be built when funds become available. However, if the most feasible project would consist of one dam, 500 ft. high, and we could improve the benefit–cost ratio by reducing the height to 300 ft., we should not do so, because that would permanently eliminate the possibility of full resources development.

Another reason why it is good policy to construct first the projects with the highest benefit–cost ratio, is to avoid below-marginal projects from slipping into the development programme. In many cases, certain projects have different benefit–cost ratios, depending on when they are added to the programme. Let us illustrate this with a flood control scheme, consisting of reservoirs, diversions and dikes. The first unit that is assumed, has relatively large benefits; the second unit with the same 'control effectiveness' has smaller benefits; the third unit still smaller benefits, etc., as shown in *Figure 11.11*.

435

We may therefore conclude that every component in a water development plan has a benefit–cost ratio that not only depends on the qualities of that component, but also on its place in the sequence of development. This means that there is one last unit that was still included because it still had an incremental benefit–cost ratio larger than one. We may call this a marginal unit, in spite of the fact that it may have a very attractive benefit–cost ratio when added first. It was noted earlier that the accuracy of benefit–cost ratios is open to question. It would therefore be no surprise that, when the whole economic analysis would be repeated 10 years later with new basic data, we would find that this marginal unit has to be removed from the plan. Since the development of a river basin plan may easily extend over several decades, during which time the designs and economic analyses of the remaining projects are periodically reviewed, it follows that the safest procedure would be to leave the marginal projects to the last, and to develop the projects with the highest benefit–cost ratio first.

CHAPTER 12

BIBLIOGRAPHY

CHAPTER 1: ENDS AND MEANS

BROWN, H. (1954). *The Challenge of Man's Future*. 290 pp. New York; The Viking Press. (The first part is devoted to reviewing statistics of human population and availability of natural resources. The last and most interesting part contains a discussion of the possible ways of living that man may adopt or drift into, in view of the dwindling supplies of readily available resources.)

CROSS, H. (1952). *Engineers and Ivory Towers*. 141 pp. New York; McGraw-Hill. (A collection of miscellaneous essays on the aims of the engineering profession, written from a somewhat philosophical viewpoint.)

FINCH, J. K. (1951). *Engineering and Western Civilization*. 374 pp. New York; McGraw-Hill. (This book presents a socioeconomic analysis of the concurrent rise of engineering and the western way of life. It discusses the effect of economic and social conditions upon technological development and progress. Some chapter headings: Ancient engineering; American engineering in the 19th century; Modern engineering; Engineering and economic change; the distribution of technological gains; Engineering and research.)

GALBRAITH, J. K. (1958). *The Affluent Society*. 368 pp. Cambridge, Boston; The Riverside Press. (This book is written by an outstanding economist and is regarded as a landmark in the economic literature. Its main theme is that in our capitalistic society we aim to produce more and more, for the sake of a healthy economy. There is almost no concern over what we produce. There is a danger that we shall produce more and more of what we need less and less, thus causing social imbalance. The author advocates the search for a substitute for production as a source of income, in order to break the present link between production and income.)

HUXLEY, A. (1932). *Brave New World*. 213 pp. London; Chatto and Windus. (The author, an outstanding novelist and philosopher, describes a future Utopian world. Science and technology have made fantastic progress, but the human mind is completely controlled and society is regimented into worker and leader units to ensure maximum efficiency and stability. In a recent essay: 'Brave new world revisited' the author concludes that much of his frightening fantasy of the thirties is already taking place at the present time.)

HUXLEY, A. (1948). *Ends and Means*. 335 pp. London; Chatto and Windus. (The author begins this essay by observing that progress of mankind should be measured in terms of his virtuous behaviour. Technological advance is rapid, but this may provide us with more efficient means for going backwards. In the remainder of the book, the author attempts to answer how existing society can be transformed into a more ideal one.)

HUXLEY, J. (1957). *New Bottles For New Wine*. 318 pp. London; Chatto and Windus. (The author, an outstanding biologist and philosopher, presents in this book his views on human destiny. In this 20th century the new vision of this destiny begins to emerge through the rapid development of the natural sciences, the social sciences and the humanities. Man is now capable of forming a new picture of himself, his place in nature, and his relations with the world around him. This will enable him to build new and more adequate beliefs. Some chapter headings: Transhumanism; Man's role in nature; Evolution; Human fulfilment. This book has recently been reprinted as a paperback in the 'Mentor Book' series under the title: *Knowledge, Morality, and Destiny*.)

HUXLEY, J. (1961). *The Humanist Frame*. 432 pp. London; George Allen. (The author has attempted to present humanism as a comprehensive system of ideas.

To this end, he has written a 48-page introductory chapter dealing with the aims of man's existence in general, and has called upon 25 contributing authors to discuss particular aspects. Some of these essays are interesting, such as: 'New horizons for under-developed peoples' by Sudhir Sen, and 'Humanist economics,' by Robin Morris.)

KUIPER, E. (1962). *The Ideals of the West*. 51 pp. Winnipeg. The University of Manitoba. (The author attempts to find a sense of purpose in our present western civilization in the hope that this could provide guidance to defining aims and objectives for social activities in general and for the engineering profession in particular.)

LILIENTHAL, D. E. (1953). *T.V.A. Democracy on the March*. 294 pp. New York; Harper Brothers. (The former administrator of the Tennessee Valley Authority during its conception and creation gives an account of the philosophy behind the project.)

ORWELL, G. (1949). *Nineteen Eighty Four*. 312 pp. London; Secker and Warburg. (The author, an outstanding novelist, has presented a frightening image of what our future society may drift into, if people have no clear concept of the way of life they wish to follow. In spite of an advanced technology, life has become a nightmare.)

PACKARD, V. (1960). *The Waste Makers*. 340 pp. New York; David McKay Co. (The author, a professional journalist, has presented a perhaps somewhat simplified, but very readable account of the present state of the U.S. economy. By means of several refined and subversive strategies, the large manufacturers force the people into ever-increasing consumption. As a result, our natural resources are vanishing rapidly, our way of life becomes permeated with commercialism, and our young people become preoccupied with materialism.)

CHAPTER 1: PLANNING

ACKERMAN, E. A., and LOF, G. O. G. (1959). *Technology in American Water Development*. 710 pp. Baltimore; The Johns Hopkins Press. (The two authors, an economist and a chemical engineer, first describe the physical and economic environment of water development in the U.S.; then they discuss how present advances in technology may affect the demand for water; this is followed by a discussion of increasing the available water supplies by such means as demineralization of water, weather control, use of ground water and evaporation control; the last section is devoted to organizational problems.)

ASCE COMMITTEE ON WATER RESOURCES PLANNING (1962). 'Basic considerations in water resources planning.' *Proc. Amer. Soc. civ. Engrs.*, Sept. 1962, HY5, pp. 23–55. (This paper discusses the principles, techniques, problems, and areas of deficiency involved in the field of water resources planning. The main topics are: political jurisdiction for planning; basic data requirements; availability of water; evaluating present and future water use and control requirements; factors influencing water resources planning; development of over-all plan.)

BANKS, H. O. (1959). 'Statewide water planning.' *Proc. Amer. Soc. civ. Engrs.*, Dec. 1958, IR4, paper 1861, 8 pp. (This paper discusses the importance of long-range planning in development of water. Particular reference is made to the California water plan and its three phase investigation: 1. An inventory of resources; 2. An inventory of requirements; and 3. A master plan for conservation and redistribution of the State's water supplies.)

CALIFORNIA WATER RESOURCES DEPARTMENT (1957). 'The California water plan.' *Bull. No. 3*, 246 pp. (A monumental water study, incorporating many features of river basin planning, and including flood control, water supply, irrigation and water power projects. The report begins with an inventory of water resources; then it proceeds to a discussion of the water problems of California; this is followed by the development of the overall plan; the report closes with a proposal for implementation of the plan. The estimated cost of the total plan is some twelve billion dollars.)

CONFERENCE (1959, 1960, 1961). *Western Resources*; conferences held in Colorado each summer, starting in 1959 and continued till the present. Papers of the conferences are published by the University of Colorado Press, Boulder. (The majority of the papers deal with water resources development. Some of these are listed under the author in this bibliography.)

CONFERENCE (1961). *Resources for Tomorrow*; conference, held in Montreal, October 23–28, 1961: Vol. I, papers on agriculture, water, regional development, 623 pp.; Vol. II, papers on forestry, wild life, recreation and fisheries, 438 pp.; Vol. III, proceedings of the conference, 519 pp.; Queen's Printer, Ottawa. (The conference included twelve papers on water resources development. Some of these are listed under the author in this bibliography.)

CURRAN, C. D., and FLEMING, F. W. (1961). 'Land and water resource planning in Texas.' *Proc. Amer. Soc. civ. Engrs.*, Nov. 1961, WW4, pp. 41–52. (This paper describes the functioning of the U.S. study commission, Texas, which is a new attempt to coordinate federal, state, and local interests for area-wide water resources planning.)

ECKSTEIN, O. (1958). *Water Resources Development*. 300 pp. Cambridge, Harvard University Press. (The author deals with the problem of river basin development primarily from an economic viewpoint. First he applies the economic concepts of market value and alternative cost to water resource development; then he reviews different aspects of water development; finally he discusses benefit–cost criteria and the allocation of joint costs.)

ENGINEERS JOINT COUNCIL (1957). 'Principles of a sound national water policy.' *Amer. Soc. civ. Engrs., Civil Engineering*, May 1957, pp. 52–57. (This is a restatement of the 1950 report on the same subject. A great number of engineers from all branches of engineering have contributed to the report. Its principal conclusions: 1. Development plans should provide for maximum utilization of resources. 2. Control of the waters of the U.S. is in the national interest, but not necessarily a function of the Federal Government. 3. Beneficiaries of federal water projects should reimburse their share of the cost. 4. Project reports should propose reimbursement plans. 5. All projects within a river basin should be coordinated within a comprehensive plan. 6. There should be a board of review for the impartial analysis of all federal water resources projects. 7. All federal agencies should conform to the same water policy.)

FOX, I. K., and PICKEN, I. (1960). 'The upstream–downstream controversy in the Arkansas–White–Red Basins survey.' University of Alabama Press, 53 pp. (Between 1950 and 1955 the Department of Agriculture, the Corps of Engineers and other agencies prepared a comprehensive river basin plan of the Arkansas–White–Red Basins. The authors analyse and evaluate the performance of the inter-agency committee. Because of basic policy differences between the agencies, the requirements for an integrated plan were never fulfilled.)

HIRSHLEIFER, J., and others. (1960). *Water Supply*. 378 pp. Chicago; The University of Chicago Press. (This book has been widely read since its publication. The three authors, two economists and a chemist, take a somewhat unorthodox standpoint with regard to the development of water resources. Although the book deals primarily with water supply for municipal, industrial and irrigation use, its principles can be extended to other areas of water development. It is contended that the real need for water is not as large as it is made appear to be by the river basin planners. A more realistic pricing of water would reduce the demand and bring it closer in balance with the supply.)

JOHNSON, W. E. (1957). 'Missouri river basin plan in operation.' *Trans. Amer. Soc. civ. Engrs.*, Vol. 122, pp. 654–65. (A brief description by the chief engineer of the project. Deals primarily with the functioning of the big reservoirs on the Missouri River.)

KRISTJANSON, K., and SEWELL, W. R. D. (1961). 'Water management problems and issues in Canada.' *Resources for Tomorrow* conference papers, Queen's Printer, Ottawa, pp. 205–10; reprinted in *Regional and Resource Planning in Canada*, Holt, Rinehart and Winston, Toronto 1963, pp. 173–79. (Comprehensive planning is required for such rivers as the Fraser, Columbia, Nelson, Ottawa and others. Neither the federal nor provincial governments have the organizational arrangements to conduct such planning. This makes it necessary to modify the existing policies and procedures for water development.)

BIBLIOGRAPHY TO CHAPTER 1: PLANNING

KRUTILLA, J. V. (1960). 'River basin development.' *Resources for the Future.* Washington, 35 pp. (The author, who is an economist, deals with the sequence and timing in river basin development with special application to Canadian–United States Columbia River Basin planning. He discusses the importance of constructing the most economic projects first, to ensure that a non-economic project is not incorporated early into a system only because benefits evaluated on a first added basis excel its cost.)

KRUTILLA, J. V., and ECKSTEIN, O. (1958). *Multiple Purpose River Development.* 301 pp. Baltimore; The Johns Hopkins Press. (The main theme is the application of economic principles to water development problems. The first part deals with a theory of efficient resource allocation and the social cost of federal financing. The second part applies the theory to such cases as the Hells Canyon Dam, the Alabama–Coosa River System, and the Willamette River Development.)

KUIPER, E. (1961). 'The water resources of the Nelson River Basin.' *Resources for Tomorrow* conference papers, Queen's Printer, Ottawa, pp. 337–44; reprinted in *Regional and Resource Planning in Canada*, Holt, Rinehart and Winston, Toronto 1963, pp. 180–89. (The author gives first a review of the present state of water development in the 440,000 square mile Nelson Basin; then he projects the economic activity of the Canadian prairie region into the future, estimates water requirements, and develops a water plan, involving the diversion of several major rivers.)

LINDNER, C. P. (1958). 'The Apalachicola river basin project.' *Trans. Amer. Soc. civ. Engrs.*, Vol. 123, pp. 973–88. (Four major dams on the Apalachicola River have been planned for navigation, flood control, water power, and stream flow regulation. Some of the planning and design aspects are presented in this paper.)

MAAS, A., and others. (1962). *Design of Water Resources Systems.* 660 pp. Cambridge; Harvard University Press. (This book, prepared by six authors, is the result of several years of study under the Harvard Water Resources Program. The authors approach the problem of river basin development from a highly theoretical viewpoint. First, they develop efficiency and economic objectives; then they apply hydrologic and engineering criteria; then they prepare a mathematical model, that is subsequently solved on a digital computer. It is believed that these techniques allow planners to compare many more alternatives than they do under present practices.)

MARTIN, R. C., and others (1960). *River Basin Administration and the Delaware.* 390 pp. Syracuse University Press. (This study represents the first major attempt to examine the physical, economic, institutional, political, and administrative factors bearing on the problem of government water-resources administration of a specific river basin, prior to development. After having analysed experience in other river basins, and having appraised the local situations, recommendations are made for an administrative structure for the Delaware Basin Development.)

RITER, J. R. (1955). 'Planning a large irrigation project.' *The Yearbook of Agriculture 1955*, U.S. Government Printing Office, pp. 328–33. (The author gives a brief summary of all phases involved in the process of planning, including the reconnaissance plan, a study of land resources, a study of water supplies, and the economic analysis.)

SCHAD, T. M. (1962). 'Perspective on national water resources planning.' *Proc. Amer. Soc. civ. Engrs.*, July 1962, pp. 17–41. (This paper traces the development of national planning for river basin development in the U.S. from the early 19th century to date; it also considers the role of the civil engineer in comprehensive water resources planning for the future.)

SPENGLER, J. J. (1960). 'Natural resources and economic growth.' *Resources for the Future.* Washington, 306 pp. (A collection of papers, presented at a conference held at Ann Arbor in 1960. Included are such topics as: the role of resources at different stages of economic development; international trade and investment; resource policies in relation to economic growth.)

THOMAS, R. O. (1960). 'Water—a limiting resource.' *Trans. Amer. Soc. civ. Engrs.*, Vol. 125, pp. 101–22. (This paper discusses the anticipated demand for water in the U.S., which is compared with the availability. The paper intends to evoke discussion on the apparent adverse balance between supply and demand.)

UNITED NATIONS ECONOMIC COMMISSION FOR ASIA AND THE FAR EAST (1955). *Multiple-purpose River Basin Development:* Part 1, 'Manual of river basin planning,' 79 pp.; Part 2A, 'Water resource development in Ceylon, Taiwan, Japan and the Philippines,' 140 pp.; Part 2B, 'Water resource development in Burma, India and Pakistan,' 135 pp., New York and Bangkok. (These three volumes contain a discussion of the principles of river basin development, such as appraisal of water resources, stages of investigation, benefit–cost analysis, and their application to problems in the Far East.)

U.S. BUREAU OF RECLAMATION (1954). *Colorado River Storage Project,* 83rd Session, House Document No. 364, 332 pp. Washington; U.S. Govt. Printing Office. (This report deals with the planning, design, and economic analysis of the Colorado River storage project, which includes the 400 million dollar Glen Canyon dam. The main purpose of this project is water power and irrigation development. Additional benefits accrue to water supply, flood control and recreation. For design features of the Glen Canyon dam see the reference: Schultz, 1962, Chapter 5.)

U.S. COMMISSION ON ORGANIZATION OF THE EXECUTIVE BRANCH OF THE GOVERNMENT (1955). *Report on Water Resources and Power.* 219 pp. Washington; U.S. Govt. Printing Office. (This report discusses primarily administrative policies regarding water resources development. Of particular interest are the three supplementary volumes of task force reports, dealing with power, reclamation, water supply, flood control, and navigation; and presenting 17 special studies by different specialists. Some of the reviews are highly critical of prevailing government practices. Abstracts of the task force reports may be found in *Proc. Amer. Soc. civ. Engrs.,* Papers 897–900 incl., Power Division, 1956.)

U.S. CORPS OF ENGINEERS (1958). *Water Resource Development of the Columbia River Basin.* A report by the Division Engineer, North Pacific, 396 pp. (This report, and six appendixes, present an economical and practical plan for water resource development in the Columbia basin to meet the future needs of the region. The major water plan includes dams, power plants, irrigation works and navigation facilities at a total cost of nearly two billion dollars.)

U.S. PRESIDENT'S WATER RESOURCES COUNCIL (1962). 'Policies, Standards, and Procedures.' 13 pp. Washington; U.S. Govt. Printing Office. (This Senate document provides an up-to-date set of uniform standards for the formulation and evaluation of water resources projects. It has been prepared by the four Departments of the Army, the Interior, Agriculture, and Health, Education and Welfare.)

U.S. PRESIDENT'S WATER RESOURCES POLICY COMMISSION (1950). Vol. I: *A Water Policy for the American People,* 445 pp.; Vol. II: *Ten Rivers in America's Future,* 801 pp.; Vol. III: *Water Resources Law;* Washington; U.S. Govt. Printing Office. (The first volume deals with the extent and character of federal government participation in major water resources programmes; and appraisal of the priority of such programmes from the standpoint of economic and social need; and criteria to evaluate the feasibility of such projects. Some of the chapter headings in Vol. I: National objectives; Unity in planning; A resources investment program; The need for basic information; Watershed management; Land reclamation; Water supply; Hydro-electric power; Flood control; Recreation; Legal aspects. The second volume deals with the present and future water development programmes in ten major river basins of the U.S.: the Columbia; the Central Valley of California; the Missouri; the Rio Grande; the Colorado; the Connecticut; the Alabama-Coosa; the Potomac; the Ohio; and the Tennessee. In each basin the topics of discussion are: regional characteristics; the river and its tributaries; present water resources development; project economics; plans for future development.)

U.S. SENATE SELECT COMMITTEE ON NATIONAL WATER RESOURCES (1959). *Report on Water Resources Activities in the United States.* 147 pp. Washington; U.S. Govt. Printing Office. (After an exhaustive study extended over several years, this committee reports on the extent to which water resources activities in the

U.S. are related to the national interest and on the extent of water resources activities that are required to provide quantity and quality of water needed by the country between 1960 and 1980. The main report is supported by 32 committee prints, dealing with different aspects of the problem, some of which contain extremely valuable material from the viewpoint of water resources planning. A few of the more important prints follow. Print No. 4: Surface water resources of the U.S. contains data on drainage area, minimum, average, and maximum discharge of more than a thousand rivers. Print No. 5: Population projections and economic assumptions. Print No. 7: Future water requirements for municipal use. Print No. 11: Future needs for navigation. Print No. 14: Future needs for reclamation. Print No. 15: Floods and flood control. Print No. 32: Water supply and demand.)

WHITE, G. F. (1959). 'A geographer's view of the problems of the South Platte,' Western Resources Conference Papers, University of Colorado, Boulder, 1960, pp. 75–84. (The author, who was chairman of the United Nations Panel on integrated river development, gives a critical review of a series of papers on a master plan for the development of the water resources of the South Platte River. His remarks are very interesting and would apply to any river basin. One of his main points is that the people, living in the basin, should be given a wide choice of the kind of resources development—water, minerals, transportation, city development, recreation—that they would like to see take place in their region.)

CHAPTER 2: HYDROLOGY

ASCE COMMITTEE ON HYDROLOGY (1949). *Hydrology Handbook. Amer. Soc. civ. Engrs.*, Manual of Engineering Practice No. 28, 184 pp. (This handbook is prepared for the hydraulic engineer who occasionally is confronted with a problem involving hydrology, but has no reason to prepare himself as a specialist in the field. Some chapter headings: Data processing; Infiltration; Run-off; Flood frequency; Evaporation and transpiration; Groundwater storage.)

ASCE SUBCOMMITTEE ON FLOODS (1953). 'Review of flood frequency methods.' *Trans. Amer. Soc. civ. Engrs.*, Vol. 118, pp. 1,220–30. (This report summarizes the available methods of predicting flood frequencies. It considers the factors affecting the accuracy of such predictions and gives the limitations of the various approaches to the problem.)

BERNARD, M. (1944). 'Primary role of meteorology in flood flow estimating.' *Trans. Amer. Soc. civ. Engrs.*, Vol. 109, pp. 311–82. (Precipitation data for 350 storms have been reviewed, about 150 are presented in this paper, including time-area-depth values. The principal meteorological elements are defined and discussed to give a basic understanding of the mechanism of storms.)

BERRY, W. M., and STICHLING, W. (1954). 'Evaporation from lakes and reservoirs in the northern plains region of North America.' *Int. Union Geod. & Geophys.*, General Assembly, Rome, Vol. 3, pp. 121–34. (This paper presents field observations as well as theoretical research to establish a method of estimating evaporation from open water surfaces for different locations and different times of the year.)

BIBLIOGRAPHY (1952). *Annotated Bibliography on Hydrology, 1941–50*. Washington 25, D.C.; U.S. Government Printing Office. (A comprehensive bibliography on hydrologic literature, published in the United States and Canada between 1941 and 1950.)

BURNS, J. I., and others (1961). 'Hydrology and flood control features of Oroville Dam.' *Trans. Amer. Soc. civ. Engrs.*, Vol. 126, Part IV, pp. 473–91. (This paper describes the hydrologic studies pertaining to one of the major components in the California Water Plan.)

BUTLER, S. S. (1957). *Engineering Hydrology*. 356 pp. Prentice Hall, Englewood Cliffs, N.J. (An elementary textbook on all hydrologic subjects, except ground water, which is discussed comprehensively in four chapters, 110 pp.)

CLARK, R. H. (1955). 'Predicting the run off from snow melt.' *Engng. J.*, EIC, April 1955, pp. 434–41. (This paper discusses the relationship between antecedent

BIBLIOGRAPHY TO CHAPTER 2

moisture conditions, snowfall, rainfall, the melting period, and resultant run-off, primarily for conditions in central Canada.)

CRAGWALL, J. S. (1955). 'Indirect methods of river discharge measurement.' *Proc. 6th Hydr. Conf., 1955*, State University of Iowa, pp. 35–59. (Often during floods, for many good reasons, current meter measurements cannot be obtained. Under such conditions, peak discharges may be approximately measured by indirect methods, as described in this paper.)

EZEKIEL, M. (1945). *Methods of Correlation Analysis*. 548 pp. New York; John Wiley. (One of the most comprehensive textbooks on this subject.)

HAZEN, A. (1930). *Flood Flows*. 160 pp. New York; John Wiley. (Although this book is somewhat out of date, Hazen's method of plotting flood flow data is still being used, and some of his writing is still of interest. Some chapter headings: Method of plotting annual floods; Drawing a smooth curve; Effect of one large flood; Seasonal distribution of flood flows; Effects of great storms.)

HICKOX, G. H. (1946). 'Evaporation from a free water surface.' *Trans. Amer. Soc. civ. Engrs.*, Vol. III, pp. 1–66. (An analogy is drawn between heat transfer and mass transfer as they occur in evaporation. The results of experiments are correlated by the use of dimensionless parameters.)

HORNER, W. W. (1956). *Hydrology*. (See reference Abbett, R. W., 1956, *American Civil Engineering Practice*, in Bibliography to Chapter 5.)

HORNER, W. W., and JENS, S. W. (1942). 'Surface run-off determination from rainfall without using coefficients.' (See in Bibliography to Chapter 6.)

HORTON, R. E. (1943). 'Evaporation maps of the U.S.' *Trans. Amer. Geophys. Union*, Vol. 24, pt. 2, pp. 743–53. (This paper presents maps of mean annual, winter, and summer evaporation, based on records.)

KUIPER, E. (1957). '100 frequency curves of North American rivers.' *Proc. Amer. Soc. civ. Engrs.*, Oct. 1957, HY5, Paper 1395, 31 pp. (This paper presents frequency curves of maximum annual flood flows of some large rivers in North America. The frequency curves are drawn as straight lines. It was found that the inclination of the lines is a function of drainage area, climate and soil conditions.)

LANGBEIN, W. B., and others (1949). 'Annual run-off in the U.S.' *Circ. 52, U.S. geol. Surv.*, U.S. Govt. Printing Office, Washington, 14 pp. (This report contains maps showing isograms of run-off, and briefly explains the climatologic, geologic, and topographic factors influencing the run-off. For more up-to-date figures on average stream flow see reference U.S. Senate Select Committee, 1959, Report on water resources, Print No. 4, in Bibliography to Chapter 1: Planning.)

LINSLEY, R. K., and others (1949). *Applied Hydrology*. 689 pp. New York, McGraw-Hill. (Although not the most recent, this is still one of the best and most comprehensive handbooks on the subject. Some chapter headings: Climate; Humidity; Winds; Precipitation; Snow; Evaporation; Groundwater; Hydrograph analysis; Run-off relations; Run-off distribution.)

MATTHES, G. H. (1956). 'River surveys in unmapped territory.' *Trans. Amer. Soc. civ. Engrs.*, vol. 121, pp. 739–58. (This paper presents short-cut methods for obtaining stream-flow data, adapted to exploratory or reconnaissance surveys. See also the writings of the same author on hydrometric surveys in Abbett, 1956, *Civil Engineering Practice*, Vol. II; for reference see Bibliography to Chapter 5.)

McCAIG, I. W., and others (1963). 'Hydrologic simulation of a river basin.' *Engng. J., EIC*, June 1963, pp. 39–43. (This paper describes the use of electronic computers for hydrologic studies in connection with flood control and power generation. The natural hydrologic processes and reservoir control procedures are represented by mathematical functions.)

MEINZER, O. E. (1942). *Hydrology: Volume IX of Physics of the Earth*. 712 pp. New York; Dover Publications. (One of the most comprehensive handbooks on the subject, prepared by 25 authors. Some of the chapter and section headings: Precipitation by M. Bernard; Evaporation by S. T. Harding; Snow by J. E. Church; Soil moisture by K. Terzaghi; Ground water by O. E. Meinzer; The unit hydrograph by L. K. Sherman; Mechanics of rivers by L. G. Straub.)

BIBLIOGRAPHY TO CHAPTER 3

REID, J. L., and BRITTAIN, K. G. (1962). 'Design concepts of the Braseau development, including river and hydrology studies.' (See in Bibliography to Chapter 7.)

RIESBOL, H. S. (1954). 'Snow hydrology for multi-purpose reservoirs.' *Trans. Amer. Soc. civ. Engrs.*, Vol. 119, pp. 595–627. (This paper describes briefly the processes of snow melting and run-off and presents criteria for determining reservoir space allocation and spillway capacities.)

SCHAFMAYER, A. J., and GRANT, B. E. (1938). 'Rainfall intensities and frequencies.' *Trans. Amer. Soc. civ. Engrs.*, Vol. 103, pp. 344–401. (Available records of excessive rainfall for 19 cities in the U.S. were plotted on semi-log paper. The graphs were straight lines of marked regularity.)

SHERMAN, L. K. (1932). 'Stream flow from rainfall by the unit-graph method.' *Engng. News Rec.*, Vol. 108, pp. 501–5. (The first publication of the unit hydrograph principle. For another discussion by the same author on the same subject see Meinzer, 1942, *Hydrology*, pp. 514–25.)

SNYDER, F. F. (1938), 'Synthetic unit hydrographs.' *Trans. Amer. geophys. Union.* Vol. 19, pp. 447–54. (This paper presents equations and empirical data that can be used to compose unit-hydrographs if no field data are available.)

SYMPOSIUM (1961). *Evaporation from Water, Snow and Ice.* 210 pp. Ottawa; National Research Council of Canada. (A collection of 11 papers by Canadian authors. Some of the titles: Evaporation from water, snow and ice by G. P. Williams; Evaporation computations for prairie reservoirs by G. A. McKay and W. Stichling; Evaporation from land surfaces by K. M. King.)

U.S. CORPS OF ENGINEERS (1946). 'Spillway design flood Garrison Reservoir.' 66 pp. Omaha, Nebr.; District Engineer's Office. (Comprehensive report on the determination of the maximum possible flood of the Missouri River at Garrison, dealing with such topics as: meteorological analysis, run-off characteristics, unit-hydrographs, flood routing, freeboard, and reservoir operations.)

U.S. GEOLOGICAL SURVEY (1945). *Stream-gaging Procedure.* 245 pp. Washington; U.S. Govt. Printing Office. (This manual describes in great detail the methods and practices of the U.S. Geological Survey to measure stream-flow discharge.)

U.S. WEATHER BUREAU (1947). 'Generalized estimates of maximum possible precipitation over the U.S.' *Hydromet. Rep. 23*, 77 pp. (A series of charts showing variation over eastern U.S. of limiting rainfall rates for areas of 10, 200 and 500 square miles and for specific duration.)

WILLIAMS, G. R. (1950). *Hydrology.* (See reference Rouse, H., 1950, Engineering Hydraulics, in Bibliography to Chapter 3.)

WILLIAMS, G. R. (1961). 'Cyclical variations in world-wide hydrologic data.' *Proc. Amer. Soc. civ. Engrs.*, Nov. 1961, HY6, pp. 71–88. (Precipitation and run-off records are analysed to discover correlation with sun spot cycles. Short-term cycles are revealed through use of 3 year moving averages. Long-term variations from the mean are indicated by use of differential mass curves.)

WILSON, W. T. (1941). 'An outline of the thermodynamics of snow melt.' *Trans. Amer. geophys. Union*, Vol. 21, pt. 1, pp. 182–95. (A comprehensive discussion of the rates of heat transfer in snow, the sources of heat, the effect of mechanical forces, and a suggested technique for computing snow melt.)

WISLER, C. O., and BRATER, E. F. (1959). *Hydrology.* 408 pp. New York; John Wiley. (An elementary textbook, reviewing the field of hydrology. Some chapter headings: The hydrograph; Factors affecting run off; Precipitation; Groundwater; Run off; Snow; Floods; Stream flow records.)

ZOLLER, J. H., and LENZ, A. T. (1958). 'Snowmelt run off.' *Proc. Amer. Soc. civ. Engrs.*, Nov. 1958, HY6, Paper 1834, 31 pp. (The primary factors relating to the melting of snow are evaluated for a river basin in Wisconsin.)

CHAPTER 3: HYDRAULICS

ABBETT, R. W. (1956). *American Civil Engineering Practice*, Vol. II. (See in Bibliography to Chapter 5.)

ALBERTSON, M. L., and others. (1960). *Fluid Mechanics for Engineers.* 567 pp. Englewood Cliffs, N.J., Prentice-Hall. (A good elementary and practical textbook on engineering hydraulics. The chapter on flow in open channels is

444

comprehensive and clear. Other chapter headings: Fluid statics; Fluid dynamics; Fluid resistance; Flow in closed conduits; Flow around submerged objects.)

ASCE TASK FORCE ON FRICTION FACTORS (1963). 'Friction factors in open channels.' *Proc. Amer. Soc. civ. Engrs.*, March 1963, HY2, pp. 97–143. (This report represents four years of work in assembling a comprehensive bibliography and in summarizing current thought on friction in open channels. It is concluded that use of Darcy-Weisbach *f* has advantages over the use of Manning's *n*.)

BROOKS, N. H. (1958). 'Mechanics of streams with movable beds of fine sand.' *Trans. Amer. Soc. civ. Engrs.*, Vol. 123, pp. 526–94. (This paper presents a laboratory study of the influence of bed configuration upon the relationship between sediment transport and channel slope. It was found that for a given discharge and slope, the depth of flow was a function of the sediment discharge. At a small sediment discharge, the bed became rough and the depth of flow became large. At a large sediment discharge, the bed became smooth and the depth of flow became small. For the laboratory flume, the sediment discharge could not be expressed as a function of depth and slope. However, when the water discharge and sediment load were used as independent variables, all other quantities were uniquely determined.)

CAREY, W. C., and KELLER, M. D. (1957). 'Systematic changes in the beds of alluvial rivers.' *Proc. Amer. Soc. civ. Engrs.* August 1957, HY4 Paper 1331, 24 pp. (This paper describes the sand wave formation and movement on the Mississippi River. The relation to hydraulic roughness is discussed.)

CHATLEY, H. (1938). 'Hydraulics of large rivers.' *J. Junior Inst. Eng. Lond.*, June 1938, pp. 401–16. (This paper discusses the characteristics of the Mississippi, the Yellow, and the Yangtze Rivers, from a viewpoint of frequency of floods, fluctuation of precipitation, deforestation and aridity, erosion and delta formation.)

CHOW, V. T. (1959). *Open-channel Hydraulics*. 680 pp. New York; McGraw-Hill. (This is by far the most comprehensive handbook on this subject, written in a clear style and dealing with practical problems. Some chapter headings: Open channel flow classifications; Energy and momentum principles; Critical flow; Design of channel for uniform flow; Gradually varied flow theory; Practical problems; Flow over spillways; Hydraulic jump; Non-prismatic channels; Unsteady flow; Flood routing.)

EINSTEIN, H. A., and BARBAROSSA, N. L. (1952). 'River channel roughness.' *Trans. Amer. Soc. civ. Engrs.*, Vol. 117, pp. 1121–88. (This paper offers a rational approach to the solution of the problem of determining the frictional losses in natural streams. An interdependent treatment of hydraulic and sedimentary characteristics is presented.)

KARMAN, T. VON. (1934). 'Turbulence and skin friction.' *J. aero. Soc.*, Vol. 1, pp. 1–18. (This paper deals with boundary layer resistance and its interpretation in terms of a friction factor.)

KENNEDY, R. J., and FULTON, J. F. (1961). 'The effect of secondary currents upon the capacity of a straight open channel.' *Trans. Engng. Inst. Canada*, Vol. 5, No. 1, pp. 12–18. (This paper describes how the pattern of secondary currents in a smooth rectangular channel may change abruptly with increasing depth, and thus affect the channel's carrying capacity by as much as 20 per cent.)

KINDSVATER, C. E., and CARTER, R. W. (1955). 'Tranquil flow through open channel constrictions.' *Trans. Amer. Soc. civ. Engrs.*, Vol. 120, pp. 955–92. (A practicable solution of the discharge equation for an open channel restriction has been achieved by the application of a systematic experimental investigation to an approximate analysis.)

KING, H. W. (1939). *Handbook of Hydraulics*. 617 pp. New York; McGraw-Hill. (This handbook presents the essentials of engineering hydraulics, followed by a wealth of empirical data, such as: coefficients of discharge for various shaped orifices, gates and weirs; roughness coefficients for pipes and open channels; permissible velocities in canals; discharge at critical depth for various shaped channels; rating table for current meters.)

LAMB, H. (1945). *Hydrodynamics*. 738 pp. New York; Dover Publications. (A reprint of the first edition published in 1879. This is one of the classical texts of the hydraulic literature. It deals with the mathematical concepts of the movement of a fluid in a highly theoretical fashion.)

LANE, E. W. (1949). 'Low temperature increases sediment transport in Colorado River.' (See in Bibliography to Chapter 4.)

MANNING, R. (1891). 'Flow of water in open channels and pipes.' *Trans. Inst. civ. Engrs. Ireland*, Vol. 20, pp. 161–207. (In this paper Manning proposes a rather complicated formula which has later been reduced to the well-known Manning formula.)

MATTHES, G. H. (1940). 'Stage transmission in the lower Mississippi River.' University of Iowa, Studies in Engineering, Bull. 20, pp. 240–47. (This paper discusses the rate of travel of flood crests and the influence of cut-offs, levees, and valley storage.)

MATTHES, G. H. (1956). 'Stream flow characteristics.' (See Abbett, R. W., 1956, *American Civil Engineering Practice*, in Bibliography to Chapter 5.)

PROCEEDINGS (1937, 1948, 1951, 1953, 1955, 1957, 1959, 1961, 1963). *Confs. Assoc. Hydr. Res.* Secretariat: Raam 61, Delft. Netherlands. (These proceedings contain a wealth of information on hydraulic engineering. Most papers deal with hydraulics laboratory techniques and the results of research; some papers deal with practical engineering problems. Some conference themes: sediment transportation; density currents; air entrainment; waves; tidal hydraulics; modern instrumentation; outlet works; hydraulics of shiplocks; ice problems; scale effect; cavitation; turbulence; groundwater flow; hydraulic problems for computers. The Association also issues an annual publication, *Hydraulic Research*, listing all important research being conducted at all hydraulics laboratories in the world, except the U.S. and Canada. Those two countries issue their own annual publication, *Hydraulic Research in the U.S. and Canada*, U.S. Govt. Printing Office, Washington.)

PROCEEDINGS (1939, 1942, 1946, 1952, 1955, 1958). *Hydr. Confs. Iowa Inst. Hydr. Res.*, published by the State University of Iowa. (These publications contain a large number of interesting papers on hydraulics, some of which are listed in this Bibliography under their author's name.)

ROUSE, H. (1946). *Elementary Mechanics of Fluids*. 376 pp. New York; Wiley. (An excellent introductory textbook on fluid motion. Although somewhat theoretical, it is fairly easy to read and has good illustrations. Some chapter headings: Fluid velocity and acceleration; Effects of gravity on fluid motion; One-dimensional method of flow analysis; Surface resistance; Form resistance.)

ROUSE, H. (1950). *Engineering Hydraulics*. 1,039 pp. New York; Wiley. (One of the most comprehensive textbooks on the subject, prepared by thirteen authors. Some chapter headings: Fundamental principles by H. Rouse; Hydrology by G. R. Williams; Flow of groundwater by C. E. Jacob; Channel transition by A. T. Ippen; Gradually varied channel flow by C. J. Posey; Flood routing by B. R. Gilcrest; Sediment transportation by C. B. Brown.)

ROUSE, H. (1959). *Advanced Mechanics of Fluids*. 444 pp. New York; Wiley. (A highly theoretical textbook on fluid motion, prepared by nine authors. Some chapter headings: Principles of irrotational flow by P. G. Hubbard; Laminar motion by Chia-Shun Yih; Turbulence by Tien-To Siao; Boundary layers by L. Laudweber; Free turbulence sheer flow by E. M. Laursen.)

SAYRE, W. W., and ALBERTSON, M. L. (1961). 'Roughness spacing in rigid open channels.' *Proc. Amer. civ. Engrs.*, May 1961, HY3, pp. 121–50. (This paper describes a series of experiments to determine the effect of the spacing of roughness elements on the total resistance of flow. Logarithmic flow formulae are developed on the basis of the observed data.)

SCHNACKENBERG, E. C. (1951). 'Slope discharge formulae for alluvial streams.' *Proc. N. Z. Inst. Eng.*, Vol. 37, pp. 340–449. (This paper presents a good deal of field data on the friction factor of open channel flow.)

SHUKRY, A. (1950). 'Flow around bends in an open flume.' *Trans. Amer. Soc. civ. Engrs.*, Vol. 115, pp. 751–88. (Presents results of a laboratory study of the spiral motion around bends. The effects of varying Reynold's number, depth-breadth ratio, radius-breadth ratio, and deflection angle are indicated. It is concluded that spiral motion exists in straight flumes as well as in curved flumes.)

446

SIMONS, D. B., and RICHARDSON, E. V. (1961). 'Forms of bed roughness in alluvial channels.' *Proc. Amer. Soc. civ. Engrs.*, May 1961, HY3, pp. 87–105. (This paper discusses the different regimes of flow in alluvial channels and their associated roughness.)

SIMONS, D. B., and RICHARDSON, E. V. (1962). 'Resistance to flow in alluvial channels.' *Amer. Soc. civ. Engrs.*, Vol. 127, pt. I, pp. 927–1006. (This paper presents the initial results of a flume study of alluvial channels. A detailed classification of the regimes of flow, the forms of bed roughness, and the basic concepts pertaining to flow resistance are discussed.)

SIMONS, D. B., and others (1962). 'Depth-discharge relations in alluvial channels.' *Proc. Amer. Soc. civ. Engrs.*, Sept. 1962, HY5, pp. 57–72. (The roughness of alluvial channels varies with the flow, the channel, the bed material, and the sediment load. A change in roughness is reflected in a change in stage, depth, velocity, and hence in stage-discharge and depth-discharge relations. These relations are examined and the results are presented.)

TULTS, H. (1956). 'Flood protection of canals by lateral spillways.' *Proc. Amer. Soc. civ. Engrs.*, Oct. 1956, HY5, Paper 1077, 17 pp. (Water levels in front of a lateral spillway crest can be computed by the Bernoulli and the continuity equations. The paper discusses different water surface profiles upstream, in front, and downstream of the spillway, depending on the flow stage and the location of the spillway.)

U.S. WATERWAYS EXPERIMENT STATION (1935). 'Studies of river bed materials and their movement.' Pap. 17 of *U.S. Wat. Exp. St.*, Vicksburg, 161 pp. (This paper reports on a study that was aimed at determining the tractive force required to move bed material of different sizes and physical characteristics.)

VANONI, V. A. (1946). 'Transportation of suspended sediment by water.' (See in Bibliography to Chapter 4.)

VANONI, V. A., and NOMICOS, G. N. (1960). 'Resistance properties of sediment laden streams.' *Trans. Amer. Soc. civ. Engrs.*, Vol. 125, pp. 1140–75. (Laboratory experiments showed that the friction factor of a stream carrying suspended sediment is less than a comparable one without sediment.)

VOGEL, H. D., and THOMPSON, P. W. (1933). 'Flow in river bends.' (See in Bibliography to Chapter 4.)

CHAPTER 4: RIVER MORPHOLOGY

BATES, C. C. (1953). 'Rational theory of delta formation.' *Bull. Amer. Assoc. Petr. Geol.*, Vol. 37, pp. 2119–62. (Interesting discussion of delta formation. A jet theory is introduced to explain the behaviour of a sediment-laden stream in its mouth. Attention is given to the density contrast between the sea or lake water and the river water.)

BELL, H. S. (1944). 'Stratified flow in reservoirs and its use in prevention of silting.' *U.S. Dept. Agr. Misc. Pub. 491*, 46 pp. (One of the important papers, discussing the transport of sediment by means of density currents. Describes their behaviour and characteristics. Presents data on the quantity of sediment transported by density currents. Pays particular attention to their possible use in the more efficient operation of reservoirs.)

BENEDICT, P. C., and others (1953). 'Total sediment load measured in turbulence flume.' *Proc. Amer. Soc. civ. Engrs.*, Vol. 79, separate 230, 48 pp. (This paper describes an interesting method of measuring the total load in a stream. Through agitating the flow in a flume with artificial roughness, all the bed load is brought into suspension. The total load can then be measured by using suspended-sediment sampling devices.)

BIBLIOGRAPHY (1950). *Annotated Bibliography on Sedimentation.* Washington 25, D.C.; U.S. Government Printing Office. (A comprehensive bibliography, compiled as a guide to the literature on the engineering aspects of sedimentation. Every effort has been made to cover all of the known sources of information in the English language. Each reference has been annotated with respect to subject matter and geographical location.)

BLENCH, T. (1957). *Regime Behaviour of Canals and Rivers.* 138 pp. London;

Butterworths. (This book discusses the regime theory, developed by Kennedy, Lindley, Lacey, Inglis and Blench, when developing irrigation systems in India and Pakistan. The regime theory is applied to the design of canals as well as the stability of river channels.)

BONDURANT, D. C. (1950). 'Sedimentation studies at Conchas Reservoir in New Mexico.' *Proc. Amer. Soc. civ. Engrs.*, Vol. 76, Sept. 29, 10 pp. (Interesting discussion of the delta formation at the upstream end of an artificial reservoir.)

BORLAND, W. M., and MILLER, C. R. (1960). 'Sediment problems of the Lower Colorado River.' *Proc. Amer. Soc. civ. Engrs.*, April 1960, HY4, pp. 61–87. (The construction of major dams on the Colorado River instigated a series of river adjustments that have required corrective measures. Some of the problems, methods and results are given.)

BROOKS, N. H. (1958). 'Mechanics of streams with movable beds of fine sand.' (See in Bibliography to Chapter 3.)

BROWN, C. B. (1944). 'The control of reservoir silting.' *U.S. Dept. Agr. Misc. Pub. 521*, 166 pp. (A comprehensive study dealing with all aspects of reservoir silting. Describes methods of sedimentation control under six headings: Selection of the reservoir site; Design of the reservoir; Control of sediment inflow; Control of sediment deposition; Removal of sediment deposits; Watershed erosion control.)

BROWN, C. B. (1950). 'Sediment transportation.' (See Rouse, H., 1950, *Engineering Hydraulics*, in Bibliography to Chapter 3.)

BUCKLEY, A. B. (1923). 'Influence of silt on the velocity of water flowing in open channels.' *Minutes of Proc. Inst. civ. Engrs.*, Vol. 216, pp. 183–211. (Discusses experiments conducted in the Nile River to determine the influence of silt on the velocity of the water. Contends that, all other factors being equal, the velocity of flow in a given section is increased by the presence of silt in the water. Followed by some critical discussions, whereby the reliability of the original data is challenged. See also Vanoni, 1946.)

CAREY, W. C., and KELLER, M. D. (1957). 'Systematic changes in the beds of alluvial rivers.' (See in Bibliography to Chapter 3.)

CHATLEY, H. (1938). 'Hydraulics of large rivers.' (See in Bibliography to Chapter 3.)

CHIEN, N. (1955). 'Graphic design of alluvial channels.' *Proc. Amer. Soc. civ. Engrs.*, separate 611, 17 pp. (An interesting paper, presenting some graphs that are made to evaluate the effects on river regime by the diversion of flows, construction of dams, contraction of channels, elimination of river forks, and cut-off of bends.)

CHIEN, N. (1955). 'The present status of research on sediment transport.' *Proc. Amer. Soc. civ. Engrs.*, separate 565, 33 pp. (A comprehensive paper on the mechanics of sediment transport. Emphasis is placed on the significance and implications of the various phases of the problem. Controversial issues are discussed, and efforts are made to unify some of the highly diversified results. Ninety-three references are presented.)

CHIEN, N. (1957). 'A concept of the regime theory.' (See in Bibliography to Chapter 8.)

COREY, H. T. (1913). 'Irrigation and river control in Colorado River Delta.' (See in Bibliography to Chapter 8.)

DAVIS, W. M. (1902). 'Base level, grade and peneplain.' *J. Geol.* 10, pp. 77–111. (One of the classical papers on the concept of a graded river. Very interesting discussion on the balance of a river system in view of the geomorphologic changes in the surface of the earth. This paper is reprinted in Davis, 1954.)

DAVIS, W. M. (1954). *Geographical Essays.* 777 pp. Dover Publications Inc. (This edition is an unabridged republication of the 1909 edition. The author is one of the outstanding geologists of all times. His writings are lucid and interesting. Many of the chapters in this book deal with river morphology.)

EAKIN, H. M. (1935). 'Diversity of current direction and load distribution on stream bends.' *Trans. Amer. geophys. Un.*, Vol. 16, pp. 467–72. (Describes field experiments of helical flow on a bend in the Mississippi River. It is concluded that a slight helical flow or, rather, 'circulation' exists. Although this experiment was carried out extensively, the results do not appear to be conclusive.)

EINSTEIN, H. A. (1950). 'The bed-load function for sediment transportation in open channel flows.' *Bull. U.S. Dept. Agr. Tech. 1026.* (One of the most important papers on sediment transport. It enables the computation of the total bed-material load in an alluvial stream, when the slope, discharge, channel cross-section and bed-material composition are given. Because of the complicated nature of the subject, the paper is difficult to read. However, the mechanics of the computations can be followed in a simple step-by-step method.)

EINSTEIN, H. A., and JOHNSON, J. W. (1950). 'The laws of sediment transportation.' In *Applied Sedimentation*, by P. D. Trask, John Wiley and Sons, pp. 62–71. (A brief and very interesting review of the laws that govern the transportation of suspended load and bed load. The discussion is understandable for the layman and contains no formulae.)

EINSTEIN, H. A., and BARBAROSSA, N. L. (1952). 'River channel roughness.' (See in Bibliography to Chapter 3.)

EINSTEIN, H. A., and CHIEN, N. (1953). 'Transport of sediment mixtures with large ranges of grain sizes.' *Univ. Calif. Inst. Eng. Res.*, M. R. D., Sediment Series No. 2. (This paper refines the Einstein, 1950, relationships and makes them applicable to sediment mixtures with large ranges of grain sizes.)

EINSTEIN, H. A., and CHIEN, N. (1956). 'Similarity of distorted river models with movable beds.' *Trans. Amer. Soc. civ. Engrs.*, Vol. 121, pp. 440–62. (Similarity conditions are derived from the theoretical and empirical equations which describe the hydraulics and sediment transport of rivers. A numerical example illustrates the method.)

ELZERMAN, J. J., and FRYLINK, H. C. (1951). 'Present state of the investigations on bed-load movement in Holland.' *Ass. Internat. d'Hydrology Scientifique*, Assemblée Générale de Bruxelles 1951, pt. 3, pp. 100–16. (Compares bed-load computations, using the formulae of Kalinske, Meyer-Peter and Muller, with actual bed-load measurements on rivers in Holland. It is concluded that both formulae can be transformed into functions that give a satisfactory agreement with the observations.)

FISK, H. M., and others (1954). 'Sedimentary framework of the modern Mississippi Delta.' *J. sediment Petrol.*, Vol. 24, pp. 76–99. (Interesting discussion of the arrangement of sedimentary units in the bird-foot delta of the Mississippi River. The shape and distribution of sedimentary units, as disclosed from borings, are shown on maps and sections. Factors controlling delta growth are the load of the river, number of main distributaries of the river, depth of water into which the delta front advances, and subsidence.)

FREEMAN, J. R. (1922). 'Flood problems in China.' (See in Bibliography to Chapter 6.)

FREEMAN, J. R. (1928). 'Flood control on the River Po in Italy.' (See in Bibliography to Chapter 6.)

FRIEDKIN, J. F. (1945). 'A laboratory study of the meandering of alluvial rivers.' Vicksburg, Miss., U.S. Waterways Experiment Station, 40 pp. (One of the most important and interesting papers on stream meandering. It contains a discussion of the basic principles of meandering and contains the results of tests and analyses on the causes of meandering and the development of the meander pattern under variable conditions of discharge and hydraulic properties of the channel.)

GILBERT, G. K. (1877). 'The geology of the Henry Mountains.' *U.S. Geog. & Geol. Survey* of the Rocky Mountain Region. (Includes a classic discussion, pp. 99–150, on the principles of fluvial erosion and debris transportation.)

GILBERT, G. K. (1890). 'Lake Bonneville.' *U.S. Geol. Survey*, Monog. 1, 438 pp. (A technical paper which deals with the geology of Lake Bonneville. Includes an interesting section, pp. 65–70, on delta formation.)

GILBERT, G. K. (1914). 'Transportation of debris by running water.' *U.S. Geol. Survey*, Prof. Paper 86, 259 pp. (A monumental work, presenting the results of laboratory investigations to determine the laws which control the movement of bed-material load in a stream. The lengthy discussions of the laboratory results are interesting and contain a wealth of information, but are of limited value for practical problems. One chapter on the behaviour of natural streams is very interesting.)

GILBERT, G. K. (1918). 'Hydraulic mining debris in the Sierra Nevada.' *U.S.*

Geol. Survey, Prof. Paper 105, 154 pp. (A comprehensive paper concerning the movement of hydraulic-mining debris from the Sierra Nevada into the Sacramento River system. River valley aggradation was at places as much as 40–70 ft.)

GRIFFITH, W. M. (1927). 'A theory of silt and scour.' *Minutes of Proc. Inst. civ. Engrs.*, Vol. 223, pp. 243–63. (Discusses the silt-transporting characteristics of streams. Assumes a set of empirical formulae, based mainly on Kennedy's formula, governing the natural shape of sediment-transporting watercourses. The paper is followed by rather critical discussions, pointing out that Kennedy's law does not always work in practice and that it has a limited range of application. An interesting illustration of the possible range in grain sizes of suspended sediment is provided by F. W. Woods, who remarks in his discussion that he saw after the great flood of 1910 in the Gomal River, India, a boulder of 4 ft. length and 3 ft. diameter, lodged in the steel latticework of a pier of a suspension bridge, at a level at least 10 ft. above low-water level.)

GRUNSKY, C. E. (1930). 'Silt transportation by Sacramento and Colorado Rivers.' *Trans. Amer. Soc. civ. Engrs.*, Vol. 94, pp. 1104–33. (Suspended-sediment transportation is shown to vary with cross-section, discharge, time of the year and location along the river. Interesting discussion of eccentric river phenomena, such as boils and pulsations.)

HARRISON, A. S. (1953). 'Deposition at the head of reservoirs.' *Bull. Univ. Iowa*, 34, Studies in Eng., pp. 199–227. (Very interesting discussion of the factors which must be taken into account in an engineering estimate of future deposition of sediment at the heads of existing or proposed reservoirs, so that action for the control or alleviation of problems which could arise therefrom can be planned before the problems become serious.)

HARRISON, A. S. (1954). 'Study of effects of channel stabilization and navigation project on Missouri River levels.' Corps of Engineers, U.S. Army, Omaha District, Memorandum 18. (Contains an interesting discussion of the effect of the sediment-transport intensity on hydraulic resistance of the river channel.)

HATHAWAY, G. A. (1948). 'Observations on channel changes, degradation and scour below dams.' *Int. Ass. Hyd. Res. Rpt.*, pp. 287–307. (Discusses the significance of sediment transportation as associated with channel changes below hydraulic structures. Interesting example of aggradation on the Arkansas River due to the use of water for irrigation.)

HJULSTROM, F. (1935). 'Studies of the morphological activity of rivers as illustrated by the River Fyris.' *Bull. Geol. Inst. Univ. Uppsala*, Vol. 25, pp. 221–528. (Discusses the mechanics of streams, in particular the changes of mass of water under the influence of gravity and its own kinetic energy; laminar movement, the transition between laminar and turbulent and the characteristics of turbulence. Considers the problem of the transportation of solid matter in water as bed load and suspended load.)

INGLIS, SIR CLAUDE (1947). 'Meanders and their bearing on river training.' *Inst. civ. Engrs., Lond.* Maritime and Waterways Paper No. 7. (An interesting paper in which an attempt is made to show that when variations in discharge and sediment load are taken into account, meandering is determined by the same basic factors as flow in straight channels. Formulae are given, expressing the meander length and meander width in terms of stream-channel width and river discharge.)

INGLIS, SIR CLAUDE (1949). 'The behaviour and control of rivers and canals.' Pub. No. 13, Central Waterway, Irrigation and Navigation Research Station, Poona Prison Press, India, pt. 1: 280 pp., pt. 2: 200 pp. (A discussion of rivers and canals, based mostly on the regime theory, developed by Kennedy and Lacey. Lengthy descriptions of laboratory experiments with river training structures. The book is mainly of interest in that it tends to develop the reader's own independent approach to the various problems.)

KUIPER, E. (1953, 1954). 'Hydrometric surveys 1953, 1954.' Saskatchewan River Reclamation Project, Interim Reports Nos. 3 and 8, P.F.R.A., Winnipeg, Man., Canada, 24 pp. and 36 pp. (An account of hydrometric survey trips on the Churchill River and Saskatchewan River, including discussions of features of delta formation.)

KUIPER, E. (1960). 'Sediment transport and delta formation.' *Proc. Amer. Soc. civ. Engrs.*, Feb. 1960, HY2, pp. 55–68. (This paper deals with sediment computations that were carried out in connection with dike design for the Saskatchewan Delta Reclamation Project.)

KUIPER, E. (1962). 'Report sedimentation South Saskatchewan reservoir.' Dept. of Agriculture, Winnipeg, Man., 45 pp. (This report presents the findings of a study to determine the formation of the delta which will develop at the upper end of the South Saskatchewan reservoir, and the stream bed aggradation which will occur upstream of the reservoir.)

LACEY, G. (1930). 'Stable channels in alluvium.' (See in Bibliography to Chapter 8.)

LACEY, G. (1933). 'Uniform flow in alluvial rivers.' (See in Bibliography to Chapter 8.)

LACEY, G. (1946). 'A general theory of flow in alluvium.' (See in Bibliography to Chapter 8.)

LANE, E. W. (1934). 'Retrogression of levels in river beds below dams.' *Engng. News Rec.*, Vol. 112, pp. 836–38. (Interesting presentation of degradation data below dams on the Sutlej River in India, Yuba River and Rio Grande River in the U.S.A., and the Saalach River in Germany. The article is followed by three supplementary statements by L. F. Harza, S. Schulits, and A. M. Shaw, giving further details of some of the degradation examples.)

LANE, E. W. (1937). 'Stable channels in erodible material.' *Trans. Amer. Soc. civ. Engrs.*, Vol. 102, pp. 123–42. (Very interesting paper, outlining the principles involved in the design of stable channels as related to the size, shape, and quantity of sediment transport.)

LANE, E. W. (1949). 'Low temperature increases sediment transportation in Colorado River.' *Civil Engng.*, Vol. 19, pp. 45–46. (An interesting presentation of sediment-transport data on the Colorado River. It shows that for a given discharge, the sediment load may be as much as two and half times as great in winter as in summer, due to the difference in water temperature.)

LANE, E. W. (1955). 'The importance of fluvial morphology in hydraulic engineering.' *Proc. Amer. Soc. civ. Engrs.*, Hydr. Div., Paper 745, 17 pp. (One of the last papers of this outstanding hydraulic engineer, reviewing the historical development of the concept of natural streams in equilibrium. Some discussion of the changes in river profile due to artificial interference.)

LANE, E. W., and BORLAND, W. M. (1953). 'River-bed scour during floods.' *Proc. Amer. Soc. civ. Engrs.*, separate 254. (Interesting discussion of the lowering of stream beds during floods. It is concluded that during floods the bed of the Rio Grande, in general, scours out at the narrow sections and that most of the material thus removed is deposited in the next wide section downstream. See on this subject also Straub, 1935, p. 1152; and Gilbert, 1914, p. 221.)

LANE, E. W., and KALINSKE, A. A. (1939). 'The relation of suspended load to bed-material in rivers.' *Trans. Amer. geophys. Un.*, Vol. 20, pp. 637–41. (This paper refers to the equation of O'Brien, 1933, which relates the suspended-material concentration at any relative distance above the bottom of a river channel to the known concentration at some relative distance. The purpose of the paper is to develop the relationship between the composition and concentration of the suspended material and the composition of the stream bed, and the hydraulic characteristics of the river. For further development of this theory, see Einstein, 1950.)

LEIGHLY, J. B. (1932). 'Toward a theory of the morphologic significance of turbulence in the flow of water in streams.' *Univ. Calif. Publ. Geogr.*, Vol. 6, pp. 1–22. (An attempt is made to apply a theory of turbulence to problems of geomorphologic activity in streams. It is shown that maximum erosion will likely occur at the base of steep slopes and not in the middle of the river bottom or in the sloping banks. This feature may be sufficient to explain meandering and an appeal to centrifugal action may be unnecessary. An interesting paper, although somewhat difficult to read.)

LELIAVSKY, S. (1955). *An Introduction to Fluvial Hydraulics.* 257 pp. London; Constable. (This book reviews the various theories, methods and facts bearing on the flow of water in erodible channels. Some chapter headings: Traction and suspension; Surface slope and particle size; Dunes and ripples; Bed-load; Side-slope stability; Sediment transport.)

LEOPOLD, L. B., and MADDOCK, T. (1953). 'The hydraulic geometry of stream channels.' *U.S. Geol. Surv., Prof.* Paper 252, 56 pp. (An interesting paper dealing with the relationship between depth, width, velocity, suspended load, and discharge of stream channels. The suspended load is shown to be an index to total load for purposes of explaining the observed average characteristics of natural channel systems.)

LINDNER, C. P. (1952). 'Diversions from alluvial streams.' *Proc. Amer. Soc. civ. Engrs.*, Vol. 78, separate 112. (Very interesting discussion of the hydraulic effects of diversion and their withdrawal of sediment from alluvial streams. Factors influencing the diversion of bed load and the variation in the quantity of diversion with the angle of diversion are developed. It is noted that branching channels carry relatively more sediment than the straight main channel. This is explained as follows: The highest velocities occur in the upper layers of the stream. These layers of water are deflected into the side channel with greater difficulty than the lower layers. These lower layers transmit most of the sediment load, and a large proportion of this load is thus moved into the branch channel. The paper is followed by interesting discussions by A. R. Thomas, D. C. Bondurant and others.)

LOBECK, A. K. (1939). *Geomorphology.* 731 pp. New York; McGraw-Hill. (A textbook containing about 100 pages of interesting discussions on the morphology of rivers. Devotes much space to the concept of graded rivers. Diagrams illustrate the change in river profile due to changes in sediment transport.)

MACKIN, J. H. (1948). 'Concept of the graded river.' *Bull. Geol. Soc. Am.*, Vol. 59, pp. 463–512. (Very interesting paper extending the theory of grade originally set forth by Gilbert and Davis. Discusses the profile of a graded stream and factors controlling slope of graded profile.)

MATTHES, G. H. (1934). 'Floods and their economic importance.' (See in Bibliography to Chapter 6.)

MATTHES, G. H. (1941). 'Basic aspects of stream meanders.' *Trans. Amer. Geophys. Un.*, Vol. 22, pp. 632–36. (Very interesting discussion, setting forth certain fundamentals relating to the dynamics of meandering streams derived from observations on streams of various sizes.)

MATTHES, G. H. (1947). 'Mississippi River cutoffs.' *Trans. Amer. Soc. civ. Engrs.*, Vol. 113, pp. 1–15. (Describes natural and artificial cut-offs in the lower Mississippi River along a 50 mile stretch north of Vicksburg, Miss. Gives the effects of cut-offs on river shortening and flood-stage lowering.)

MATTHES, G. H. (1956). 'River engineering.' Section 15 of Abbett, Vol. II (see reference in Bibliography to Chapter 5). (This section contains 61 pages on river characteristics and sediment transport.)

MEYER-PETER, E., and MULLER, R. (1948). 'Formulas for bed-load transport.' *Internatl. Ass. for Hydr. Res. Rpt.* 2nd Mtg., Stockholm, pp. 39–64. (One of the important papers on the transport of bed load. Describes the development of the Swiss experimental bed-load transportation formula from its first simple form in 1934 to its later more complicated form in 1948.)

NEDECO (1959). 'River studies of the Niger and Benue.' North Holland Publishing Co., Amsterdam, 1,000 pp. (This is a report by the Netherlands Engineering Consultants to the Government of Nigeria regarding the navigation potential of the Niger and Benue Rivers. The report is of general interest inasmuch as it presents extensive discussion of river morphology and sediment transport.)

O'BRIEN, M. P. (1933). 'Review of the theory of turbulent flow and its relation to sediment transportation.' *Trans. Amer. Geophys. Un.*, Vol. 14, pp. 487–91. (One of the first papers demonstrating that if the average concentration of suspended sediment at any point in a wide, straight stream was known, the average concentration at any other point could be determined. For further development of this theory, see Lane and Kalinske, 1939.)

PIERCE, R. C. (1916). 'The measurement of silt-laden streams.' *U.S. Geol. Surv.*, Water-Supply Paper 400-C, pp. 39–51. (Contains an interesting description of so-called 'sand waves' that occur during flood stages. They are believed to be caused by anti-dune formations of the river bed.)

POWELL, J. W. (1961). *The Exploration of the Colorado River*. 176 pp. New York; Doubleday. (A reprint of Major Powell's diary from 1870. An account of the first trip down the Colorado River by boat, entertaining in style, and including discussions of fluvial erosion.)

RIPLEY, H. C. (1927). 'Relation of depth to curvature of channels.' *Trans. Amer. Soc. civ. Engrs.*, Vol. 91, pp. 207–65. (Gives formulae for the computation of the cross-section of river bends as a function of the mean depth, width, and radius of curvature. Followed by an interesting and critical discussion. It is mentioned that the South Platte in Denver tore out a railroad bridge and carried away a locomotive. After the waters had subsided to the usual small trickle, the sandy bottom appeared precisely as before but the locomotive was never seen again.)

RUBEY, W. W. (1933). 'Equilibrium conditions for debris-laden streams.' *Trans. Amer. Geophys. Un.*, Vol. 14, pp. 497–505. (Derives a general energy equation of stream work, including the energy consumed in friction and in supporting suspended sediment. The latter is in some cases only a few per cent of the first.)

RUBEY, W. W. (1952). 'The geology of Hardin Quadr., Illinois.' *U.S. Geol. Surv.*, Prof. Paper 218. (Contains an interesting section—Concept of adjusted cross-section of stream channels, pp. 129–36—dealing with a quantitative analysis of stream characteristics.)

SALISBURY, R. D., and ATWOOD, W. W. (1908). 'The interpretation of topographic maps.' *U.S. Geol. Surv.*, Prof. Paper 60, 84 pp. 170 plates. (Contains a very interesting discussion, pp. 26–40, of stream erosion, alluviation, and the meaning of young, mature and old streams, illustrated by sections of full-size topographic maps.

SCHAANK, E. M. H., and SLOTBOOM, G. (1937). 'Enkele mededeelingen betreffende de zandbeweging op den Neder-Rijn.' *De Ingenieur*, 52e Jaargang, pp. B167–71. (A report in Dutch on the transport of sand on the Rhine River in Holland.)

SHUKRY, A. (1950). 'Flow around bends in an open flume.' (See in Bibliography to Chapter 3.)

SIMONS, D. B., and RICHARDSON, E. V. (1962). 'Resistance to flow in alluvial channels.' (See in Bibliography to Chapter 3.)

STANLEY, J. W. (1951). 'Retrogression on the lower Colorado River.' *Trans. Amer. Soc. civ. Engrs.*, Vol. 116, pp. 943–57. (An interesting survey of bed and bank erosion, sediment deposition and sediment loads in the river, subsequent to the closure of the Hoover Dam in 1935.)

STEVENS, J. C. (1936). 'The silt problem.' *Trans. Amer. Soc. civ. Engrs.*, Vol. 101, pp. 207–250. (A comprehensive treatise pertaining to various subjects related to transportation and deposition of sediment. Gives the amount of sediment deposits for 25 reservoirs in the U.S. and 7 foreign reservoirs.)

STEVENS, J. C. (1947). 'Future of Lake Mead and Elephant Butte Reservoir.' *Trans. Amer. Soc. civ. Engrs.*, Vol. III, pp. 1231–54. (A report presenting an analysis of sediment records for Elephant Butte Reservoir. From these data, the useful lives of Lake Mead and Elephant Butte Reservoir are estimated. To prolong the life of Lake Mead, considers upstream storage of sediment. No practical method of removing the enormous amounts of sediment in the reservoirs has yet been found.)

STRAUB, L. G. (1935). 'Silt investigations in the Missouri Basin.' In *Missouri River*, 73rd Congr., 2nd sess., H. Doc. 238, pp. 1032–183. (Comprehensive study of the sedimentary characteristics of the Missouri River system. Discusses variation of silt discharge from its source to its mouth; relation of silt discharge to water discharge and to velocity; bed-load transportation; theoretical basis for the measurement of the amount of bed material transported; relation between bed load and suspended load; interesting miscellaneous observations concerning the behaviour of the stream channels.)

STRAUB, L. G. (1942). 'Mechanics of rivers.' (Interesting discussion of river regime, including such topics as character of sediment in rivers, sorting of bed-material, analysis of changes in river beds, channel contraction works, river gradients and river-bed changes. See reference in Bibliography to Chapter 2: Meinzer, O. E., 1942, *Hydrology*, pp. 614–36.)

SUNDBORG, A. (1956). *The River Klaralven: A Study of Fluvial Processes.* 316 pp. Esselte Aktiebolag; Stockholm. (A Swedish book, published in English, presenting a synopsis of recent facts and theories regarding fluvial processes, and applying some of these to the investigation of an actual river.)

THORNBURY, W. D. (1954). *Principles of Geomorphology.* 618 pp. New York; John Wiley. (A geological textbook that contains an excellent description of river morphology. Some chapter headings: The fluvial geomorphic cycle; Complications of the fluvial cycle; Stream deposition; The peneplain concept.)

TODD, O. J., and ELIASSEN, S. (1940). 'The Yellow River problem.' *Trans. Amer. Soc. civ. Engrs.,* Vol. 105, pp. 346–416. (Interesting paper dealing with problems of controlling the Yellow River. Section on the regime of the stream and the sediment-transport characteristics. Interesting discussions by H. van der Veen, E. W. Lane, and others.)

TURNBULL, W. M., and others (1950). 'Geology of Lower Mississippi Valley.' In *Applied Sedimentation* by P. D. Trask, New York; John Wiley, pp. 210–29. (Interesting description of the morphologic history of the Lower Mississippi River. Discussion of natural river levees, point bar deposits and crevasses.)

TWENHOFEL, W. H. (1950). *Principles of Sedimentation.* 673 pp. New York; McGraw-Hill. (Primarily a geological textbook on sedimentation, which covers the subject rather concisely. Contains an interesting chapter on sediment transportation and deposition.)

UNITED NATIONS (1951). 'Methods and problems of flood control in Asia and the Far East.' *U.N. Publ.,* Sales No. 1951. II.F.5., Bangkok, 45 pp. (An interesting review of flood-control problems with emphasis on diking problems. Contains the following chapters: 1. Flood-control methods in Asia, discussing different local methods of laying out and constructing river dikes; 2. Problems of flood control, discussing stability of rivers and flood control by dikes, reservoirs, and land conservation methods.)

UNITED NATIONS (1953). 'River training and bank protection.' *U.N. Publ.,* Sales No. 1953, II.F.6., Bangkok, 100 pp. (A comprehensive review of different river training and bank protection methods, with emphasis on problems in Asia and the Far East. Contains the following chapters: 1. River training, discussing the theory of river flow and river training, methods of river training in general, and in Asia in particular; 2. Bank protection, discussing bank erosion and theory of erosion and scour, methods of bank protection in general, and in Asia in particular; 3. River-work practice in other regions, discussing Mississippi River problems and river works in Europe, Australia and New Zealand.)

UNITED NATIONS (1953). 'The sediment problem.' *U.N. Publ.,* Sales No. 1953. II.F.7., Bangkok, 92 pp. (A summary report of literature on transport and deposition of sediment, with emphasis on sediment problems in Asia and the Far East. Contains the following chapters: 1. Soil erosion, providing data on silt yield of various drainage areas; 2. Transportation of sediment, discussing the mechanics of sediment transport in detail. See on this subject also a publication by Chien, 1955, on sediment transport; 3. Silting and scouring of channels, summarizing the regime theory of irrigation channels; 4. Silting of reservoirs, mostly a summary of a study by Brown, 1944; 5. Action of sediment on the regime of rivers, discussing briefly some aspects of river regime; 6. Sampling and analysis of sediment, mostly a summary of reports by the U.S. Government, 1952.)

U.S. GOVERNMENT (1952). 'A study of methods used in measurement and analysis of sediment loads in streams.' St. Paul Dist. Sub-office, Corps of Engrs.; Hydraulics Lab., Univ. of Iowa. (The most comprehensive study on this subject, published in the following nine reports: 1. Field practice and equipment used in sampling suspended sediment, 1940; 2. Equipment used for sampling bed load and bed-material, 1940; 3. Analytical study of methods of sampling suspended sediments, 1941; 4. Methods of analysing sediment samples, 1941; 5. Laboratory investigations of suspended-sediment samplers, 1941; 6. The design of improved types of suspended-sediment samplers, 1952; 7. A study of new methods for size analysis of suspended-sediment samples, 1943; 8. Measurement of the sediment discharge of streams, 1948; 9. Density of sediments deposited in reservoirs, 1943.)

BIBLIOGRAPHY TO CHAPTER 5

VANONI, V. A. (1946). 'Transportation of suspended sediment by water.' *Trans. Amer. Soc. civ. Engrs.*, Vol. 111, pp. 67–133. (One of the important papers on laboratory measurements of sediment and velocity distributions for various values of slope of channel, rate of flow and sediment transport. It was found that the measured distributions have the same form as the theoretical distributions, but do not agree quantitatively. It was also found that suspended sediment tends to reduce the turbulent transfer of momentum and, hence, the resistance to flow, allowing the sediment-laden water to flow more rapidly than a comparable clear-water flow. See also on this subject, Buckley, 1923. Interesting discussions by R. W. Powell and P. F. Nemenyi on double spiral motion in straight open conduits.)

VANONI, V. A., and NOMICOS, G. N. (1960). 'Resistance properties of sediment laden streams.' (See Bibliography to Chapter 3.)

VETTER, C. P. (1949). 'The Colorado River Delta.' 'Reclamation Era, Vol. 35, Nos. 10 and 11, pp. 216–20 and 217–19. (Very interesting brief description of the formation of a large river delta.)

VETTER, C. P. (1953). 'Twenty years of sediment work on the Colorado River.' *Univ. Iowa, Studies in Engng., Bull. 34*, pp. 5–33. (Interesting description of the change in the regime of the Colorado River subsequent to the construction of the Hoover Dam, Parker Dam, Imperial Dam, and others. Quantitative data on stream-channel aggradation and degradation are included.)

VOGEL, H. D., and THOMPSON, P. W. (1933). 'Flow in river bends.' *Civ. Engng.*, Vol. 3, pp. 266–68. (Experiments on a Mississippi River model seem to discredit helicoidal theory of flow. It was found that bed load moves from the concave towards the convex bank, but that the bottom current is parallel to the banks. Reference is made to Leighly, 1932, and his theory of 'turbulence in the direction of decreased velocity.')

WOODFORD, A. O. (1951). 'Stream gradients and Monterey Sea Valley.' *Bull. Geol. Soc. Amer.*, Vol. 62, pp. 799–852. (Includes an interesting discussion of river profiles and the various factors that influence the shape of the profile.)

CHAPTER 5: HYDRAULIC STRUCTURES

ABBETT, R. W. (1956). *American Civil Engineering Practice*, Volume II (*Hydraulic Engineering*). 935 pp. New York; Wiley. (One of the most recent and comprehensive handbooks on hydraulic engineering, prepared by 10 authors. The 181 pp. chapter on river engineering by the outstanding hydraulic engineer G. H. Matthes merits special attention. Other chapter headings: Hydrology by W. W. Horner; Dams by W. P. Creager and B. O. McCoy; Hydroelectric power by W. P. Creager and E. B. Strowger; Irrigation by D. P. Barnes; Harbor engineering by R. W. Abbett and E. E. Halmos; Public water supply by R. Hazen.)

ASCE COMMITTEE ON MASONRY DAMS (1951). 'Final report of the subcommittee on uplift in masonry dams.' *Trans. Amer. Soc. civ. Engrs.*, Vol. 117, pp. 1218–52. (This is the result of several years of study by a panel of experts to 'study the possibility of clarifying the uplift problem'.)

ASCE TASK FORCE ON HYDRAULIC DESIGN OF SPILLWAYS (1963). Bibliography on same. *Proc. Amer. Soc. civ. Engrs.*, July 1963, HY4, pp. 117–39.

BEACH EROSION BOARD. (See U.S. Beach Erosion Board.)

BERRYHILL, R. H. (1963). 'Experience with prototype energy dissipators.' *Proc. Amer. Soc. civ. Engrs.*, May 1963, HY3, pp. 181–201. (This paper contains data on the actual performance of stilling basins. Some of the projects that have sustained damage or erosion, are reported. Comparisons with preceding hydraulic model tests are made.)

BRADLEY, J. H., and PETERKA, A. J. (1957). 'Hydraulic design of stilling basins.' *Proc. Amer. Soc. civ. Engrs.*, Oct. 1957, HY5, Papers 1401–6 incl., 129 pp. (This is a group of six papers on the hydraulic design of stilling basins and their associated appurtenances for a variety of conditions. The papers emphasize practical design procedures. Sample problems are included.)

BRETSCHNEIDER, C. L. (1959). 'Revisions in wave forecasting in deep and shallow water.' *Proc. 6th Conf. Coastal Engng.*, University of California. (This paper presents some mathematical derivations and empirical relationships between wind velocity, fetch length, water depth, wave height, and wave period. Some of the more important findings are quoted in U.S. Beach Erosion Board, 1961, *Shore Protection Planning and Design.*)

CONGRESS (1933, 1936, 1948, 1951, 1955, 1958, 1961). *Large Dams.* Secretariat, 51 Rue Saint Georges, Paris. (The proceeding papers of these seven congresses contain a wealth of information on the design and construction of large dams and all its associated aspects. Some conference themes: research on construction materials; stability of earth dams; uplift in dams; methods to avoid piping; maximum flood discharge; spillway design; reservoir sedimentation; economics of concrete dams; aggregates for concrete.)

CREAGER, W. P., JUSTIN, J. D., and HINDS, J. (1945). *Engineering For Dams* (in 3 volumes). 929 pp. New York; Wiley. (One of the most comprehensive handbooks on the subject. Volume I deals with general design; investigation of dam sites; choice of type of dam; preparation of foundation; spillway design floods. Volume II deals with forces acting on dams; design procedures; gravity dams; arch dams; buttress dams. Volume III deals with earth dams; rock-fill dams; timber dams; headwater control.)

DAVIS, C. V. (1942). *Handbook of Applied Hydraulics.* 1083 pp. New York; McGraw-Hill. (One of the most comprehensive handbooks on the subject, prepared by 18 authors. Some chapter headings: Hydrology by C. S. Jarvis; River regulation by reservoirs by T. T. Knappen and C. V. Davis; Gravity dams by C. V. Davis; Arch dams by I. E. Houk; Earth dams by T. T. Knappen; Rock-fill dams by I. C. Steele; Spillways by E. W. Lane; Spillway Gates by J. S. Bowman; Outlet works by P. A. Kinzie; Hydroelectric plants by J. C. Stevens; Irrigation by I. E. Houk; Drainage by G. W. Pickels.)

ELEVATORSKI, E. A. *Hydraulic Energy Dissipators.* 214 pp. New York; McGraw-Hill. (An up-to-date and comprehensive textbook on the theory of the hydraulic jump and the design of stilling basins.)

HARZA, L. F. (1949). 'The significance of pore pressure in hydraulic structures.' *Trans. Amer. Soc. civ. Engrs.*, Vol. 114, pp. 193–289. (It is shown that hydrostatic uplift acts over the entire horizontal area, instead of one half of the area, as sometimes assumed. It is also shown that the hydrostatic force against a dam is applied progressively along the seepage route, instead of the upstream face.)

HERBICH, J. B., and others (1963). 'Effect of berm on wave run-up on composite beaches.' *Proc. Amer. Soc. civ. Engrs.*, May 1963, WW2, pp. 55–72. (The objective of this study was to evaluate the wave run-up on a composite slope, where the width of the horizontal berm is considerable. It is found that the run-up greatly depends on the backwash of the previous wave.)

HUFFT, J. C. (1958). 'Laboratory study of wind waves in shallow water.' *Proc. Amer. Soc. civ. Engrs.*, Sept. 1958, WW4, Paper 1765, 19 pp. (This paper shows experimental relationships between wind velocity, fetch, and wave parameters for two different water depths.)

HUNT, I. A., and BAJORUNAS, L. (1959). 'The effect of Seiches at Cormeaut Harbour.' *Proc. Amer. Soc. civ. Engrs.*, June 1959, WW2. pp. 31–41. (This paper analyses the wind set-up oscillations on Lake Erie which may reach a magnitude of several feet, extending over several hours.)

KEENER, K. B. (1951). 'Uplift pressures in concrete dams.' *Trans. Amer. Soc. civ. Engrs.*, Vol. 116, pp. 1218–64. (This paper presents a valuable addition of basic data regarding uplift pressures under masonry dams, founded on rock. A case history of the Hoover Dam is included.)

KUIPER, E. (1951). 'Construction of Harlinger breakwater.' The Dock and Harbour Authority, London, March 1951, pp. 341–47. (This paper describes the design and construction of a harbour dam, 30 ft. high and 3,000 ft. long, composed of a sand core and asphalt cement cover. Several experiments were conducted to test the suitability of asphalt cement for application above and below mean sea level.)

KUIPER, E. (1959). 'Practical flood frequency analysis.' *Proc. symp. on spillway design floods*, Queen's Printer, Ottawa, pp. 87–100. (This paper presents a discussion of the rational method and the frequency analysis method for determining spillway design floods. It is concluded that the two methods are supplementary rather than opposed to one another.)

LANE, E. W. (1935). 'Security from under-seepage, masonry dams on earth foundations.' *Trans. Amer. Soc. civ. Engrs.*, Vol. 100, p. 1235. (This paper deals with the required length of the path of percolation under concrete dams to prevent piping. It is one of Lane's classic papers on hydraulic engineering, still widely being used in design practices.)

LELIAVSKY, S. (1958). *Uplift in Gravity Dams.* 267 pp. New York; Frederick Ungar. (The author, who is professor in Egypt, and was formerly hydraulic engineer in Russia, reviews the entire subject of uplift in gravity dams.)

LINSLEY, R. K., and FRANZINI, J. B. (1955). *Elements of Hydraulic Engineering.* 582 pp. New York; McGraw-Hill. (This textbook deals mostly with hydrology and applied hydraulics. Some chapter headings: Hydrology; Water law; Dams; Spillways; Open channels; Hydraulic machinery; Irrigation; Water supply; Hydroelectric Power; River navigation; Flood control; Multiple-purpose projects.)

MACDONALD, D. H., and others (1960). 'Kelsey generating station dam and dikes.' (See in Bibliography to Chapter 7.)

MACKENZIE, G. L. (1960). 'The South Saskatchewan river dam.' *Engng. J.*, EIC, May 1960, pp. 50–55. (This dam is the principal structure in a 200 million dollar water power and irrigation project in Central Canada. This paper gives a general description. The following papers deal with particular aspects: Hydrologic investigations by W. M. Berry and others, *Engng. J.*, April 1961, pp. 61–68; Geology of the project by D. H. Pollack, *Engng. J.*, April 1962, pp. 37–46; Structural design of tunnels by C. Booy, *Engng. J.*, Oct. 1961, pp. 83–91. See also under Peterson, 1957, in Bibliography to Chapter 5. For a general description of the project see the magazine *Water Power*, Sept. and Oct. 1963.)

McCAIG, I. W., and others (1962). 'Selection of dike freeboard and spillway capacity for Grand Rapids generating station.' *Engng. J.*, EIC, Nov. 1962, pp. 46–52. (This paper discusses selection of design wind, computation of wind set-up, wave height and wave uprush on the Cedar Lake reservoir. The required freeboard on the dam and dikes was in the order of 10 ft.)

MIDDLEBROOKS, T. A. (1953). 'Earth dam practice in the United States.' *Trans. Amer. Soc. civ. Engrs.*, Vol. CT, pp. 697–722. (This paper, by an outstanding soil mechanics engineer, reviews the history of earth dams, including a discussion of the types of failure that have been experienced. Best current practice in design and construction is described. Predictions of future advances are set forth. The paper concludes with an extensive bibliography.)

MIDDLEBROOKS, T. A., and JERVIS, W. H. (1947). 'Relief wells for dams and levees.' *Trans. Amer. Soc. civ. Engrs.*, Vol. 112, pp. 1321–402. (The writers have studied the problem of relieving uplift pressures for a number of years and have developed certain design criteria, which have been quite successfully applied. This paper presents the theoretical and empirical background.)

MONFORE, G. E. (1954). 'Ice pressure against dams.' (See Symposium, 1954.)

PETERSON, R. (1957). 'Design and construction of earth dams in Western Canada.' *Engng. J.*, EIC, Feb. 1957, pp. 129–37. (This paper reviews general principles of earth dam design and discusses how these are applied to the typical glacial till, shale and clay foundations, found in the river valleys of the prairie provinces.)

ROSE, E. (1947). 'Thrust exerted by expanding ice sheet.' *Trans. Amer. Soc. civ. Engrs.*, Vol. 112, pp. 871–900. (This paper presents a rational procedure for estimating the magnitude of ice pressures, and a review of available data on the subject.)

SAVAGE, R. P. (1959). 'Wave run-up on roughened and permeable slopes.' *Trans. Amer. Soc. civ. Engrs.*, Vol. 124, pp. 852–70. (Laboratory tests pertaining to the subject are described. Curves relating the run-up to wave steepness, beach slope, slope roughness, and slope permeability are presented.)

SAVILLE, T. (1958). 'Wave run-up on shore structures.' *Trans. Amer. Soc. civ. Engrs.*, Vol. 123, pp. 139–50. (Laboratory tests determining run-up on shore structures as a result of wave action are described. Curves relating the run-up to wave steepness, structure type, and depth at structure toe are presented.)

SAVILLE, T., and others (1962). 'Freeboard allowances for waves in inland reservoirs.' *Proc. Amer. Soc. civ. Engrs.*, May 1962, WW2, pp. 93–124. (The selection of freeboard allowance for wind-generated waves is reviewed. Hydraulic model studies applicable to the subject are examined. Sample computations are presented.)

SCHULTZ, E. R. (1962). 'Design features of Glen Canyon dam.' *Proc. Amer. Soc. civ. Engrs.*, July 1962, PO2, pp. 113–42. (This paper discusses the geology of the foundation, stress studies of the dam, foundation treatment, design of the penstocks, river outlets, spillway, and diversion procedures. See also the reference: U.S. Bureau of Reclamation, 1954, Colorado River Storage Project; in Bibliography to Chapter 1: Planning)

SIBUL, O. J., and JOHNSON, J. W. (1957). 'Laboratory study of wind tides in shallow water.' *Proc. Amer. Soc. civ. Engrs.*, April 1957, WW1, Paper 1210, 32 pp. (Experiments were conducted to determine the wind set-up with smooth and rough bottom conditions.)

SYMPOSIUM (1950). 'Multiple purpose reservoirs.' (See in Bibliography to Chapter 6.)

SYMPOSIUM (1954). 'Ice pressure against dams.' *Trans. Amer. Soc. civ. Engrs.*, Vol. 119, pp. 1–42. (The magnitude of pressure exerted by an ice sheet remains one of the major uncertainties in the design of dams. This Symposium presents data from Sweden, Canada, and the U.S.)

SYMPOSIUM (1956). 'Arch dams.' *Proc. Amer. Soc. civ. Engrs.*, April 1956, PO2, Papers 959, 960; June 1956, PO3, Papers 990–997; Aug. 1956, PO4, Paper 1045; Dec. 1956, PO6, Paper 1134; Feb. 1957, PO1, Papers 1182, 1183; April 1957, PO2, Paper 1217; June 1957, PO3, Papers 1267, 1286; Aug. 1957, PO4, Paper 1351. (Some titles: The philosophy of arch dams by A. Coyne; Portuguese experience by A. C. Xerez; Development in Italy by C. Semenza; Design and construction of Ross Dam by C. E. Sherling and L. R. Scrivner.)

SYMPOSIUM (1958). 'Rock fill dams.' *Proc. Amer. Soc. civ. Engrs.*, Aug. 1958, PO4, Papers 1733–49. (Some titles: Performance of T.V.A. central core dams by G. K. Leonard and O. H. Raine; The Bersimis sloping core dams by F. W. Patterson and D. H. MacDonald; The Derbendi Khan dam by G. V. Davis; Review and statistics by J. B. Snethlage and others.)

SYMPOSIUM (1958). 'McNary dam.' *Proc. Amer. Soc. civ. Engrs.*, Symp. Series No. 2, 154 pp. (A collection of five papers on the design and construction of the McNary dam, power house, spillway, and shiplock on the Columbia River. Contains an interesting account of the closure of the river.)

SYMPOSIUM (1959). *Spillway Design Floods.* 304 pp. National Research Council of Canada, Queen's Printer, Ottawa. (A collection of papers on the hydrology and flood frequency analysis associated with determining the design capacity of spillways.)

SYMPOSIUM (1963). 'Spillway design floods.' *Int. Assn. for Hydraulic Research*, 10th Congress, in London, Vol. 2, 235 pp. (A collection of 23 papers, from 14 different countries, on various aspects of design flood criteria and spillway design.)

TERZAGHI, K. (1948). *Soil Mechanics in Engineering Practice.* 565 pp. New York; Wiley. (This book, written by the most outstanding soil mechanics engineer, is surprisingly easy to read. Some sections discuss the stability of slopes, the design of dikes and earth dams, and the foundation of dams.)

TERZAGHI, K., and LEPS, T. M. (1960). 'Design and performance of Vermillion dam.' *Trans. Amer. Soc. civ. Engrs.*, Vol. 125, pp. 63–100. (The foundation conditions of this earth dam were so complicated that it was necessary to check on the design assumptions during construction and to modify some details in accordance with the findings.)

THOM, H. C. S. (1960). 'Distribution of extreme winds in the U.S.' *Proc. Amer. Soc. civ. Engrs.*, April 1960, ST4, pp. 11–24. (Extreme wind speed data are given for 30 ft. elevation and with recurrence intervals of 2, 50, and 100 years.)

THYSSE, J. TH., and SCHYF, J. B. (1949). 'Report on waves.' *17th Int. Navigation Congr.*, Section II, Communication 4, Lisbon. (Presents the results of observations on lakes and inland seas, and experiments in the Delft Hydraulics Laboratory, relating length of fetch, depth of water, velocity of wind, and height of waves. Some of these results are quoted in U.S. Beach Erosion Board, 1961, *Shore Protection Planning and Design*, pp. 25 and 27.)

U.S. BEACH EROSION BOARD (1954). 'The effect of fetch widths on wave generation.' *Techn. Memo 70*, U.S. Govt. Printing Office, Washington. (Presents a procedure for taking into account the configuration of a reservoir in determining the effective fetch of the reservoir. A summary of this memo is presented in U.S. Beach Erosion Board, 1961, *Shore Protection Planning and Design*, pp. 16h and 16i.)

U.S. BEACH EROSION BOARD (1961). *Shore Protection Planning and Design*. 400 pp. Washington; U.S. Govt. Printing Office. (This is a comprehensive manual for the design of dikes, seawalls, breakwaters and groins. The first part of the book deals with deep water and shallow water waves; tables and graphs are presented to estimate the wave lengths for given conditions; also discussed is the magnitude of wind set-up. The second part of the book discusses littoral transport. The third part deals with wave uprush and overtopping. The last part deals with design and construction.)

U.S. BUREAU OF RECLAMATION (1950–60). Manual. (See in Bibliography to Chapter 8.)

U.S. BUREAU OF RECLAMATION (1960). *Design of Small Dams*. 611 pp. Washington; U.S. Govt. Printing Office. (This book is intended to serve as a guide to safe practices for the design of small dams in public works programmes. Some chapter headings: Project planning; Flood studies; Selection of type of dam; Foundations; Earth fill dams; Rock fill dams; Concrete gravity dams; Spillways; Outlet works.)

U.S. COMMITTEE ON LARGE DAMS (1958). *Register of Dams in the U.S.* 429 pp. New York; McGraw-Hill. (This publication presents statistical data and photographs of practically all dams of any importance in the U.S. The statistics include: height, length, volume of dam, reservoir capacity, purpose, and installed power capacity if any.)

U.S. CORPS OF ENGINEERS (1946). 'Spillway design flood Garrison Reservoir.' (See in Bibliography to Chapter 2.)

U.S. CORPS OF ENGINEERS (1950–60). *Engineering and Design Manual*. (See in Bibliography to Chapter 9.)

U.S. CORPS OF ENGINEERS (1955). 'Waves and wind tides in shallow lakes and reservoirs.' Office of the District Engineer, Jacksonville, Fla., 46 pp. (This is the summary report of a series of studies made to determine the wind set-up, wave height and wave uprush for shallow lakes. A good deal of the studies have been carried out on Lake Okeechobee in Florida.)

U.S. TENNESSEE VALLEY AUTHORITY (1940–50). *Technical Reports*, U.S. Govt. Printing Office, Washington. (Twenty-four reports, with a total of some 3,000 pages, present the engineering aspects of the T.V.A. projects. Some titles of reports: No. 1 the Norris project; No. 12 the Fontana project; No. 22 Geology and foundation treatment; No. 24 design of T.V.A. projects.)

CHAPTER 6: FLOOD CONTROL

AMERICAN INSURANCE ASSOCIATION (1956). *Studies of Floods and Flood Damage*. 296 pp. New York; American Insurance Association. (An interesting study carried out by Parsons, Brinckerhoff, Hall and Macdonald, to determine if there were facts that might warrant a modification of the traditional position of the insurance business that the writing of flood insurance is not a feasible undertaking.)

BAINES, W. D. (1961). 'On the transfer of heat from a river to an ice sheet.' *Trans. Eng. Inst. Canada*, Vol. 5, No. 1, pp. 27–32. (A conventional heat formula is applied and examples are given of its use. It is shown that very small temperature differences must exist between the water and the ice sheet.)

BARROWS, H. K. (1948). 'Floods, their hydrology and control.' 432 pp. New York; McGraw-Hill. (This book presents a good deal of statistical data on flood control projects, constructed in the U.S.)

BEARD, L. R. (1963). 'Flood control operation of reservoirs.' *Proc. Amer. Soc. civ. Engrs.*, Jan. 1963, HY1, pp. 1–23. (This paper deals with multi-purpose reservoir operation. Procedures and criteria are developed to operate most effectively for flood control while minimizing the interference with other interests. Actual operation in California is cited.)

BURNS, J. I., and others (1961). 'Hydrology and flood control features of Oroville dam.' (See in Bibliography to Chapter 2.)

CHOW, V. T. (1959). *Open Channel Hydraulics*. (See in Bibliography to Chapter 3.)

ERSKINE, H. M. (1957). 'Losses due to ice storage in Heart River, N.D.' *Proc. Amer. Soc. civ. Engrs.*, June 1957, HY3, Paper 1261, 17 pp. (Flow records indicate water losses due to storage as ice is relatively low for small discharges; and the opposite for high discharges.)

FREEMAN, J. R. (1922). 'Flood problems in China.' *Trans. Amer. Soc. civ. Engrs.*, Vol. 85, pp. 1405–60. (Very interesting discussion of river training and flood-control problems on the Yellow River, with particular reference to the transportation and deposition of sediment. Various features of delta formation and river behaviour are discussed.)

FREEMAN, J. R. (1928). 'Flood control on the River Po in Italy.' *Proc. Amer. Soc. civ. Engrs.*, Vol. 54, pp. 957–92. (Description of the Po River, its floods and control works. It is concluded that the widely believed concept that the bed of a river is gradually raised as a result of dike construction appears to be untrue. Interesting discussion by H. P. Davis, who points out that excluding the river from its natural flood-plains must result in the deposition of sediment in the river bed or in the river delta. In the long run, both must occur. R. D. Goodrich mentions that he saw a whole village destroyed by bank undercutting in three days in 1924 when the Pei Ho was in flood.)

HORNER, W. W., and JENS, S. W. (1942). 'Surface run off determination from rainfall without using coefficients.' *Trans. Amer. Soc. civ. Engrs.*, Vol. 107, pp. 1039–117. (This paper outlines a method for estimating infiltration and retention to determine surface–run-off volume. This method is used for storm-drain design.)

KIRPICH, P. Z., and OSPINA, C. S. (1959). 'Flood control aspects of Cauca Valley development.' *Proc. Amer. Soc. civ. Engrs.*, Sept. 1959, HY9, pp. 1–34. (The Cauca Valley project in Colombia includes flood control as well as power and irrigation projects. This paper discusses the hydrology of the region and the flood control development as affected by economic and political considerations.)

KUIPER, E. (1952). 'Report conservation and flood control Assiniboine river.' Dept. of Agriculture, Winnipeg, Man., 61 pp. (This summary report presents the findings of a three-year study of flood control and conservation in the Assiniboine Basin, extending over 60,000 square miles. It was found that construction of the Russell Reservoir in the headwaters and diversion of the lower Assiniboine into lake Manitoba had more merits than any alternative scheme.)

KUIPER, E. (1956). 'Report Saskatchewan river reclamation project.' Dept of Agriculture, Winnipeg, Man., 48 pp. (This summary report presents the findings of a three-year study of reclamation possibilities in the Saskatchewan delta. It was found that one million acres of fertile farmland can be made available at a cost of 20 dollars per acre. This cost includes the construction of dikes along the river banks to prevent flooding, and the cost of drainage canals and pumping plants in the interior to maintain a suitable groundwater level.)

LANE, E. W. (1955). 'Design of stable channels.' *Trans. Amer. Soc. civ. Engrs.*, Vol. 120, pp. 1234–79. (The method of designing stable channels which is presented in this paper is based on the tractive force along the perimeter of the channel. This force should be sufficiently large to prevent sedimentation and small enough to prevent scour. Limiting values of the tractive force are offered.)

LEOPOLD, L. B., and MADDOCK, T. (1954). *The Flood Control Controversy*. 350 pp. A discussion. New York; The Ronald Press Co. (Of flood control by small dams and upstream watershed management, versus flood control by large downstream reservoirs. The book also discusses the public interest, the allocation of the costs and benefits, and the economic and political means for carrying this development forward.)

LUTHIN, J. N. (1957). *Drainage of Agricultural Lands*. 620 pp. Madison, Wisconsin; American Society of Agronomy. (An American textbook on land drainage with important contributions from agronomists in Holland and England. Some chapter headings: Water table in equilibrium with rainfall; Drainage by pumping from wells; Drainage of irrigated land; Soil permeability; Soil conditions related to drain depth.)

MATTHES, G. H. (1934). 'Floods and their economic importance.' *Trans. Amer. Geophys. Un.*, Vol. 15, pp. 427–32. (Interesting discussion of the function of floods in the regime of natural rivers. Discusses creative functions of floods, such as the periodic removal of debris accumulations in stream channels, the deposition of silt on flood plains, and the building up of delta lands. Describes the results on stream equilibrium if man interferes with natural flood conditions.)

MATTHES, G. H. (1948). 'Mississippi river cutoffs.' *Trans. Amer. Soc. civ. Engrs.*, Vol. 113, pp. 1–39. (Natural and man-made cut-offs on the lower Mississippi in the 50-mile stretch north of Vicksburg are described with respect to their effect on river shortening and flood-stage lowering.)

MATTHES, G. H. (1956). 'River engineering.' Section 15 of Abbett, Vol. II. (See in Bibliography to Chapter 5). (This section contains 30 pages on flood control.)

NEWTON, C. T. and HEDGER, H. E. (1959). 'Los Angeles County flood control and water conservation.' *Proc. Amer. Soc. civ. Engrs.*, June 1959, WW2, pp. 81–97. (This paper discusses the planning and coordination of the flood control works for the Los Angeles metropolis. Arrangements are made to conserve a major portion of the limited run-off for replenishment of a dwindling ground water supply.)

PAFFORD, R. J. JR., (1957). 'Operation of Missouri river main stem reservoirs.' *Proc. Amer. Soc. civ. Engrs.*, Vol. 83, paper 1370, 16 pp. (Description of the six main stem multiple purpose reservoirs system on the Missouri River. Outlines the main features of operation of these reservoirs in the interests of flood control, irrigation, water supply, and power. Discussion of operating plans and criteria.)

PARISET, E., and HAUSSER, R. (1961). 'Formation and evolution of ice covers on rivers.' *Trans. Engng. Inst. Canada*, Vol. 5, No. 1, pp. 41–49. (This paper presents the results of experimental and theoretical research on the formation of ice, and the blocking of rivers by ice jams.)

PICKELS, G. W. (1941). *Drainage and Flood Control Engineering.* 476 pp. New York; McGraw-Hill. (Comprehensive discussion of drainage work from the drainage of small areas by tile underdrains to the protection of large areas from floods, and drainage by open ditches. See also Section 22, pp. 1013–30, by the same author, in Davis, C. V., 1942, *Handbook of Applied Hydraulics*, under Chapter 5 of this Bibliography.)

RUTTER, E. J. (1951). 'Flood control operation of T.V.A. reservoirs.' *Trans. Amer. Soc. civ. Engrs.*, Vol. 116, pp. 671–707. (This paper describes actual flood conditions and alternative hypothetical conditions during the severe storms of 1946, 1947 and 1948. The effectiveness of the reservoirs is discussed.)

SYMPOSIUM (1938). 'National aspects of flood control.' *Trans. Amer. Soc. civ. Engrs.*, Vol. 103, pp. 551–719. (This is a collection of eight papers and extensive discussion on this subject. Some titles: Federal responsibility by J. J. Davis; Economic aspects by N. B. Jacobs; Floods in the upper Ohio by E. K. Morse; Federal plans by W. E. R. Covell.)

SYMPOSIUM (1949). 'Panama canal—the sea level project.' *Trans. Amer. Soc. civ. Engrs.*, Vol. 114, pp. 607–908. (A collection of nine papers with extensive discussion on this three billion dollar prospective project. Some titles of papers: Flood control by F. S. Brown; Tidal currents by J. S. Meyers and E. A. Schultz; Design of channel by J. E. Reeves and E. H. Bourguard.)

SYMPOSIUM (1950). 'Multi-purpose reservoirs.' *Trans. Amer. Soc. civ. Engrs.*, Vol. 115, pp. 789–908. (This symposium reveals the lack of agreement among the philosophies of the federal agencies primarily responsible for multiple reservoirs and the conflicts between the several functions to be served by such reservoirs. Two of the papers deal with reservoir operation for flood control: 1. Relation to flood control and navigation by M. Elliot. 4. Use for flood control by A. L. Cochran.)

SYMPOSIUM (1958). 'Old River diversion control.' *Trans. Amer. Soc. civ. Engrs.*, Vol. 123, pp. 1129–81. (The Atchafalaya River, a distributary of the Lower

Mississippi River, provides a route to the sea, that is only one-half as long as the course along the main channel. To prevent the Atchafalaya from capturing the main flow of the Mississippi, and thus to upset established navigation and water supply for New Orleans, the bifurcation must be controlled. Four different papers deal with the general problems: the hydraulic requirements; the foundation design; and the hydraulic structures.)

SYMPOSIUM (1960). 'Delta project.' *Trans. Amer. Soc. civ. Engrs.*, Vol. 125, pp. 1267–1307. (Three papers in this symposium describe the principles of controlling floods in the estuary of the Rhine River in Holland. One of the subjects is bottom erosion during enclosure operations.)

SYMPOSIUM (1961). 'Underseepage and its control.' *Trans. Amer. Soc. civ. Engrs.*, Vol. 126, Part I, pp. 1427–1568. (Seepage and sand boils, landward of dikes, are often a problem during flood stages. The purpose of the three papers in this symposium is to develop a better understanding of the phenomena and to present possible methods of control.)

U.S. TENNESSEE VALLEY AUTHORITY (1961). 'Floods and flood control.' *Techn. Rep. No. 26*, Knoxville, Tennessee, 302 pp. (This report covers the development, operation, costs and benefits of the Tennessee River system for flood control, based upon the integrated multiple-purpose system of reservoirs. Some chapter headings: Flood-producing storms; Design flood flows; Reservoir operation for flood control; Effect of changes in land use on floods; Benefits from flood control.)

WHITE, G. F., and others (1958). *Changes in Urban Occupance of Flood Plains in the U.S.* 235 pp. Chicago; The University of Chicago Press. (This is a report on changes in flood plain use since the enactment of the Flood Control Act of 1936. It has become apparent that Federal large-scale flood control measures have reduced the flood hazard, but also encouraged further encroachment upon the flood plains. The extent to which the net annual flood losses for the nation as a whole were reduced, is unclear.)

WHITE, G. F., and others (1961). *Papers on Flood Problems.* 228 pp. Chicago; The University of Chicago Press. (This publication presents 30 papers on non-engineering aspects of flood problems. Some titles: The strategy of using flood plains by G. F. White; Composition of flood losses; Flood hazard evaluation by R. W. Kates; Flood plain regulation by W. E. Akin.)

WILLIAMS, G. P. (1959). 'Frazil ice.' *Engng. J., EIC*, Nov. 1959, pp. 55–60. (This paper reviews the theory of frazil ice formation; then follows a discussion of remedial measures; the paper is concluded with an extensive bibliography.)

WILLIAMS, G. P. (1963). 'Probability charts for predicting ice thickness.' *Engng. J., EIC*, June 1963, pp. 31–35. (These charts, based on observations, provide information on the probable rate of increase of ice thickness and the probable maximum thickness that will be attained under a wide variety of conditions in North America.)

CHAPTER 7: WATER POWER

BROWN, J. G. (1958). *Hydroelectric Engineering Practice* (3 volumes). 2,400 pp. London; Blackie. (This comprehensive handbook on water power engineering has been prepared by 11 British authors. Vol. I deals with Civil Engineering. Some chapter headings: Rainfall and Run off; Hydraulics; Hydraulic models; Planning of the scheme; Gravity dams; Embankment dams; Arch dams; Spillways; Intake works; Canals; Tunnels; Surge chambers; Steel pipe lines; Power station design. Vol. II deals with mechanical and electrical engineering. Vol. III deals with economics, operation and maintenance.)

BRUDENELL, R. N., and GILBREATH, J. H. (1959). 'Economic complementary operations of hydro storage and steam power in the integrated T.V.A. system.' *Trans. Amer. Inst. Elect. Engrs.*, Vol. 78, Part III-A, pp. 136–56. (The first part of the paper deals with the theory behind the preparation of economy guide lines for the operation of power systems composed of thermal plants and hydro plants with reservoir storage. The second part of the paper deals with the application of the theory to the T.V.A. system.)

CONFERENCE (1924, 1930, 1936, 1950, 1956, 1962). *World Power.* Central Office,

201–2 Grand Buildings, Trafalgar Square, London. (The proceeding papers of the above six plenary meetings, plus the additional 12 sectional meetings in intervening years, provide a wealth of information on power development in general. A good portion of the papers deal with water power. Some conference themes: Power resources of the world; Recent technical and economic developments; The utilization of water resources; Planning of hydro-electric developments; Power for under-developed countries.)

COOKE, J. B. (1959). 'Haas hydroelectric power project.' *Trans. Amer. Soc. civ. Engrs.*, Vol. 124, pp. 989–1026. (The Haas Power project in California presents some unusual features, such as an underground powerhouse and the use of crushed rock as concrete aggregate. These and other features are described in the paper.)

CREAGER, W. P., and JUSTIN, J. P. (1950). *Hydroelectric Handbook.* 1151 pp. New York; Wiley. (A monumental work, prepared by two outstanding hydraulic engineers and 19 contributors. This comprehensive treatise deals with American water power engineering practice. Some chapter headings: Rainfall; Runoff; Flood flows; Investigation of sites; Hydraulics; Head, power, efficiency; Pondage and storage; Cost of hydro power; Market requirements and load studies; Capacity of the development; Gravity dams; Arch dams; Earth dams; Rock fill dams; Spillways; Intakes; Conduits; Tunnels; Water hammer; Surge tanks; Powerhouse; Turbines; Generators; Transmission; Operations.)

DARIN, K. and others (1959). *Principles of Power Balance Calculations for Economic Planning and Operation of Integrated Power Systems.* 66 pp. Stockholm; Eklunds and Vasatryck. (This publication, in English, deals with Swedish practices to determine optimum utilization of long-term storage reservoirs for co-ordination of hydro-electric power and thermal power.)

FATHY, A., and SHUKRY, A. S. (1956). 'The problem of reservoir capacity for long-term storage.' *Proc. Amer. Soc. civ. Engrs.*, Oct. 1956, HY5, Paper 1082, 27 pp. (This paper deals with the same problem as Hurst, 1951, only from a different angle. Deviations in the arithmetic mean for groups of observations are used instead of the deviation of one observation.)

GIRAND, J. (1941). 'Water supply on upper Salt River Arizona.' *Trans. Amer. Soc. civ. Engrs.*, Vol. 106, pp. 398–409. (This paper discusses methods used to determine economical storage requirements and power outputs of a proposed hydro-electric plant, on a river where only a few years of flow records were available. Probable future hydrographs were constructed, based on well-defined trends of long-period variations in river flow, studied from tree ring measures, rainfall records and adjacent stream gauge records.)

GUNTER, G. H., and BRUCE, J. F. (1962). 'John Day lock and dam: planning and site selection.' *Proc. Amer. Soc. civ. Engrs.*, July 1962, PO2, pp. 35–56. (This project, now under construction, is a multi-purpose navigation, flood control, and power project on the lower Columbia River. Notable features are: the highest single lift lock in the world (113 ft.), 3,100 MW hydro-electric capacity, and 2,250,000 cusec spillway capacity. This paper describes aspects of site selection and plant lay-out.)

HURST, H. E. (1951). 'Long term storage capacity of reservoirs.' *Trans. Amer. Soc. civ. Engrs.*, Vol. 116, pp. 770–808. (This paper presents a solution to the problem of determining the reservoir storage required on a given stream, to guarantee a given draft. Probability methods are applied. The results have been used in the Nile River Basin. See also Fathy, 1956.)

KREUGER, R. E. (1959). 'Multiple-purpose power plant capacity.' *Proc. Amer. Soc. civ. Engrs.*, Aug. 1959, PO4, pp. 51–64. (This paper outlines the major factors determining the total plant capacity of hydro-electric plants attached to multiple-purpose projects of the Bureau of Reclamation.)

KUIPER, E. (1958). 'Report on the control of lakes Winnipeg and Manitoba.' Govt. Printing Office, Winnipeg, Man., 58 pp. (This summary report presents the findings of a two-year study of power development on the Nelson River and multi-purpose control of the large Manitoba lakes. It was found that Lake Manitoba should be regulated for flood control, while a total of some 30 million acre-ft. of storage capacity could be made available on Lakes Winnipeg and

Winnipegosis, to enhance the potential 4,000 MW power development on the Nelson River.)

MacDonald, D. H., and others (1960). 'Kelsey generating station dam and dikes.' *Engng. J.*, *EIC*, Oct. 1960, pp. 87–98. (This paper describes the first power plant on the Nelson River. Notable features of this project are the construction of dams and dikes on both unfrozen and permanently frozen foundations. This paper describes perma frost investigations, dam and dike design, and initial results.)

Martin, T. J. (1960). 'Warsak hydro electric project.' *Water Power*, Nov. 1960, pp. 431–36; Dec. 1960, pp. 457–62. (This scheme in north-west Pakistan has been designed and built by Canadian engineers in collaboration with the Government of Pakistan, under the auspices of the Colombo Plan. The head on the dam is 140 ft.; the installed capacity is 160 MW. A notable feature of the project is the partial river diversion during construction, with the remainder of flood flow being allowed to pass over the unfinished dam.)

Matthias, F. T., and others (1960). 'Planning and construction of the Chute-des-Passes hydro-electric power project.' *Engng. J.*, EIC, Jan. 1960, pp. 39–51. (This development is located on the Saguenay River, a tributary of the St. Lawrence River. The total plant capacity is 750 MW. Notable features are two miles of 38 ft. diam. supply tunnels, 50 ft. diam. discharge tunnels, and an underground powerhouse. This project is also described in the *Water Power*, May 1959, pp. 166–74.)

McMordie, R. C. (1962). 'Aspects of Columbia River projects in Canada.' *Engng. J.*, EIC, Oct. 1962, pp. 47–54. (This paper reviews the Columbia River treaty between Canada and the U.S.; the storage and power projects on the Columbia River in Canada; and the methods of determining and dividing the resultant benefits.)

McQueen, A. W. F., and Simpson, C. N. (1958). 'Underground power plants in Canada.' *Proc. Amer. Soc. civ. Engrs.*, June 1958, PO3, Paper 1670, 22 pp. (This paper presents a review of the factors affecting design practice in Canada. Reference is made to the Bersimis, Chute-des-Passes, and Kemano projects.)

Mosonyi, E. (1957). *Water Power Development* (2 volumes). 1,800 pp. Budapest; Publishing house of the Hungarian Academy of Sciences. (This comprehensive handbook has been prepared by one author. It deals primarily with water power engineering practice in Europe and Russia. Some chapter headings: History and types of water power development; Low-head power plants; High-head power plants; Pumped-storage systems; Water power economics.)

Reid, J. L., and Brittain, K. G. (1962). 'Design concepts of the Brazeau development, including river and hydrology studies.' *Engng. J.*, *EIC*, Oct. 1962, pp. 60–66. (The Brazeau River is a tributary of the Saskatchewan River. The power plant will operate under a gross head of 390 ft.; its ultimate capacity will be 600 MW. One of the notable features is a pumping plant between the reservoir and the headrace canal. The second half of the paper is devoted to the hydrology of the spillway design flood.)

Stall, J. B., and Neill, J. C. (1963). 'Calculated risks of impound reservoir yield.' *Proc. Amer. Soc. civ. Engrs.*, Jan. 1963, HY1, pp. 25–34. (By developing an array of probabilities of occurrence of various droughts during various design periods, the design engineer can gain an improved understanding of the risks associated with adopting dependable yields from reservoirs.)

Symposium (1954). 'The Nechako-Kemano-Kitimat development.' *Engng. J.*, *EIC*, April 1953, Nov. 1954, reprinted in one issue, 112 pp. (A collection of 12 papers on different aspects of this water power development in British Columbia. One of the notable features of this project is that a small river, with an average flow of 6,500 cusec, normally flowing in easterly direction, was dammed off and diverted via tunnels in a westerly direction towards sea-level, providing a gross head of 2,600 ft. The total capacity of the plant is 450 MW. Its principal load is the smelting of aluminium at Kitimat.)

Symposium (1961). 'Penstocks.' *Proc. Symp. Amer. Soc. civ. Engrs.*, Series No. 4. (A collection of reprints from the following proceeding papers in the power division: 1216, 1284, 1285, 1344, 1396, 1397, 1398, 1457, 2291. Some titles:

Water hammer design criteria by J. Parmakian; Penstock design and construction by G. R. Latham; Determination of stresses on anchor blocks.)

SYMPOSIUM (1961). 'Pumped storage.' *Proc. Amer. Soc. civ. Engrs.*, July 1962, PO2, pp. 183–251, and PO4, pp. 83–142. Some titles: General planning of pumped storage by R. D. Ley and E. S. Loane; Run-off river pumped storage project by M. J. Hroncich and J. M. Mullarkey; Dike investigations of pumped storage facilities by M. G. Salzman; Reversible pump-turbines by M. Braikevitch and others; Foreign pumped storage projects by R. D. Harza.)

U.S. BUREAU OF RECLAMATION (1954). 'Selecting hydraulic reaction turbines.' *Engng. Monograph No. 20*, Technical Information Branch, Denver Federal Center, 45 pp. (This publication has been prepared to permit rapid selection of the proper unit, determination of its major dimensions, and prediction of its performance.)

VOETSCH, C., and FRESEN, M. H. (1938). 'Economic diameter of steel penstocks.' *Trans. Amer. Soc. civ. Engrs.*, Vol. 103, pp. 89–132. (Methods are developed to determine the minimum total annual cost of steel penstocks for hydro-electric plants.)

WHIPPLE, W., JR. (1955). 'Principles of federal hydro-electric power development.' *Proc. Amer. Soc. civ. Engrs.*, Vol. 81, paper 739, 20 pp. (Discusses the controversy over federal hydro-electric development and its economic utilization in conjunction with privately owned thermal electric systems.)

WING, L. S., and GRIFFIN, R. H. (1955). 'Selection of installed capacity at hydro-electric power plants.' *Proc. Amer. Soc. civ. Engrs.*, May 1955, PO, Paper 697, 36 pp. (This paper presents a method for determining the capacity of hydro plants by selecting the lowest-cost increments of hydro capacity and energy.)

CHAPTER 8: IRRIGATION

ABBETT, R. W. (1956). *American Civil Engineering Practice*, Vol. II. (See in Bibliography to Chapter 5.)

BARNES, D. P. (1956). 'Irrigation and land drainage.' Section 17, 95 pp., of Abbett, Vol. II (see reference in Bibliography to Chapter 5). (Sub-titles of this section: Project development; Availability of water; Suitability of land; System design; Land drainage.)

BERG, P. H. (1962). 'Methods of applying irrigation water.' *Trans. Amer. Soc. civ. Engrs.*, Vol. 127, Part III, pp. 61–74. (The factors influencing the selection of a method of applying irrigation water on a farm are described in general. The various methods of irrigation and the associated conveyance and distribution systems are presented.)

BLANEY, H. F. (1952). 'Consumptive use of water—definition, methods, and research data.' *Trans. Amer. Soc. civ. Engrs.*, Vol. 117, paper 2525, 24 pp. (Presents a very interesting method of estimating the consumptive use of various classes of agricultural crops on the basis of available climatological data.)

BLENCH, T. (1957). *Regime Behaviour of Canals and Rivers*. (See in Bibliography to Chapter 4.)

CHIEN, N. (1957). 'A concept of the regime theory.' *Trans. Amer. Soc. civ. Engrs.*, Vol. 122, pp. 785–805. (The regime theory is analysed on the basis of the bed load functions. Under certain conditions, agreement between the two is found. The use of the regime theory for conditions other than those in India and Pakistan should be approached with great care.)

CONGRESS (1951, 1954, 1957, 1960, 1963), *on Irrigation and Drainage*. Secretariat 184, Golf Links Area, New Delhi 3, India. (The transaction of these five congresses and the annual bulletins of the International Commission on Irrigation and Drainage contain a wealth of information on irrigation development all over the world. Some conference themes: Economics of irrigation works; Control of water table; Reclamation of saline lands; Multi-purpose reservoir operation.)

COREY, H. T. (1913). 'Irrigation and river control in Colorado River Delta.' *Trans. Amer. Soc. civ. Engrs.*, Vol. 76, pp. 1204–453. (Aspects of delta formation and the effects of erosion and sedimentation in the stream channel on changes in the longitudinal and cross-sectional profiles are discussed. Interesting account of how the Colorado River broke through towards the Salton Sea in 1905, due to an artificial cut in the natural river bank. Six attempts were made to close the gap, the last one succeeded in 1906.)

GOODRICH, R. D. (1957). 'Methods for determining consumptive use of water in irrigation.' *Trans. Amer. Soc. civ. Engrs.*, Vol. 122, pp. 806–22. (Standard methods of determining rates of consumptive use are described, and the use of the results are demonstrated.)

HOUK, I. E. (1942). 'Irrigation and irrigation structures.' Sections 20 and 21, pp. 947–1012, of Davis, *Handbook of Applied Hydraulics* (see in Bibliography to Chapter 5). (Sub-titles of these two sections: Land classification; Water supply; Water requirements; Conveyance losses and waste; Irrigation losses and waste; Consumptive use; Irrigation methods; Diversion weirs; Distribution system; Conveyance structures.)

HOUK, I. E. (1951). *Irrigation Engineering* (2 vols.). 1,076 pp. New York; Wiley. (This is one of the most comprehensive and most recent textbooks on the subject. Some chapter headings: Soil fertility; Climatic factors; Run-off and stream flow; Irrigation and water requirements; Irrigation water supplies; Planning irrigation projects; Conveyance of irrigation water; Diversion dams and intakes; Storage dams.)

KUIPER, E. (1964). 'Ebro Basin water resources development.' Madrid, Spain, 26 pp. (This report presents an outline of the studies that must be undertaken to ensure that Spain's water resources will be developed to the greatest advantage of its economic progress. Particular emphasis is placed on irrigation development.)

LACEY, G. (1930). 'Stable channels in alluvium.' *Minutes Proc. Inst. civ. Engrs.*, *Lond.*, Vol. 229, pp. 259–92. (The first of a series of three parts on the so-called 'regime theory', developed by a number of English engineers engaged on Indian irrigation problems. Gives Kennedy's and other formulae for the critical velocity as a function of depth and type of silt transported. Gives principles governing stable channels in alluvium. Some of the following discussions are rather critical about the reliability of the original data.)

LACEY, G. (1933). 'Uniform flow in alluvial rivers and canals.' *Minutes Proc. Inst. civ. Engrs.*, Vol. 237, pp. 421–53. (The second article on regime theory. The first article is not summarized and has to be read first. Some of the following discussions contain interesting remarks about river behaviour in general.)

LACEY, G. (1946). 'A general theory of flow in alluvium.' *J. Inst. civ. Engrs.*, Vol. 27, pp. 16–47. (The third article on regime theory. The first two articles have to be read first. As a whole, the articles are rather difficult to understand and call for concentrated reading. The equations that are presented should only be applied when means are available to check their validity for the case under consideration. An interesting review of Lacey's regime theory is given by Chien, 1957.)

LANE, E. W. (1955). 'Design of stable channels.' (See in Bibliography to Chapter 6.)

LOWRY, R. L., and JOHNSON, A. F. (1942). 'Consumptive use of water for irrigation.' *Trans. Amer. Soc. civ. Engrs.*, Vol. 107, pp. 1243–302. (Average annual consumptive use in 20 selected areas is shown to have a linear relation to accumulated daily temperature above 32°F during the growing season. Discusses the cause of deviations from average requirements.)

MULDROW, W. C. (1948). 'Forecasting productivity of irrigable lands.' *Trans. Amer. Soc. civ. Engrs.*, Vol. 113, pp. 562–76. (From a broad water-resource planning standpoint, it is desirable that available water be used where it will do the most good. A simple extension of the heat-unit method from consumptive use of water to tonnage of crop produced will provide a method of estimating the productivity of lands on a proposed project.)

RITER, J. R. (1955). 'Planning a large irrigation project.' (See in Bibliography to Chapter 1: Planning.)

SIMONS, D. B., and ALBERTSON, M. L. (1960). 'Uniform water conveyance channels in alluvial material.' *Proc. Amer. Soc. civ. Engrs.*, May 1960, HY5, pp. 33–71. (Methods of designing stable irrigation channels are developed. The theory includes a modification of the regime theory as well as a modification of the tractive-force theory.)

U.S. BUREAU OF RECLAMATION (1950–60). *Manual*. Federal Center, Denver. (The U.S. Bureau of Reclamation has prepared numerous documents that deal with its design practices. The complete manual includes hydrology, water studies, irrigation, dams, canals, and power plants.)

CHAPTER 9: NAVIGATION

BRUCE, J. W., and others (1957). 'Modern facilities for Ohio River navigation.' *Proc. Amer. Soc. civ. Engrs.*, May 1957, WW2, Paper 1239, 16 pp. (This paper describes the general requirements and criteria of the new canalization works on the Ohio River.)

BURPEE, L. H. (1961). 'Canadian section of the St. Lawrence Seaway.' *Trans. Amer. Soc. civ. Engrs.*, Vol. 126, Part IV, pp. 184–209. (This paper describes the St. Lawrence Seaway project and summarizes the navigation features of the Canadian share of the project. It also discusses operational aspects.)

BUSH, J. L. (1962). 'Channel stabilization on the Arkansas River.' *Proc. Amer. Soc. civ. Engrs.*, May 1962, WW2, pp. 51–67. (The Arkansas River in Oklahoma is a meandering stream with unstable channel; the banks are subject to extensive caving; sand and gravel bars change with every rise. Various types of structures were to be installed to establish a stable channel suitable for navigation.)

FORSTON, E. P., and FENWICK, G. B. (1961). 'Navigation model studies of new Ohio River locks.' *Trans. Amer. Soc. civ. Engrs.*, Vol. 126, Part IV, pp. 171–83. (This paper describes the successful utilization of hydraulic models in designing a proper lay-out of the locks and approaches, as well as methods of operation of the dam gates.)

FRANZIUS, O. (1936). *Waterway Engineering*. 527 pp. Cambridge; The Technology Press M.I.T. (First published in Germany in 1927 and translated by L. G. Straub. This textbook deals comprehensively with such topics as river regulation; tidal rivers; sea-shore development; weirs; shiplocks; canals. Most of the prevailing techniques of European waterway engineering are discussed in this book.)

GOMEZ, A. (1960). 'Sacramento River deep water ship channel.' *Proc. Amer. Soc. civ. Engrs.*, Nov. 1960, WW4, pp. 53–67. (This paper gives a description of the fifty million dollar project, its functional requirements, the hydraulic studies, estimates of benefits, and economic analysis.)

GROTHAUS, W., and RIPLEY, D. M. (1958). 'St. Lawrence Seaway, 27-ft. canals and channels.' *Proc. Amer. Soc. civ. Engrs.*, Jan. 1958, WW1, Paper 1518, 22 pp. (A discussion of the basic design criteria adopted. The dams, locks and channels are then described in terms of how they meet these criteria.)

GUNTER, G. H., and BRUCE, J. F. (1962). 'John Day lock and dam: planning and dike selection.' (See in Bibliography to Chapter 7.)

HAAS, R. H., and WELLER, H. E. (1953). 'Bank stabilization by revetments and dikes.' *Trans. Amer. Soc. civ. Engrs.*, Vol. 118, pp. 849–70. (This paper describes river training works on the lower Mississippi River. The main problem is one of arresting bank recession. Revetments are the principal type of defence.)

HICKSON, R. E. (1961). 'Columbia River ship channel improvement.' *Proc. Amer. Soc. civ. Engrs.*, Aug. 1961. WW3, pp. 71–93. (This paper presents the design of channel improvements and control works on the lower Columbia River. The beneficial effects of these works are described.)

KABELAC, O. W. (1961). 'The Great Volga waterway.' *Proc. Amer. Soc. civ. Engrs.*, Feb. 1961, WW1, pp. 151–63. (This paper reviews one of the largest Russian water resource developments, including navigation, water power, water supply and reclamation aspects.)

KUIPER, E. (1951). 'Construction of the Harlingen works.' World Construction, Chicago, Jan. 1951. pp. 11–16. (This paper describes the design and construction of two shiplocks, connecting the Dutch inland canal system with the open sea. The sea side of the locks was protected by a harbour dam, covered with asphalt cement. Special measures had to be taken to reduce salt water intrusion to a minimum. The lock chambers were built on 75 ft. reinforced concrete piles.)

KUIPER, E. (1958). 'Navigation Saskatchewan River.' Winnipeg, Man. (This summary report presents the findings of a preliminary cost–benefit analysis of potential navigation facilities between Empress on the South Saskatchewan River and Selkirk on the Red River.)

McINTYRE, D. (1958). 'The Beauharnois canal locks.' *Proc. Amer. Soc. civ. Engrs.* Sept. 1958, WW4, Paper 1781, 13 pp. (This paper describes the hydraulic features of the St. Lawrence Seaway structures designed for the Beauharnois site in Canada.)

NEDECO (1959). 'River studies of the Niger and Benue.' (See in Bibliography to Chapter 4.)

RICHARDSON, G. C., and WEBSTER, M. J. (1960). 'Hydraulic design of Columbia River navigation locks.' *Trans. Amer. Soc. civ. Engrs.*, Vol. 125, pp. 345–64. (This paper describes shiplocks with lifts of about 100 ft. The designers were forced to develop improvements with the aid of hydraulic model studies. As a result there is hardly any turbulence in the lock chambers of the more recent projects.)

SENOUR, C. (1947). 'New project for stabilizing and deepening lower Mississippi River.' *Trans. Amer. Soc. civ. Engrs.*, Vol. 112, pp. 277–97. (This paper discusses the river training works that have been undertaken to obtain reduction in flood heights, to increase the flood carrying capacity of the river channel, and to maintain a navigable depth of 12 ft. at low water conditions.)

STRATTON, J. H. (1956). 'Canalized rivers and lock canals.' Section 15, pp. 148–72 of Abbett, Vol. II (see in Bibliography to Chapter 5.)

SYMPOSIUM (1951). 'Lock systems in the U.S.' *Trans. Amer. Soc. civ. Engrs.*, Vol. 116, pp. 829–90. (This Symposium presents the latest thoughts regarding the design of locks and other navigation structures.)

TWAIN, MARK (1875). *Life on the Mississippi.* 527 pp. New York; Harper and Brothers. (Contains very interesting and humorous accounts of the navigation difficulties on a large alluvial river in the state of nature.)

U.S. CORPS OF ENGINEERS (1950–60). *Engineering and Design Manual.* Washington; U.S. Govt. Printing Office. (The Corps has prepared several documents that deal with its design practices. The following topics are included: run-off from snow melt; flood hydrograph analysis; reservoir regulation; navigation dams and locks; hydro-electric power plants.)

WOODSON, R. C. (1961). 'Stabilization of the Middle Rio Grande in New Mexico.' *Proc. Amer. Soc. civ. Engrs.*, Nov. 1961, WW4, pp. 1–15. (In the Cochito to Rio Puerco reach, the Rio Grande has an average level which is normally above the level of the land behind the levees. Channel stabilization was required to protect the levees, along 105 miles of channel. The use of the Kellner jetty system is described.)

CHAPTER 10: WATER SUPPLY

ASCE COMMITTEE ON GROUNDWATER (1961). 'Groundwater basin management.' *Amer. Soc. civ. Engrs.*, *Manual of Engineering Practice No. 40*, 160 pp. (The purpose of this publication was to accumulate knowledge regarding all phases of existing and potential utilization of underground capacity for storage of water and its withdrawal for irrigation and other uses. Some chapter headings: Occurrence of water; Movement of water; Groundwater reservoirs; The hydrologic equation; The groundwater inventory; Planning for artificial recharge; Extraction of groundwater; Design and operation of projects; Legal considerations. The book concludes with an extensive bibliography.)

BABBIT, H. E., and DOLAND, J. J. (1955). *Water Supply Engineering.* 608 pp. New York; McGraw-Hill. (This textbook deals primarily with the problems of municipal water supply. Some chapter headings: Demand for water; Location of wells; Intakes; Aqueducts; Pumping stations; Distributing reservoirs; Distributing systems; Quality of water supplies; Water purification.)

CALIFORNIA DEPARTMENT OF WATER RESOURCES (1957). 'California water plan.' (See in Bibliography to Chapter 1: Planning.)

CHURCHILL, M. A. (1958). 'Effects of storage impoundments on water quality.' *Trans. Amer. Soc. civ. Engrs.*, Vol. 123, pp. 419–64. (When stream flow water is impounded in a large storage reservoir, changes in the physical, bacteriological, sanitary-chemical, and mineral quality of the water are produced. Although most of these changes result in a generally improved water quality, certain qualities may be adversely affected. This paper presents a wide variety of observations in large reservoirs.)

CONKLING, H. (1946). 'Utilization of groundwater storage in stream system development.' *Trans. Amer. Soc. civ. Engrs.*, Vol. III, pp. 275–354. (This paper considers underground reservoirs into which water is charged naturally, or can be charged artificially, and from which water is available when needed, either as gravity flow or from wells.)

HAZEN, R. (1956). 'Public water supply.' Section 18, 78 pp. of Abbett, Vol. II (see in Bibliography to Chapter 5). (Sub-titles of this section: Water requirements; Collection of water; Purification of water; Transmission and distribution of water.)

HIRSHLEIFER, J., and others (1960). *Water Supply.* (See in Bibliography to Chapter 1: Planning.)

KRUL, W. F. J. M., and others (1963). *Re-use of Water in Industry.* 256 pp. London; Butterworths. (This book contains papers on water economy, pollution, and waste treatment by seven experts from England, France and Germany.)

SYMPOSIUM (1957). 'Ground water development.' *Trans. Amer. Soc. civ. Engrs.*, Vol. 122, pp. 422–517. (This Symposium contains papers on: the planned utilization of underground storage; irrigation aspects; basin recharge; and the hydraulics of wells.)

SYMPOSIUM (1962). *Groundwater.* National Research Council of Canada, Sub-committee on Hydrology, Ottawa. (A collection of 13 papers on different ground-water problems in Canada.)

THOMAS, R. O. (1961). 'Legal aspects of water utilization.' *Trans. Amer. Soc. civ. Engrs.*, Vol. 126, Part III, pp. 633–61. (This paper discusses historical basis of water rights; their application to rights to the use of ground water; and various other legal aspects of ground water utilization.)

U.S. SENATE SELECT COMMITTEE ON NATIONAL WATER RESOURCES (1959). Report on water resources activities in the United States. (See in Bibliography to Chapter 1: Planning.)

WATER POLLUTION CONTROL (1951). A Symposium. *Trans. Amer. Soc. civ. Engrs.*, Vol. 116, pp. 1–30. (This symposium deals primarily with legislative measures to control river pollution in New England, New York, the Delaware Basin, and the Tennessee Valley.)

CHAPTER 11: ECONOMIC ANALYSIS

ADAMS, H. W. (1956). 'Economic aspects of flood plain zoning.' *Proc. Amer. Soc. civ. Engrs.*, Vol. 82, paper 882, 5 pp. (A brief discussion on the merits of flood plain zoning as an alternative consideration to flood prevention. The paper points out that zoning has not been extensively used as a primary means, but there have been numerous cases where it was applied as a supplement to structural flood control measures.)

BENNETT, N. B., JR. (1956). 'Cost allocation for multi-purpose water projects.' *Proc. Amer. Soc. civ. Engrs.*, Vol. 82, paper 961, 10 pp. (Discussion of the separable costs, remaining benefits, method of cost allocation, and the meanings attached to single-purpose alternate costs, specific and separable costs, joint costs, and remaining project costs.)

CLARENBACH, F. A. (1955). 'Reliability of estimates of agricultural damages from floods.' (A critical discussion of costs and benefits of four rural flood control projects, proposed by the Corps of Engineers and the Soil Conservation Service. It is concluded that federal construction agencies cannot be trusted to produce objective benefit-cost ratios. For reference, see in Bibliography to Chapter 1: Planning—U.S. Commission on Organization of the Executive Branch of the Government, 1955, Report on Water Resources and Power, Task Force reports, Vol. III.)

CLOUGH, D. J. (1961). 'Measures of value and statistical models in the economic analysis of flood control and water conservation schemes.' *Trans. Engng. Inst. of*

Can., Vol. 5, No. 1, pp. 33–40. (This paper discusses the need for common scale values in water development projects; it also focuses on the construction of statistical models of flood damages.)

CURRAN, C. D. (1955). 'Evaluation of federal navigation projects.' (A critical discussion of costs and benefits of several navigation projects constructed or proposed by the Corps of Engineers. For reference see in Bibliography to Chapter 1: Planning—U.S. Commission on Organization of the Executive Branch of the Government, 1955, Report on Water Resources and Power, Task Force reports, Vol. III.)

ECKSTEIN, O. (1958). *Water Resources Development.* (See in Bibliography to Chapter 1: Planning.)

ENGINEERS JOINT COUNCIL (1957). 'Principles of a sound national water policy.' (This report contains a critical discussion of prevailing methods of determining benefits and costs of water resources projects. For reference, see in Bibliography to Chapter 1: Planning.)

FAISON, H. R. (1955). 'Some economic aspects of waterway projects.' *Trans. Amer. Soc. civ. Engrs.*, Vol. 120, pp. 1480–549. (Evidence is presented that the development of inland waterways in the U.S. has generally resulted in traffic increases that more than justify the cost of improvements.)

FOGARTY, E. R. (1956). 'Benefits of water development projects.' *Proc. Amer. Soc. civ. Engrs.*, Vol. 82, paper 981, 8 pp. (A brief statement of federal policy on establishing benefits.)

GOMEZ, A. (1960). 'Sacramento River deep water ship channel.' (See in Bibliography to Chapter 9.)

GRANT, E. L., and TRESON, W. G. (1960). *Principles of Engineering Economy.* 574 pp. New York; The Ronald Press Co. (One of the most comprehensive textbooks on this subject. Some chapter headings: Interest formulas; Present worth; A pattern for economy studies; Dealing with multiple alternatives; Economy studies for governmental activities; Criteria for investment decisions.)

KRUTILLA, J. V., and ECKSTEIN, O. (1958). *Multiple Purpose River Development.* (See in Bibliography to Chapter 1: Planning.)

KUIPER, E. (1958). 'Lake Manitoba regulation.' Govt. Printing Office, Winnipeg, Man., 24 pp. (This report contains an economic analysis of the regulation of Lake Manitoba for the purpose of reducing flood damages around the perimeter of the Lake. The benefit–cost ratio was found to be 1·8. The required control works were to have been constructed shortly afterwards.)

KUIPER, E. (1961). 'Benefit–cost analysis Assiniboine River flood control.' Govt. Printing Office, Winnipeg, Man., 62 pp. (This report contains an economic analysis of a number of proposed Assiniboine River flood control projects, making allowance for water conservation aspects. It was found that a combination of the Portage Diversion and the Shellmouth Dam, at a total cost of 20 million dollars, had more merits than any other scheme.)

LI, S., and BOTTOMS, E. E. (1961). 'Economic evaluation of inland waterways projects.' *Proc. Amer. Soc. civ. Engrs.*, Aug. 1961, WW3, pp. 29–57. (Basic principles are summarized and some of the current evaluation practices are given. An extensive bibliography on the subject is included.)

McGAUHEY, P. H., and ERLICH, H. (1959). 'Economic evaluation of water.' *Proc. Amer. Soc. civ. Engrs.*, June 1959, IR2, pp. 1–21. (This paper describes the broader policy that is needed to allocate unappropriated water to various beneficial uses in such a way as to produce the greatest economic yield.)

RASMUSSEN, J. J. (1956). 'Economic criteria for water development projects.' *Proc. Amer. Soc. civ. Engrs.*, Vol. 82, paper 977, 14 pp. (An interesting paper by an economist on the optimum use of water resources, including a discussion on benefit-cost ratios.)

ROYAL COMMISSION (1952). *Report on the South Saskatchewan River Project.* 423 pp. Ottawa; Queen's Printer. (In 1951 a Royal Commission was appointed to conduct an inquiry 'Whether the economic and social returns to the Canadian people on the investment in the proposed South Saskatchewan River Project would be commensurate with the cost thereof'. This report presents the answer of the Commission, which was negative. Some chapter headings: Historical setting; Irrigation in Western Canada; The South Saskatchewan River Project; Long run economic and social benefits; The immediate impact on the economy.)

ROYAL COMMISSION (1958). *Flood Cost–Benefit*. 129 pp. Winnipeg, Man.; Govt. Printing Office. (This report contains an economic analysis of the proposed greater Winnipeg flood control projects. It is concluded that a combination of the Red River Floodway, the Assiniboine Diversion, and the Russell Reservoir, at a total cost of 73 million dollars, will have more merits than any other flood control scheme. Some chapter headings: The nature of flooding; Flood protection measures; Flood damages; Benefit–cost analysis; Flood insurance.)

SENOUR, C. (1961). 'Economics of river bank stabilization.' *Proc. Amer. Soc. civ. Engrs.*, May 1961, WW2, pp. 17–26. (This paper discusses the difficult problem of estimating the benefits of a bank stabilization project.)

SEWELL, W. R. D., and others (1962). 'Guide to benefit-cost analysis.' Queen's Printer, Ottawa, 49 pp. (This paper has been prepared by four Canadian economists for the 'Resources for Tomorrow' Conference in Montreal, in 1961. It reviews the subject of economic analysis and project evaluation and applies the principles to numerical examples.)

THE ENGINEER SCHOOL (1950). 'The Bonneville project.' The Engineer Center, Fort Belvoir, 32 pp. (This publication describes the Bonneville dam, power plant and shiplock on the Columbia River, its engineering features, its costs and benefits and the cost allocation.)

UNITED NATIONS (1955). *Multi-purpose River Basin Development*. (See in Bibliography to Chapter 1: Planning.)

U.S. INTER-AGENCY COMMITTEE ON WATER RESOURCES (1958). 'Proposed practices for economic analysis of river basin projects.' U.S. Govt. Printing Office, Washington, 56 pp. (This is one of the first, and by far the most influential document on this subject, commonly referred to as the 'green book'. The present edition is a revision of the original 1950 issue. Chapter headings are: Basic principles and concepts; Project and program formulation; Measurement of benefits and costs; Application of principles to project purposes; Cost allocation for multiple-purpose projects.)

WOLLMAN, N., and others (1962). *The Value of Water in Alternative Uses*. 426 pp., University of New Mexico Press. (This book presents a study of patterns of water use and their effects on the development of a drainage basin. Eight different schemes, based on varying degrees of agricultural, industrial, domestic and recreational use, are analysed with respect to the optimum economic benefits that can be obtained.)

INDEX

473